Earth Science

Second Edition

Edward J. Tarbuck and Frederick K. Lutgens
Illinois Central College

Charles E. Merrill Publishing Company
A Bell & Howell Company
Columbus Toronto London Sydney

Cover photo by Four By Five, Inc.

Published by
Charles E. Merrill Publishing Company
A Bell & Howell Company
Columbus, Ohio 43216

This book was set in Times Roman.
The production editors were Linda Johnstone and JoEllen Gohr.
The cover was prepared by Larry Hamill.

Library of Congress Catalog Card Number: 78–70071
International Standard Book Number: 0–675–08303–6
3 4 5 6 7 8 9 10—85 84 83 82 81 80 79

Printed in the United States of America

preface

Increased awareness and concern for our physical environment, probes to the moon and neighboring planets as well as to the bottom of the sea, development of a revolutionary theory in geology that has made scientists rewrite explanations of the earth's long and complex history—in recent years, these events and many more have helped make the earth sciences among the most dynamic and exciting of disciplines. To help the beginning student share in this excitement and gain a basic understanding of the physical environment, we have made every effort to write a text that is highly usable—a tool for learning the basic concepts of earth science. To achieve this goal, the book is written concisely, contains useful learning aids, and is well illustrated. Each chapter begins with a brief overview, important terms are indicated in boldface type, key phrases are italicized for emphasis, and review questions and a list of key terms conclude each chapter. Furthermore, a glossary serves as a ready reference for significant terms and ideas. Diagrams and photographs constitute an important part of the text and serve to make the concepts less abstract.

Earth Science is a broad and nonquantitative survey at the introductory level of topics in geology, oceanography, meteorology, and astronomy. It includes an up-to-date chapter on plate tectonics and integrates the concepts from this important and exciting theory into the chapters on volcanic activity, mountain building, and oceanography. The atmosphere unit, in addition to surveying weather elements and patterns, contains a chapter on the topical subject of climatic change as well as sections on air pollution and intentional weather modification. In the astronomy portion of the text, we have when possible incorporated current information from the space program.

While student use of the text is a primary concern, the book's adaptability to the needs and desires of the instructor is equally important. Realizing the broad diversity of earth science courses in both content and approach, we chose a relatively nonintegrated format to allow maximum flexibility for the instructor. Each of the four major units stands alone. Hence, they can be taught in any order, or a unit can be omitted entirely, without appreciable loss of continuity. Moreover, portions of some chapters may be interchanged or excluded at the instructor's discretion. For example, some instructors may wish to include the section on the sun, which appears as a portion of Chapter 17, with the discussion of stars in Chapter 20. Others may wish to incorporate those portions of Chapter 16 dealing with theories of climatic change into discussions of such topics as glaciation, volcanic activity, and astronomy.

In addition to the very useful comments provided by the reviewers of the first edition of *Earth Science*, we benefitted greatly from the constructive input of many professors who used the first edition. We are also grateful for the careful and detailed review of the entire manuscript of the second edition by Professor William S. McLoda. His insights and comments were very helpful. Finally, a special debt of gratitude goes to our wives and children for their encouragement and patience.

contents

introduction

A view of the earth from space affords us a unique perspective of our planet (Figure I.1). At first, it may strike us that the earth is a somewhat fragile-appearing sphere surrounded by the blackness of space. In fact, it is just a speck of matter in an infinite universe. As we look more closely, it becomes apparent that the earth is much more than just rock and soil. Indeed, the most conspicuous features are not the continents, but the swirling clouds suspended above the surface and the vast global ocean. From such a vantage point, we can appreciate why the earth is traditionally divided into three major parts: the solid lithosphere, the liquid hydrosphere, and the gaseous atmosphere. However, the earth is not dominated by rock, water, or air alone. Rather, it is characterized by continuous interaction as air comes in contact with rock, rock with water, and water with air.

The solid earth, or **lithosphere,** may be divided into three principal units: the dense **core;** the less dense **mantle;** and the **crust,** which is the light and very thin outer skin of the earth (Figure I.2).* The crust is not a layer of uniform thickness; rather, it is characterized by many irregularities. It is thinnest beneath the oceans and thickest where continents exist. Although the crust may seem insignificant when compared with the other units of the solid earth, it was created by the same general processes that were responsible for the earth's present structure. Thus, the crust is important in understanding the history and nature of our planet.

The **hydrosphere** is the water portion of our planet. This dynamic mass of liquid is continuously on the move, from the oceans to the air,

*In recent years, since the development of the theory of plate tectonics (Chapter 6), the term *lithosphere* is also used more specifically to denote the rigid outer layer of the earth which includes the crust and upper mantle.

A.

B.

Figure I.1 A. *View of the earth that greeted the* Apollo 8 *astronauts as their spacecraft came from behind the moon.* **B.** *A closer look. In the view, North and South America are largely obscured by clouds. The dark areas represent the ocean. (A. and B. courtesy of NASA)*

to the land, and back again. The global ocean is obviously the most prominent feature of the hydrosphere, blanketing 71 percent of the earth's surface and accounting for about 97 percent of the earth's water. However, the hydrosphere also includes the fresh water found in streams, lakes, and glaciers, as well as that found in the ground. Although these latter sources constitute just a tiny fraction of the total, they are much more important than their meager percentage indicates, for they are responsible for sculpturing and creating many of our planet's varied landforms.

The earth is surrounded by a thick, life-giving gaseous envelope called the **atmosphere.** This blanket of air, hundreds of kilometers thick, is an

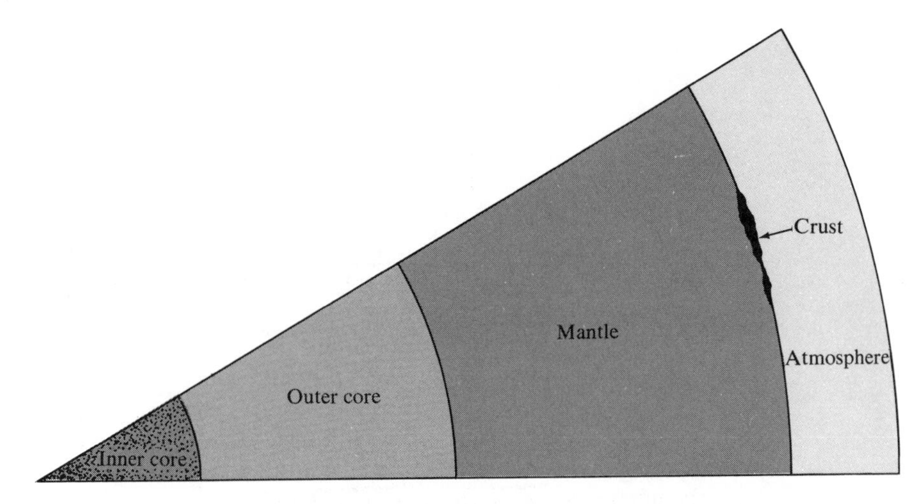

Figure I.2 *Cross-sectional view of the earth's layered structure. The inner core, outer core, and mantle are drawn to scale, but the thickness of the crust is exaggerated by about three times.*

integral part of the planet. It not only provides the air that we breathe but also acts to protect us from the sun's intense heat and dangerous radiation. The energy exchanges that continually occur between the atmosphere and the earth's surface and between the atmosphere and space produce the effects we call weather.

The shape of the earth is essentially spherical, a fact that we all take for granted. To reinforce our belief, all we need to do is view one of the thousands of photographs taken of the earth from space (Figure I.1). For most everyday purposes, it is sufficient to assume that our planet is a sphere with a diameter of about 12,900 kilometers (8000 miles). However, the earth's true form is *not* a perfect sphere. Refined measurements have revealed that the shape of the earth is that of an **oblate spheroid.** That is, *the earth is a sphere that is flattened at the poles and bulged at the equator* (Figure I.3). This not-quite spherical shape is attributed to the effect of the prolonged stress on the somewhat plastic earth by the rotation of our planet about its axis. When the dimensions are rounded off to the nearest whole kilometer, the earth's equatorial

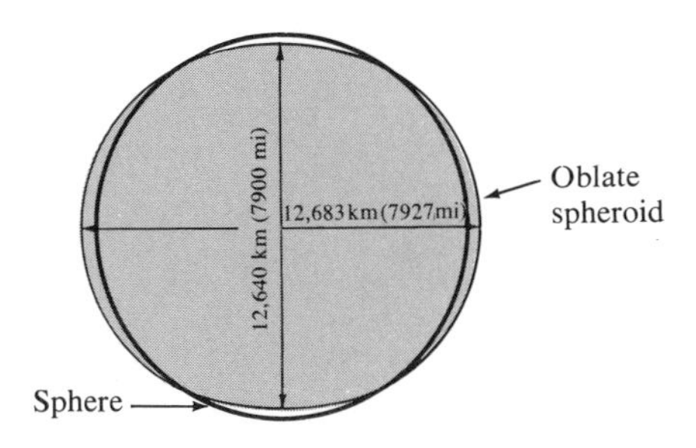

Figure I.3 *Comparison of a sphere and an oblate spheroid.*

diameter is 12,757 kilometers (7927 miles), and the polar diameter is 12,714 kilometers (7900 miles). The difference is only 43 kilometers (27 miles). The degree of oblateness is slight, the polar diameter being only 27/7927 or about 1/300 less than the equatorial diameter. On a standard 16-inch globe, this would amount to an imperceptible difference of only 1/20 inch.

Earth science is a name for all the sciences that collectively seek to understand the earth and its neighbors in space. *Geology*, which literally means "the study of the earth," examines the origin and development of the solid earth, as well as its composition and the processes that operate beneath and upon its surface. In many instances, it requires significant consideration of the hydrosphere and atmosphere. *Meteorology* and *climatology* are concerned with the activities and structure of the atmosphere, and *oceanography* deals with the dynamics of the oceans. At the boundary between the ocean and the atmosphere, oceanography and meteorology merge, and in a similar manner, oceanography merges with geology at the ocean floor.

The study of the earth is not confined to investigations of the lithosphere, hydrosphere, and atmosphere and to their many interactions and interrelationships. The earth sciences also attempt to relate our planet to the larger universe. Because the earth is related to all of the other objects in space, the science of *astronomy* is very useful in probing the origins of our own environment. Since we are so closely acquainted with the planet on which we live, it is easy to forget that the earth is just a tiny object in a vast universe. Indeed, the earth is subject to the same physical laws that govern the great many other objects that populate the infinite expanses of space. Thus, to understand the theories of the earth's origin, it is necessary to learn something about the other members of our solar system. Moreover, it is necessary to view the solar system as a part of the great assemblage of stars that comprise our galaxy, which, in turn, is but one of many galaxies.

part 1

the solid earth

Quartz crystals. (Courtesy of Ward's Natural Science Establishment, Inc., Rochester, N.Y.)

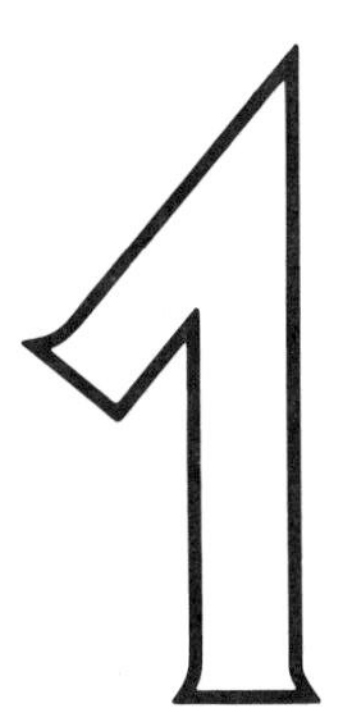

Minerals and Rocks

The study of geology appropriately begins with the study of rocks and minerals, since a knowledge of the materials that make up the earth is basic to our understanding of much of the rest of geology. In the pages that follow we shall examine atoms and elements, the building blocks of minerals, and then investigate the three major groups of rocks. Our inquiry shall attempt to answer some important questions: How do rocks form? How are minerals and rocks identified? What characteristics differentiate one group of rocks from another? What can the study of rocks and minerals tell us about the nature of our planet?

Some Historical Notes about Geology

The nature of our earth—its materials and processes—has been a focus of study for centuries. Writings about such topics as fossils, gems, earthquakes, and volcanoes date back to the Greeks, more than 2300 years ago. Certainly the most influential of the Greek philosophers was Aristotle. Because Aristotle was a philosopher, his explanations were not always based on keen observations and experiments but often were arbitrary pronouncements. He believed that rocks were created under the "influence" of the stars, and that earthquakes occurred when air crowded into the ground, was heated by central fires, and escaped explosively. When confronted with a fossil fish, he explained that, "a great many fishes live in the earth motionless and are found when excavations are made." Although Aristotle's explanations may have been adequate for his day, they unfortunately continued to be expounded for many centuries, thus thwarting the acceptance of more up-to-date accounts. Frank D. Adams states in *The Birth and Development of the Geological Sciences* (New York: Dover, 1938) that, "throughout the Middle Ages Aristotle was regarded as the head and chief of all philosophers; one whose opinion on any subject was authoritative and final."

Catastrophism

During the seventeenth and eighteenth centuries the doctrine of **catastrophism** was the prevailing philosophy guiding explanations about the earth. Briefly stated, catastrophists believed that the earth's landscape had been modeled primarily by great world-shaking catastrophes. Features such as mountains and canyons, which today we know take great periods of time to form, were explained as having been produced rapidly, that is, catastrophically. This philosophy was an attempt to fit the rate of earth processes to the then-current ideas on the age of the earth. In 1654, Archbishop James Usher, a scholar of the Bible, concluded that the earth was approximately 6000 years old, having been created in 4004 B.C. Later, another biblical scholar named Lightfoot was even more specific, declaring that the earth had been created at 9:00 A.M. on October 26, 4004 B.C.

The relationship between catastrophism and the age of the earth has been summarized nicely as follows:

> That the earth had been through tremendous adventures and had seen mighty changes during its obscure past was plainly evident to every inquiring eye; but to concentrate these changes into a few brief millenniums required a tailor-made philosophy, a philosophy whose basis was sudden and violent change.*

The Birth of Modern Geology

The late eighteenth century is generally regarded as the beginning of modern geology, for it was during this period that James Hutton, a Scottish scientist, put forth the concept of **uniformitarianism.** *Uniformitarianism means that the processes which we can observe acting upon and within the earth today are the same processes that have acted in the past.* Simply stated, uniformitarianism means that the present is the key to the past. Today this rather simple idea is one of the foundations of the science of geology.

The acceptance of the concept of uniformitarianism, however, meant the acceptance of a very long history for the earth, for although processes vary in their intensity, they still take a very long time to create or destroy major features of the landscape.

For example, rocks containing fossils of organisms that lived in the sea more than 15 million years ago are now part of mountains that stand 3000 meters (9800 feet) above sea level. This means that the mountains were uplifted 3000 meters in about 15 million years, which works

---

*H.E. Brown, V.E. Monnett, and J.W. Stovall. *Introduction to Geology* (New York: Blaisdell, 1958).

A.

B.

Figure 1.1 *Geologic processes often act so slowly that changes may not be visible during an entire lifetime. These two photographs were taken from the same vantage point nearly one hundred years apart. Photograph A was taken by J.K. Hillers in 1872 and photograph B was taken in 1968 by E.M. Shoemaker. The photos reveal practically no visible signs of erosion. (Photos courtesy of U.S. Geological Survey)*

out to a rate of only 0.2 millimeter per year! Rates of erosion (the processes that wear away land) are equally slow (Figure 1.1). Estimates indicate that the North American continent is being lowered at a rate of just 3 centimeters per 1000 years. Thus, as you can see, it takes tens of millions of years for nature to build mountains and wear them down again. But even these time spans are relatively short on the time scale of earth history, for the rock record contains evidence that shows the earth has experienced many cycles of mountain building and erosion. Concerning the everchanging nature of the earth through great expanses of geologic time, Hutton stated: "We find no sign of a beginning, no prospect of an end." A quote from William L. Stokes sums up the significance of Hutton's basic concept:

> In the sense that uniformitarianism implies the operation of timeless, changeless laws or principles, we can say that nothing in our incomplete but extensive knowledge disagrees with it.*

In the pages of this part, entitled "The Solid Earth," we shall be examining the materials that compose our planet and the processes that modify it. It will be important to remember that although

---

**Essentials of Earth History* (Englewood Cliffs, New Jersey: Prentice-Hall, 1966), p. 34.

many features of our physical landscape may seem to be unchanging in terms of the tens of years we might observe them, they are nevertheless changing, but on time scales of hundreds, thousands, or even many millions of years.

Minerals

The outer layer of the earth—the crust—is only as thick when compared to the remainder of the earth as a peach skin is to a peach, yet it is of supreme importance to us. We depend on it for fossil fuels and as a source of such diverse minerals as talc for baby powder and gold for world trade. In fact, on occasion, the availability or absence of certain earth materials has altered the course of history. The crust is a veritable storehouse of rocks, which in turn are made up of minerals. First we will turn our attention to minerals.

Structure of Minerals

Minerals, like all matter, are made up of **elements.** At present, 103 elements are known, a dozen of which have been produced only in the laboratory. Some minerals like gold are made entirely of one element, but most are a combination of two or more elements joined chemically to form a **compound.**

In order to better understand how elements combine into compounds, we will first consider the smallest part of matter that still retains the characteristics of an element—the **atom**—because it is this extremely small particle that does the combining. Each atom has a central region, called the **nucleus,** which contains very dense positively charged **protons** and neutral particles called **neutrons** (Figure 1.2). The number of protons determines the **atomic number** and name of the element. For example, all atoms with six protons are carbon atoms, all those with eight protons are oxygen atoms, and so forth. Orbiting the nucleus are negatively charged particles called **electrons,** which are equal in number to the protons. As a result, the opposite charges of an atom balance each other.

When an atom combines chemically, it either gains, loses, or shares electrons with another

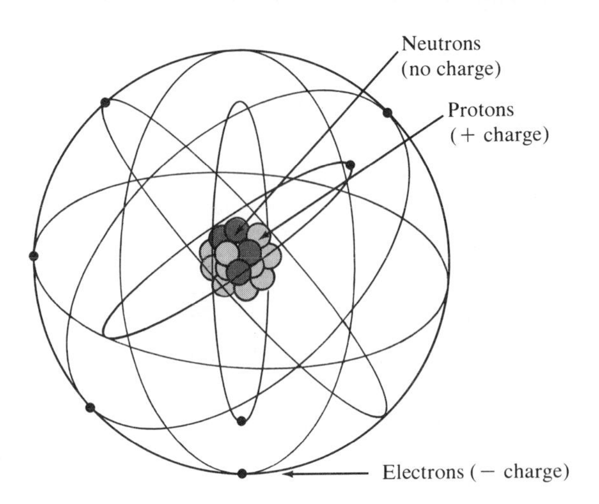

Figure 1.2 *Structure of an atom. Atoms consist of a central nucleus composed of protons and neutrons which is encircled by electrons.*

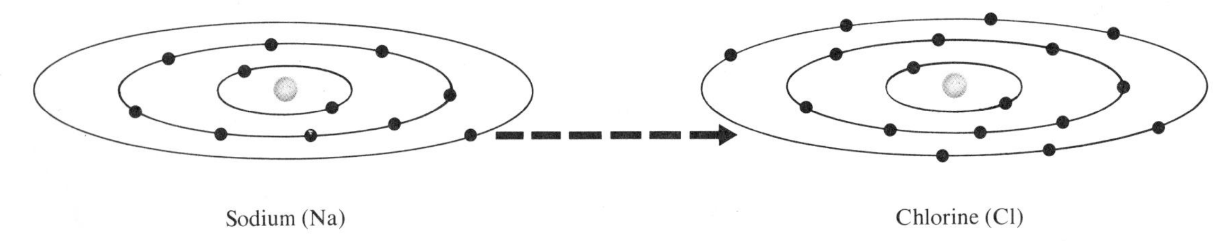

Figure 1.3 *Chemical bonding of sodium and chlorine to produce sodium chloride. By transferring one electron from sodium to chlorine, sodium becomes a positive ion, and chlorine a negative ion.*

atom, making one a positively charged atom and the other negatively charged. Atoms that have an unequal charge because of a gain or loss of electrons are called **ions.** Simply stated, *oppositely charged ions attract one another and produce a neutral chemical compound.*

An example of chemical bonding using sodium (Na) and chlorine (Cl) to produce sodium chloride (NaCl, commonly called table salt) is shown in Figure 1.3. When the sodium atom loses one electron it becomes a positive ion, and when the chlorine atom gains one electron it becomes a negative ion. These opposite charges act as a bond, holding the atoms together. Here, it is interesting to note that chlorine is a green, poisonous gas and sodium is a silvery metal, which, if held in your hand, could burn a hole in it. Together, however, these atoms produce the compound sodium chloride, which looks nothing like either and is a requirement for human life. This example can also serve to illustrate a difference between a rock and a mineral. The mineral sodium chloride (called halite) is a chemical compound and has properties unique to it and grossly different from the elements which make it up. A **rock,** on the other hand, is a *mixture* of minerals, with each mineral retaining its own characteristics.

Although electrons have the active role in a chemical reaction, they do not contribute significantly to the weight (mass) of an atom. It follows then that the weight of an atom is centered in the nucleus. The **atomic weight** is obtained by totaling the number of neutrons and protons in the nucleus. It is not uncommon, however, for atoms of the same element to have varying numbers of neutrons, and therefore to have different atomic weights. Such atoms are called **isotopes** of that element. For example, carbon has two common isotopes, one having a weight of 12 (C_{12}), the other having a weight of 14 (C_{14}). Recall that all atoms of the same element must have the same number of protons (atomic number) and that carbon always has six. Hence, carbon-12 has six neutrons to give it a weight of 12, and carbon-14 has eight to give it a weight of 14.

Most isotopes are stable, but a few like carbon-14 go through a process of natural disintegration called **radioactivity,** which occurs when the forces that bind the nucleus together are not strong enough. The rate at which the unstable nuclei of these elements break apart (decay) is measurable and makes such elements useful "clocks" in dating the events of earth history. (A more complete discussion of radioactivity and dating may be found in Chapter 9.)

Physical Properties of Minerals

Minerals are solids formed by inorganic processes. *Each mineral has an orderly arrangement of atoms (crystalline structure) and a definite chemical composition which give it a unique set of physical properties.* Since the internal structure and chemical composition of a mineral are difficult to determine without the aid of sophisticated tests and apparatus, the more easily recognized physical properties are used in identification. A discussion of some diagnostic physical properties follows.

Figure 1.4 *Galena crystals. (Courtesy of Ward's Natural Science Establishment, Inc., Rochester, N.Y.)*

Crystal Form. When the external expression of a mineral reflects its orderly internal arrangement of atoms, the mineral is demonstrating its **crystal form.** Figure 1.4 illustrates the characteristic crystal form of the lead-bearing mineral galena. *Any time a mineral is permitted to form without space restrictions, it will develop individual crystals.* However, most of the time crystal growth is interrupted because of competition for space, resulting in an intergrown mass of crystals, none of which exhibit their crystal form.

Color. Although color is the most obvious feature of a mineral, it is possibly the least reliable. Slight impurities in the common mineral quartz, for example, give it a variety of colors, including pink, purple (amethyst), white, and even black. For other minerals like sulfur, color is an excellent diagnostic property.

Streak. **Streak** is the color of a mineral in its powdered form and is obtained by rubbing the mineral across a plate of unglazed porcelain. While the color of a mineral often varies from sample to sample, the streak usually does not, and is therefore the more reliable property.

Figure 1.5 *Smooth surfaces produced when a mineral with cleavage is broken. This sample exhibits three planes of cleavage (six sides), and all planes meet at 90-degree angles.*

Luster. **Luster** is the appearance or quality of light reflected from the surface of a mineral. Minerals that have the appearance of metals, regardless of color, are said to have a *metallic luster.* Minerals with a *nonmetallic luster* are described by various adjectives, including *vitreous (glassy), pearly, silky, resinous,* and *earthy (dull).*

Hardness. One of the most useful diagnostic properties of a mineral is **hardness,** the resistance of a mineral to abrasion or scratching. This is a relative property which is determined by rubbing a mineral of unknown hardness against one of known hardness, or vice versa. A numerical value can be obtained by using **Mohs scale** of hardness, which consists of ten minerals arranged in order from 1 (softest) to 10 (hardest) as follows:

Hardness	*Mineral*
1	Talc
2	Gypsum
3	Calcite
4	Fluorite
5	Apatite
6	Orthoclase
7	Quartz
8	Topaz
9	Corundum
10	Diamond

Any mineral of unknown hardness can be compared to these or to other objects of known hardness. A fingernail has a hardness of 2.5, a copper penny 3, and a piece of glass 5.5.

Cleavage. **Cleavage** is the tendency of a mineral to break along planes of weak bonding. The cleavage of a mineral is described by the number of planes exhibited and the angles at which they meet (Figure 1.5). Some minerals have several cleavage planes which produce smooth surfaces when broken, while others exhibit poor cleavage, and still others have no cleavage at all.

Cleavage should not be confused with crystal form. When a mineral exhibits cleavage, it will break into pieces which have the same configuration as the parent. By comparison, the quartz crystals shown on the opening page of this chapter do not have cleavage and, if broken, would shatter into shapes that do not resemble each other or the original crystals.

Fracture. Minerals that do not exhibit cleavage are said to **fracture** when broken. Those that break into smooth curved surfaces like broken glass have a *conchoidal fracture* (Figure 1.12). Others break into splinters or fibers, but most fracture irregularly.

Specific Gravity. **Specific gravity** is a number which represents the ratio of the weight of a mineral to the weight of an equal volume of water. For example, if a mineral weighs three times as much as an equal volume of water, its specific gravity is 3. With a little practice, you can estimate the specific gravity of minerals by hefting them in your hand. The average specific gravity for minerals is around 2.6, but some metallic minerals have a specific gravity two or three times greater.

The Silicates

Over 2000 minerals are presently known to exist. Fortunately for those studying minerals, fewer than 20 are abundant. Collectively, these few make up most of the earth's crust and are classed as the **rock-forming minerals.** It is also interesting to note that only eight elements compose the bulk of these minerals and over 98 percent (by weight) of the continental crust (Table 1.1). The two most abundant elements are silicon and oxygen, which combine to form the mineral group known as the **silicates.** Every silicate mineral contains oxygen and silicon, and all except quartz contain one or more additional elements to complete their structure.

All silicate minerals have the same basic building block, the **silicon-oxygen tetrahedron.** This structure is composed of four oxygen atoms with a smaller silicon atom positioned in the space between them (Figure 1.6). In some minerals the tetrahedra are joined into chains, sheets, or three-dimensional networks by sharing oxygen atoms (Figure 1.7). These larger silicate structures are then connected to each other by other elements. The primary elements that join silicate structures are iron (Fe), magnesium (Mg), potassium (K), sodium (Na). and calcium (Ca). Because iron and magnesium are nearly the same size, they can readily substitute for each other without changing the structure of the mineral. This also holds true for calcium and sodium, which can occupy the same site in a crystal structure. Aluminum (Al) is unique as a substitute for silicon in the tetrahedron.

The main groups of silicate minerals and common examples of each are given in Figure 1.8. The feldspars are by far the most adundant

Table 1.1 ***Relative Abundance of Major Elements in the Continental Crust***

Element	Approximate % by Weight
Oxygen (O)	46
Silicon (Si)	27
Aluminum (Al)	8
Iron (Fe)	6
Calcium (Ca)	4
Potassium (K)	3
Sodium (Na)	3
Magnesium (Mg)	3
Total	100

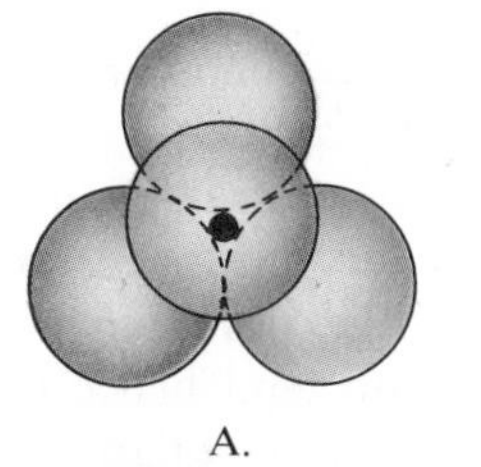

Figure 1.6 *Top view of the oxygen-silicon tetrahedron.* **A.** *The four large spheres represent oxygen atoms, and the one dark one represents a silicon atom.* **B.** *Diagrammatic representation of the tetrahedron using four points to represent the positions of the oxygen atoms.*

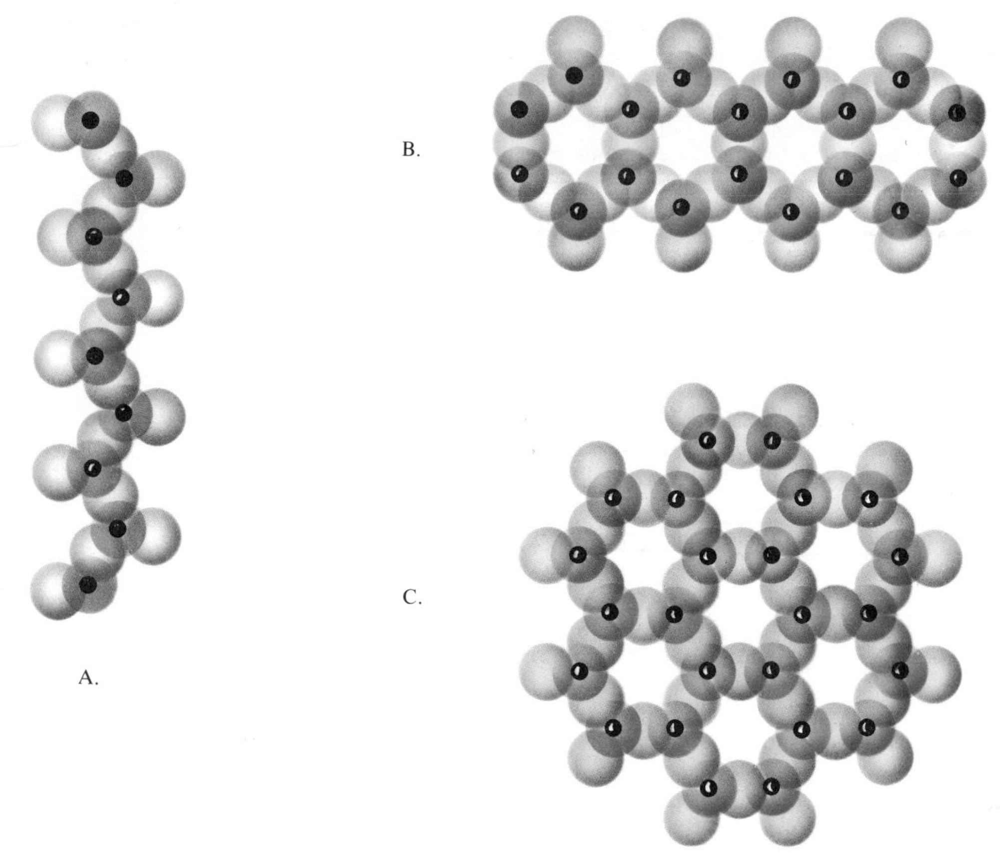

Figure 1.7 *Three types of silicate structures.* **A.** *Single chains.* **B.** *Double chains.* **C.** *Sheet structures.*

group, comprising over 50 percent of the crust. Quartz, the second most common mineral in the continental crust, is the only one made completely of silicon and oxygen.

Because the silicon-oxygen bonds are strong, silicate minerals tend to cleave (break) along the silicate structure rather than across it. Each group has a particular silicate structure and, consequently, a particular cleavage. For example, the micas have a sheet structure and hence cleave into flat plates (Figure 1.9), while quartz, which has equally strong bonds in all directions, has no cleavage.

Nonsilicate Minerals

Although other mineral groups can be considered scarce when compared with the silicates, many are important from an economic standpoint. Table 1.2 lists some examples of oxides, sulfides, sulfates, halides, and native elements that are of economic value. Another important rock-forming group is the carbonates, which includes the mineral calcite ($CaCO_3$). This mineral is the major constituent of limestone and marble.

The Rock Cycle

The **rock cycle** (Figure 1.10) is one means of viewing many of the interrelationships of geology. By studying the rock cycle we may ascertain the origin of the three basic rock types and gain some insight into the role of various geologic processes in transforming one rock type into an-

Mineral		Idealized Formula	Cleavage	Silicate Structure
Olivine		$(Mg,Fe)_2SiO_4$	None	Single tetrahedron
Pyroxene group		$(Mg,Fe)SiO_3$	Two planes at right angles	Chains 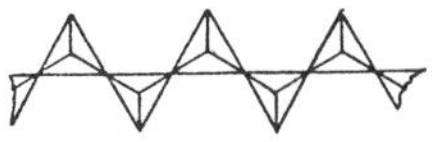
Amphibole group		$(Ca_2Mg_5)Si_8O_{22}(OH)_2$	Two planes at 60° and 120°	Double chains 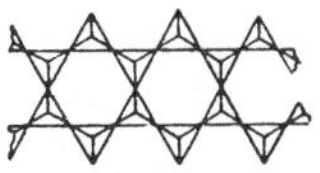
Micas	Muscovite	$KAl_3Si_3O_{10}(OH)_2$	One plane	Sheets 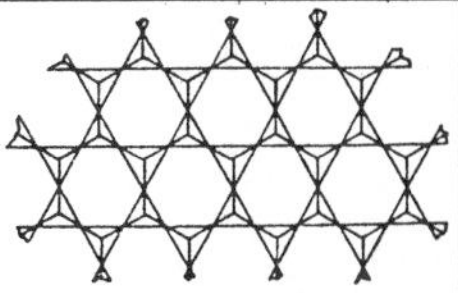
	Biotite	$K(Mg,Fe)_3Si_3O_{10}(OH)_2$		
Feld-spars	Orthoclase	$KAlSi_3O_8$	Two planes at 90°	Three-dimensional networks 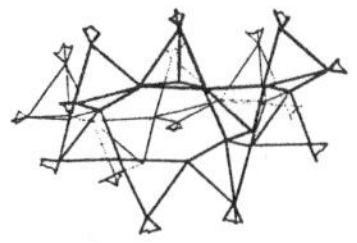
	Plagioclase	$(Ca,Na)AlSi_3O_8$		
Quartz		SiO_2	None	

Figure 1.8 *Common silicate minerals. Note that the complexity of the silicate structure increases down the chart.*

other. The concept of the rock cycle, which may be considered as a basic outline of physical geology, was initially proposed by James Hutton. This rock cycle, shown in Figure 1.10, indicates processes by arrows and materials in boxes.

The first rock type, **igneous rock,** originates when molten material called **magma** cools and solidifies. This process, called **crystallization,** may occur either deep beneath the earth's surface or, following a volcanic eruption, at the surface. Initially, or shortly after forming, the earth is believed to have been molten. Therefore igneous rocks were the first rocks to compose the earth's crust.

If igneous rocks are exposed at the surface of the earth, they will undergo **weathering,** in which the day-in-and-day-out influences of the atmosphere slowly break the rocks into tiny bits. These bits will be picked up, transported, and deposited by any of a number of erosional agents—gravity, running water, glaciers, wind, or waves. Once these bits of rock, called **sediment,** are deposited, usually as horizontal beds in the ocean, they will undergo **lithification,** a term meaning "conversion into rock." Sediment is lithified when compacted by the weight of overlying layers or when cemented as percolating groundwater fills the pores with mineral matter. If the resulting **sedimentary rock** is buried deep within the earth or involved in the dynamics of mountain

Figure 1.9 *Sheet-type cleavage common to the micas. (Courtesy of Ward's Natural Science Establishment, Inc., Rochester, N.Y.)*

Table 1.2 ***Common Nonsilicate Mineral Groups***

Group	Member	Formula	Description
Oxides	Hematite	Fe_2O_3	Ore of iron
	Magnetite	Fe_3O_4	Ore of iron
	Corundum	Al_2O_3	Used as an abrasive
	Ice	H_2O	Solid form of water
Sulfides	Galena	PbS	Ore of lead
	Sphalerite	ZnS	Ore of zinc
	Pyrite	FeS_2	Fool's gold
	Chalcopyrite	$CuFeS_2$	Ore of copper
Sulfates	Gypsum	$CaSO_4 \cdot 2\,H_2O$	Used for plaster
	Anhydrite	$CaSO_4$	Used for plaster
Native elements	Gold	Au	
	Copper	Cu	
	Diamond	C	
	Sulfur	S	
	Graphite	C	
Halides	Halite	NaCl	Common salt
	Fluorite	CaF_2	Used in steel making, chemicals, ceramics

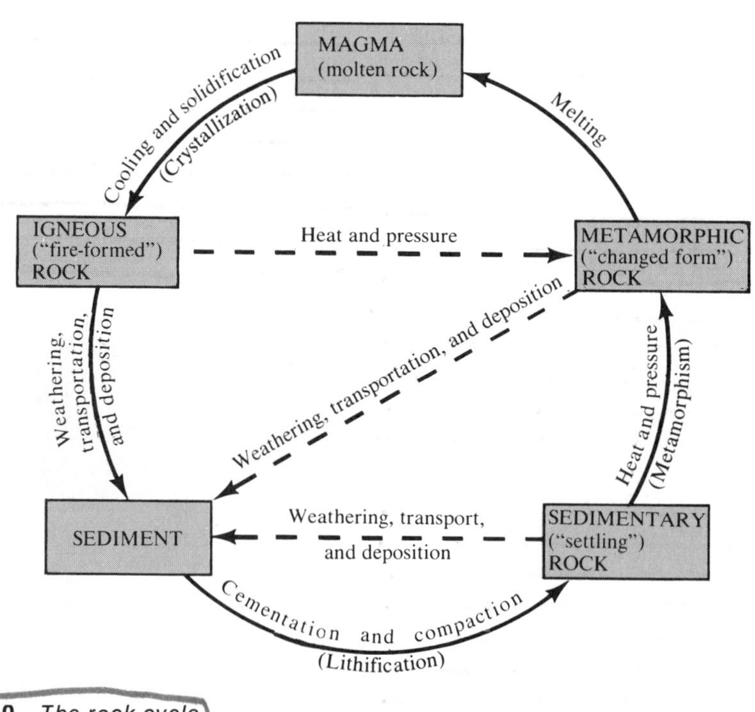

Figure 1.10 *The rock cycle.*

building, it will be subjected to great pressures and heat. The sedimentary rock will react to the changing environment and turn into the third rock type, **metamorphic rock.** When metamorphic rock is subjected to still greater heat and pressure, it will melt, creating magma, which will eventually solidify as igneous rock.

The full cycle just described does not always take place. "Shortcuts" in the cycle are indicated by dashed lines in Figure 1.10. Igneous rock, for example, rather than being exposed to weathering and erosion at the earth's surface, may be subjected to the heat and pressure found far below and change to metamorphic rock. On the other hand, metamorphic and sedimentary rocks, as well as sediment, may be exposed at the surface and turned into new raw materials for sedimentary rock.

As you study the remaining pages of this chapter, which discuss details of each of the three basic rock types, remember the rock cycle. Although rocks may seem to be unchanging masses, the rock cycle shows us that they are not. The changes, however, take time—great amounts of time.

Igneous Rocks

Igneous rocks form from the cooling and crystallization of hot, molten magma. This molten rock consists primarily of the eight abundant elements shown in Table 1.1, along with some gasses, particularly water vapor, which are contained within the magma by the pressure of the confining rocks. The magma body, which is lighter than the surrounding rocks, works its way toward the surface, and on occasion breaks through, producing a volcanic eruption. The spectacular explosions which sometimes accompany an eruption are produced by the gasses (volatiles) escaping as the confining pressure is reduced near the surface. Sometimes blockage of the vent and surface water seepage into the magma chamber can produce catastrophic explosions. Along with ejected rock fragments, a volcanic eruption generates extensive lava flows. **Lava** is similar to magma, except most of the gasses have escaped. The rocks which result when lava solidifies are classified as **extrusive,** or **volcanic.** The magma which does not reach the surface must eventually crystallize at depth. The igneous rocks produced in this manner are termed **intrusive,** or **plutonic,** and they would never be observed if erosion did not strip away the overlying rocks.

Formation and Classification of Igneous Rocks

A large variety of igneous rocks exist and are differentiated on the basis of their texture and mineral content. The **texture,** or size of the crystals composing the rock, is determined primarily by the rate at which the magma cools. *Rapid cooling yields small crystals, while slow cooling results in the formation of larger crystals.* Igneous rocks which form at the earth's surface or as small tabular masses below demonstrate a fine-grained texture, with the individual crystals being too small to be seen with the unaided eye. Intrusive rocks, especially those associated with massive features such as batholiths and laccoliths (see Chapter 7), exhibit a coarse-grained texture, a result of the slow cooling that produced them.

A large mass of magma at depth may require tens of thousands, even millions, of years to solidify. Since all minerals within a magma do not crystallize at the same rate, it is possible for some to become quite large before others even start to crystallize. If magma containing some large crystals should change environments by erupting at the surface, for example, the molten portion of the lava would cool quickly. The resulting rock, having large crystals embedded in a matrix of fine crystals, is called a **porphyry** (Figure 1.11). Some igneous rocks, like obsidian, are **glassy** (Figure 1.12). When lava cools very rapidly, there is not sufficient time for the ions in the molten solution to arrange themselves in an orderly fashion. The igneous rock that results has no crystalline structure and is termed a volcanic glass. Finally, some igneous rocks contain numerous holes (vesicles) which were produced as gasses escaped from the upper por-

A.

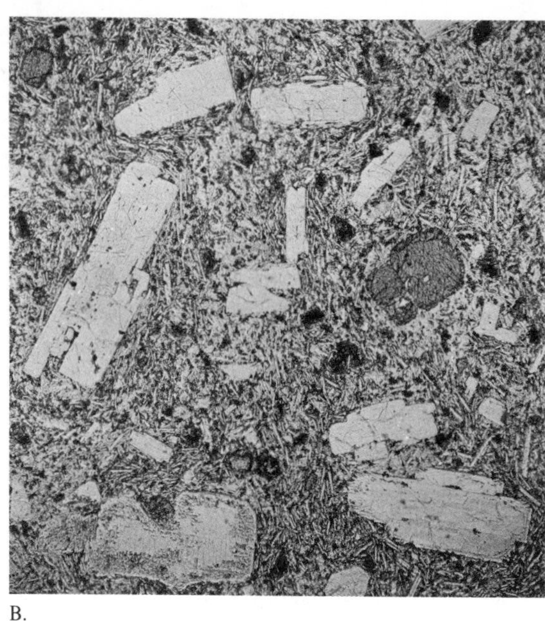

B.

Figure 1.11 *Porphyritic texture.* **A.** *Hand sample exhibiting two grain sizes.* **B.** *Microscopic view of a thin slice of a porphyry. (Courtesy of U.S. Geological Survey)*

tion of a lava flow. Such a rock is said to be **vesicular** (Figure 1.13). In summary, it is quite apparent that *the texture of an igneous rock provides numerous clues that can be used to decipher the circumstances of the rock's formation.*

The mineral makeup of an igneous rock is determined by the chemical composition of the magma from which it crystallized. Because such a large variety of igneous rocks exists, it seems logical to assume that an equally large variety of magmas must also exist. Geologists have found that various eruptive stages of the same volcano often have magmas exhibiting somewhat different mineral compositions, particularly if an extensive period of time separated eruptions. Evidence of this type led them to look into the possibility that a single magma might produce rocks of varying mineral content. The initial investigation into the crystallization of magma was done by N.L. Bowen in the first quarter of this century. Bowen found that as magma was cooled in a laboratory, certain minerals always crystallized first. Since those that crystallized were no longer a part of the **melt** (liquid portion of a magma, excluding any solid material), the concentration of each element remaining in the liquid was increased. At successively lower temperatures, other minerals would begin to crystallize as shown in the sequence in Figure 1.14. It is possible to visualize how the earliest-

Figure 1.12 *Obsidian, a glassy volcanic rock. (Courtesy of Ward's Natural Science Establishment, Inc., Rochester, N.Y.)*

formed minerals could settle to the bottom of the magma chamber to form a rock (Figure 1.15A). When the remaining melt crystallizes, either in place or in a new location, which can happen if it is squeezed out of the magma chamber, it will form a rock with a mineral composition much different from the first (Figure 1.15B). This process which involves the segregation of minerals by crystallization and settling is called **fractional crystallization.**

Bowen also demonstrated that if a mineral remained in the melt after it crystallized it would react with the remaining melt and produce the next mineral in the sequence. The two reaction series discovered by Bowen are shown in Figure 1.14. In the left-hand series, olivine is the first mineral to form, and if it is not removed, it will react and become pyroxene. As the temperature drops, the minerals continually change, until biotite, the last mineral in this series, is formed. The right-hand series is a continuum where calcium-rich feldspar will become more and more sodium rich as crystallization continues. At any given stage in the crystallization process the remaining melt has a slightly different chemical composition, and consequently a different rock type will be produced when it crystallizes. Dur-

Figure 1.13 *Vesicular texture. Vesicles form as gas bubbles escape near the top of a lava flow.*

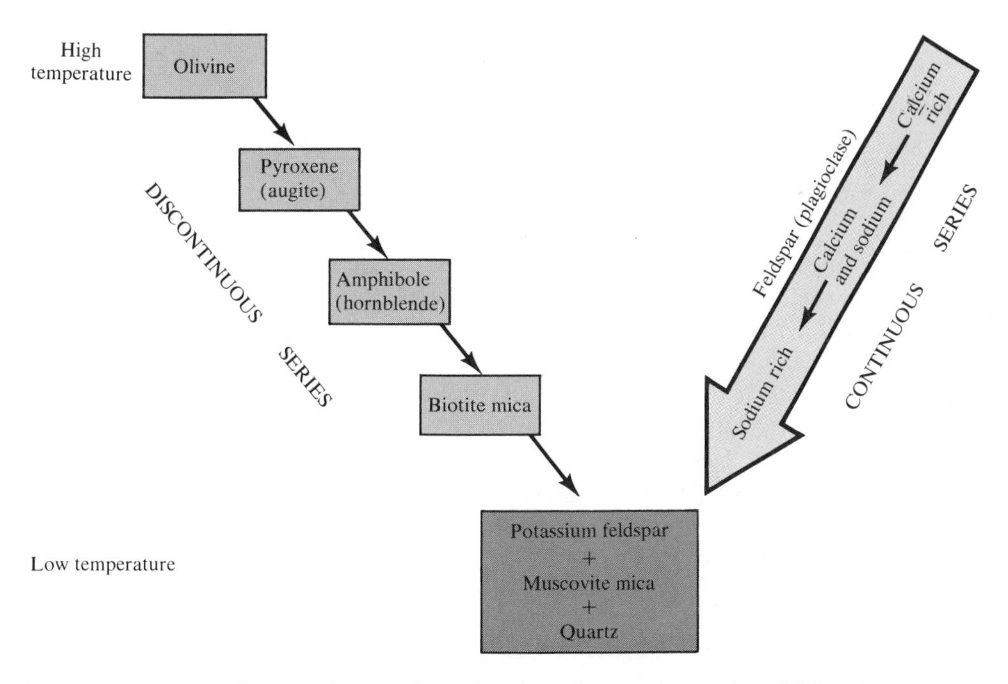

Figure 1.14 *Bowen's reaction series, showing the sequence in which minerals crystallize from a magma. Compare this figure to the mineral composition of the rock groups in Table 1.3. Note that each rock group consists of minerals that crystallize at the same time.*

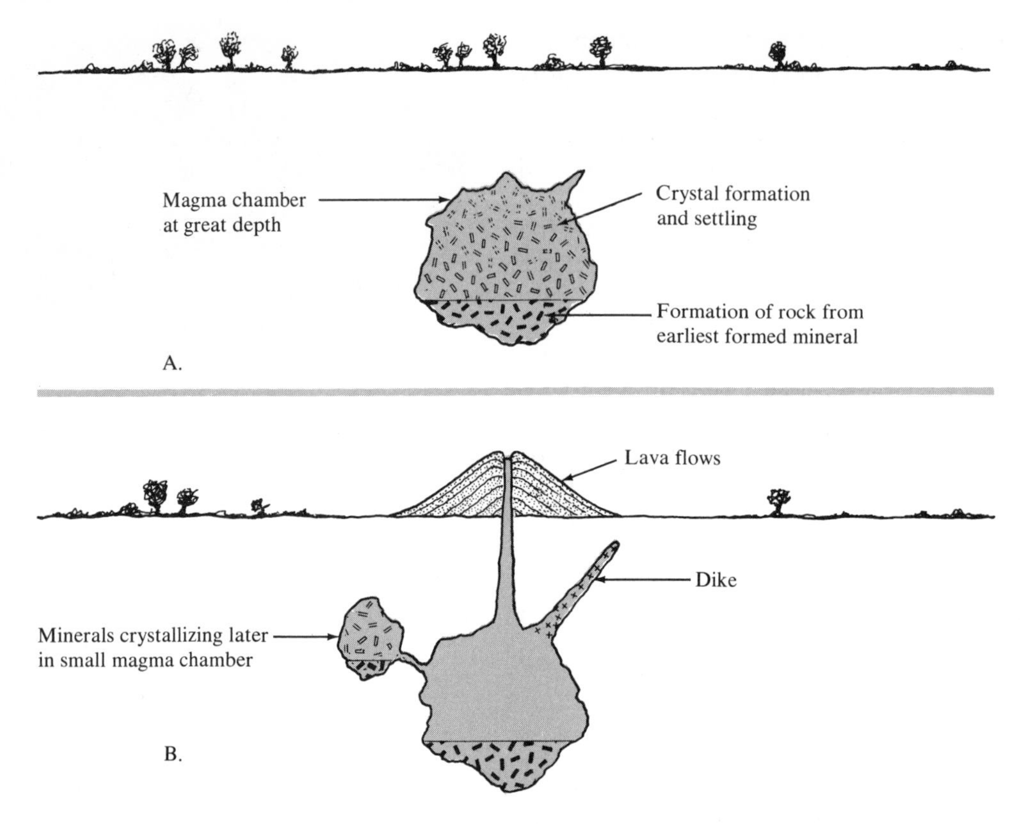

Figure 1.15 *Separation of minerals by fractional crystallization.* **A.** *Illustration of how the earliest-formed minerals can be separated from a magma by settling.* **B.** *The remaining melt could be squeezed into a number of different locations and, upon further crystallization, generate rocks having a composition much different from the first.*

ing the last stage of crystallization, after about 85 percent of the magma has solidified, the remaining liquid will form the minerals quartz, potassium feldspar, and muscovite.

Minerals that crystallize from a magma at about the same time are most often found together composing the same igneous rock. Hence the classification of igneous rocks closely approximates Bowen's reaction series. A general classification scheme based on texture and mineral composition is provided in Table 1.3. The rocks on the right side of the chart consist of the first minerals to crystallize and are higher in iron (Fe) and magnesium (Mg) and lower in silica (SiO_2) than later ones. Their iron content makes them darker in color and slightly heavier than the other rocks. The term **mafic,** derived from **Ma**gnesium and **Fe**, the chemical symbol for iron, is used for rocks of this composition. The rocks on the left are the last to crystallize and are made mainly of **fel**dspar and **si**lica (quartz); consequently the term **felsic** is applied to them. The **intermediate** igneous rocks are made up of minerals from the middle region of the reaction series, and the **ultramafic** rocks from the very earliest minerals to crystallize, mainly olivine.

Because more felsic rock (granite) exists than can be accounted for by the fractional crystallization of magma, other sources have been explored to explain its formation. It has been demonstrated that partial melting of a rock will generate a magma with a composition closer to

Table 1.3 ***Common Igneous Rocks***

	Felsic	Intermediate	Mafic	Ultramafic
Intrusive	Granite	Diorite	Gabbro	Peridotite
Extrusive	Rhyolite	Andesite	Basalt	None
Mineral composition	Quartz Potassium feldspar	Hornblende Sodium feldspar Calcium feldspar	Calcium feldspar Pyroxenes	Olivine Pyroxenes
Minor mineral constituents	Sodium feldspar Muscovite Biotite Hornblende	Biotite Pyroxenes	Olivine Hornblende	Calcium feldspar

the felsic end of the spectrum (Table 1.3). Melting, like crystallization, occurs in stages, but in exactly the opposite order. The felsic minerals, which are the last to crystallize, are the first to melt. For example, partial melting of an ultramafic rock like peridotite would yield a melt of mafic composition similar to the rock basalt. The volcanic islands of Hawaii are basaltic in composition and are believed to have been produced by the partial melting of ultramafic rocks at depth. Further, partial melting of intermediate rocks is undoubtedly a source of some felsic rocks. Another mechanism thought to generate felsic magma is the melting of sediments which were derived from preexisting granitic rock. Simply stated, the magma is a recycled product—a part of the rock cycle described earlier.

Sedimentary Rocks

The weathering process attacks rocks at the earth's surface by either chemical or mechanical means and breaks them into bits and pieces of all sizes. The end product, sediment, is transported, most often by running water, deposited, usually in the ocean, and eventually lithified. *Sedimentary rocks, therefore, are formed when the weathered products of preexisting rocks are lithified.* The word *sedimentary* gives some clue to the nature of these rocks, for it is derived from the Latin *sedimentum,* which means "settling," referring to the settling out of material from a fluid. Not all sediment is deposited in this fashion, but the great bulk of it is. Although sedimentary rocks represent less than 10 percent of the outer 16 kilometers (10 miles) of the earth, they account for about 75 percent of the rocks at the earth's surface. Therefore we may think of sedimentary rocks as comprising a very thin and somewhat discontinuous layer in the uppermost portion of the crust. This fact is readily understood when we consider that these rocks form at or very near the surface of the earth.

Classification

The sediments from which sedimentary rocks are composed may be grouped into two basic categories: clastic and nonclastic. We shall deal with clastic sediments first.

Clastic Sediments. **Clastic sediments** are broken fragments of preexisting rock ranging in size from microscopic clay particles to very large boulders. Sedimentary rocks composed of these materials are classified according to the size of the particles which predominate, Wentworth's scale being our guide (Table 1.4). Where gravel-sized particles predominate, the rock is called *conglomerate* if the sediment is rounded, and *breccia* if the pieces are angular (Figure 1.16). *Sandstone* is the name given rocks where sand-sized grains prevail, while *shale,* the most common sedimentary rock, is made of very fine-grained sediment. *Siltstone,* another

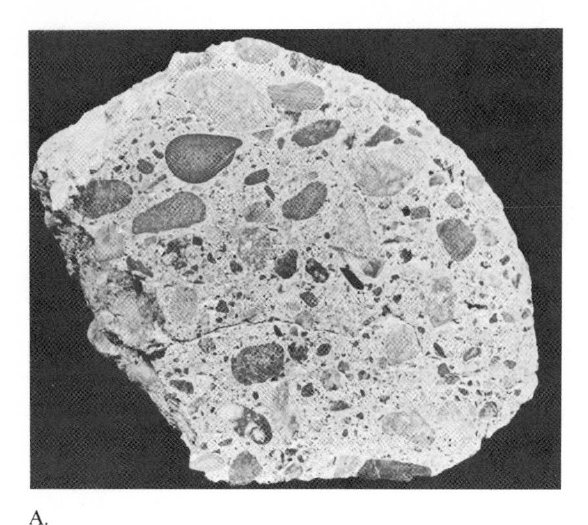

A.

B.

Figure 1.16 **A.** *Conglomerate.* **B.** *Breccia. (Courtesy of Ward's Natural Science Establishment, Inc., Rochester, N.Y.)*

rather fine-grained sedimentary rock, may sometimes be difficult to differentiate from rocks such as shale that are composed of even smaller, clay-sized sediment (Figure 1.17). One test if you are in doubt is to "taste" the sample in question. If the sediment "tastes" gritty, it is likely silt; if not, it is clay.

Clastic sedimentary rocks are classified primarily on the basis of the size of the rock and mineral fragments which compose them. There are some exceptions to this statement; that is, in some cases the mineral composition of a clastic sedimentary rock plays a part in its classification. For example, most sandstones are predominantly quartz, but when appreciable quantities of feldspar are present, the rock is called *arkose.* When the sandstone includes a quantity of clay as well as angular quartz grains, the rock is *graywacke.* In addition, rocks composed of clastic sediments are rarely composed of grains of just one size. Consequently, we may correctly classify a rock containing quantities of both sand and silt as *sandy siltstone* or *silty sandstone,* depending upon which particle size dominates.

Nonclastic Sediments. The second major category of sediments are the **nonclastics,** which may be subdivided as follows: chemical precipitates, biochemical sediment, and organic sediment. The first of these, **chemical precipitates,** forms when mineral matter in solution is precipitated from water, usually because of changes in water temperature or in the chemical makeup of the water. For example, when ocean currents carry cool water into a warm region, the change in temperature causes some calcite (calcium carbonate) to precipitate from the solution. Cal-

Table 1.4 ***Wentworth's Scale of Clastic Sediments***

Sediment Name	Size Range (mm)	Clastic Rock Name
Gravel	> 2	Conglomerate and breccia
Sand	1/16–2	Sandstone
Silt	1/256–1/16	Siltstone
Clay	$< 1/256$	Shale

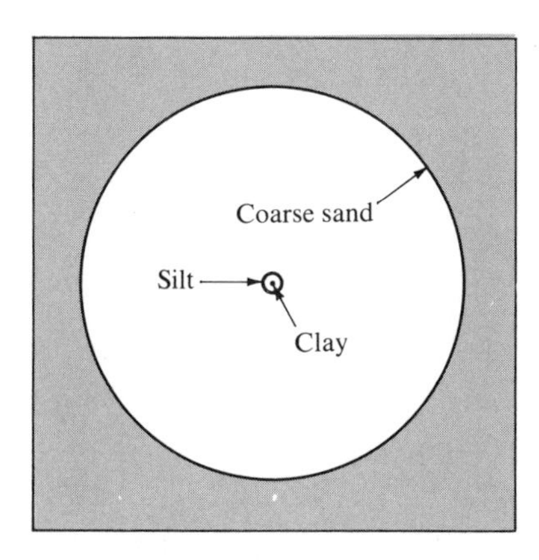

Figure 1.17 *Diagram of the relative sizes of three kinds of clastic sediments. Even though the sediments have been enlarged about 60 times, the clay particle can barely be seen.*

cite is the main constituent of the sedimentary rock *limestone,* which is produced primarily in tropical waters. When silica (SiO_2) is precipitated, the result is various forms of quartz, including *flint, chert, jasper,* and *agate.* Often the mechanism triggering deposition of chemical precipitates is evaporation. Minerals commonly precipitated in this fashion include halite (sodium chloride), the chief component of *rock salt,* and gypsum (hydrous calcium sulphate), the main ingredient of *rock gypsum.*

In the geologic past, many places that are now dry land were covered by shallow arms of the sea that eventually became cut off from the main body of water (much as the Caspian Sea is today). As the water evaporated, the salts were left behind as **evaporite deposits.** Today these deposits serve as an important source of many chemicals. Similar deposits may be seen in such places as Death Valley, California. Here, following rains or periods of snowmelt in the mountains, streams flow from the surrounding mountains into an enclosed basin. As the water evaporates, it leaves its dissolved materials behind as a white crust on the ground, called salt flats (Figure 1.18).

Biochemical sediment is produced when water-dwelling plants and animals extract from the water dissolved mineral matter, usually calcite, and less often silica, to form shells or other hard parts. These shells, many of which are microscopic, collect in great quantities on the ocean bottom, forming the raw material for sedimentary rocks. The white chalk cliffs of Dover in England, for example, are composed of such material, which has been lithified and uplifted above sea level (Figure 1.19). By far the most common sedimentary rock formed from biochemical sediments is calcite-rich *limestone.* By closely examining a piece of limestone you may often see the shells and shell fragments which compose it.

Coal is the most important sedimentary rock composed of **organic sediment** and is the end product of the burial of large amounts of plant material for extended periods of time. Under the right circumstances a great volume of plants may accumulate rather than totally decompose, which is normally the case. One such environment is an oxygen-poor swamp. At various times during earth history such environments have been relatively common. With each successive stage in coal formation, added pressure drives off impurities and volatiles as follows:

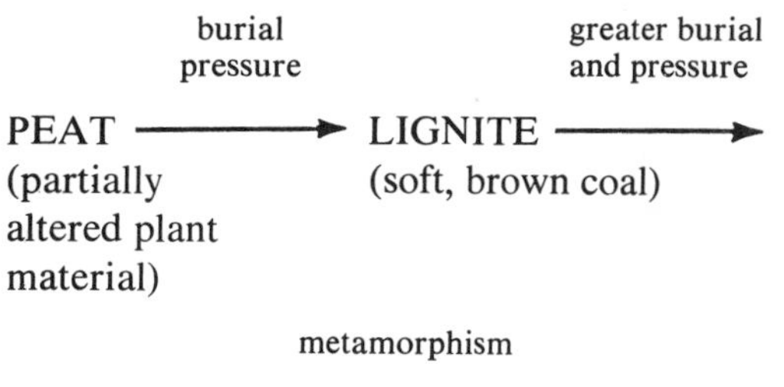

Bituminous coal is by far the most important of our coal resources. *Anthracite* is formed when bituminous coal undergoes metamorphism, and although it is cleaner burning, it is not as widespread and is more expensive to mine.

Organic matter is present in small quantities in many other sedimentary rocks, both clastic and nonclastic. When present, it usually colors the rock black. In such cases the term **carbona-**

Figure 1.18 *These salt flats (composed of gypsum and rock salt) in Death Valley, California, are an example of evaporite deposits. (Photo by C.B. Hunt, U.S. Geological Survey)*

ceous is often applied to describe the rock. Thus a black, organic-rich shale would be called carbonaceous shale.

In summary, the classification of sedimentary rocks is a difficult matter. As was pointed out earlier, many clastic sedimentary rocks are a mixture of more than one particle size. Furthermore, *even though sedimentary rocks are classified as clastic or nonclastic, most are a combination of both.*

Lithification

Sediments are most commonly lithified, that is, turned into rock, through compaction and/or cementation. Compaction occurs when the weight of overlying layers compresses the

Figure 1.19 *The White Chalk Cliffs of Dover. (Courtesy of British Tourist Authority)*

sediment below. As the grains of sediment are pressed closer and closer together there is a considerable reduction in pore space and volume. Compaction is most significant as a lithifier in fine-grained sedimentary rocks, such as shale. Often sediment is cemented together as the openings between particles become filled with mineral matter that has precipitated from percolating underground water. Calcite, silica, and iron oxide are the most common cements. The identification of the cementing material is a relatively simple matter. Calcite cement will effervesce with dilute hydrochloric acid, while iron oxide gives the rock a characteristic red, orange, or yellow color. Silica, the hardest of the cements, produces the hardest sedimentary rocks.

Although most sedimentary rocks are lithified by compaction, cementation, or a combination of both, some are made of interlocking crystals. Chemical precipitates, for the most part, fit in this latter category.

Figure 1.20 *Strata. This outcrop of sedimentary strata illustrates the characteristic layering of this group of rocks. (Photo by G.K. Gilbert, U.S. Geological Survey)*

Features of Sedimentary Rocks

Sedimentary rocks are particularly important in the interpretation of earth history. These rocks form at the earth's surface, and as layer upon layer of sediment accumulates, each records the nature of the environment at the time the sediment was deposited. *These layers, called* **strata,** *or* **beds,** *are probably the single most characteristic feature of sedimentary rocks* (Figure 1.20).

As geologists examine sedimentary rocks, much can be deduced. A conglomerate, for example, may indicate a high-energy environment, like a rushing stream, where only the coarse materials can settle out. If the rock is arkose, it may signify a dry climate, where little chemical alteration of feldspar is possible. Carbonaceous shale is a sign of a low-energy, organic-rich environment such as a swamp or lagoon. Other features found in some sedimentary rocks also give clues to past environments (Figure 1.21).

Fossils, the evidence or remains of prehistoric life, are perhaps the most important inclusions found in sedimentary rock. Knowing the nature of the life forms that existed at a particular time may help to answer many questions about the environment. Was it land or ocean? A lake or swamp? Was the climate hot or cold, rainy or dry? Was the ocean water shallow or deep, turbid or clear? Fossils are important tools used in interpreting the geologic past (more on this in Chapter 9).

Metamorphic Rocks

Metamorphic rocks are formed from sedimentary and igneous rocks by the action of heat, pressure, and chemically active fluids—the *agents of change.* The process of metamorphism takes place when rock is subjected to conditions unlike those in which it formed, becomes unstable, and attempts to achieve a state of equilibrium in the new environment. The changes occur deep beneath the earth's surface, below the zone of weathering and above the zone of melting. Since the formation of metamorphic rocks is completely hidden from view, which is not the case for many

A.
B.
C.

Figure 1.21 **A.** *Ripple marks may indicate a beach or stream-channel environment. (Photo by J.R. Stacy, U.S. Geological Survey)* **B.** *Mud cracks form when wet mud or clay dries out and shrinks, perhaps signifying a tidal flat or a desert basin. (Photo by L.W. Stephenson, U.S. Geological Survey)* **C.** *The cross-bedding in this sandstone indicates that it was once a sand dune. (Photo by H.E. Gregory, U.S. Geological Survey)*

sedimentary and some igneous rocks, metamorphism is undoubtedly one of the most difficult processes for geologists to study.

Metamorphism most often occurs in one of two settings. First, during mountain building great quantities of rock are subjected to the intense pressures and temperatures associated with such large-scale deformation. The end result may be extensive areas of metamorphic rocks that are said to have undergone **regional metamor-**

phism. The greatest volume of metamorphic rocks is produced in this fashion. Second, when rock is in contact or close proximity to a mass of magma, **contact metamorphism** takes place. The changes are caused primarily by the high temperatures of the molten material, which tend to "bake" the surrounding rock.

Since the word *metamorphic* literally means "changed form," what sort of changes are we talking about? Among the most distinguishing features of many metamorphic rocks is something called foliation. Because of the great pressures associated with metamorphism, some minerals like the micas, hornblende, and augite become reoriented and aligned at right angles to the pressure (Figure 1.22). The resulting linear orientation of minerals may give the rock a banded or layered appearance, and the rock is said to be **foliated.** On the other hand, some minerals are not easily realigned, and as a result, rocks composed of such minerals are **nonfoliated,** or **massive.** Table 1.5 lists and briefly describes some common representatives of each category.

Because of the high temperatures, great pressures, and presence of circulating solutions from nearby magma, new minerals may form. The type and size of the new minerals will depend upon the degree or intensity of metamorphism. Heat and pressure will also cause small mineral grains to recrystallize into larger ones, generally resulting in a denser rock than before.

The results of the metamorphic process are very complex. In fact, metamorphic rocks may undergo so much change that the original parent

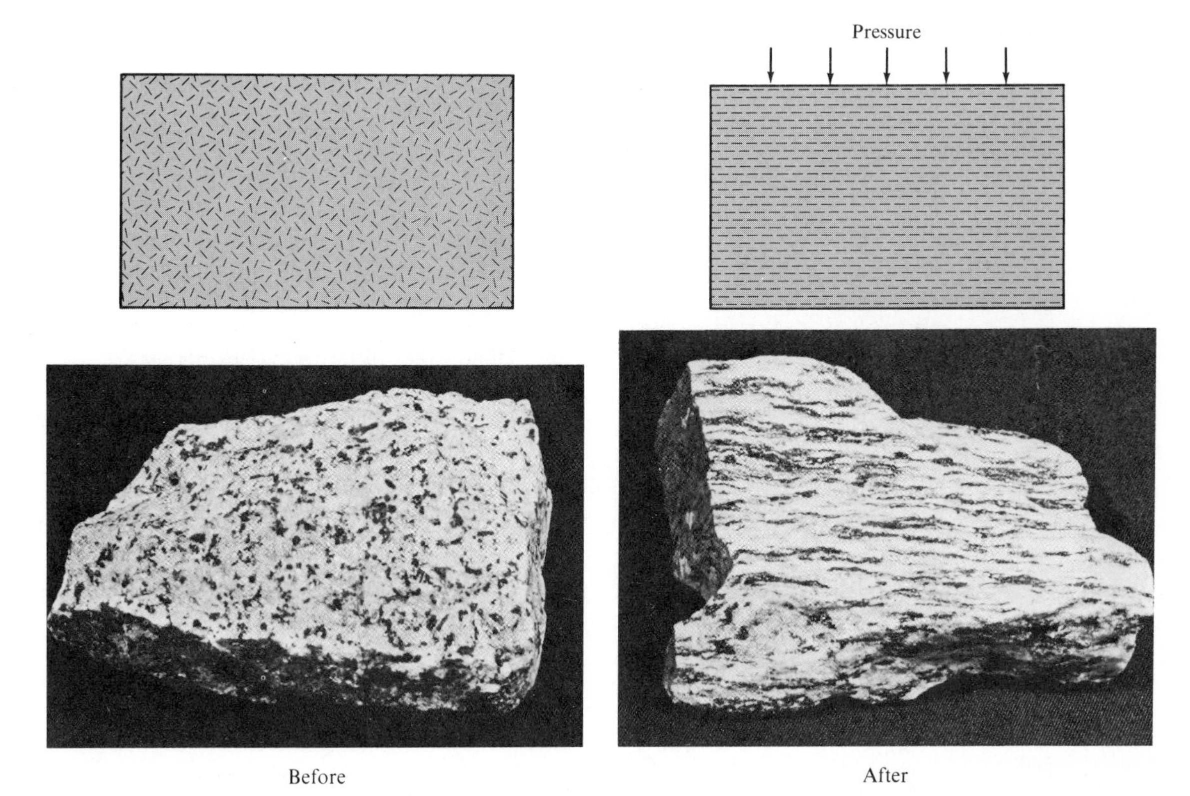

Figure 1.22 *Under the pressures of metamorphism, some mineral grains become reoriented and aligned at right angles to the pressure. The resulting linear orientation of mineral grains gives the rock a foliated texture. If the coarse-grained igneous rock (granite) on the left underwent intense metamorphism, it could end up closely resembling the sample on the right (gneiss).*

Table 1.5 ***Description of Common Metamorphic Rocks***

FOLIATED TEXTURE

Slate Very fine-grained rock composed primarily of microscopic flakes of mica. Slate is produced by the low-grade metamorphism of shale.

Schist Among the most abundant foliated metamorphic rocks, *schists* are composed in large part of visible flakes of platy minerals. As with slate, schist often forms from shale, but in the case of schist, the metamorphism was more intense. Often included are new minerals formed during metamorphism and unique to metamorphic rocks. Schists are named based upon their mineral composition. Figure 1.23 is a photo of garnet-mica schist.

Gneiss (pronounced "nice") Most often has the composition of granite, although other compositions are possible. The outstanding feature of gneiss is a coarsely banded or streaked appearance, reflecting the characteristic alternating layers of light and dark minerals (Figure 1.22).

NONFOLIATED TEXTURE

Marble Results when limestone is metamorphosed. The large interlocking crystals of calcite so characteristic of marble form when the much smaller grains in limestone are recrystallized. When pure, marble is white, but it is often colored by impurities.

Quartzite Common metamorphic rock formed from quartz sandstone. Quartzite may closely resemble marble in appearance but is much harder.

---

rock may be unrecognizable or change so slightly that the new rock is hard to differentiate from its predecessor. As compared to the igneous or sedimentary rocks from which they formed, metamorphic rocks: (1) may have a different mineral composition; (2) may contain minerals that have a linear orientation (foliation); (3) may have larger mineral grains; and (4) will likely be denser.

Figure 1.23 *Garnet-mica schist. (Courtesy of the American Museum of Natural History)*

REVIEW

1. Why did Aristotle have a "bad" influence on the science of geology?
2. Contrast the philosophies of catastrophism and uniformitarianism. How did the proponents of each perceive the age of the earth?
3. List the three main particles of an atom and tell how they differ from one another.
4. If the number of electrons in an atom is 35 and the atomic weight is 80, calculate the following:
   a. the number of protons.
   b. the atomic number.
   c. the number of neutrons.
5. What occurs in an atom to produce an ion?
6. What is an isotope?
7. Although all minerals have an orderly internal arrangement of atoms (crystalline structure), most mineral samples do not demonstrate their crystal form. Explain.
8. Why might it be difficult to identify a mineral by its color?
9. If you found a glassy-appearing mineral while rock hunting and had hopes that it was a diamond, what simple test might help you make a determination?

10. Explain the use of corundum as given in Table 1.2 in terms of Mohs hardness scale.
11. Gold has a specific gravity of almost 20. If a 25-liter pail of water weighs about 25 kilograms, how much would a 25-liter pail of gold weigh?
12. What is the difference between silicon and silicate?
13. Explain the statement "One rock is the raw material for another" using the rock cycle.
14. If a lava flow at the earth's surface had a mafic composition, what rock type would the flow likely be (see Table 1.3)? What igneous rock would form from the same magma if it did not reach the surface but instead crystallized at great depth?
15. What does a porphyritic texture indicate about an igneous rock?
16. How are granite and rhyolite different (see Table 1.3)? The same?
17. Relate the classification of igneous rocks to Bowen's reaction series.
18. How are sediments converted to sedimentary rock?
19. If you had a sample of limestone, how could you determine whether it was composed of biochemical sediment or was a chemical precipitate?
20. For what purpose is Wentworth's scale used?
21. Why do you suppose fossils are found primarily in sedimentary rocks?
22. Why are sedimentary rocks very important in the interpretation of earth history?
23. What is metamorphism? What are the *agents of change*?
24. Distinguish between regional and contact metamorphism.
25. What feature would make schist and gneiss easily distinguished from quartzite and marble?
26. How do metamorphic rocks differ from the igneous and sedimentary rocks from which they formed?

KEY TERMS

Review your understanding of important terms in this chapter by defining and explaining the importance of each term listed below. Terms are listed in their order of occurrence in the chapter.

catastrophism
uniformitarianism
element
compound
atom
nucleus
proton
neutron
atomic number
electron
ion
rock
atomic weight
isotope
radioactivity
crystal form
streak
luster
hardness
Mohs scale
cleavage
fracture
specific gravity
rock-forming minerals
silicates
silicon-oxygen tetrahedron
rock cycle
igneous rock
magma
crystallization
weathering
sediment

lithification
sedimentary rock
metamorphic rock
lava
extrusive (volcanic)
intrusive (plutonic)
texture
porphyry
glassy
vesicular
melt
fractional crystallization
mafic
felsic
intermediate
ultramafic
clastic sediments
nonclastics
chemical precipitate
evaporite deposits
biochemical sediment
organic sediment
carbonaceous
strata (beds)
fossils
regional metamorphism
contact metamorphism
foliated
nonfoliated (massive)

Air view of the Madison Canyon landslide. The slide, triggered by an earthquake, dammed a river, creating a lake. (Photo by J.R. Stacy, U.S. Geological Survey)

Weathering
Soil
Mass Wasting

2

Weathering, Soils, and Mass Wasting

The earth's surface is constantly changing. Rock is disintegrated and decomposed, moved to lower elevations by gravity, and carried away by water, wind, or ice. In this manner the earth's physical landscape is sculptured. This chapter focuses on the first two steps of this never-ending process—weathering and mass wasting—probing into how and why rock breaks up and what mechanisms act to move it downslope. Soil, an important product of the weathering process and a vital resource to people, is also examined.

At a casual glance the face of the earth appears to be without change, unaffected by time. Not too many years ago most people believed that mountains, lakes, and deserts were permanent features of an earth which was thought to be no more than a few thousand years old. However, today we know that mountains must eventually succumb to erosion and be washed into the sea, that lakes will fill with sediment or be drained by streams, and that deserts come and go as minor climatic changes occur. The earth is indeed a dynamic body. Volcanic and tectonic activities are elevating parts of the earth's surface, while opposing processes are continually removing materials from higher elevations and moving them to lower elevations.* The latter processes include:

(1) **Weathering**—the disintegration and decomposition of rock at or near the surface of the earth.
(2) **Erosion**—the incorporation and transportation of material by a mobile agent, usually water, wind, or ice.
(3) **Mass wasting**—the transfer of rock material downslope under the influence of gravity.

Weathering cannot be separated easily from the other two processes, because as weathering breaks apart rock, it encourages the processes of erosion and mass wasting, and the transport of material by erosion and mass wasting in turn furthers the disintegration and decomposition of rock.

Weathering

All materials are susceptible to weathering. You are undoubtedly familiar with the fabricated product concrete, which closely resembles the sedimentary rock conglomerate. A new concrete sidewalk has a smooth, fresh look. Not too many years later that same sidewalk will appear chipped, cracked, and rough, with pebbles exposed at the surface. If a tree is nearby, its roots may heave and buckle the concrete as well. The same processes which will eventually destroy a concrete sidewalk will also act to break down rock.

Weathering occurs when rock is mechanically broken (disintegrated) and chemically altered (decomposed). Although we shall discuss these two processes separately, remember that *they work simultaneously in nature.*

Mechanical Weathering

When a rock undergoes **mechanical weathering** it is broken into smaller pieces, each retaining the characteristics of the original material. The end result is many small pieces from a single large one. Figure 2.1 shows how breaking a rock into smaller pieces increases the amount of surface area available for chemical attack. An analogous situation occurs when sugar dissolves in water. Here, a large single crystal will dissolve much slower than an equal volume of granules because of the difference in surface area. *Hence, by breaking rocks into smaller pieces, mechanical weathering increases the amount of surface area available for chemical attack.*

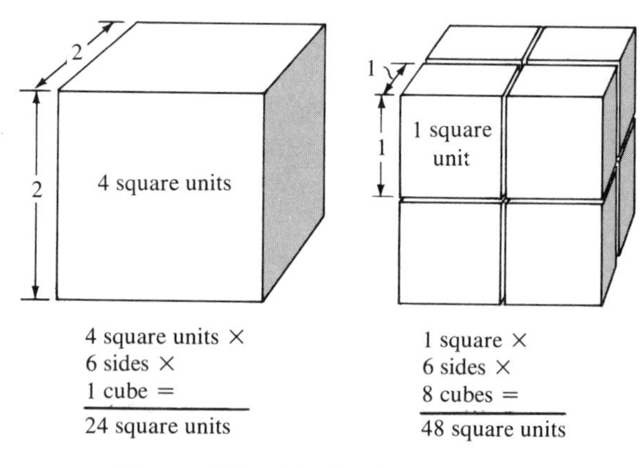

Figure 2.1 *Mechanical weathering increases the amount of surface area available for chemical attack.*

*__Tectonic activities__ are activities that result in the deformation of the earth's crust.

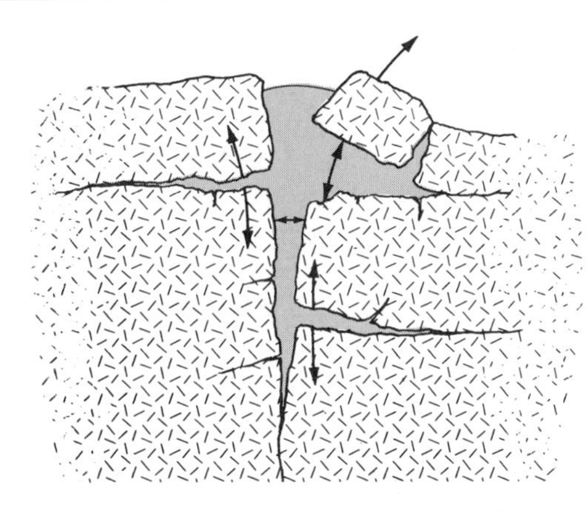

Figure 2.2 *Frost wedging. As water freezes it expands, exerting a force great enough to break rock. (From Robert J. Foster,* Physical Geology, *3rd ed., Columbus, Ohio: Charles E. Merrill, 1979)*

Freezing contributes most to the mechanical weathering of rocks. When water freezes it expands about 9 percent, exerting a tremendous force. Persons who have forgotten to put antifreeze in their cars during freezing weather or have put a full beverage bottle into the freezer for a quick chill and forgotten about it can verify this fact. In nature, water works its way into cracks and, upon freezing, expands and wedges the rock apart. This process is appropriately called **frost wedging** (Figure 2.2). Much frost wedging occurs in mountainous regions where a daily freeze-thaw cycle exists. Here, sections of rock are wedged loose and may tumble into large piles called **talus slopes,** which are found at the base of cliffs (Figure 2.3).

When large igneous bodies, particularly those composed of granite, are exposed by erosion, parallel slabs begin to break loose. The process which produces these onionlike layers is called **exfoliation** and is thought to occur, at least in part, because of the great reduction in pressure when the overlying rock is removed. The outer layers expand most and separate from the remaining rock body. Usually the granite body will take on a dome shape like that of Stone Mountain, Georgia, or Liberty Cap in Yosemite National Park (Figure 2.4). Mine shafts provide another view of how rocks behave if the confining pressure is removed. Rocks have been known to explode off the walls of mine shafts because of the reduced pressure.

Although many cracks in rocks are caused by a reduction in pressure, others are produced by contraction during the crystallization of magma and by tectonic forces during mountain building. Cracks like these are called **joints.** Joints are important rock structures which allow water to penetrate to depth and start the process of weathering long before the rock ever reaches the surface.

The daily cycle of temperature change is thought to weaken rocks, particularly in hot, dry regions where daily variations often exceed 30° C (50° F). Heating a rock causes it to expand, and cooling causes it to contract. Although this process was once thought to be of major importance in the disintegration of rock, laboratory experiments have not been able to substantiate this idea. In one test rocks were heated to temperatures much higher than those normally experienced on the earth's surface and then were

Figure 2.3 *Talus slope made of weathered debris at the base of a cliff. (Photo by H.E. Malde, U.S. Geological Survey)*

Figure 2.4 *Liberty Cap, an exfoliation dome in Yosemite National Park. (Courtesy of National Park Service, U.S. Department of the Interior)*

cooled. This procedure was repeated many times to simulate hundreds of years of weathering, but the rocks showed little apparent change. Additional work is needed before the impact of daily temperature variations on weathering can be determined.

Weathering is also accomplished by the activities of plants, burrowing animals, and people. Plant roots in search of minerals grow into rock joints, and as the roots grow, they wedge the rock apart. Burrowing animals are active in furthering disintegration and moving material to the surface, where chemical weathering can attack. Further, many dead organisms produce acids as they decay, another aid to chemical weathering. Where rock has been blasted in search of minerals or for road construction, the impact of people is quite noticeable, but on a worldwide scale, we probably rank behind burrowing animals in earth-moving accomplishments.

Chemical Weathering

During **chemical weathering** the internal structure of a mineral is altered by the removal and/or the addition of elements. *Water is the main agent of chemical weathering.* Although water in a pure form is inactive, a small amount of some dissolved material is all that is needed to activate it. Oxygen dissolved in water will *oxidize* some materials. For example, when an iron nail is found in the soil it will have a coating of rust (iron oxide), and if the time of exposure has been long, the nail will be so weak that a person can break it like a toothpick. Mafic rock, like basalt, contains some iron. When these rocks are exposed to weathering, they begin to oxidize, and a yellow to reddish brown rust will appear on the surface.

Carbon dioxide (CO_2) dissolved in water (H_2O) forms carbonic acid (H_2CO_3), the same weak acid produced when soft drinks are carbonated. Rain dissolves some carbon dioxide as it falls through the atmosphere, and additional amounts released by decaying organic matter are acquired as the water percolates through the soil. Carbonic acid ionizes to form the very reactive hydrogen ion (H^+) and the bicarbonate ion (HCO_3^-).

To illustrate how a rock chemically weathers when attacked by carbonic acid, we will examine granite, the most abundant continental rock. Granite consists mainly of quartz and potassium feldspar. The idealized reaction between carbonic acid and potassium feldspar is as follows:

$$\underset{\text{Potassium feldspar}}{2\,KAlSi_3O_8} + \underset{\text{Carbonic acid}}{2\,(H^+ + HCO_3^-)} + \underset{\text{Water}}{H_2O} \longrightarrow$$

$$\underset{\text{Clay mineral}}{Al_2Si_2O_5(OH)_4} + \underset{\text{Potassium bicarbonate}}{2\,KHCO_3} + \underset{\text{Silica}}{4\,SiO_2}$$

The hydrogen ion (H^+) attacks and replaces potassium (K) in the feldspar, thereby disrupting its crystalline structure. Once removed, the potassium is available as "food" for plant growth or becomes the soluble product potassium bicarbonate ($KHCO_3$), which may be incorporated into other minerals or carried to the ocean, where it eventually precipitates to form a sedimentary rock. The most abundant by-products of the chemical breakdown of feldspar are residual clay minerals. Because clay minerals are the end product of weathering, they are very stable under surface conditions. Consequently, clay minerals make up a high percentage of the inorganic material in soils. The most abundant sedimentary rock, shale, is also composed of clay minerals. Some of the silica that is removed from the feldspar structure will go into solution with the groundwater (water beneath the earth's surface). This silica will eventually precipitate, producing nodules of chert or flint, fill in the pore space of such things as buried wood to produce petrified wood, or be carried to the ocean, where microscopic animals like diatoms will remove it to build silica shells.

Quartz, the other main component of granite, is very resistant to chemical weathering; hence it remains substantially unaltered. As granite weathers, the feldspar crystals dull and slowly turn to clay, releasing the once-interlocked quartz grains, which still retain their fresh, glassy appearance. Some of the quartz remains in the soil, but much is washed to the sea, where it becomes the main constituent of sandy beaches and in time is often converted into the sedimentary rock sandstone.

Table 2.1 lists the weathered products of quartz, feldspar, and other silicate minerals. Recall that silicate minerals make up most of the earth's crust and that only eight elements are the dominant constituents of the silicate minerals. Upon chemical weathering, sodium, calcium, potassium, and magnesium produce soluble products which can be removed by groundwater. The element iron combines with oxygen, producing insoluble iron oxides, most notably limonite and hematite, which remain in the soil, giving it a reddish brown or yellowish color. Under most conditions the three remaining elements, aluminum, silicon, and oxygen, join with water to produce stable clay minerals.

Table 2.1 ***Products of Weathering***

Mineral	Residual Products	Material in Solution
Quartz	Quartz grains	Silica
Feldspars	Clay minerals	Silica K^+, Na^+, Ca^{++}
Hornblende	Clay minerals Limonite Hematite	Silica Ca^{++}, Mg^{++}
Olivine	Limonite Hematite	Silica Mg^{++}

Figure 2.5 *Successive shells are loosened from rock as the weathering process continues to penetrate deeper and deeper into the rock. (Photo by C.A. Kaye, U.S. Geological Survey)*

As chemical weathering alters the internal structure of minerals, physical changes are also occurring. For example, when cube-shaped rock fragments produced by regular joint systems are attacked by chemical weathering, the fragments take on a spherical shape. Any process that tends to give the weathered rock a spherical shape is called **spheroidal weathering.** One type of spheroidal weathering resembles exfoliation, except that much smaller rocks are involved. As the minerals in the rock weather to clay, they increase in size because of the addition of water into their structure. This increase in bulk exerts an outward force that is thought to cause concentric layers of rock to break loose and fall off (Figure 2.5). Hence chemical weathering can produce forces great enough to cause mechanical weathering.

Rates of Weathering

The rate at which rock weathers depends on many factors. We have already seen how the particle size influences the rate of weathering. The mineral makeup of a rock is also a very important factor, which can be demonstrated by comparing headstones carved from different rock types. Those made of slate, which is composed primarily of clay minerals, are very resistant to chemical weathering. This is not true of marble headstones (Figure 2.6). *The order in which the silicate minerals weather is the same as their order of crystallization.* The minerals that crystallized first formed under much higher temperatures and pressures than those that crystallized last. Consequently, these early-formed minerals are not as stable at the earth's surface, where the temperature and pressure are radically different from those in the environment where they formed. By examining Bowen's reaction series (Figure 1.14), we can see that olivine crystallizes first and is consequently the least resistant to weathering, while quartz, which crystallizes last, is the most resistant.

Climatic factors, particularly temperature and

A.

B.

Figure 2.6 A. *An examination of headstones reveals the rate of chemical weathering on diverse rock types. The slate headstone (left) was erected 125 years before the marble headstone (right), whose inscription date, 1818, is illegible.* **B.** *These two marble headstones further illustrate the rate of deterioration of marble. (Photos courtesy of Sheldon Judson)*

Figure 2.7 *Chemical weathering of Cleopatra's Needle, a granite obelisk.* **A.** *Before it was removed from Egypt. (Courtesy of The Metropolitan Museum of Art)* **B.** *After a span of 75 years in New York City's Central Park. After having survived intact for about 35 centuries in Egypt, the windward side has been almost completely defaced in less than a century. (Courtesy of New York City Parks)*

moisture, are of primary significance to the rate of rock weathering. *The optimum environment for chemical weathering is a combination of warm temperatures and abundant moisture.* In polar regions chemical weathering is ineffective because frigid temperatures keep the available moisture locked up as ice, while in arid regions there is insufficient moisture to foster much chemical weathering. A classic example of how climate affects the rate of weathering was provided when Cleopatra's Needle, a granite obelisk, was moved from Egypt to New York City. After withstanding approximately 3500 years of exposure in the dry climate of Egypt, the hieroglyphics were almost completely removed from the windward side in less than 75 years in the wet and chemical-laden air of New York City (Figure 2.7).

Soil

Soil has accurately been called "the bridge between life and the inanimate world." All life owes its existence to the dozen or so elements that must ultimately come from the earth's crust. First, weathering converts the rock to soil. Then, plants carry out their role as the great intermediary, assimilating the necessary elements and making them available to animals and people.

With few exceptions, the earth's land surface is covered by **regolith,** the layer of rock and mineral fragments produced by weathering. Some would call this material soil. However soil is more than an accumulation of loose rock debris. **Soil** *is a combination of mineral and organic matter, water, and air, that portion of the regolith that supports the growth of plants.* Although the

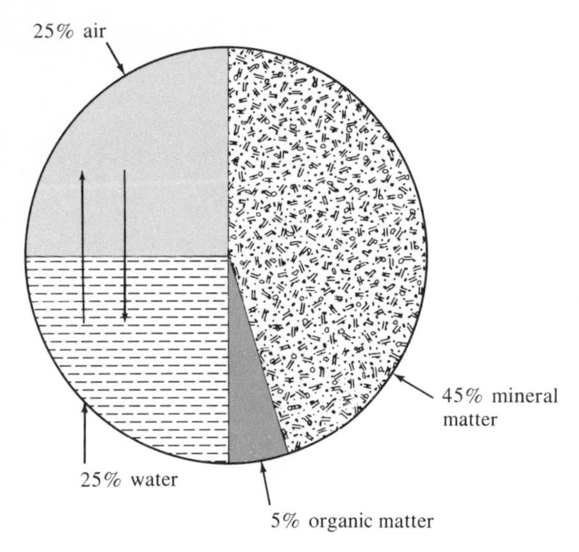

Figure 2.8 *Composition (by volume) of a soil in good condition for plant growth. Although the percentages vary, each soil is composed of mineral and organic matter, water, and air.*

proportions may vary, the major components do not (Figure 2.8). About one-half of the total volume of a good-quality surface soil is a mixture of disintegrated and decomposed rock (mineral matter) and the remains of animal and plant life (organic matter). The remaining half consists of pore spaces, where air and water circulate.

The Soil Profile

If you were to dig a trench, you would see that its walls consisted of a series of horizontal layers. These layers, called **horizons,** collectively make up the **soil profile** (Figure 2.9). Three basic horizons are identified and from top to bottom are labeled *A, B,* and *C,* respectively. The *A* and *B* horizons may be subdivided further.

The uppermost layer in a soil profile, the *A* horizon, is often called the **surface soil.** It is the part of the soil with the greatest biological activity and is therefore the horizon where organic matter is most plentiful. Since it lies at the surface, the *A* horizon is also the part of the soil that falling rain reaches first. As a consequence, soluble materials and tiny particles such as clay are **leached** (washed out) by percolating water.

Lying immediately below the surface soil is the *B* horizon, or **subsoil.** It is in this zone that much of the material removed from the *A* horizon is deposited. Because of this, the *B* horizon is referred to as the *zone of accumulation*. Since the *B* horizon has an intermediate position in the soil profile, it may be considered, at least in part, a transitional zone. For example, living organisms and organic matter are more abundant in the *B* than in the *C* horizon, but considerably less so than in the *A* horizon. The *A* and *B* horizons together constitute the **solum,** or "true soil." It is in the solum that the soil-forming processes are active and that living roots and other plant and animal life are largely confined.

Below the solum is the *C* horizon, a layer characterized by partially altered parent material and little if any organic matter (Figure 2.9). While the parent material may be so dramatically altered in the solum that its original character is not recognizable, it is easily identifiable in the *C* horizon.

The boundaries between soil horizons may be very sharp, or the horizons may blend gradually from one to another. Furthermore, some soils lack horizons altogether. Such soils are called **immature,** because soil building has been going on for only a short time. Immature soils are also characteristic of steep slopes where erosion continually strips away the soil, preventing full development.

Controls of Soil Formation

Soil is the product of the complex interplay of several factors, including parent material, time, climate, plants and animals, and slope. Although all of these factors are interdependent, it will be helpful to examine their roles separately.

Parent Material. The **parent material** from which a soil has evolved may either be underlying bedrock or a layer of unconsolidated deposits. Soils formed on bedrock are termed **residual soils,** while those developed on unconsolidated deposits are called **transported soils** (Figure 2.10).

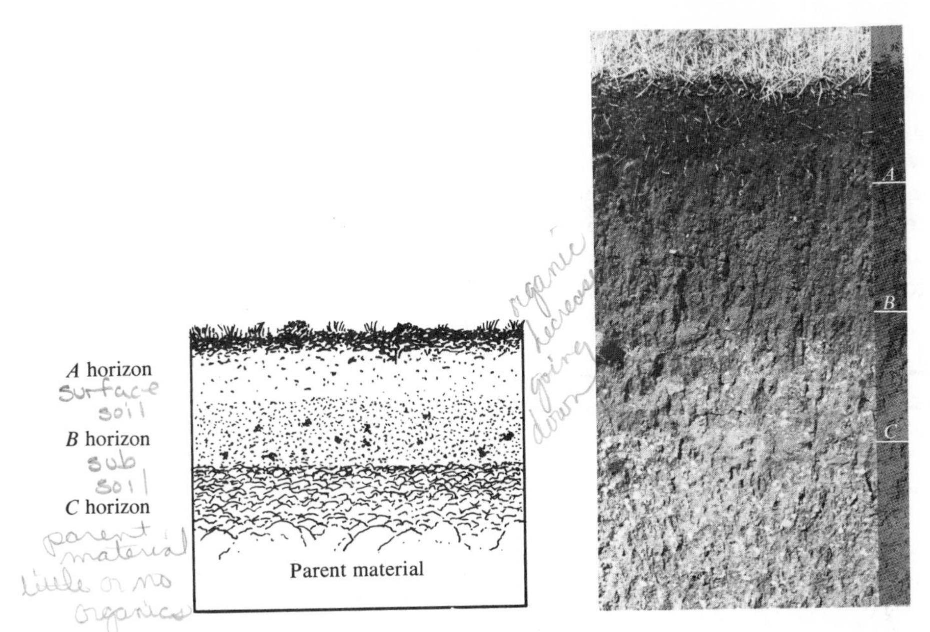

Figure 2.9 *Soil profile. Mature soils are characterized by a series of horizontal layers called horizons, which comprise the soil profile. (Photo by R.W. Simonson, Soil Conservation Service, U.S. Department of Agriculture; drawing from Foster,* Physical Geology, *3rd ed., Columbus, Ohio: Charles E. Merrill, 1979)*

The nature of the parent material influences soils in two ways. First, the type of parent material to some degree will affect the rate of weathering, and thus the rate of soil formation. For example, the mineral composition of the parent material will influence the rate of chemical weathering. Also, since unconsolidated deposits are already partly weathered, soil development on such material will likely progress more rapidly than when bedrock is the parent material. Second, the chemical makeup of the parent material will affect the fertility of the soil. If it lacks elements that are necessary to plant growth, its usefulness is obviously diminished.

At one time it was thought that the parent material was the primary factor causing differences between soils. Today soil scientists realize that other factors, especially climate, are more important. In fact, it has been found that *similar soils are often produced from different parent materials and that dissimilar soils have developed from the same parent material.* Such discoveries reinforce the importance of the other soil-forming factors.

Time. If weathering has been going on for a comparatively short time, the character of the parent material determines to a large extent the characteristics of the soil. As the weathering process continues, the influence of parent material on soil is overshadowed by the other soil-forming factors. Therefore time is an important control of soil formation. It is not possible to list the length of time it takes for various soils to evolve, because the soil-forming processes act at varying rates under different circumstances. However it is safe to say that *the longer a soil has been forming, the thicker it becomes and the less it resembles the parent material from which it formed.*

Climate. Climate is considered to be the most important control of soil formation. It determines whether chemical or mechanical weathering will predominate and also greatly influences the rate and depth of weathering. For instance, a hot and wet climate may produce a thick layer of chemically weathered soil in the same amount of time that a cold and less humid climate produces a thin mantle of mechanically weathered debris. Furthermore, the amount of precipitation influences the degree to which various materials are leached from the soil, thereby affecting soil fertility. Finally, climatic conditions are an important control on the type of plant and animal life present.

Plants and Animals. The chief function of plants and animals is the furnishing of organic matter to the soil. Certain bog soils are composed almost entirely of organic matter, while desert soils may contain only a small fraction of a percent. Although the quantity of organic matter present varies substantially among soils, no soil completely lacks it.

The primary source of organic matter is plants, although animals and the uncountable millions of microorganisms also contribute. When organic matter is decomposed, it supplies important nutrients for plants, as well as food for the animals and microorganisms living in the soil. Consequently, soil fertility is in part related to the amount of organic matter present. Furthermore, the decay of plant and animal remains causes the formation of various organic acids. These complex acids hasten the weathering process. Organic matter also has a high water-holding ability and thus aids water retention in a soil.

Microorganisms, including fungi, bacteria, and the single-celled protozoa, play the active role in the decay of plant and animal remains. The end product is **humus,** a jellylike material that no longer resembles the plants and animals from which it formed. In addition, certain microorganisms aid soil fertility because they have the ability to *fix* (change) atmospheric nitrogen into soil nitrogen.

Earthworms and other burrowing animals act to mix the mineral and organic portions of a soil. Earthworms, for example, feed on the organic matter in the soil and thoroughly mix soils in which they live, often moving and enriching many tons per acre each year. Burrows and holes also aid the passage of water and air through the soil.

Slope. Slope has a significant impact on the

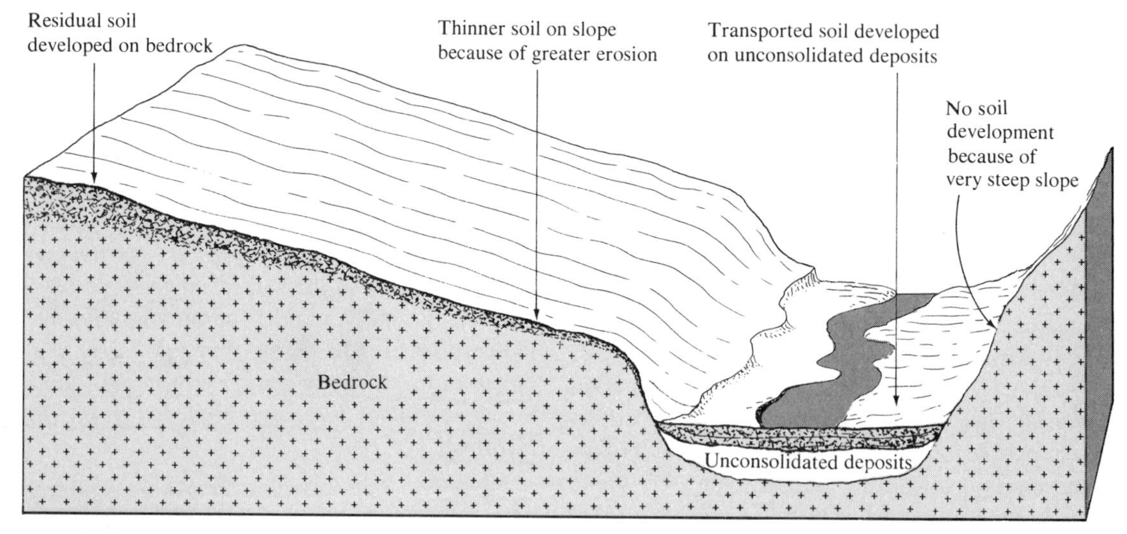

Figure 2.10 *Parent material. The parent material for residual soils is the underlying bedrock, while transported soils form on unconsolidated deposits. Also note that soils are thinner, or non-existent, on the slopes.*

amount of erosion and the water content of soil. On steep slopes soils are often poorly developed. In such situations the quantity of water soaking in is slight, and as a result, the moisture content of the soil may not be sufficient for vigorous plant growth. Further, because of accelerated erosion on steep slopes, the soils are thin, or in some cases nonexistent (Figure 2.10). On the other hand, poorly drained and waterlogged soils found in bottomlands have a much different character. Such soils are usually very thick and very dark, the dark color resulting from the large quantity of organic matter that accumulates because saturated conditions retard the decay of vegetation. The optimum slope for soil development is a flat-to-undulating upland surface. Here we find good drainage, minimum erosion, and sufficient infiltration of water into the soil.

Slope orientation, the direction the slope is facing, is another aspect worthy of mention. In the mid-latitudes a south-facing slope will receive a great deal more sunlight than a north-facing slope. In fact, a steep north-facing slope may receive no direct sunlight at all. The difference in the amount of solar radiation received will cause differences in soil temperature and moisture, which in turn may influence the nature of the vegetation and the character of the soil.

Although this section has dealt separately with each of the soil-forming factors, remember that all of them work together to form soil. No single factor is responsible for a soil being as it is, but rather the combined influence of parent material, time, climate, plants and animals, and slope.

Soil Types

In the discussion which follows we shall briefly examine some common soil types. As you read, notice that the characteristics of each of the soil types are primarily manifestations of the prevailing climatic conditions. A summary of the characteristics of the soils discussed in this section is provided in Table 2.2.

The term **pedalfer** gives a clue to the basic characteristic of this soil type. The word is derived from the Greek **ped***on,* meaning "soil," and the chemical symbols **Al** (aluminum) and **Fe** (iron). Pedalfers are characterized by an accumulation of iron oxides and aluminum-rich clays in the *B* horizon. In the mid-latitude areas where the annual rainfall exceeds 63 centimeters (25 inches) most of the soluble materials, such as calcium carbonate, are leached from the soil and carried away by underground water. The less soluble iron oxides and clays are carried from the *A* horizon and deposited in the *B* horizon, giving it a brown to red brown color. These soils are best developed under forest vegetation where large quantities of decomposing organic matter provide the acid conditions necessary for leaching. In the United States pedalfers are found east of a line extending from northwestern Minnesota to south-central Texas (Figure 2.11).

Pedocal is derived from the Greek **ped***on,* meaning "soil," and the first three letters of **cal***cite* (calcium carbonate). As the name implies, pedocals are characterized by an accumulation of calcium carbonate. This soil type is found in the drier western United States in association with grassland and brush vegetation (Figure

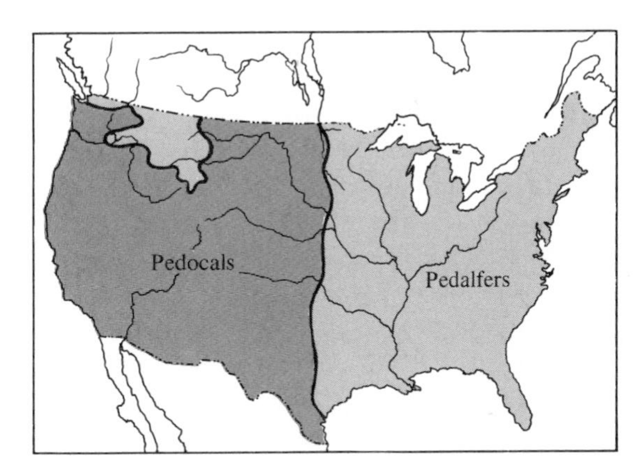

Figure 2.11 *Pedalfers and pedocals. In the United States pedalfers characterize the more humid eastern part of the country, while pedocals are found in the drier western portion. The boundary is the 63-centimeter (25-inch) rainfall line. (From Foster,* Physical Geology, *3rd ed., Columbus, Ohio: Charles E. Merrill, 1979)*

Table 2.2 ***Summary of Soil Types***

Climate	Temperature Humid (> 63 cm rainfall)	Temperature Dry (<63 cm rainfall)	Tropical (heavy rainfall)	Extreme Arctic Desert (hot or cold)
Vegetation	Forest	Grass and brush	Grass and trees	Almost none, so no humus develops
Typical area	Eastern U.S.	Western U.S.		
Soil type	Pedalfer	Pedocal	Laterite	
Topsoil	Sandy; light colored; acid	Commonly enriched in calcite; whitish color	Zones not developed: Enriched in iron (and aluminum); brick red color	No real soil forms because there is no organic material. Chemical weathering is very slow
Subsoil	Enriched in aluminum, iron, and clay; brown color	Enriched in calcite; whitish color	All other elements removed by leaching	
Remarks	Extreme development in conifer forests, because abundant humus makes groundwater very acid. Produces light gray soil because of removal of iron	*Caliche* is name applied to the calcite-rich soils	Apparently bacteria destroy humus, so no acid is available to remove iron	

2.11). Here, rainwater percolating through the soil often evaporates before it can remove soluble materials, chiefly calcium carbonate. The result is a whitish accumulation called **caliche.** In addition, since chemical weathering is less intense in drier areas, pedocals generally contain a smaller percentage of clay minerals than pedalfers.

In the hot and wet climates of the tropics soils called **laterites** develop. Since chemical weathering is intense under such climatic conditions, these soils are usually deeper than soils developing over a similar period of time in the mid-latitudes. Not only does leaching remove the soluble materials like calcite, but the great quantities of percolating water also remove much of the silica, with the result that oxides of iron and aluminum become concentrated in the soil. The iron gives the soil a distinctive red color. When dried, laterites are very hard. In fact, some people use this soil for making bricks. If the parent rock contained little iron, the product of weathering is an aluminum-rich accumulation called **bauxite.** Bauxite is the primary ore of aluminum.

Since bacterial activity is very high in the tropics, laterites contain practically no humus. This fact, coupled with the highly leached and bricklike nature of these soils, makes laterites poor for growing crops. The infertility of these soils has been borne out repeatedly in tropical countries where cultivation has been expanded into such areas.

In cold or dry climates soils are generally very thin and poorly developed. The reasons for this are fairly obvious. Chemical weathering progresses very slowly in such climates, and the scanty plant life yields very little organic matter.

Mass Wasting

Periodically newspapers carry stories relating the terrifying and often grim details of landslides. On Friday, June 28, 1974, the papers carried such an account when a massive landslide occurred along a steep hillside in the Andes Mountains of Columbia. A peasant described it this way:

> The earth began to rumble like when thousands of horses gallop on the range." Witnesses said the rock and debris covered more than 875 meters (2400 feet) of the twisting highway at the base of the slope and "pushed around tractors and other heavy equipment like toys.

Officials estimated the dead at more than 200. Since the landslide also dammed a river, there was also the threat of flooding. Certainly for those in the vicinity of the landslide it was a major catastrophe. Fortunately, mass movements such as this are infrequent and seldom affect many people.

Landslides are spectacular examples of a normal geologic process called **mass wasting.** Mass wasting refers to the downslope movement of rock, regolith, and soil under the direct influence of gravity. Once weathering breaks up rock, gravity pulls the material to lower elevations, where streams usually carry it away, eventually depositing it in the ocean. In this manner the earth's landscape is slowly being shaped.

Although gravity is the controlling force of mass wasting, other factors play an important part in bringing about the downslope movement of material. Water is one of these factors. When the pores in sediment become filled with water, the cohesion between the particles is destroyed, allowing them to slide past one another with relative ease. For example, when sand is slightly moist, it may stick together quite well. However, if more water is added, filling the openings between the grains, the sand will ooze out in all directions. Thus saturation reduces the internal resistance of materials, which are then easily set in motion by the force of gravity. When clay is wetted, it becomes very slick—another example of the "lubricating" effect of water. Water also adds considerable weight to a mass of material. The added weight in itself may be enough to cause the material to slide or flow downslope.

Oversteepening of slopes is another cause for many mass movements. There are many situations in nature where this takes place. A stream undercutting a valley wall and waves pounding against the base of a cliff are but two common examples. Furthermore, through their activities, people often create oversteepened and unstable slopes that become prime sites for mass wasting.

The various types of mass wasting have been classified using a number of schemes, some of them quite complex. For our purposes, we shall divide them into two categories based on the rate of movement. Some mass movements are rapid, as the landslide described earlier. An extreme example is a landslide that occurred at Elm, Switzerland, in 1881. The rock and debris in this slide moved over 2500 meters (8000 feet) in less than one minute—a mean velocity of about 160 kilometers (100 miles) per hour. In most instances, however, velocities are considerably less than this. On the other hand, some types of mass wasting are so slow that the movement is imperceptible. These slow movements, which include creep and solifluction, are measured in terms of centimeters or meters per year.

Rapid Movements

Landslides. Landslides may be divided into two basic types: slump and rockslide. **Slump** refers to the downward slipping of a mass of rock or unconsolidated material moving as a unit along a curved surface (Figure 2.12). This movement usually results because a slope has been greatly steepened, either naturally or by people, and, as expected, the material reacts to the pull of gravity. As Figure 2.12 illustrates, there is usually a backward rotation of the material that has slumped.

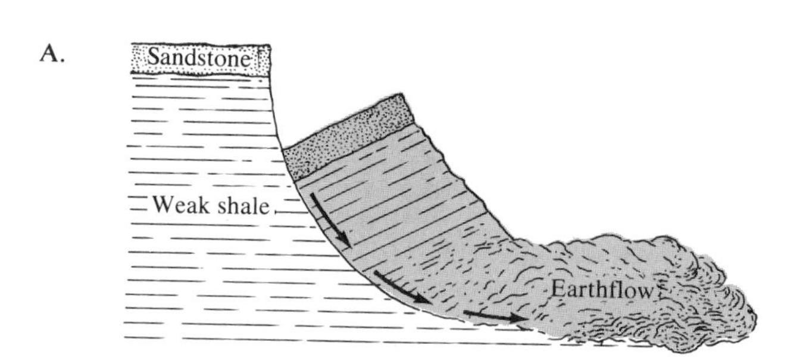

Figure 2.12 **A.** *Slump, one of the two basic types of landslide, occurs when rock or regolith slips downward en masse along a curved surface of rupture.* **B.** *Slump at Anchorage, Alaska, triggered by the devastating 1964 earthquake. (Photo by W.R. Hansen, U.S. Geological Survey; drawing from Foster,* Physical Geology, *3rd ed., Columbus, Ohio: Charles E. Merrill, 1979)*

Rockslides occur when blocks of bedrock break loose and slide down a slope. Such events are among the fastest and most destructive of mass movements. Usually rockslides take place in a geologic setting where the rock strata are inclined, or joints and fractures exist, parallel to the slope. If the rock is undercut at the base of the slope it loses support, and the rock eventually gives way. Sometimes the rockslide is triggered when rain or melting snow lubricates the underlying surface to the point where friction is no longer sufficient to hold the rock unit in place. As a result, rockslides tend to be most prevalent during the spring, when heavy rains and melting snow are generally greatest. Earthquakes are another mechanism which often triggers rockslides and other mass movements. The 1811 earthquake at New Madrid, Missouri, for example, caused landslides in an area of more than 13,000 square kilometers (5000 square miles) along the Mississippi River valley. In terms of property damage, the most devastating effects of the famous 1964 Alaska earthquake were the landslides in Anchorage, some 130 kilometers (80 miles) from the center of the quake.

The Gros Ventre Landslide. The Gros Ventre River flows west from the Wind River Range in northwestern Wyoming, through Grand Teton National Park, eventually emptying into the Snake River. On June 23, 1925, a classic landslide took place in its valley, just east of the small town of Kelly. In the span of just a few minutes a great mass of sandstone, shale, and soil crashed down the south side of the valley, carrying with it a dense pine forest. The volume

Figure 2.13 *The scar, 2.4 kilometers (1.5 miles) long and 0.8 kilometer (0.5 mile) wide, left on the side of Sheep Mountain by the Gros Ventre landslide. (Courtesy of Wyoming Travel Commission)*

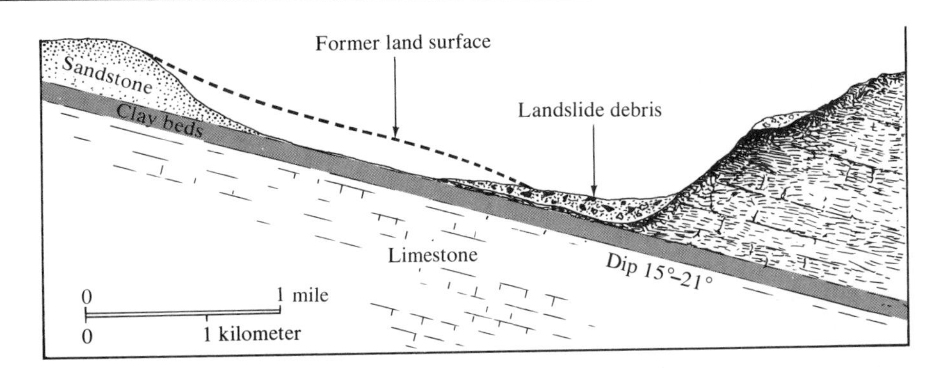

Figure 2.14 *Cross-sectional view of the Gros Ventre landslide. The slide occurred when the tilted and undercut sandstone bed could no longer maintain its position atop the saturated bed of clay. [After W.C. Alden, "Landslide and Flood at Gros Ventre, Wyoming."* Transactions *(AIME) 76(1928): 348*]

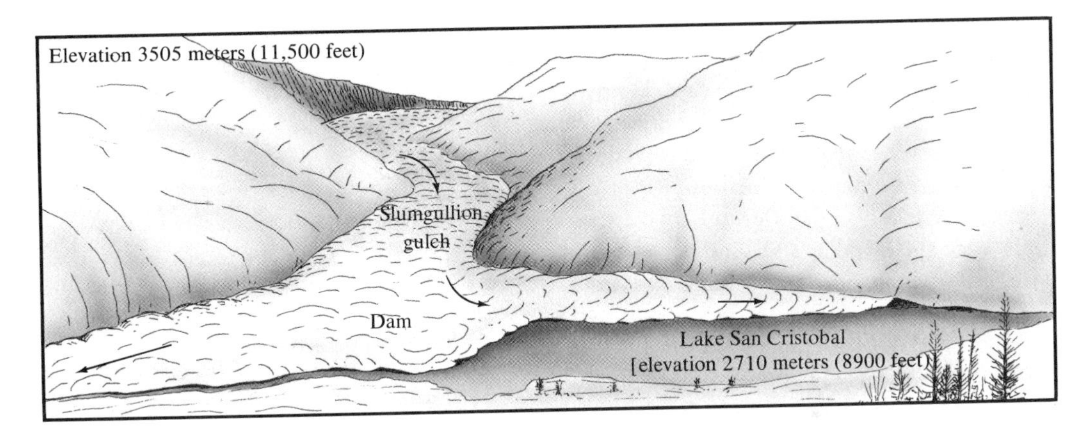

Figure 2.15 *View of the great Slumgullion mudflow from its source to Lake San Cristobal, a vertical distance of about 780 meters (2600 feet). Since it occurred in prehistoric times, there is no direct information about its rate of movement. (Drawing after Ward's Natural Science Establishment, Inc., Rochester, N.Y.; photo by W. Cross, U.S. Geological Survey)*

of debris, estimated at 38 million cubic meters (50 million cubic yards), created a 67–75-meter (225–250-foot) high dam on the Gros Ventre River (Figure 2.13). The river was completely blocked, creating a lake, which filled so quickly that a house that had been 18 meters (60 feet) above the river was floated off its foundation 18 hours after the slide. In 1927 the lake overflowed the dam, partially draining the lake and resulting in a devastating flood downstream.

Why did the Gros Ventre landslide take place? Figure 2.14 is a diagrammatic cross-sectional view of the geology of the valley. Notice the following points: (1) The sedimentary strata in this area dip (tilt) 15–21 degrees; (2) Underlying the bed of sandstone is a relatively thin layer of clay; and (3) At the bottom of the valley the river had cut through much of the sandstone layer. During the spring of 1925 water from the heavy rains and melting snow seeped through the sandstone, saturating the clay below. Since much of the sandstone layer had been cut through by the Gros Ventre River, the layer had virtually no support at the bottom of the slope. Eventually the sandstone could no longer hold its position on the wetted clay, and gravity pulled the mass down the side of the valley. The circumstances at this location were such that a landslide was inevitable.

Mudflow. **Mudflow** is a type of mass wasting which involves a flowage of debris containing a large amount of water. Mudflows are most characteristic of canyons and gullies in arid mountainous regions. When such an area experiences a heavy rain, large amounts of sediment are washed into the channel from the valley walls, which usually have little or no vegetation to anchor the loose material. The end product is a rapidly moving mass of mud having the consistency of wet concrete. Because of its high density the mudflow can carry large boulders with ease. Upon reaching the mouth of the canyon the mudflow spreads out. Figure 2.15 illustrates the well-known Slumgullion mudflow which occurred in the San Juan Mountains of Colorado and dammed the Lake Fork of the Gunnison River, creating Lake San Cristobal.

Earthflow. Unlike mudflows, **earthflows** are most common in humid regions. When water

Figure 2.16 *Earthflow. (Photo by G.K. Gilbert, U.S. Geological Survey)*

Figure 2.17 *Creep in vertical beds of shale, Washington County, Maryland. Although creep is an imperceptibly slow movement, it results in visible effects. (Photo by G.W. Stose, U.S. Geological Survey)*

saturates clay-rich regolith on a hillslope, the material may break away and flow a short distance downslope, leaving a scar on the hillside (Figure 2.16). Depending upon the steepness of the slope and the consistency of the material, the speed of the earthflow may vary from a few meters per hour to many meters per minute. Earthflows also commonly occur in association with slumps, because debris often flows out at the base of the slumped material (Figure 2.12A).

Slow Movements

The rapid movements are certainly the most spectacular and catastrophic forms of mass wasting. *However, in spite of their large size and spectacular nature, rapid movements are of less importance overall than slow movements.* While rapid types of mass wasting are characteristic of mountainous and hilly regions, the slower forms can take place on gentle slopes and are thus much more widespread.

Creep. **Creep** is the slow downhill movement of soil and regolith. Although the movement is imperceptibly slow, its effects are recognizable. Creep causes tilted fences and telephone poles, and tree trunks will often be bent as a consequence of this movement (Figure 2.17). Creep usually results from the alternate expansion and contraction of surface material caused by freezing and thawing or wetting and drying. As shown in Figure 2.18, freezing or wetting lifts the soil at right angles to the slope (solid lines), and thawing or drying allows the particles to fall back to a slightly lower level (dashed lines). Each cycle therefore moves the material a short distance downhill.

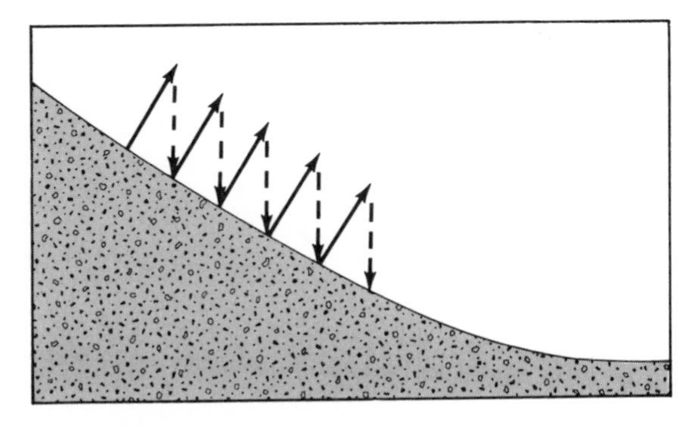

Figure 2.18 *Creep.*

Solifluction. In the frigid zones found at high latitudes **solifluction** is of major importance. In such regions only the ice in the upper few meters of regolith melts during the spring and summer, while the ground below remains permanently frozen. Since the water in the upper zone has nowhere to go, this layer remains saturated and slowly flows down even the gentlest slopes. In this manner the overburden is removed and the unweathered rock below is exposed. When the newly exposed rock is eventually weathered, it too will be removed by solifluction.

REVIEW

1. If two identical rocks were weathered, one mechanically and the other chemically, how would the products of weathering for the two rocks differ?
2. How does mechanical weathering add to the effectiveness of chemical weathering?
3. Granite and basalt are exposed at the surface in a hot and wet region.
   a. Which type of weathering will predominate?
   b. Which of the rocks will weather most rapidly? Why?
4. Heat speeds up a chemical reaction. Why then does chemical weathering proceed slowly in a hot desert?
5. How is carbonic acid (H_2CO_3) formed in nature? What results when this acid reacts with potassium feldspar?
6. What is the difference between soil and regolith?
7. List the characteristics associated with each of the horizons in a well-developed soil profile. Which of the horizons constitute the solum? Under what circumstances do soils lack horizons?
8. What factors might cause different soils to develop from the same parent material, or similar soils to form from different parent materials?

9. Which of the controls of soil formation is most important? Explain.
10. How can slope affect the development of soil? What is meant by the term *slope orientation*?
11. Distinguish between pedalfers and pedocals.
12. Soils formed in the humid tropics and the Arctic both contain little organic matter. Do both lack humus for the same reasons?
13. What role does mass wasting play in sculpturing the earth's landscape?
14. What is the controlling force of mass wasting? What other factors play important roles?
15. What factors led to the massive landslide at Gros Ventre, Wyoming (Figure 2.14)?
16. Since creep is an imperceptibly slow process, what evidence might indicate to you that this phenomenon is affecting a slope? Describe the mechanism that creates this slow movement (Figure 2.18).
17. Why is solifluction only a summertime phenomenon?

KEY TERMS

weathering
erosion
mass wasting
mechanical weathering
frost wedging
talus slope
exfoliation
joints
chemical weathering
spheroidal weathering
regolith
soil
horizons
soil profile
surface soil
subsoil
solum
immature soil
parent material
residual soil
transported soil
humus
pedalfer
pedocal
caliche
laterite
bauxite
slump
rockslide
mudflow
earthflow
creep
solifluction

Scenic valley with stream meandering on its floodplain. (Photo by W.T. Lee, U.S. Geological Survey)

3

Running Water and Groundwater

All the rivers run into the sea; yet the sea is not full; unto the place from whence the rivers come, thither they return again. (Ecclesiastes 1:7)

As the perceptive writer of Ecclesiastes indicated, water is continually on the move, from the ocean to the land and back again. This chapter deals with that part of the water cycle which involves the return of water to the sea, with some water traveling quickly via a rushing stream, and some moving more slowly below the surface. We shall examine the factors that control the movement of water, as well as look at how water sculptures the landscape and at the features that result. To a great extent, the Grand Canyon, Niagara Falls, Old Faithful, and Mammoth Cave all owe their existence to the activities of water on its way to the sea.

The Hydrologic Cycle

The amount of water on earth is immense, an estimated 1360 million cubic kilometers (326 million cubic miles). Of this total, the vast bulk—97.2 percent—is part of the world ocean. Icecaps and glaciers account for another 2.15 percent, leaving only 0.65 percent to be divided among lakes, streams, subsurface water, and the atmosphere. Although the percentage of the earth's total water found in each of the latter sources is but a small fraction of the total inventory, the absolute quantities are great.

An adequate supply of water is vital to life on earth. With increasing demands on this finite resource, science has given a great deal of attention to the continuous exchanges of water between the oceans, the atmosphere, and the continents. This unending circulation of the earth's water supply has come to be called the **hydrologic cycle.** *It is a gigantic system powered by energy from the sun in which the atmosphere provides the vital link between the oceans and continents.* Water from the oceans, and to a much lesser extent from the continents, is constantly evaporating into the atmosphere. Winds transport the moisture-laden air, often great distances, until the complex processes of cloud formation are set in motion that eventually result in precipitation. The precipitation that falls into the ocean has ended its cycle and is ready to begin another. The water that falls on the continents, however, must still make its way back to the ocean.

What happens to precipitation once it has fallen on the land? A portion of the water soaks into the ground, some of it moving downward, then laterally, finally seeping into lakes, streams, or directly into the ocean. When the rate of rainfall is greater than the earth's ability to absorb it, the additional water flows over the surface into lakes and streams. Much of the water which soaks in **(infiltration)** or runs off **(runoff)** eventually finds its way back to the atmosphere because of evaporation from the soil, lakes, and streams. Also, some of the water that infiltrates the ground surface is absorbed by plants, which

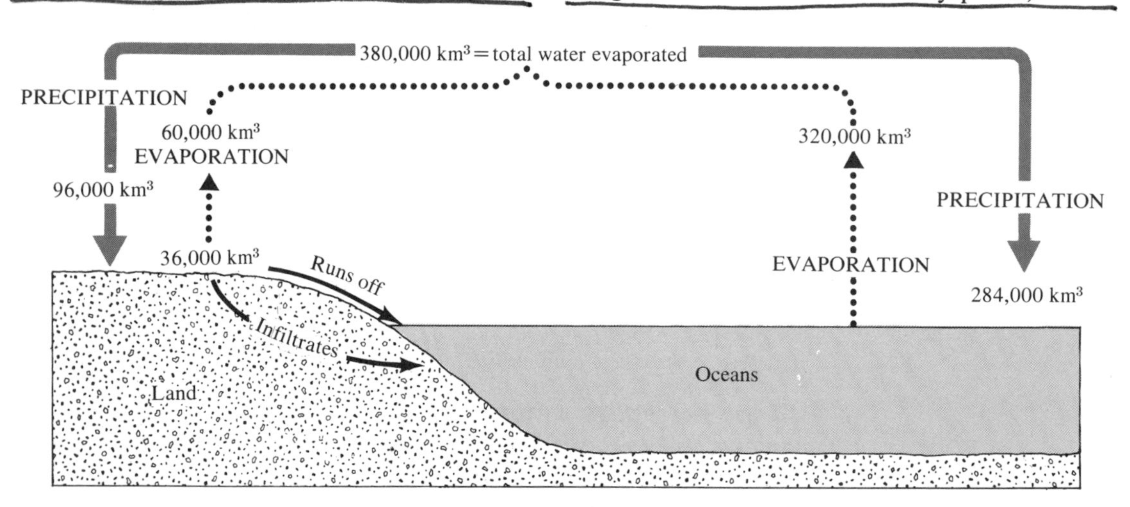

Figure 3.1 *The earth's water balance. About 320,000 cubic kilometers of water are evaporated each year from the oceans, while evaporation from the land (including lakes and streams) contributes 60,000 cubic kilometers of water. Of this total of 380,000 cubic kilometers of water, about 284,000 cubic kilometers fall back to the ocean, and the remaining 96,000 cubic kilometers fall on the earth's land surface. Since 60,000 cubic kilometers of water evaporate from the land, 36,000 cubic kilometers of water remain to erode the land during the journey back to the oceans.*

then release it into the atmosphere. This process is called **transpiration.** When precipitation falls at high elevations or high latitudes, the water may not immediately soak in, run off, or evaporate. Instead it may become part of a snowfield or a glacier. Although the water can be tied up in this fashion for many years, it will eventually melt and continue its path to the sea.

A diagram of the earth's water balance, a quantitative view of the hydrologic cycle, is shown in Figure 3.1. While the amount of water vapor in the air at any one time is but a minute fraction of the earth's total water supply, the absolute quantities that are cycled through the atmosphere over a one-year period are immense —some 380,000 cubic kilometers—enough to cover the earth's surface to a depth of about 100 centimeters. Since the total amount of water vapor in the atmosphere remains about the same, average annual precipitation over the earth must be equal to the quantity of water evaporated. However, for all of the continents taken together, precipitation exceeds evaporation. Conversely, over the oceans, evaporation exceeds precipitation. Since the level of the world ocean is not dropping, runoff from land areas must balance the deficit of precipitation over the oceans.

The hydrologic cycle represents the continuous movement of water from the oceans to the atmosphere, from the atmosphere to the land, and from the land back to the sea. *The wearing down of the earth's land surface is primarily attributable to the last of the steps in this cycle.*

Running Water

Of all the geologic processes, running water may have the greatest impact on people. We depend upon rivers for energy, travel, and irrigation; their fertile floodplains have fostered human progress since the dawn of civilization. As the dominant agent of landscape alteration, streams have shaped much of our physical environment.

Although people have always depended to a great extent on running water, its source eluded them for centuries. It was not until the sixteenth century that they first realized that streams were supplied by runoff and underground water, which ultimately had their sources as rain and snow. Runoff initially flows in broad sheets, that enter small rills, which carry it to a stream. The word *stream* is used to denote channelized flow of any size, from the smallest brook to the mighty Amazon. Although the terms *river* and *stream* are used synonymously, the term *river* is often preferred when describing a main stream into which several tributaries flow.

Streamflow

Flowing water makes its way to the sea under the influence of gravity. The time required for the journey depends upon the velocity of the stream. Some streams travel at less than 0.8 kilometer (0.5 mile) per hour, while a few rapid ones reach speeds as high as 32 kilometers (20 miles) per hour. Velocities are determined at gaging stations where measurements are taken at several locations across the channel and then averaged. Along straight stretches the highest velocities are near the center of the channel just below the surface, where friction is lowest. But when a stream curves, its zone of maximum speed shifts toward its outer bank.

Since the velocity of a stream is directly related to its ability to erode and transport materials, it is a very important characteristic. Those factors which determine velocity include gradient, channel shape and size, load, and discharge.

Gradient. The **gradient,** or slope, of a stream is defined as the drop per unit distance. The steeper the slope, the more energy for streamflow. Parts of the lower Mississippi have gradients of less than 0.3 meter per kilometer (1 foot per mile), while some mountain streams drop hundreds of meters per kilometer.

Channel Shape and Size. The channel shape determines the amount of water in contact with the channel and hence affects the frictional drag. The most efficient channel is one with the least perimeter for its cross-sectional area. Figure 3.2 compares three types of channels and shows why a semicircular channel has the least frictional drag. In addition, an increase in channel size also reduces the ratio of perimeter to cross-sectional area and therefore ups the efficiency of flow.

Load. The solid particles carried by a stream are termed its **load.** Generally, large boulders and pebbles make up the load of mountain streams, while the load of large rivers near their mouths consists mainly of gravel, sand, and silt. The type of load affects the roughness of the bed, and consequently the frictional drag.

Discharge. The **discharge** of a stream is the amount of water flowing past a certain point in a given unit of time (usually measured in cubic meters or cubic feet per second). By way of comparison, the Nile, one of the world's longest rivers, is dwarfed by the slightly shorter Amazon, which has a discharge equal to that of 50 rivers the size of the Nile.

The relationship between discharge and velocity can be illustrated by comparing streamflow at different times of the year. During floodstage the amount of water entering a river via its tributaries increases significantly. In order to handle this additional discharge the stream will increase the size of its channel by widening and deepening it. This makes for more efficient flow and thereby increases the velocity. Consequently, *an increase in discharge results in an increase in velocity.* During a period of low water the stream refills its channel with sediment. Because the stream is now rather shallow, most of its water is affected by the frictional drag imposed by the

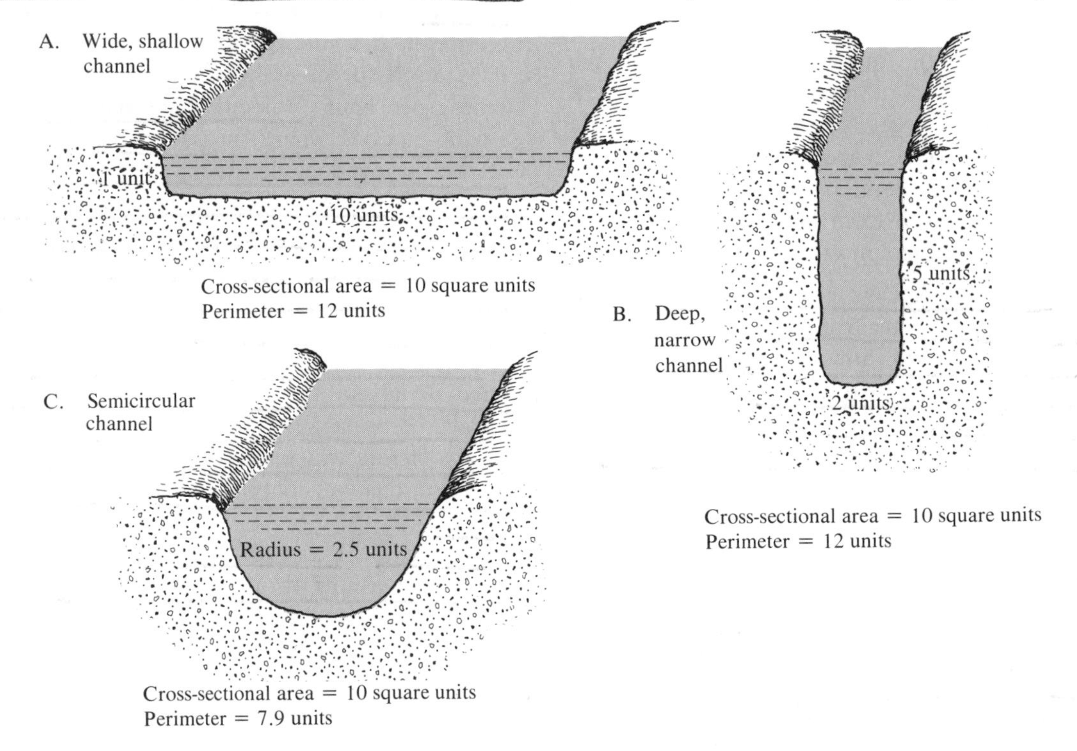

Figure 3.2 *Influence of channel shape on velocity. Although the cross-sectional area of all three channels is the same, the semicircular channel has less water in contact with the channel, and hence less frictional drag. As a result, the water will flow more rapidly in this channel, all other factors being equal.*

stream bed. Therefore, its velocity is reduced significantly.

In summary, *the velocity of a stream is affected by numerous characteristics, including gradient, channel size and shape, load, and discharge.*

Changes Downstream

Streams attempt to maintain a balance between all factors which control their flow, whether at a single point or along their entire course. Examine a stream along its full length and observe how it systematically adjusts to changes. Note that tributaries contribute additional water, which must consequently increase the discharge of the main stream (Figure 3.3). To absorb this discharge the stream increases its width, depth, and velocity. The increased discharge could produce extremely high velocities if the stream did not adjust its gradient to maintain a balance. Figure 3.4B shows a stream profile that illustrates the gradual flattening of the gradient which typically occurs downstream. Characteristically, the portion of a stream near its mouth maintains a lower gradient than its headwaters because of the increased discharge downstream.

The observed increase in velocity which occurs downstream contradicts the idea of rushing mountain streams and wide, sluggish rivers. The mental picture that many have of "old man river just rollin' along" is just not so. While a mountain stream may have the appearance of a raging torrent, its average velocity often is less than for the river near its mouth. The difference is attributable to the greater efficiency of the larger channel. Another factor which contributes to the increased velocity downstream is the change in particle size along the stream course. The smaller sediment transported near the mouth of a stream reduces the frictional drag along the bed in that region.

In summary, *a stream can maintain a higher velocity near its mouth even though it has a lower gradient than upstream because of the greater discharge, larger channel, and smoother bed.*

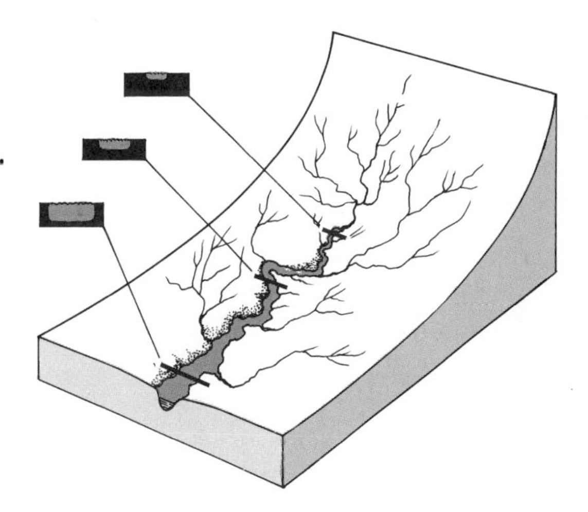

Figure 3.3 *Changes in channel size downstream. As discharge increases downstream, so do the width and depth of the stream channel.*

Base Level and Graded Streams

An important control over streamflow is base level. The **base level** is the lowest point to which a stream can erode its channel. Two general types of base level exist. Sea level is considered the **ultimate base level**, since it represents the lowest level to which stream erosion could lower the land. **Temporary,** or **local, base levels** include lakes, resistant rock, and main streams which act as base levels for their tributaries. All have the capacity to limit a stream at a certain level. For example, when a stream enters a lake, its velocity quickly approaches zero and its ability to erode decreases. Thus the lake prevents the stream from eroding below its level at any point upstream from the lake (Figure 3.4A). However, since the outlet of the lake can cut downward and drain the lake, the lake is only a temporary hindrance to the stream's ability to downcut its channel.

Any change in base level will cause a corresponding readjustment of stream activities. When a dam is built along a stream course, the reservoir which forms behind it raises the base level of the stream (Figure 3.4B). Upstream from

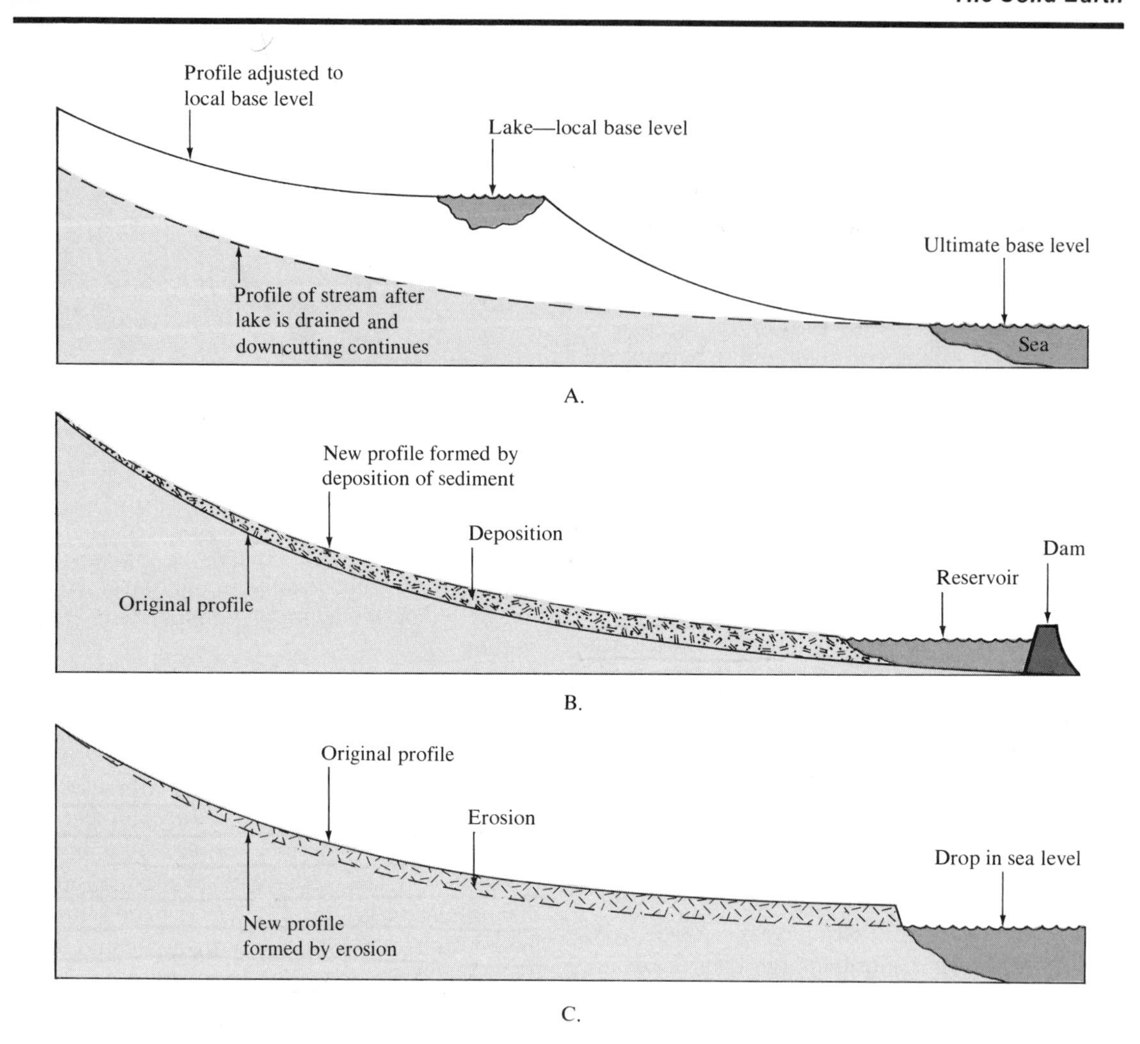

Figure 3.4 **A.** *Effect of a local base level on a stream profile.* **B.** *Adjustment of a stream profile to a change in base level.* **C.** *Readjustment of a stream to a lower base level.*

the dam the stream gradient is reduced, lowering its velocity and, hence, its sediment-transporting ability. The stream, now unable to transport all of its load, will deposit material, thereby building up its channel. This process continues until the stream again has a gradient sufficient to carry its load. The profile of the new channel would be similar to the old, except that it would be somewhat higher.

If on the other hand the base level should be lowered, either by uplifting of the land or by a drop in sea level, the stream would again readjust. The stream, now above base level, would have excess energy and would downcut its channel to establish a balance with its new base level (Figure 3.4C). Erosion would first progress near the mouth, then work upstream until the stream profile was adjusted along its full length.

The observation that streams adjust their profile for changes in base level led to the concept of a graded stream. A **graded stream** has the correct slope and other channel characteristics necessary to maintain just the velocity required to transport the material supplied to it. On the average, *a graded system is not eroding or depositing material but is simply transporting it.* Once a stream has reached this state of equilibrium, it becomes a self-regulating system in which a change in one characteristic causes an adjustment in the others to counteract the effect. Referring again to our example of a stream adjusting to a lowering of its base level, the stream would not be graded while it was cutting its new channel but would achieve this state after downcutting had ceased.

Work of Streams

The work of streams includes erosion, transportation, and deposition. These activities go on simultaneously in all stream channels, even though they are presented individually here.

Erosion

Although much of the material carried by streams has been brought in by underground water, overland flow, and mass wasting, streams do contribute to their load by eroding their own channels. If a channel is composed of bedrock, most of the erosion is accomplished by the abrasive action of water armed with sediment, a process analogous to sandblasting. Pebbles caught in eddies serve as cutting tools and bore circular holes into the channel floor, called **potholes.** In channels consisting of unconsolidated material considerable lifting can be accomplished by the impact of water alone.

Transportation

Once streams acquire their load of sediment, they transport it in three ways: (1) in solution **(dissolved load)**; (2) in suspension **(suspended load)**; and (3) along the bottom **(bed load).**

The dissolved load is brought to the stream by groundwater and to a lesser degree is acquired directly from soluble rock along the stream's course. The dissolved load makes up an estimated 20 percent of the total load of streams in the United States.

Most, but not all, streams carry the bulk of their load as suspended load (Figure 3.5). Usually only fine sand-, silt-, and clay-size particles can be carried this way, but during floodstage much larger particles are carried as well. Also during floodstage, the total quantity of material carried in suspension increases dramatically, as can be verified by persons whose homes have been sites for the deposition of this material. During floodstage the Hwang Ho (Yellow River) of China is reported to carry an amount of sediment equal in weight to the water that carries it. Rivers like this are often appropriately described as "too thick to drink, but too thin to cultivate."

The bed load, which consists of material that moves along the channel floor by rolling and sliding, is occasionally lifted for a few seconds by a swift eddy, a process termed **saltation** (Figure 3.5). Since the bed load travels much slower than both the suspended and dissolved loads, usually the least amount of material is transported in

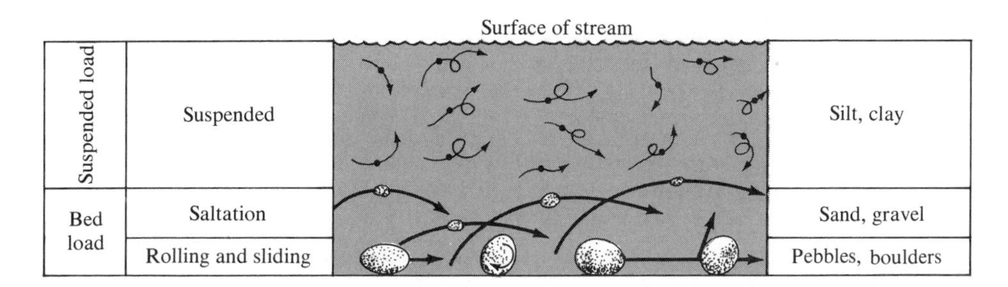

Figure 3.5 *Transportation of a stream's load.*

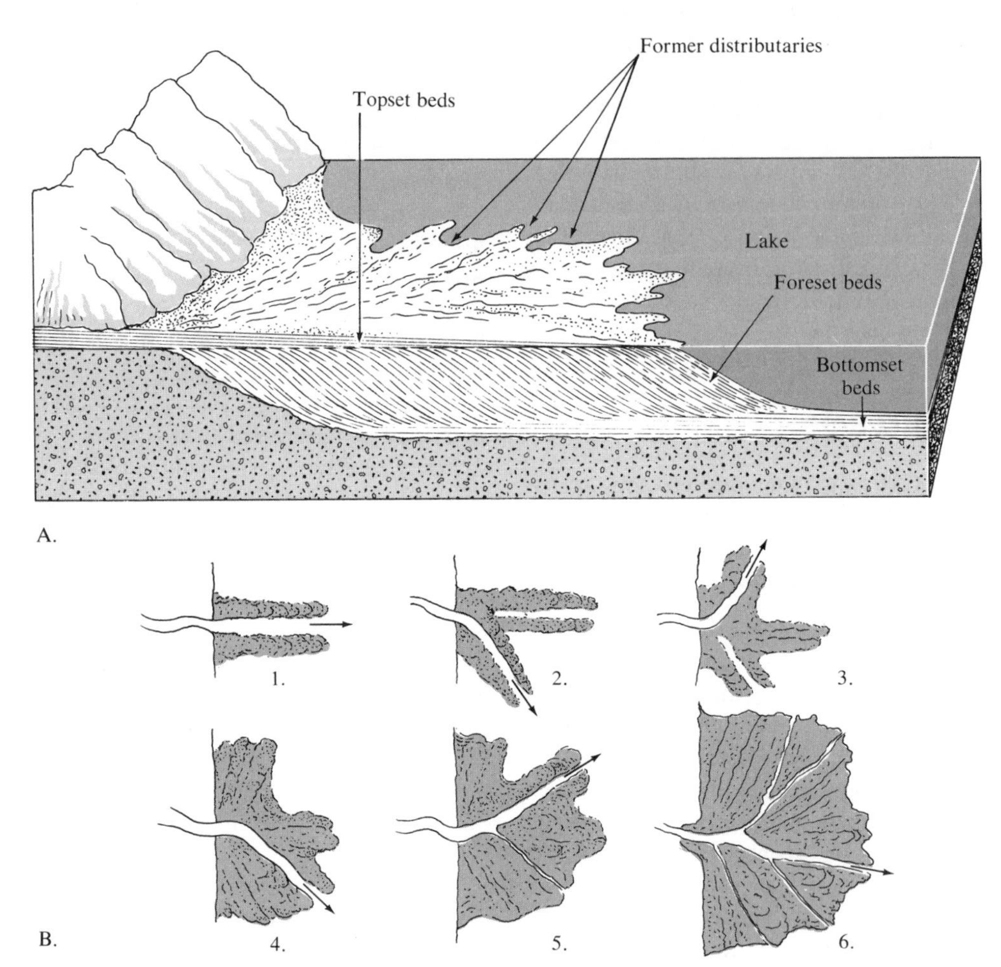

Figure 3.6 **A.** *Structure of a simple delta.* **B.** *Growth of a simple delta. Once a stream has extended its channel, the reduced gradient causes it to find a shorter distance to its base level. (A. and B. after Arthur N. Strahler,* Introduction to Physical Geography, *3rd ed., p. 455. Copyright © 1973 by John Wiley & Sons, Inc. Reprinted by permission.)*

this fashion. For illustration, consider the distribution of the 750 million tons of material carried to the Gulf of Mexico by the Mississippi River each year. Of this total, 500 million tons are carried in suspension, 200 million tons in solution, and the remaining 50 million tons as bed load.

A stream's ability to carry its load is established using two criteria. First, the **competence** of a stream is a measure of the maximum size of particles it is capable of transporting. The stream's velocity determines it competence. If the velocity of a stream doubles, the impact force of water increases four times; if the velocity triples, the force increases nine times; and so forth. Hence the huge boulders which are visible during the low-water stage and seem immovable can, in fact, be transported during floodstage

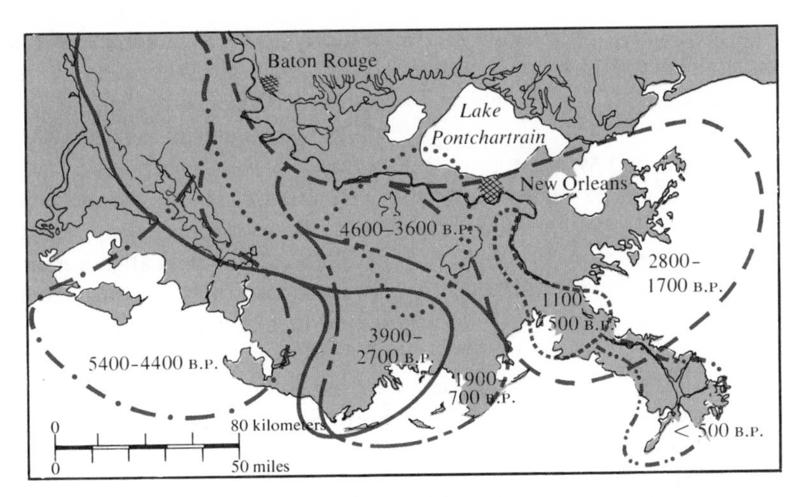

Figure 3.7 *Growth of the Mississippi Delta. The stages of development are dated in years "before present" (B.P.). The present bird's foot delta (lower right) represents the activity of the past 500 years. (After C. Kolb and J.R. Van Lopik,* Depositional Environments of the Mississippi River Deltaic Plain, *p. 22. Copyright © 1966 by the Houston Geological Society)*

because of the stream's increased velocity. Second, the maximum load a stream can carry is termed its **capacity.** The capacity of a stream is directly related to its discharge. The greater the amount of water flowing in the stream, the greater the stream's capacity for hauling sediment.

Deposition

Whenever a stream's velocity is reduced, its competence is also lowered. Consequently, some of the suspended particles begin to settle out. Material deposited in this manner is termed **alluvium.** Although some material temporarily settles in the channel, eventually it reaches its final destination, the ocean. When a stream enters the relatively still waters of an ocean or lake, its forward motion is quickly lost, and the resulting deposits form a triangular-shaped wedge called a **delta.** The finer silts and clays will settle out some distance from the mouth into nearly horizontal layers called **bottomset beds** (Figure 3.6A). Prior to the accumulation of bottomset beds, **foreset beds** begin to form. These beds are composed of coarse sediment which is dropped almost immediately upon entering a lake or ocean, forming sloping layers. The foreset beds are usually covered by thin, horizontal **topset beds** deposited during floodstage. As the delta grows outward, the gradient of the river is continually lowering, causing the stream to search for a shorter route to base level, a process il-

Figure 3.8 *Alluvial fans in the Mohave Desert. These fan-shaped structures develop where the gradient of a stream changes abruptly, such as at the foot of a mountain. (Photo by J.R. Balsley, U.S. Geological Survey)*

lustrated in Figure 3.6B. This figure shows how a simple delta grows into the idealized triangular shape of the Greek letter delta (Δ), for which it was named.

Large rivers like the Nile and Mississippi have deltas extending over thousands of square kilometers. The Mississippi delta began forming millions of years ago near the present-day town of Cairo, Illinois, and has since advanced nearly 1600 kilometers (1000 miles) to the south. New Orleans rests where there was ocean less than 5000 years ago, and the present bird's foot delta shown in Figure 3.7 has been built in the last 500 years. Figure 3.7 also shows the main channel dividing into several smaller ones called **distributaries,** which are found on most large deltas.

Alluvial fans are features similar to deltas which form on land (Figure 3.8). When mountain streams reach a plain, their gradient is abruptly lowered, and hence they immediately dump much of their load. Usually the coarse material is dropped near the base of the slope, while finer material is carried farther out on the plain.

Rivers that occupy valleys with broad, flat valley floors on occasion build a landform that parallels its channel called a **natural levee.** Natural levees are built by successive floods over a period of many years. When a stream overflows its banks, its velocity immediately diminishes, leaving coarse sediment deposited in strips bordering the channel (Figure 3.9). As the water spreads out over the valley a lesser amount of fine sediment is deposited over the valley floor. This uneven distribution of material produces the very gentle slope of the natural levee. The natural levees of the lower Mississippi rise 6 meters (20 feet) above the valley floor. The area behind the levee is characteristically

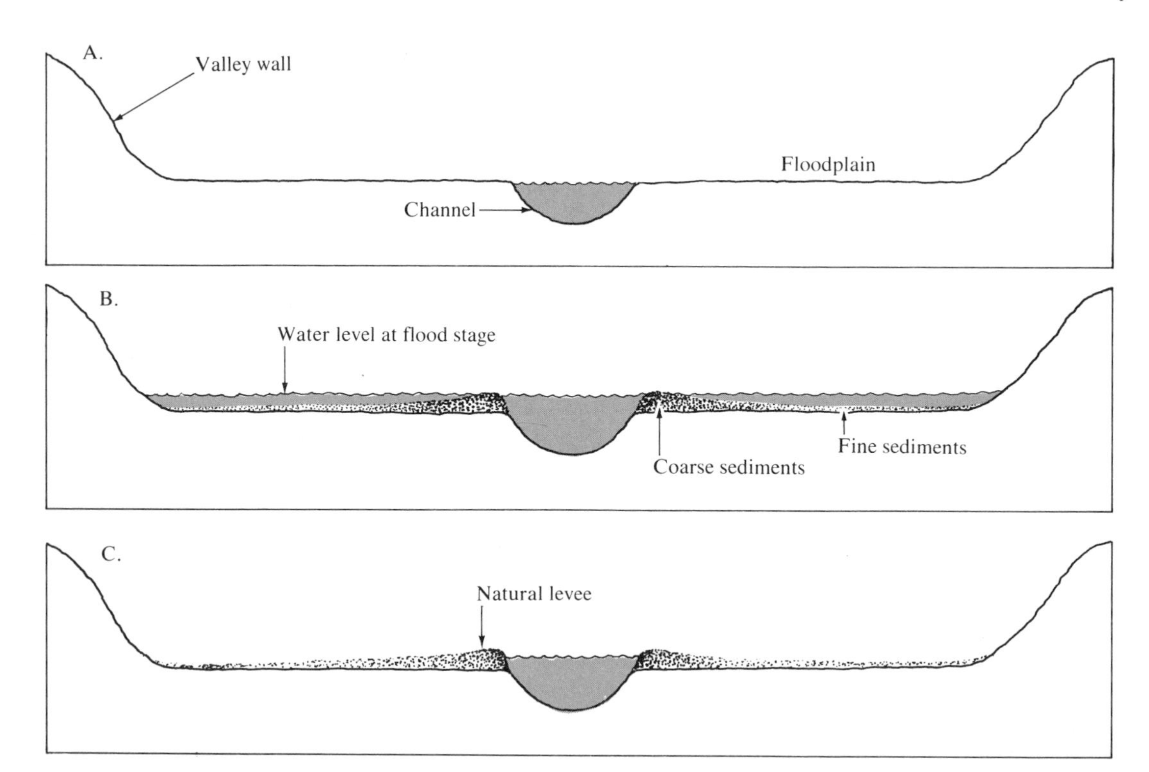

Figure 3.9 *Formation of natural levees. After many years of flooding, streams which carry some coarse sediment can build very gently sloping levees.*

Figure 3.10 *V-shaped valley of the Yellowstone River. The rapids and waterfalls of this river indicate that it is vigorously downcutting. (Courtesy of National Park Service, U.S. Department of the Interior)*

poorly drained for the obvious reason that water cannot flow up the levee and into the river. Marshes called **back swamps** result. A tributary stream that attempts to enter a river with natural levees often has to flow parallel to the main stream until it can breach the levee. Such streams are called **yazoo tributaries** after the Yazoo River, which parallels the Mississippi for over 300 kilometers.

Stream Valleys

Stream valleys can be divided into two general types. Narrow V-shaped valleys and wide valleys with flat floors exist as the ideal forms, with many gradations between. In arid regions narrow valleys often have nearly vertical walls, while in humid regions the effect of mass wasting and slope erosion caused by heavy rainfall produce the typical V-shaped valley (Figure 3.10). The type of valley gives an indication of what the stream has been doing. A narrow V-shaped valley indicates that the primary work of the stream has been downcutting toward base level. On the other hand, streams with flat-floored valleys have been widened by lateral (side-to-side) erosion (see chapter-opening photo).

The most prominent features of a narrow valley are **rapids** and **waterfalls.** Both occur where the stream profile drops rapidly, a situation which is usually caused by variations in the erodibility of the bedrock into which the stream channel is cutting. Figure 3.11 shows how a resistant bed produces a rapids by acting as a temporary base level upstream while allowing downcutting to continue downstream. Once erosion has eliminated the resistant rock, the stream profile

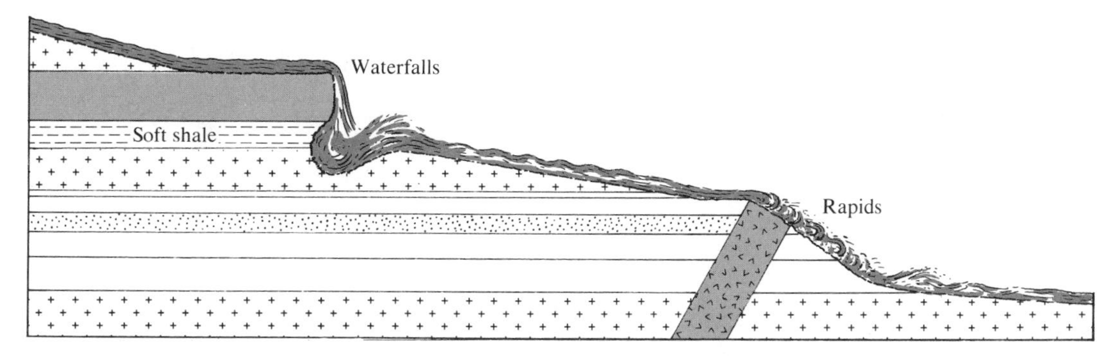

Figure 3.11 *Formation of rapids and falls on resistant rock.*

Figure 3.12 *Bird's-eye view of Niagara Falls, with the larger Horseshoe Falls on the right and the American Falls on the left. (Courtesy of U.S. Army Engineer District, Buffalo, New York)*

smooths out again. Waterfalls are places where the stream profile makes a vertical drop. One type of waterfall is exemplified by Niagara Falls (Figure 3.12). Here, the falls are supported by a resistant bed of dolomite that is underlain by a less resistant shale (Figure 3.11). As the water

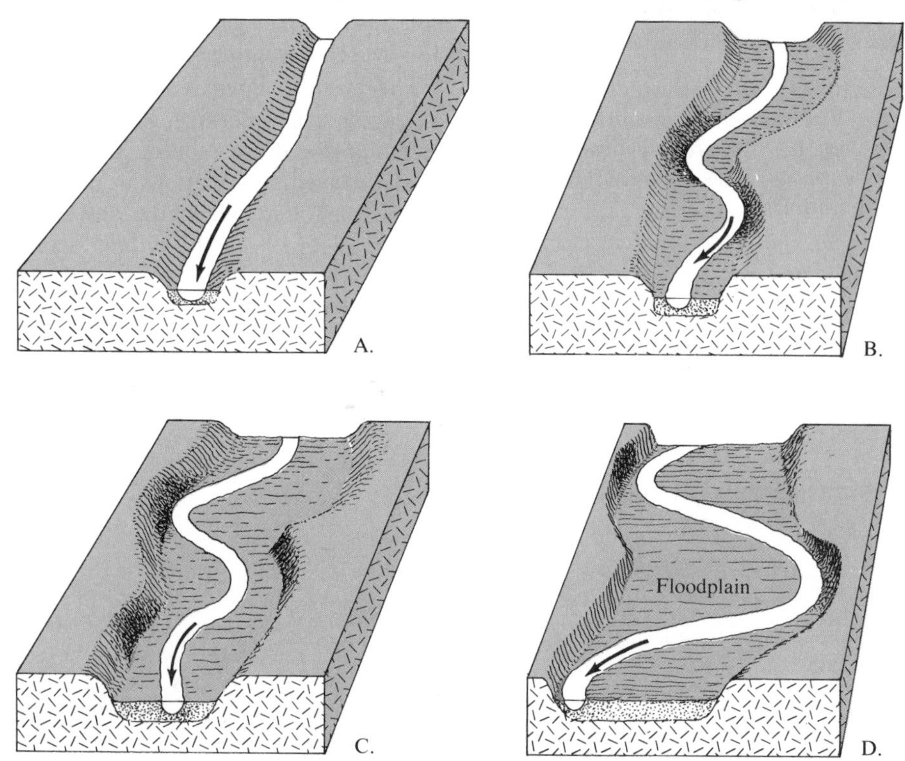

Figure 3.13 *Stream eroding its floodplain.*

plunges over the lip of the falls it erodes the less resistant shale, undermining a section of dolomite, which eventually breaks off. In this manner the waterfall retains its vertical cliff while slowly but continually retreating upstream. Since its formation Niagara Falls has retreated approximately 11 kilometers (7 miles) upstream.

Once a stream has cut its channel closer to base level it begins to reach a graded condition, and downward erosion becomes less dominant. At this point more of the stream's energy is directed from side to side. The reason for this change is not fully understood, but the reduced gradient probably is an important factor. Nevertheless it does occur, and the result is a widening of the valley as the river cuts away first at one bank and then at the other (Figure 3.13). In this manner the flat valley floor, or **floodplain,** is produced. It is appropriately named, because the river is confined to its channel except during floodstage, when it overflows its banks and inundates the floodplain.

When a river erodes laterally, creating a floodplain as just described, it is called an *erosional floodplain.* Floodplains can be depositional in nature as well. *Depositional floodplains* are produced by a major fluctuation in conditions, such as a change in base level. The floodplain in Yosemite Valley is such a feature, which was produced when a glacier gouged the former stream valley about 300 meters (1000 feet) deeper than it had been. After the glacial ice melted, the stream readjusted itself to its former base level by refilling the valley with alluvium.

Streams that flow upon floodplains, whether erosional or depositional, move in sweeping bends called **meanders.** Meanders continually change position by moving sideways and slightly downstream (Figure 3.14). The sideways movement occurs because the maximum velocity of the stream shifts toward the outside of the bend, causing erosion of the outer bank (Figure 3.15). At the same time the reduced current at the inside of the meander results in the deposition of coarse sediment, especially sand. Thus by eroding its outer bank and depositing material along its inner bank a stream moves sideways and slightly downstream without changing its channel size. On occasion one meander moves downstream faster than another and eventually erodes the neck of land between them (Figure 3.16). When this happens, the meander is said to be *cut off,* and if abandoned, it is called an **oxbow.** Because some oxbows contain water, the term **oxbow lake** is given to them (Figure 3.17). The process of meander cutoff has the effect of shortening the river and was described quite humorously by Mark Twain in *Life on the Mississippi:*

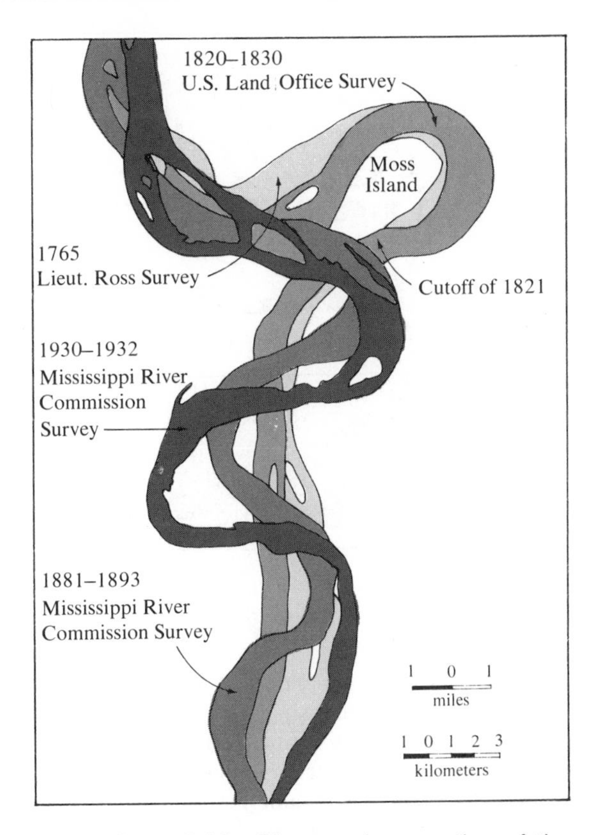

Figure 3.14 *Changes in a section of the Mississippi River above Memphis between 1765 and 1932. (After Arthur N. Strahler,* Introduction to Physical Geography, *3rd ed., p. 451. Copyright © 1973 by John Wiley & Sons, Inc. Reprinted by permission.)*

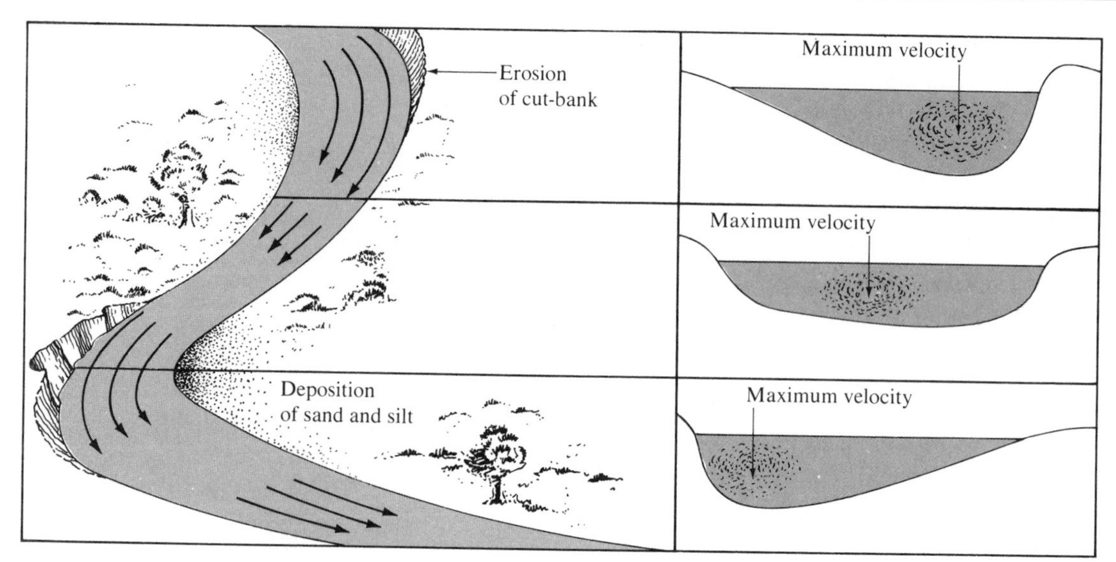

Figure 3.15 *Lateral movement of meanders. By eroding its outer bank and depositing material on the inside of the bend, a stream is able to shift its channel.*

> In the space of one hundred and seventy-six years the lower Mississippi has shortened itself two hundred and forty-two miles. This is an average of a trifle over one mile and a third per year. Therefore, any calm person, who is not blind or idiotic, can see that in the Old Oolitic Silurian Period, just a million years ago next November, the Lower Mississippi River was upwards of one million three hundred thousand miles long, and stuck out over the Gulf of Mexico like a fishing rod. And by the same token any person can see that seven hundred and forty-two years from now the Lower Mississippi will be only a mile and three quarters long, and Cairo and New Orleans will have joined their streets together, and be plodding comfortably along under a single mayor and a mutual board of aldermen. There is something fascinating about science. One gets such wholesale returns of conjecture out of such a trifling investment of fact.

Although the data used by Mark Twain may be reasonably accurate, he purposely forgot to include the fact that the Mississippi created many new meanders, thus lengthening its course by a similar amount. In fact, with the growth of the delta, the Mississippi is actually getting longer, not shorter.

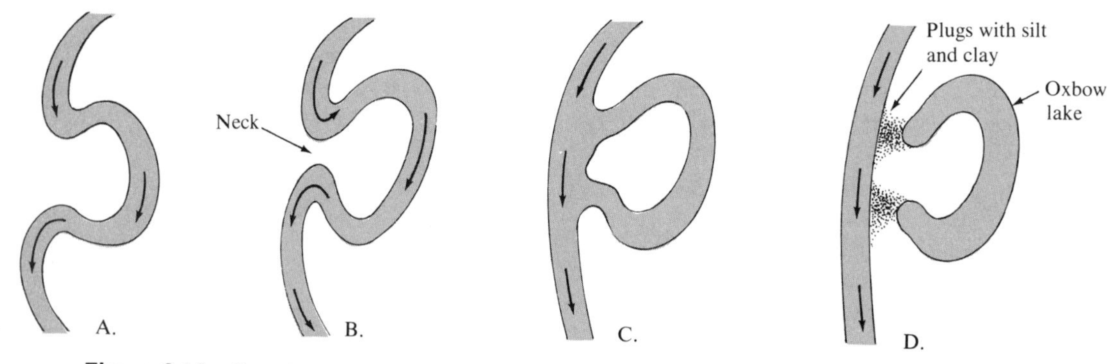

Figure 3.16 *Development of a cutoff.*

Figure 3.17 *Aerial view of oxbow lakes along the White River in Arkansas. (Photo by J.R. Balsley, U.S. Geological Survey)*

Not all streams meander placidly on their floodplains. Those that have more load than their channel can handle tend to become clogged, causing the stream to break into many intertwining channels. Because such a stream has a braided appearance, the term **braided stream** is applied.

Drainage Systems and Patterns

A stream is just a small component in a much larger system. Each system consists of a **drainage basin,** the land area that contributes water to the stream. The drainage basin of one stream is separated from another by an imaginary line called a **divide** (Figure 3.18). Divides range in size from a ridge separating two small gullies to *continental divides,* which split continents into enormous drainage basins. For example, the continental divide that runs somewhat north-south through the Rocky Mountains separates the drainage which flows west to the Pacific Ocean from that which flows to the Atlantic. Although divides separate the drainage of two streams, if they are tributaries of the same river, they are both a part of that larger drainage system.

All drainage systems are made up of an interconnected network of streams which form par-

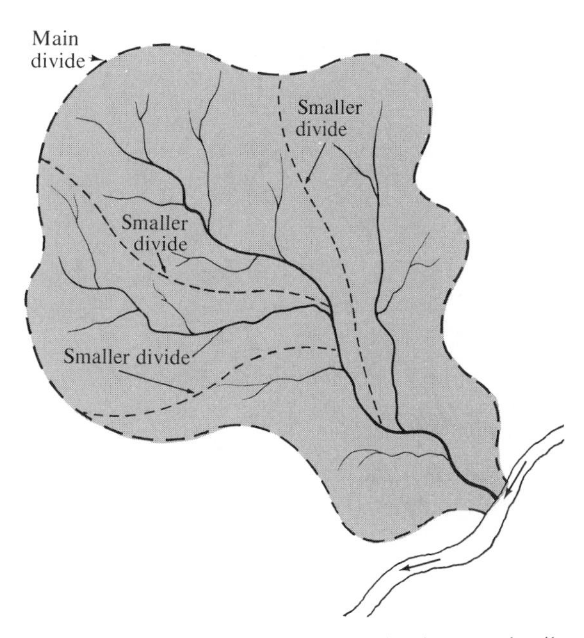

Figure 3.18 *Drainage basins and divides. Divides separate the drainage basins of each stream. Drainage basins and divides also exist for the smallest tributary but are not shown.*

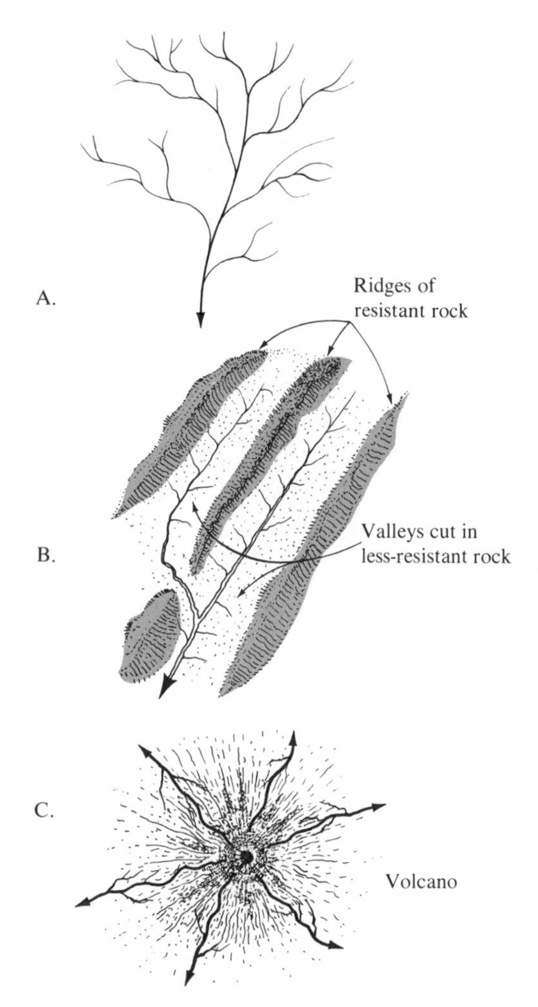

Figure 3.19 *Drainage patterns.* **A.** *Dendritic.* **B.** *Trellis.* **C.** *Radial.*

ticular patterns depending on the material over which they flow. In regions where the surface material is relatively uniform a branching tree-like arrangement develops and is referred to as a **dendritic pattern** (Figure 3.19A). Where bands of alternately resistant and less-resistant rocks are exposed at the surface, a **trellis pattern** forms (Figure 3.19B). Here the main stream occupies a valley cut into the soft rock, while tributaries flow down the resistant ridges, entering the main stream at nearly right angles. Where an elevated structure such as a volcano or dome exists, the drainage often takes on a **radial pattern** (Figure 3.19C).

Stages of Valley Development

Contrary to the popular belief of his day, James Hutton proposed that streams were responsible for cutting the valleys in which they flowed. Later geologic work conducted in stream valleys substantiated Hutton's pronouncement and further revealed that the development of stream valleys progresses in a somewhat orderly fashion. The evolution of a valley has been arbitrarily divided into three sequential stages: youth, maturity, and old age.

As long as the stream is downcutting to establish a graded condition with its base level, it is considered youthful. Rapids, an occasional waterfall, and a narrow V-shaped valley are all visible signs of the vigorous downcutting that is going on. Other features of a youthful stream include a steep gradient, little or no floodplain, and a rather straight course without meanders (Figure 3.20A).

When a stream reaches maturity downward erosion diminishes and lateral erosion dominates. Thus the mature stream begins actively cutting its floodplain and meandering upon it (Figure 3.20B). During the mature stage cutoffs occur, producing oxbows, and a few streams may even produce low natural levees. In contrast to the gradient of a youthful stream, the gradient of a mature stream is much lower, and the profile is much smoother, since all rapids and waterfalls have been eliminated.

A stream enters old age after it has cut its floodplain several times wider than its **meander belt,** which is the width of the meander (Figure 3.20C). When this stage is reached the stream is rarely near the valley walls; hence it ceases to enlarge the floodplain. Thus the primary work of a river in an old-age valley is the reworking of unconsolidated floodplain deposits. Because this task is easier than cutting bedrock, a stream in an old-age valley shifts more rapidly than a stream in a mature valley. For example, some meanders

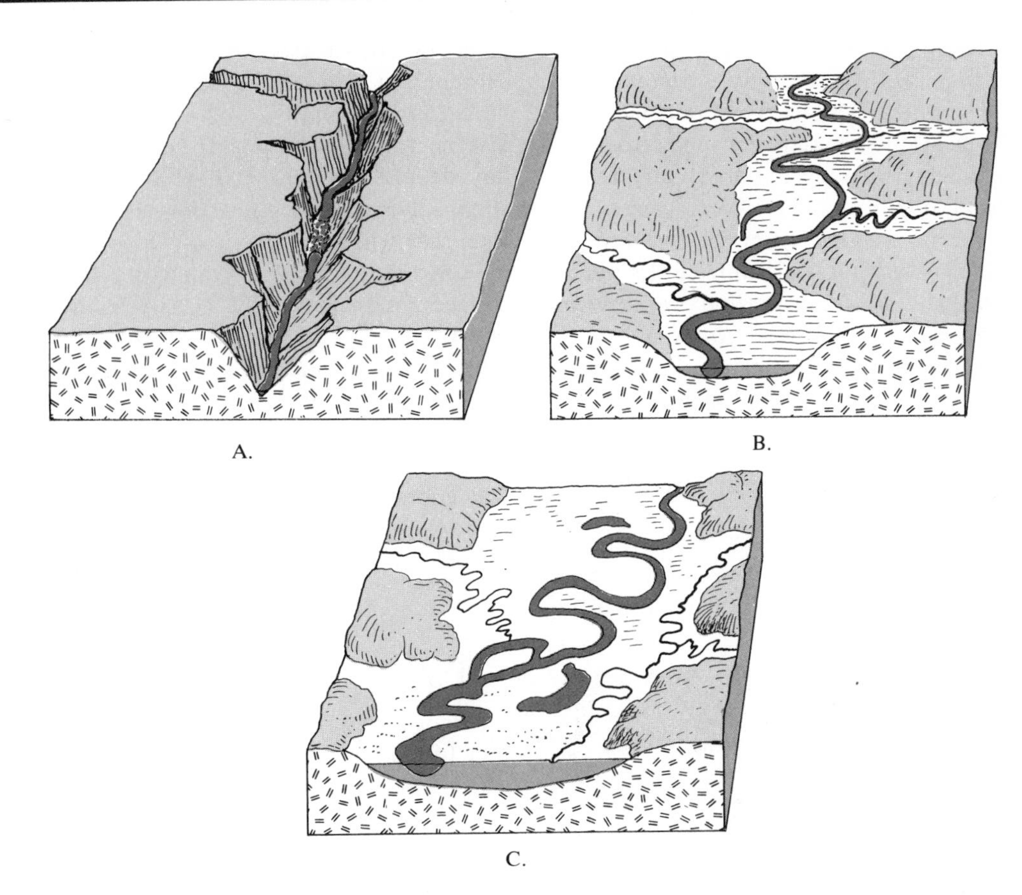

Figure 3.20 *Stages of valley development.* **A.** *Youth. The youthful stage is characterized by downcutting and a V-shaped valley.* **B.** *Maturity. Once a stream has sufficiently lowered its gradient, it begins to erode laterally, producing a wide valley.* **C.** *Old age. After the valley has been cut several times wider than the zone of the meandering stream, it has entered old age. (B. and C. courtesy of Ward's Natural Science Establishment, Inc., Rochester, New York)*

of the lower Mississippi move 20 meters (60 feet) a year, and its large floodplain is dotted with oxbow lakes and old cutoffs. Natural levees are also common features of old-age valleys and, when present, are accompanied by back swamps and yazoo tributaries.

Thus far we have assumed that the base level of a stream remains constant as a river progresses from youth to old age. On many occasions, however, the land is uplifted. The effect of uplifting on a youthful stream is to increase its gradient and accelerate its rate of downcutting. However, uplifting of a mature stream would cause it to abandon lateral erosion and revert to downcutting. Rivers of this type are said to be **rejuvinated,** and the meanders are known as **entrenched meanders** (Figure 3.21). Mature streams may eventually readjust to uplift by cutting a new floodplain at a level below the old one. The remnants of the old floodplain are often present in the form of flat surfaces called **terraces.**

Two additional points concerning valley development should be made. First, *the time required for a stream to reach any given stage depends on several factors, including the erosive ability of the stream, the nature of the material through which the stream must cut, and its height above base level.* Consequently, a stream which starts out very near base level and only has to cut through unconsolidated sediments may reach maturity in a matter of a few hundred years. On the other hand, the Colorado River, where it is actively cutting the Grand Canyon, has retained its youthful nature for an estimated 15 million years. Second, *individual portions of a stream reach each stage at different times.* Often the lower reaches of a stream attain old age while the headwaters are still youthful in character.

Cycle of Landscape Evolution

While streams are cutting their valleys they simultaneously sculpture the land. To describe this unending process we will need a starting point. For this reason only, we will assume the existence of a relatively flat upland area in a humid region. Until a well-established drainage system forms, lakes and ponds will occupy any depressions that exist (Figure 3.22A). As streams form and begin to downcut toward base level they will drain the lakes. During the youthful stage the landscape retains its relatively flat surface, interrupted only by narrow stream valleys (Figure 3.22B). As downcutting continues, relief increases, and the flat, youthful landscape is transformed into one consisting of the hills and valleys which characterize the mature stage (Figure 3.22C). Eventually some of the streams will approach base level, and downcutting will give way to lateral erosion. As the cycle nears the old-age stage the effects of overland flow and mass wasting, coupled with the lateral erosion by streams, will reduce the land to a **peneplain** ("near plain"), an undulating plain near base level (Figure 3.22D). Although no peneplains are known to exist today, there is evidence that they formed in the past and have since been uplifted. Once a peneplain has formed, uplifting starts the cycle over again. Most often, uplifting interrupts the cycle before it reaches old age.

Figure 3.21 *Entrenched meanders of the Colorado River in Canyonlands National Park, Utah. (Photo by S.W. Lohman, U.S. Geological Survey)*

Water beneath the Surface

Many chronicles show that people have known from ancient times that much water lies underground. The Bible includes references to "the water under the earth" (Exodus 20:4) and to "the fountains of the great deep" (Genesis 7:11). How much water is underfoot? The U.S. Geological Survey estimates that the volume of water in the upper 0.8 kilometer (0.5 mile) of the continental crust is about 3000 times greater than the volume of water in all rivers at any one time, and nearly 20 times greater than the combined volume in all lakes and rivers. Thus this vast reservoir of underground water represents a significant part of the world's water resources. In addition, it is of importance as an equalizer of streamflow and as an agent of erosion. It is the

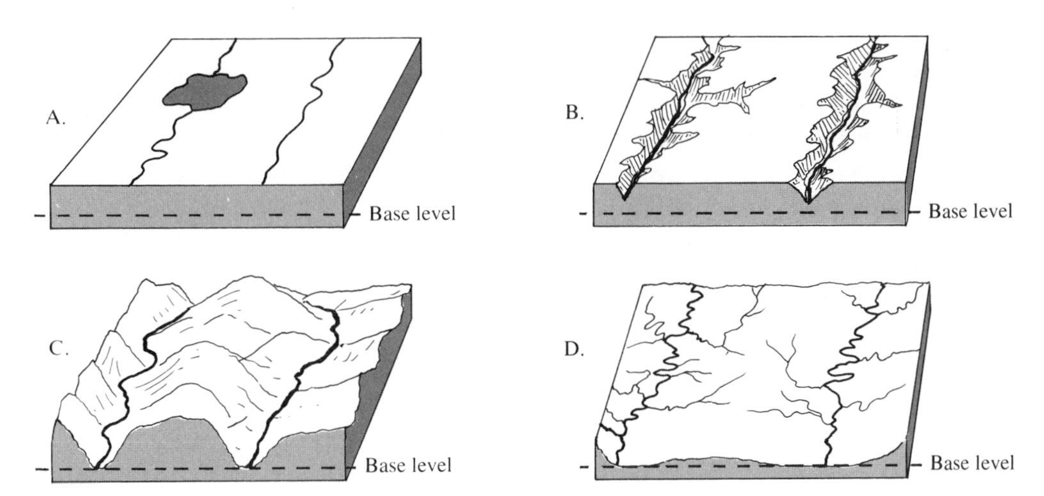

Figure 3.22 *Idealized cycle of landscape evolution.* **A.** *Initial stage. The land is poorly drained and situated well above base level.* **B.** *Youthful stage. Streams have cut downward and drained the lakes. The landscape is still relatively flat between stream valleys.* **C.** *Mature stage. All of the area between the initial streams has been eroded by running water. Most of the landscape is in slope, and maximum relief exists.* **D.** *Old-age stage. Lateral erosion, mass wasting, and slope wash have lowered most of the hills to the level of the floodplain. The entire surface has become an undulating plain near base level. (B.–D. from Foster,* Physical Geology, *3rd ed., Columbus, Ohio: Charles E. Merrill, 1979)*

work of subsurface water which creates caves, caverns, and other related features.

Distribution and Movement of Groundwater

Beneath most land areas there is a zone where the spaces in sediments and rocks are completely saturated with water. The water in this saturated zone is called **groundwater,** and the upper limit of the zone is known as the **water table.** The configuration of the water table is not always perfectly flat. Rather it tends to coincide with the surface topography but is more subdued. Swamps occur when the water table is right at the surface, while lakes may result when the land is below the level of the water table (Figure 3.23). In desert regions the water table may lie hundreds of me-

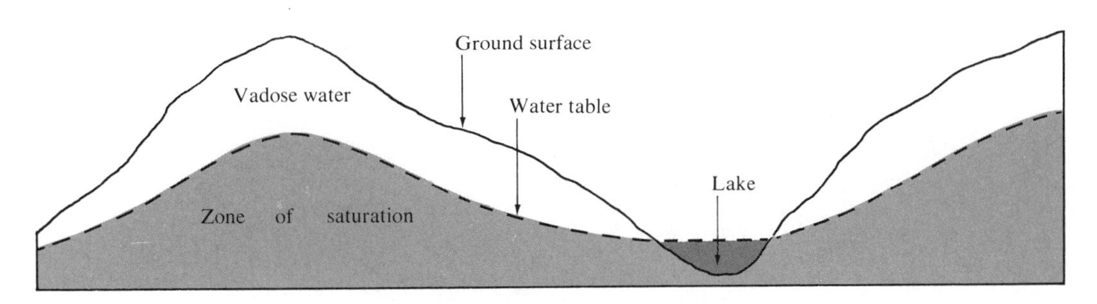

Figure 3.23 *The configuration of the water table generally coincides with the surface topography but is more subdued. When there is a depression below the water table, a lake will result.*

ters below the surface. The depth of the zone of saturation depends upon the local geology. Its lower limit is the point where rock becomes so dense that water cannot penetrate further.

Water in the unsaturated zone above the water table is called **vadose water.** Water in this zone is percolating downward toward the saturated zone. A certain amount of vadose water is always present because of surface tension with rock and soil particles.

The quantity of groundwater that can be stored in rocks depends upon the volume occupied by open spaces, that is, the **porosity** of the material. The openings may be pores in sediment or sedimentary rock, or fractures in crystalline rocks. Porosity alone, however, is not a satisfactory measure of a material's capacity to yield groundwater. The **permeability** of a material, its ability to transmit water, is of prime importance. In material like clay, which has very small openings, most of the water may be held by molecular attraction and be unable to move. Therefore, even though some clays have high porosities, they yield little or no water. On the other hand, material with large openings may yield much water, although the porosity may be small. Permeable rock strata or unconsolidated materials which transmit groundwater are called **aquifers.**

The rate of groundwater movement is usually very slow, especially when compared to the movement of streams. Rather than measuring velocity in meters per second or kilometers per hour, the movement of groundwater is measured in meters or centimeters per day. Although rates of tens of meters per day have been measured, a rate of 12–15 meters (40–50 feet) per day is considered high.

Wells and Springs

The most common device used by humans for removing groundwater is the **well,** an opening bored into the zone of saturation. Wells serve as reservoirs into which groundwater moves and from which it can be pumped to the surface. Digging for water dates back many centuries and continues to be an important method of obtaining water. Today in many parts of the world, including the United States, water from wells irrigates more land than does water from streams.

The level of the water table may fluctuate considerably during the course of a year, dropping during dry seasons and rising following periods of rain. Therefore, to insure a continuous supply of water, a well should penetrate many meters below the water table. When water is pumped from a well, it produces a depression

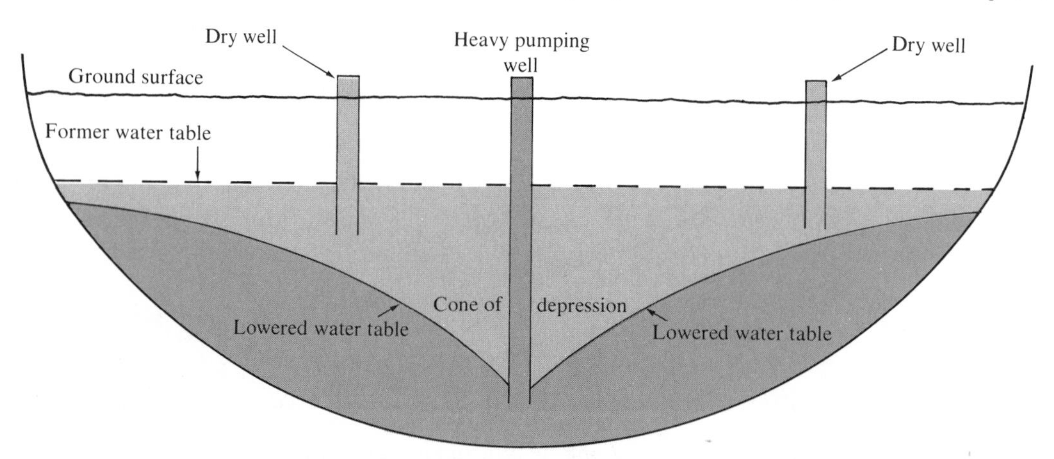

Figure 3.24 *A cone of depression in the water table often forms around a pumping well. If heavy pumping lowers the water table, some wells may be left high and dry.*

Figure 3.25 *Thousand Springs, Gooding County, Idaho. (Photo by I.C. Russell, U.S. Geological Survey)*

in the water table, roughly conical in shape, known as a **cone of depression.** If pumping is heavy, the water table may not only be lowered immediately around the well but may extend over a large area (Figure 3.24). This has been the case in portions of the western United States. Under these circumstances, it can be said that the groundwater is literally being "mined," for even if pumping were to cease immediately, it could take hundreds of years for the groundwater to be replenished. The following example illustrates this point:

> Consider for example, a location in the dry southwestern United States where annual recharge to an aquifer is on the order of only two-tenths of an inch of water. In such areas it is not uncommon to pump two feet or more of water per year for irrigation or other uses. In this over-simplified example if the entire aquifer were pumped at that rate, yearly pumpage would be equivalent to 120 years' recharge, and ten years of pumping would remove a 1200-year accumulation of water. New recharge during the pumping period would be negligible. Mechanical problems and economic factors prevent complete dewatering of an aquifer, but the example is valid in principle.*

*U.S. Geological Survey, *Water of the World,* 1968, p. 16.

When the water table intersects the ground surface, **springs** result (Figure 3.25). There are many circumstances which might create springs. Two such situations are illustrated in Figure 3.26. In some places the main zone of saturation is overlain by unsaturated material that contains an impervious layer, above which a local zone of saturation may occur. If this **perched water table** intersects a hillside a spring or series of springs are found (Figure 3.26A). Springs also occur when fractures or solution channels (subterranean channels created when soluble rock, such as limestone, is dissolved by groundwater) intersect the surface (Figure 3.26B).

Artesian Wells

The term **artesian** is applied to any situation in which groundwater rises in a well above the level where it was initially encountered. For such a situation to occur, two conditions must exist (Figure 3.27): (1) Water must be confined to an aquifer that is inclined so that one end is exposed at the surface, where it can receive water; and (2) Impermeable layers, both above and below the aquifer, must be present to prevent the water from escaping. When such a layer is tapped, the pressure created by the weight of the water above will cause the water to rise. If there were no friction the water in the well would rise to the level of the water at the top of the aquifer. However friction reduces the height of this pressure surface. The greater the distance from the recharge area (area where water enters the inclined aquifer), the greater the friction and the less the rise of water. In Figure 3.27 Well 1 is a **nonflowing artesian well,** because at this location the pressure surface is below ground level. When the pressure surface is above the ground and a well is drilled into the aquifer, a **flowing artesian well** is created (Well 2, Figures 3.27A and B).

Artesian systems act as conduits, transmitting water from remote areas of recharge great distances to the points of discharge. In this manner water which fell in central Wisconsin years ago is now taken from the ground and used by com-

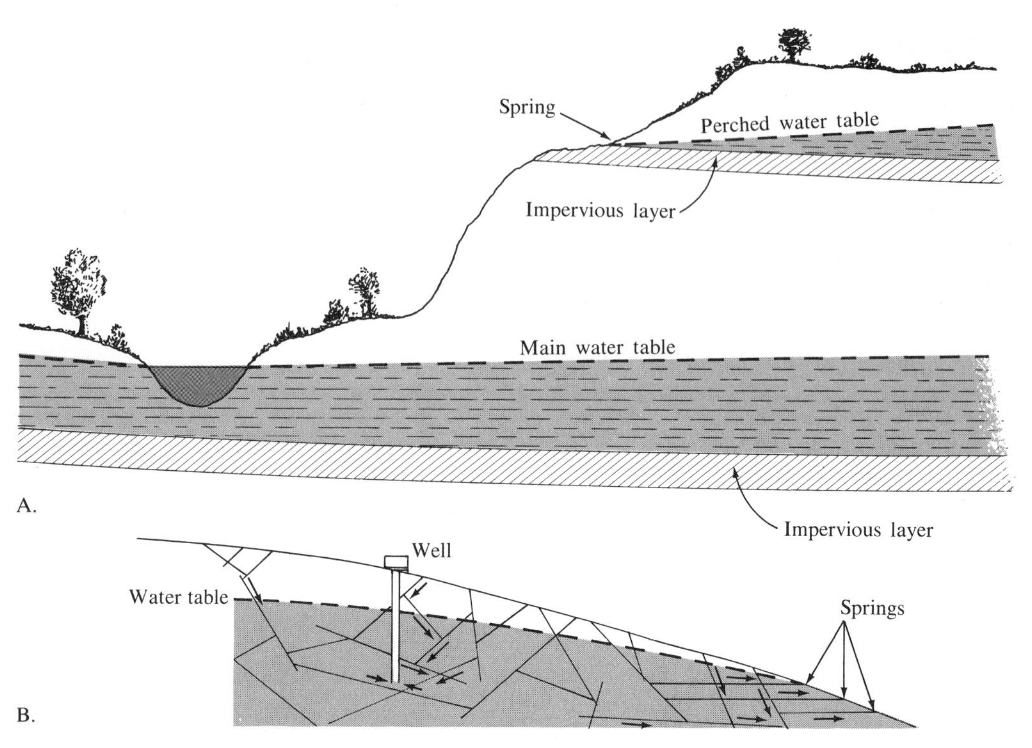

Figure 3.26 *Springs occur when the water table intersects the ground surface. (A. after Water, The 1955 Yearbook of Agriculture, p. 51; B. from Foster,* Physical Geology, *3rd ed., Columbus, Ohio: Charles E. Merrill, 1979)*

munities many kilometers away, in Illinois. In South Dakota such a system has brought water from the Black Hills in the west, eastward across the state. On a different scale, city water systems may be considered examples of artificial artesian systems (Figure 3.28). The water tower, into which water is pumped, may be considered the area of recharge, the pipes the confined aquifer, and the faucets in homes the flowing artesian wells.

Hot Springs, Geysers, and Geothermal Energy

Hot Springs

By definition, the water in hot springs is 6–9° C (10–15° F) warmer than the mean annual air temperature for the localities where they occur. In the United States alone, there are well over 1000 such springs.

Mineral explorations over the world have shown that temperatures in deep mines and oil wells usually rise with an increase in depth below the surface. Temperatures in such situations increase an average of about 2° C per 100 meters (1° F per 100 feet). Therefore when groundwater circulates at great depths, it becomes heated, and if it rises to the surface, the water emerges as a hot spring. The water of some hot springs in the United States, particularly in the East, is heated in this manner. The great majority (over 95 percent) of the hot springs (and geysers) in the United States are found in the West. The reason for such a distribution is that the source of heat for most hot springs is cooling igneous rock, and it is in the West that igneous activity has been most recent.

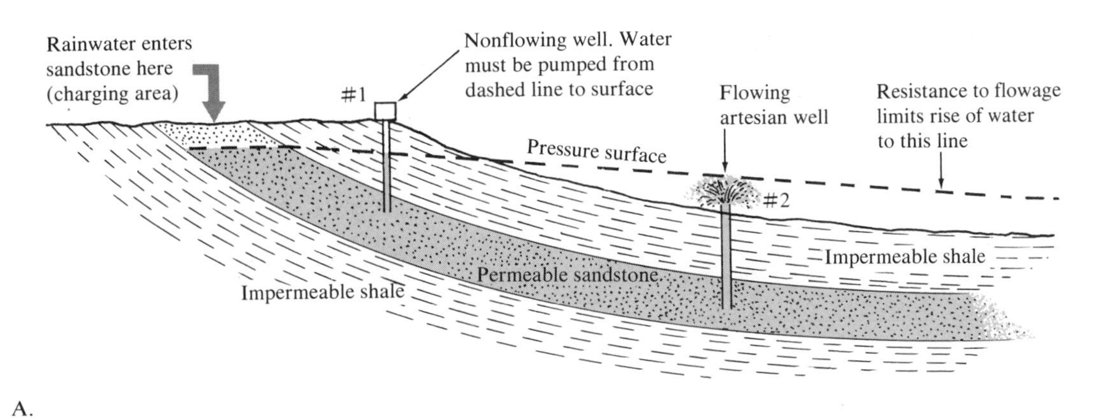

A.

B.

Figure 3.27 **A.** *Artesian systems occur when an inclined aquifer is surrounded by impermeable beds. (From Foster,* Physical Geology, *3rd ed., Columbus, Ohio: Charles E. Merrill, 1979)* **B.** *A flowing artesian well, Gallatin County, Montana. (Photo by J.R. Stacy, U.S. Geological Survey)*

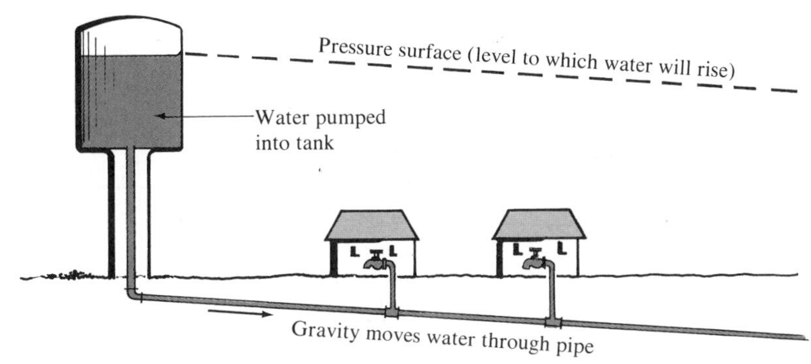

Figure 3.28 *City water systems may be considered to be artificial artesian systems.*

Figure 3.29 *A geyser in action. Old Faithful in Yellowstone National Park emits up to 45,000 liters of hot water and steam about once each hour.* **A.** *Recharge stage.* **B.** *Preliminary eruption stage.* **C.** *Full eruption stage.* **D.** *Steam stage. (Photos by J.R. Stacy, U.S. Geological Survey)*

Geysers

Geysers are intermittent hot springs or fountains where columns of water are ejected with great force at various intervals, often rising 30–60 meters (100–200 feet). After the jet of water ceases, a column of steam rushes out, usually with a thundering roar. Perhaps the most famous geyser in the world is Old Faithful in Yellowstone National Park, which erupts about once each hour (Figure 3.29). Geysers are also found in other parts of the world, including New Zealand and Iceland, where the term *geyser,* meaning "spouter" or "gusher," was coined.

Geysers occur when groundwater is heated in underground chambers. At the bottom of the chamber, the water is under great pressure because of the weight of the overlying water. Consequently, a temperature above 100° C (212° F) is required before it will boil. For example, at the bottom of a 300-meter (1000-foot) chamber water must attain a temperature of nearly 230° C (450° F) before it will boil. The heating causes the water to expand, with the result that some flows out at the top. This decreases the pressure, and the water quickly turns to steam and causes the geyser to erupt (Figure 3.30).

Groundwater from hot springs and geysers usually contains more material in solution than groundwater from other sources because hot water is a more effective dissolver than cold. When the water contains much dissolved silica, *geyserite* is deposited around the spring. *Travertine,* a form of calcite, is a characteristic deposit at hot springs in limestone regions (Figure 3.31). Some hot springs contain sulfur. In addition to making the water taste bad, sulfur emits an unpleasant odor. Undoubtedly Rotten Egg Spring, Nevada, is such a situation.

Geothermal Energy

Many natural geyser areas around the world are potential sites for tapping **geothermal energy,** that is, natural steam used for power generation. In New Zealand, Italy, the Soviet Union, and the United States underground supplies of superheated steam are now being used to provide power for generating electricity. In the United States the first commercial geothermal power plant was built in 1960 at "The Geysers," north of San Francisco (Figure 3.32). The most favorable geologic factors for a geothermal reservoir of commercial value include:

(1) A potent source of heat, such as a large magma chamber. The chamber should be deep enough to ensure adequate pressure and a slow rate of cooling, and yet not be so deep that the natural circulation of water is inhibited. Magma chambers of this type are most likely to occur in regions of recent volcanic activity;

(2) Large and porous reservoirs with channels connected to the heat source, near which water can circulate and then be stored in the reservoir;

(3) Capping rocks of low permeability that inhibit the flow of water and heat to the surface. A deep and well-insulated reservoir is likely to contain much more stored energy than an uninsulated, but otherwise similar, reservoir.*

It is too early to judge whether natural steam has the potential to satisfy an important part of the world's requirements for electrical power, but with the need to develop new sources of energy, the possibilities are definitely worth exploring.

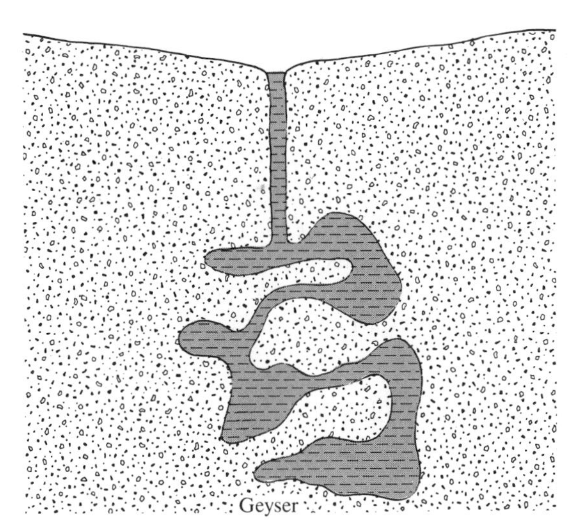

Figure 3.30 *Idealized diagram of a geyser. A geyser can form if the heat is not distributed by convection. In this figure the water near the bottom is heated to near its boiling point. The boiling point is higher there than at the surface because the weight of the water above increases the pressure. The water higher in the geyser system is also heated and so expands and flows out at the top, reducing the weight of the water on the bottom. At the reduced pressure on the bottom boiling occurs. The bottom water flashes into steam, and the expanding steam causes an eruption. (From Foster,* Physical Geology, *3rd ed., Columbus, Ohio: Charles E. Merrill, 1979)*

The Geologic Work of Groundwater

The primary erosional work carried on by groundwater is solution (dissolving soluble rock). Rainwater becomes slightly acid as carbon dioxide dissolves in it, producing carbonic acid (Chapter 2). As the water percolates through the soil, decaying plants add complex organic acids. In regions where easily soluble rock, especially limestone, underlies the surface, groundwater

*Adapted from U.S. Geological Survey, *Natural Stream for Power,* 1968.

Figure 3.31 *Mineral deposits surrounding Mammoth Hot Springs, Yellowstone National Park. (Photo by J.K. Hillers, U.S. Geological Survey)*

has its greatest impact. Many areas of the world have landscapes that are to a large extent the result of underground solution. Such areas are said to exhibit **karst topography,** a term derived from a region in Yugoslavia. Generally, arid and semiarid areas do not develop karst topography. When solution features exist in such regions, they are likely remnants of a time when more humid conditions prevailed.

The most common and widespread features in karst areas are depressions called **sinkholes,** or **sinks** (Figure 3.33). In the limestone areas of Kentucky and southern Indiana there are an estimated 600,000 of these depressions, varying in depth from just a meter or two to a maximum of more than 30 meters (100 feet). A sinkhole may form when a cavern roof collapses, or may develop slowly without physical disturbance of the rock as solution gradually carries away material. Following a rainfall in regions with many sinkholes, most runoff is funneled below ground, so streams are often absent. When streams are present, they eventually flow into a sink and are diverted underground. The names of such streams often give a clue to their fate. In the Mammoth Cave area of Kentucky, for example, there is Sinking Creek, Little Sinking Creek, and Sinking Branch. Other sinkholes become plugged with clay and debris, creating ponds.

Limestone caverns are believed to form at or just below the water table along joints in the rock. Limestone which is not highly jointed tends

Figure 3.32 *"The Geysers," north of San Francisco, California, is the site of the first commercial geothermal power plant in the United States. (Courtesy of U.S. Geological Survey)*

Figure 3.33 *A recently created sinkhole in Alabama known as the "December Giant," formed when the roof of a cavern collapsed. (Courtesy of T.V. Stone, Geological Survey of Alabama)*

to have a minimum of solution features, because joints are natural avenues through which groundwater moves. In the course of time, solution enlarges the joints, often resulting in a complex system of tunnels and chambers. Since caves are created in the zone of saturation, it is only possible to enter them after the water table has dropped. In the United States, Carlsbad Caverns, New Mexico, and Mammoth Cave, Kentucky, are probably the best-known areas of cavern development. The latter locality includes an elaborate system comprised of at least 240 kilometers (150 miles) of subterranean passageways.

Probably the most striking features found in caves are the bizarre stone formations which often give them a wonderland appearance (Figure 3.34). **Stalactites** appear to be large stone icicles hanging from the roof of the cavern, while **stalagmites** jut up from the floor. As water seeps through the roof of a cave evaporation leaves a small deposit of calcite behind. Over the centuries a stalactite will gradually form. When the water drops to the floor of the cave some more evaporates, again depositing calcite, and a stalagmite begins growing upward. When a stalagmite and a stalactite meet, a **column** is formed.

Figure 3.34 *Stalactites, stalagmites, and columns in Carlsbad Caverns, New Mexico. (Photo by Grant Kennicott, National Park Service, U.S. Department of the Interior)*

REVIEW

1. Describe the movement of water through the water cycle. Is there more than one path which precipitation may take after it has fallen?
2. A stream starts out 2000 meters above sea level and travels 250 kilometers to the ocean. What is its average gradient in meters per kilometer?
3. Suppose that the stream mentioned in question 2 developed extensive meanders so that its course was lengthened to 500 kilometers. Calculate its new gradient. How does meandering affect gradient?
4. Why is the Amazon considered the largest river on earth when the Nile is actually longer?
5. When the discharge of a stream increases, what happens to the stream's velocity?
6. Why does the downstream portion of a river have a gentle gradient when compared to the headwater region?
7. Define *base level.* Name the main river in your area. For what streams does it act as base level? What is the base level for the Mississippi River?
8. In what three ways does a stream transport its load?
9. If you collect a jar of water from a stream, what part of its load will settle to the bottom of the jar? What portion will remain in the water?
10. Differentiate between competency and capacity.
11. In what way is a delta similar to an alluvial fan? In what way are they different?
12. What is a divide?
13. Why is it possible for a youthful valley to be older (in years) than a mature valley?
14. Do mature and old-age valleys make good political boundaries? Explain.
15. What is groundwater and how is it related to the water table?
16. Explain the difference between porosity and permeability.
17. Under what circumstances can a material have a high porosity but not be a good aquifer?
18. Why is the pumping of water in some areas of southwestern United States a serious problem?
19. What is meant by the term *artesian*? Under what circumstances do artesian wells form?
20. What is the source of heat for most hot springs and geysers? How is this reflected in the distribution of these features?
21. List two conditions that are required for the development of karst topography.
22. Differentiate between stalactites, stalagmites, and columns. How do these features form?

KEY TERMS

hydrologic cycle
infiltration
runoff
transpiration
gradient
load
discharge
base level
ultimate base level
temporary (local) base level
graded stream
potholes
dissolved load
suspended load
bed load
saltation
competence
capacity
alluvium
delta
bottomset beds
foreset beds
topset beds
distributaries
alluvial fans
natural levee

back swamp
yazoo tributaries
rapids
waterfalls
floodplain
meanders
oxbow
oxbow lake
braided stream
drainage basin
divide
dendritic pattern
trellis pattern
radial pattern
meander belt
rejuvinated
entrenched meanders
terraces
peneplain
groundwater
water table
vadose water
porosity
permeability
aquifer
well
cone of depression
spring
perched water table
artesian
nonflowing artesian well
flowing artesian well
geyser
geothermal energy
karst topography
sinkholes (sinks)
stalactites
stalagmites
column

Yentna Glacier in south-central Alaska. (Courtesy of U.S. Geological Survey)

Glacers

Glacial Movement

Glacial Erosion

Glacial Deposits

Causes of Glaciation

Deserts

The Evolution of a Desert Landscape

Wind Erosion

Wind Deposits

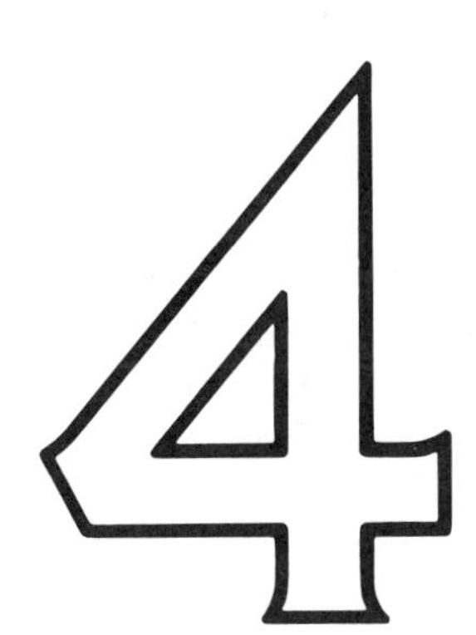

Glaciers, Deserts, and Wind

Today glaciers cover nearly 10 percent of the earth's land surface; however in the recent geologic past ice sheets were three times as large, covering vast areas with ice thousands of meters thick. Many present-day landscapes still bear the mark of these glaciers. The first part of this chapter examines glaciers and the erosional and depositional features they create. The second part is devoted to dry lands and the geologic work of wind. Since desert and near-desert conditions prevail over an area as large as that affected by the massive glaciers of the Ice Age, the nature of such landscapes is indeed worth investigating.

Since glacial ice is a naturally occurring solid composed of interlocking crystals of the mineral ice, it must be considered as a rock. Although it is most often classified as a metamorphic rock, because it is formed by recrystallization under pressure, it also shares some characteristics with the two other major rock groups. Like igneous rocks, glacial ice is a frozen fluid. Further, it shares a common feature with sedimentary rocks: Both are deposited in layers at the earth's surface and may accumulate to great thicknesses.

Certainly the geologic effects of glacial erosion and deposition, which are the primary focus of this chapter, have been and continue to be of considerable scientific interest. However, the study of glaciers has recently gained attention for other reasons. For example, the movement of ice sheets may act as an early warning system for worldwide climatic change. Glaciers are still abundant on the earth today and, whether they are advancing or retreating, may be an important indicator of otherwise subtle changes in global climate. Moreover, since glacial ice accumulates in layers year after year, the study of these layers gives indications of past climatic variations. The composition of the ice itself and the nature of the air trapped in the ice have proven to be valuable tools in examining past changes in the earth's atmospheric environment.

Glaciers

A glacier is a thick mass of ice that originates on land from the compaction and recrystallization of snow which shows evidence of past or present flow. Glaciers form in places where more snow accumulates each year than melts away. The resulting accumulation is a **snowfield.** If the accumulation is thick enough, the pressure of overlying snow transforms the snow below into glacial ice. The boundary separating the snowfield from zones without permanent snow cover is the **snowline.** The elevation of the snowline varies considerably. In polar regions it may be sea level, while in equatorial areas the snowline may rest at 6000 meters (20,000 feet).

Although glaciers are found in many parts of the world today, they are usually located in areas remote from populous regions. Literally thousands of relatively small glaciers exist in mountainous regions. Such glaciers are generally confined to mountain valleys and are most often termed **alpine,** or **valley, glaciers.** The total volume of all alpine glaciers is about 210,000 cubic kilometers (50,000 cubic miles), comparable to the combined volume of the world's large saline and freshwater lakes.

On a different scale, **continental glaciers** are massive accumulations of ice covering extensive areas. Two such glaciers, the Greenland and Antarctic ice sheets, exist on the earth today. The glacier on Greenland covers 80 percent of this large island's land area, occupying about 1.7 million square kilometers (667,000 square miles) and averaging nearly 1500 meters (5000 feet) thick (Figure 4.1A). When compared to the glacier covering the continent of Antarctica, the Greenland ice sheet seems quite small. Eighty percent of the world's ice and nearly two-thirds of the earth's fresh water are represented by Antarctica's glacier, which covers an area almost one and one-half times that of the United States (Figure 4.1B). If this ice were melted, sea level would rise an estimated 60 meters (200 feet), inundating many densely populated coastal areas. The hydrologic importance of the continent and its ice can be illustrated in another way. If Antarctica's ice sheet were melted at a suitable rate, it could feed (1) the Mississippi River for more than 50,000 years, (2) all the rivers in the United States for about 17,000 years, (3) the Amazon River for approximately 5000 years, (4) all the rivers of the world for about 750 years.

As the foregoing example illustrates, the quantity of glacial ice on the earth today is great; however, *glaciers today cover only slightly more than one-third the area they once did.* If Antarctica is excluded, glaciers once occupied an area about 13 times greater than present-day

The Lower Falls of the Grand Canyon of the Yellowstone. An excellent example of a valley in the youthful stage of development. (Photo by I.J. Witkind, U.S. Geological Survey)

This huge scar on the side of Sheep Mountain resulted from the famous Gros Ventre landslide of June 23, 1925, near Kelly, Wyoming. (Courtesy of Ward's Natural Science Establishment, Inc., Rochester, N.Y.)

Salt flats are common features in intermontane basins that are characterized by interior drainage. The Inyo Mountains are in the background. (Courtesy of Ward's Natural Science Establishment, Inc., Rochester, N.Y.)

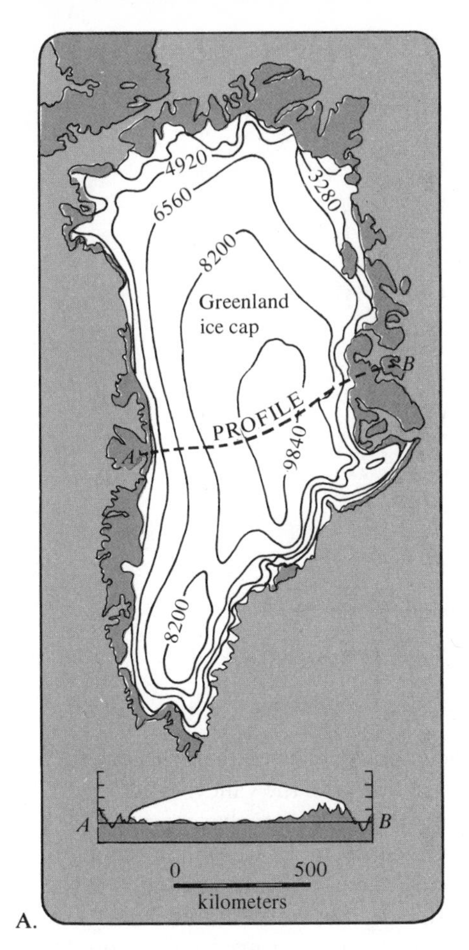

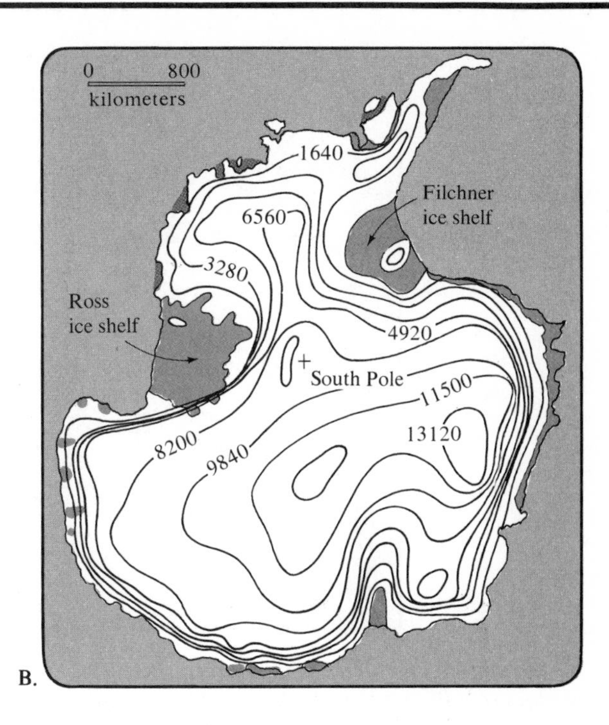

Figure 4.1 *The ice sheets of* **A.** *Greenland and* **B.** *Antarctica are the only continental glaciers on earth today. The contours indicate the depth of the ice in feet. (A. and B. courtesy of Ward's Natural Science Establishment, Inc., Rochester, N.Y.)*

glaciers. During the **Pleistocene epoch,** or **Ice Age,** glaciers left their imprint on 10 million square kilometers (4 million square miles) of North America, 5 million square kilometers (2 million square miles) of Europe, and about 4 million square kilometers (1.5 million square miles) of Siberia. Figure 4.2 illustrates the maximum extent of Pleistocene glaciation in North America.

The Ice Age began at least 2.5 million years ago, while the last continental glaciers receded from North America and Eurasia between 10,000 and 15,000 years ago. The Pleistocene epoch, however, was not a period of continuous glaciation. Four separate stages of glaciation have been identified in North America, and up to six in Europe. *Thus the Ice Age was characterized by alternating periods of glacial advance and retreat.* The time spans between glacial advances, called **interglacial periods,** often exceeded the amount of time when glaciers were present (Figure 4.3).

Glaciers have not been ever-present features throughout the earth's long 4½–5-billion-year history. In fact, for most of geologic time glaciers

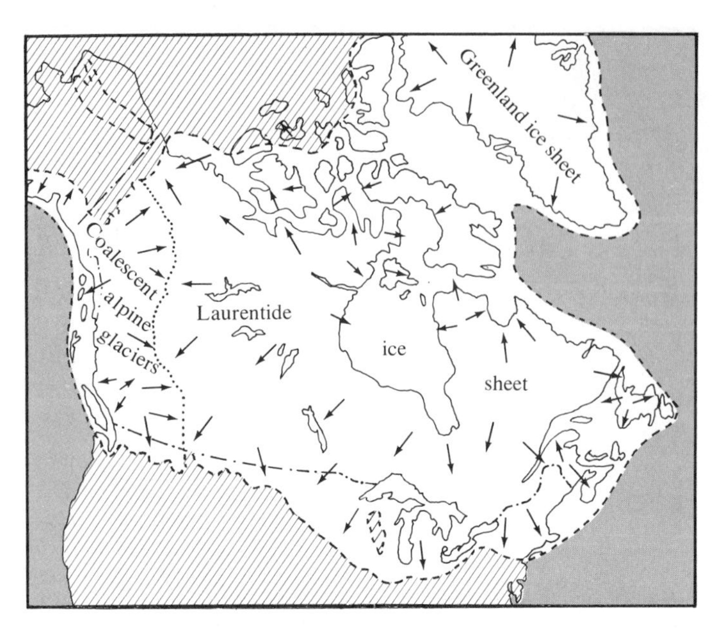

Figure 4.2 *At their maximum extent, Pleistocene glaciers covered about 10 million square kilometers (4 million square miles) of North America.*

have been absent. Evidence does indicate that in addition to the Pleistocene epoch there were at least three other periods of glacial activity: 2000 million, 600 million, and 250 million years ago. However, it is the most recent period of glaciation, the Ice Age, which is of greatest interest, for the features of many present-day landscapes are a reflection of the work of Pleistocene glaciers.

Glacial Movement

Unlike streamflow, the movement of glaciers is not readily apparent to the casual observer. More than 100 years ago the first measurements of glacial movement were made. In this experiment stakes were carefully placed in a straight line across the top of an alpine glacier with their positions marked on the valley walls. Periodically the positions of the stakes were noted. This experiment, and many others conducted since, have shown that all of the stakes moved down the valley, but not at the same rate. In fact, the position of the stakes indicated that the center of the glacier moved faster than the edges, where there is friction with the valley walls (Figure 4.4).

Today, the fact that glaciers move is easily demonstrated by time-lapse photography, which can capture the movement on film. The speed of glacial flow varies considerably, from less than a centimeter to several meters per day. At these rates, it may take hundreds or even thousands of years for ice to travel from one end of the glacier to the other.

The movement or flow of a glacier occurs because of the pull of gravity on the mass of ice. While the mechanisms of movement are quite complex and not fully understood, investigations have shown that two zones can be identified (Figure 4.5). The lower portion of the glacier, 30–60 meters (100–200 feet) below the surface, is the **zone of flow.** Here the weight of the overlying ice is great, and the ice behaves plastically, undergoing a slow and continuous change in shape. Ice in the upper portion of the glacier, however, does not flow but is carried along. Since it is not under great pressure, the ice is brittle and often develops cracks known as

crevasses (Figure 4.6). This upper zone is appropriately called the **zone of fracture.**

Whether the margins of a glacier are advancing, retreating, or stationary depends upon the rate of snow accumulation and ice formation on the one hand, and the rate of melting and evaporation on the other. Glaciers can be divided into two areas, one where there is a net accumulation of ice, and another where there is a net loss. If the amount of ice accumulating exceeds the quantity wasting away, the glacial front advances until a state of equilibrium is reached. At this point the ice front is stationary. At a later time when melting and evaporation exceed accumulation, the ice front will recede. *Whether the margins of the glacier are advancing, retreating, or stationary, the ice within the glacier continues to flow forward.* In the case of a receding glacier, the ice simply does not flow forward rapidly enough to offset wastage.

From the preceding discussion it seems logical to conclude that *climate is the primary factor that controls a glacier.* When the climate grows cooler and/or precipitation increases, glaciers grow and advance; if the temperature warms or precipitation diminishes, glaciers shrink.

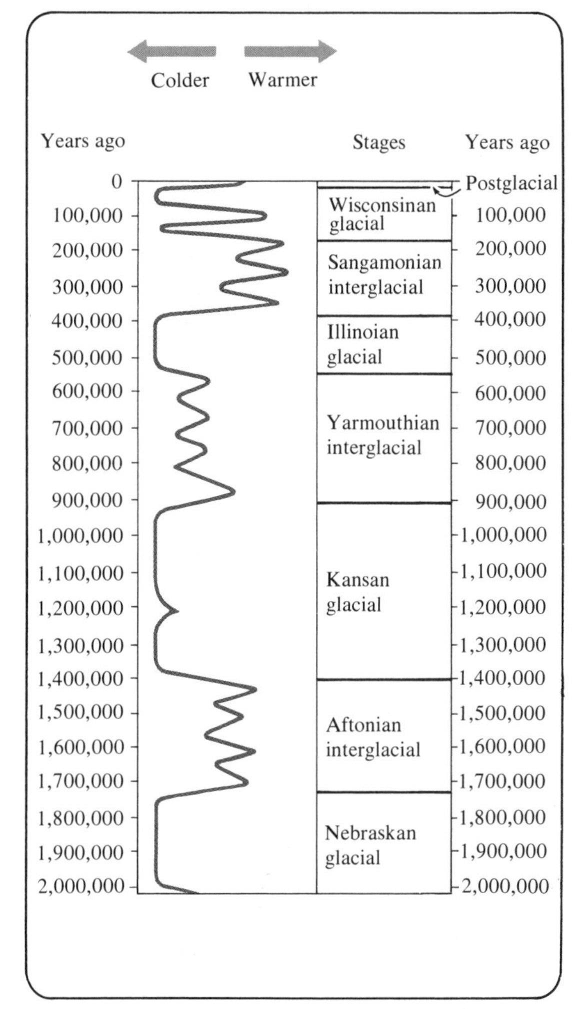

Figure 4.3 *The Pleistocene epoch was marked by fluctuating climatic conditions that led to alternating glacial and interglacial periods. Four separate stages of glaciation are recognized for North America. [After D.B. Ericson and G. Wollin, "Pleistocene Climates and Chronology in Deep-sea Sediments,"* Science *162 (1968):1233. Copyright 1968 by the American Association for the Advancement of Science]*

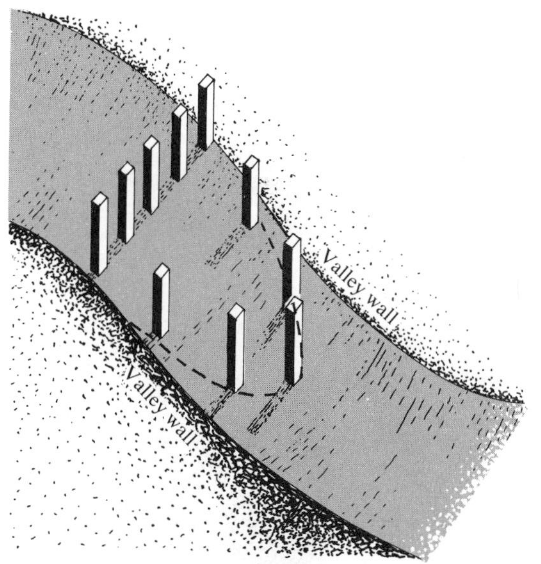

Figure 4.4 *Diagrammatic aerial view of an alpine glacier. Movement of the glacier is shown by the displacement of stakes driven into the surface of the ice. The stakes indicate that the center of the glacier moves faster than the edges.*

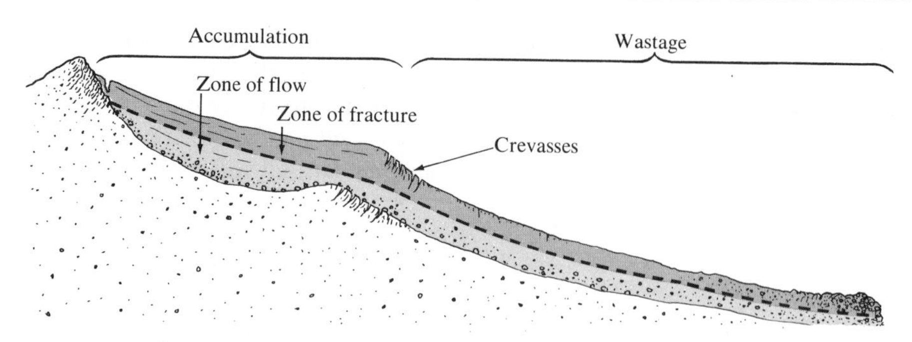

Figure 4.5 *Cross-sectional view of an alpine glacier. Note the zone of flow and the brittle zone of fracture, where crevasses develop. The advance or retreat of the ice front depends upon imbalances between accumulation and wastage.*

Glacial Erosion

Glaciers are capable of carrying on great amounts of erosional work. Although glaciers are of limited importance as an erosional agent today, many landscapes modified by the widespread glaciers of the Pleistocene epoch still reflect to a high degree the work of ice.

Glacial erosion is primarily of two types. As a glacier flows over a fractured bedrock surface it lifts blocks of jointed rock and carries them off, a process known as **plucking.** In this manner rock fragments of all sizes are incorporated into the glacier. A second erosional process is **abrasion** (or grinding). As the ice, armed with its load of rock fragments, moves along it acts as a giant rasp or file which abrades the surface. The results of glacial abrasion include scratches called **glacial striations** and deep gouges, or **grooves,** in the bedrock surface (Figure 4.7). Abrasion by

Figure 4.6 *Blue Glacier, Olympic Park, Washington. Crevasses form in the brittle ice of the zone of fracture. They do not continue down into the zone of flow. (Courtesy of U.S. Geological Survey)*

Figure 4.7 *Results of glacial abrasion. Glacial striations in bedrock, Worcester County, Massachusetts. (Photo by W.C. Alden, U.S. Geological Survey)*

fine silt-sized sediment in the glacier yields a polished surface. The ground-up rock produced by the grinding effect of the glacier is called **rock flour.** If you were to collect a jar of water from a melting glacier and let it stand for a period of time, the fine rock flour would settle to the bottom of the jar.

In regions eroded by continental ice sheets extensive, glacially scoured surfaces are the rule, whereas in mountainous areas glacial erosion yields many spectacular features. Prior to glaciation alpine valleys are characteristically V-shaped because streams are well above base level and are therefore downcutting. However in mountainous regions that have been glaciated the valleys are no longer narrow. As a glacier moves down a valley once occupied by a stream the ice modifies it in three ways: The glacier widens, deepens, and straightens the valley, so that what was once a youthful V-shaped valley is transformed into a U-shaped **glacial trough** (Figure 4.8).

Since the magnitude of the glacial erosion depends upon the thickness of the ice, main or trunk glaciers cut their valleys deeper than their tributaries are able to do. After the glaciers have receded, the tributary valleys stand high above the main trough and are termed **hanging valleys.** Hanging valleys often produce spectacular cascading waterfalls, such as those in Yosemite National Park, California (Figure 4.8).

Figure 4.8 *Bridalveil Fall in Yosemite National Park cascades from a hanging valley into the U-shaped glacial trough below. (Photo by F.E. Matthes, U.S. Geological Survey)*

Figure 4.9 *Aerial view of typical bowl-shaped depressions called cirques along the Continental Divide in Colorado. (Photo by T.S. Lovering, U.S. Geological Survey)*

At the head of a glacial valley is a bowl-shaped basin, or **cirque** (Figure 4.9), which represents the area of snow accumulation and glacial ice formation. Such a depression is thought to be produced by a combination of frost wedging and plucking. After the glacier forms, frost action along the margins of the ice enlarges the cirque. When the ice eventually disappears, the steep-walled cirque may contain a small lake or pond called a **tarn.**

When a series of cirques are present along either side of a mountain range, a knifelike ridge may be created as the cirques enlarge. Such features were named **arêtes** by mountaineers. When three or more cirques surround a mountain summit, a pyramidlike peak called a **horn** is formed, as exemplified by the famous Matterhorn, in the Alps of Switzerland (Figure 4.10). If two cirques are back-to-back along a mountain ridge and their common headwall (back wall) is breached, a pass, or **col,** through the ridge is created. The landforms carved by alpine glaciers are summarized in Figure 4.11.

Mountainous areas adjacent to the ocean that have been subjected to glaciation are often characterized by **fiords.** These steep-sided inlets of the sea are common features along many coasts, including those of Norway, British Columbia, Alaska, and southern Chile (Figure 4.12). The

Figure 4.10 *The Matterhorn. (Courtesy of Swiss National Tourist Office)*

Figure 4.11 *Landforms carved by alpine glaciers.* **A.** *Period of maximum glacial activity.* **B.** *Landscape as it appears following glaciation. (A. and B. after Ward's Natural Science Establishment, Inc., Rochester, N.Y.)*

Figure 4.12 *Like other fiords, the spectacular Geiranger fiord in Norway is a drowned glacial trough. (Courtesy of Norwegian National Travel Office)*

fiords were created when glacial troughs were partially submerged as sea level rose following the Ice Age.

Glacial Deposits

Glaciers are capable of acquiring a huge load of debris by plucking and abrasion. Further, weathering and mass wasting along valley walls add substantial quantities of material to the sides of an alpine glacier. Ultimately the glacially transported sediments are deposited when the ice melts. **Drift,** the general term for any glacial deposit, is an archaic term that was coined by persons who believed that large boulders and other debris of a rock type unlike that found in the area had been "drifted" in on icebergs during Noah's Flood. Today we know these materials are glacial deposits, but the term *drift* remains.

Glacial deposits are of two types: (1) those deposited directly by the glacier, which are known as **till;** and (2) materials deposited by glacial meltwater, called **outwash.** Till is deposited as glacial ice melts and drops its load of rock fragments. Deposits of till are characteristically unsorted, a mixture of many different sizes of sediment (Figure 4.13). When boulders are found in till, they are called **glacial erratics,** indicating that they were derived from a source outside the area where they are found. In many areas erratics may be seen dotting pastures and farm fields. In fact, sometimes these large rock fragments are cleared from the fields and piled into fences (Figure 4.14). By examining the erratics as well as the mineral composition of the remaining till, geologists are often able to trace the path that the glacier took.

Outwash deposits, by contrast, are usually well-sorted and stratified accumulations of silt, sand, and gravel. Such deposits reflect the sorting action of running water. Outwash and till deposited by the continental glaciers during the Pleistocene blanket vast areas in many parts of the world, including the northern United States. Thicknesses often reach to many tens, or even hundreds, of meters. *The general effect of these deposits has been to reduce the local relief and to level the topography.*

Figure 4.13 *Glacial till is an unsorted mixture of many different sediment sizes. (Photo by W.C. Alden, U.S. Geological Survey)*

Figure 4.14 *Land cleared of glacial erratics which were then piled into walls about a field near Whitewater, Wisconsin. (Photo by W.C. Alden, U.S. Geological Survey)*

Moraines, Outwash Plains, and Kettle Holes

Perhaps the most widespread features created by glacial deposition are *moraines,* which are simply layers or ridges of till. Several types of moraines are identified; some are common only to mountain valleys, and others are associated with areas affected by either continental or alpine glaciers. Lateral and medial moraines fall in the first category, while end moraines and ground moraines are in the second.

As Figure 4.15 illustrates, the sides of an alpine glacier accumulate large quantities of debris from the valley walls. When the glacier wastes away, these materials are left as ridges, or **lateral moraines,** along the sides of the valley. **Medial moraines** are formed when two alpine glaciers coalesce (Figure 4.15).

As the name implies, **end moraines** form at the terminus of a glacier. Here, while the ice front is stagnant, the glacier continues to carry in and deposit large quantities of rock debris, creating a ridge of till tens to hundreds of meters high. The end moraine marking the farthest advance of the glacier is called the **terminal moraine,** while those moraines formed as the ice front periodically stagnated during retreat are

Figure 4.15 *Yentna Glacier in south-central Alaska is fed by many tributary glaciers. The dark stripes within the glacier are medial moraines that were created when the lateral moraines of merging alpine glaciers coalesced. (Courtesy of U.S. Geological Survey)*

termed **recessional moraines** (Figure 4.16). As the glacier recedes, a layer of till is laid down, forming a gently undulating surface of **ground moraine.** Ground moraine has a leveling effect, filling in low spots and clogging old stream channels, often leading to a disruption of drainage.

In front of the end moraines glacial meltwater drops sediment, producing stratified deposits of sand, silt, and gravel called **outwash plains.** Numerous depressions, some containing water, may be seen in many outwash plains and moraines. These depressions were created when blocks of stagnant ice became lodged in the glacial deposits and melted, leaving a pit called a **kettle hole** (Figure 4.17). Figure 4.18 depicts a hypothetical area which shows some of the depositional features described in this section as they would appear in relation to one another.

Other Depositional Features

Drumlins are streamlined asymmetrical hills composed of till (Figure 4.19). They range in height from 15 to 60 meters (50 to 200 feet) and average 0.4–0.8 kilometer (0.25–0.50 mile) in length. The steep side of the hill faces the direction from which the ice advanced, while the gentler slope points in the direction the ice moved. Drumlins are not found singly, but rather in clusters, sometimes called drumlin

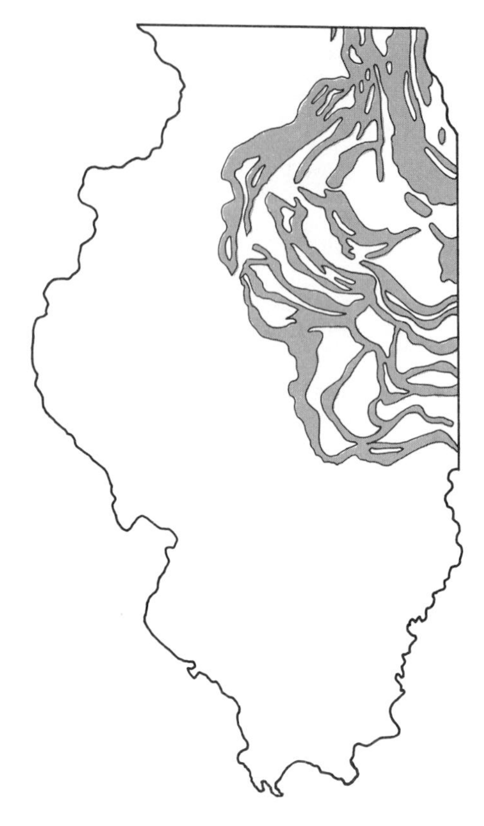

Figure 4.16 *End moraines deposited by the most recent (Wisconsinan age) continental ice sheets in Illinois. (After Illinois State Geological Survey)*

Figure 4.17 *Kettle-hole pond in gravel near the terminus of Baird Glacier in southeastern Alaska. (Photo by A.F. Buddington, U.S. Geological Survey)*

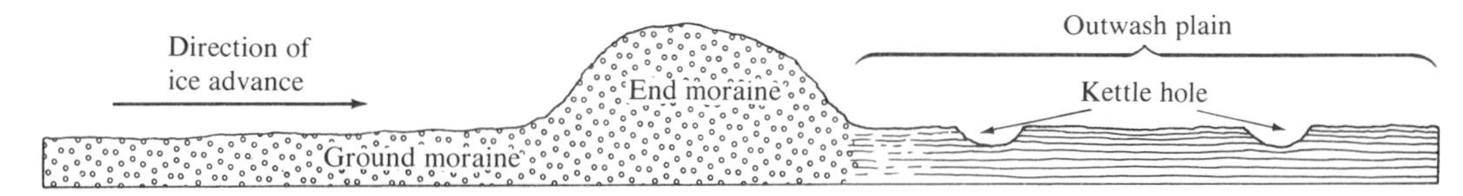

Figure 4.18 *The outwash plain, formed from sediments deposited by glacial meltwater, lies in front of the end moraine, which was created while the ice front was stationary. A layer of ground moraine is deposited as the ice front retreats.*

fields. Although drumlin formation is not well understood, their streamlined shape seems to indicate that they were molded in the zone of flow within an active glacier. Some geologists believe drumlin fields were created when a glacier advanced and reshaped an end moraine.

In areas once occupied by continental ice sheets, sinuous ridges composed largely of sand and gravel may be found. These ridges, called **eskers,** are thought to be deposits made by streams flowing in tunnels beneath the ice, near the terminus of a glacier. They may be several meters high and extend for many kilometers. In many areas they are mined for sand and gravel, and for this reason, eskers are disappearing in some localities. **Kames** are steep-sided hills, which, like eskers, are composed of sand and gravel (Figure 4.20). Kames are believed to have originated when sediment collected in openings in stagnant ice.

Figure 4.19 *Aerial view of a drumlin. (Courtesy of Ward's Natural Science Establishment, Inc., Rochester, N.Y.)*

Figure 4.20 *White Kame in Kettle Moraine State Forest, Wisconsin. (Photo by G.J. Knudson, Wisconsin Department of Natural Resources)*

Causes of Glaciation

A great deal is known about glaciers and glaciation. Much has been learned about glacier formation and movement, the extent of glaciers past and present, and the features created by glaciers, both erosional and depositional. However, to find a widely accepted theory for the causes of ice ages is a very difficult task, for the literature of science is permeated with a great array of theories, all of which appear to have shortcomings. R.F. Flint summarized the situation this way:

> Although nearly 150 years have elapsed since the Glacial Theory was proposed, the causes remain elusive because insufficient attention has been concentrated on relevant geophysical factors, particularly as regards the atmosphere.*

* *Glacial and Quaternary Geology* (New York: Wiley, 1971), p. 789.

To some degree, Flint's statement is not quite as true today as it was when it was written, for science is beginning to take a serious new look at the possible mechanisms that may cause glacial ages. However, a complete answer still appears to be many years in the future.

Any theory that attempts to explain the causes of the Pleistocene Ice Age and other periods of glacial activity must answer two questions: (1) What caused the onset of glacial conditions? (2) What caused the alternation of glacial and interglacial stages? Most investigators agree that in order for continental ice sheets to have formed, average temperatures must have been lower than at present, although not a great deal lower. Furthermore, precipitation must have been higher. This latter assumption is reinforced when we consider that many areas in the high latitudes such as northern Canada and Siberia have average temperatures below freezing, but meager precipitation and no glaciers.

Although none of the theories on the causes for the Ice Age have won the approval of the entire scientific community, several quite plausible explanations are popular. These include the possible climatic effects of such geologic phenomena as continental drift (Chapter 6) and explosive volcanic activity. Further, other proposed mechanisms of climatic change rely upon phenomena that occur away from the planet earth but nevertheless may influence its climate. More specifically, some scientists believe that variations in solar output might be the cause, while others link the Ice Age to various changes in the earth's orbit. In reality, any or all of these mechanisms may be (at least in part) responsible for the extensive glaciations of the Pleistocene. In Chapter 16, "The Changing Climate," all of these theories are examined in some detail.

Deserts

The desert (arid) and steppe (semiarid) regions of the world encompass about 42 million square kilometers (16 million square miles), almost 30 percent of the earth's land surface. The major deserts listed in Table 4.1 alone account for one kilometer out of every seven on earth. These areas are characterized by scanty and highly irregular precipitation. Because of the lack of water, weathering is reduced and soils are thin, as compared to more humid areas. In contrast to the rounded slopes and rock edges in humid regions, *the topography of arid landscapes is characteristically angular because of the diminished chemical weathering.*

There are a number of common misconceptions which many people have about deserts. One common fallacy about deserts is that they are lifeless, or practically so. However, although reduced in amount and different in character, vegetation and animal life are indeed present. In addition, many visualize deserts as consisting of kilometer after kilometer of drifting sand. It is true that sand accumulations do exist in some areas and may be striking features, but they represent only a small percentage of the total desert area. In the Sahara, the world's largest desert, accumulations of sand cover only one-tenth of its area. The sandiest of all deserts is the Arabian, where one-third consists of sand. Also, deserts are not always hot. Tropical deserts do lack a cold season, but deserts in the mid-latitudes experience seasonal temperature changes. Among the most common misconceptions concerning deserts is the seemingly logical idea that wind is the most important agent of erosion. *Although wind is relatively more significant in dry areas than anywhere else, most desert landforms are created by running water.**

When rains come to desert areas, they usually take the form of thunderstorms. The heavy rain associated with such storms cannot all soak in, so rapid runoff results. Without a thick vegetative cover to protect the ground, great quantities of debris are carried into the previously dry stream channels, known as **washes** in the U.S. Southwest. Within a few minutes a wash is trans-

* William D. Thornbury, *Principles of Geomorphology*, 2nd ed., Wiley, p. 269. This section on common misconceptions about deserts is based upon a similar discussion in Thornbury's Chapter 11.

Table 4.1 ***Major Deserts of the World***

Desert	Approximate Area Thousands of square kilometers	Thousands of square miles
Sahara	9100	3500
Australian	3400	1300
Arabian	2600	1000
Turkistan	2000	750
North American	1300	500
Patagonian	680	260
Indian	600	230
Kalahari-Namib	570	220
Gobi-Takla Makan	520	200
Iranian	390	150
Atacama-Peruvian	360	140

formed into a torrent of sediment-choked water. Not long after the storm ceases the streamflow ends. Because of heavy losses to seepage and evaporation, desert streams are typically short and intermittent. The few permanent streams which do traverse arid regions, like the Colorado and Nile rivers, have their sources in well-watered areas outside the desert and are termed **exotic streams.**

The Evolution of a Desert Landscape

Since arid regions typically lack permanent streams, they are characterized as having interior drainage. In the United States, the dry Basin and Range region is the best example of this. This region includes southern Oregon, all of Nevada, western Utah, southeastern California,

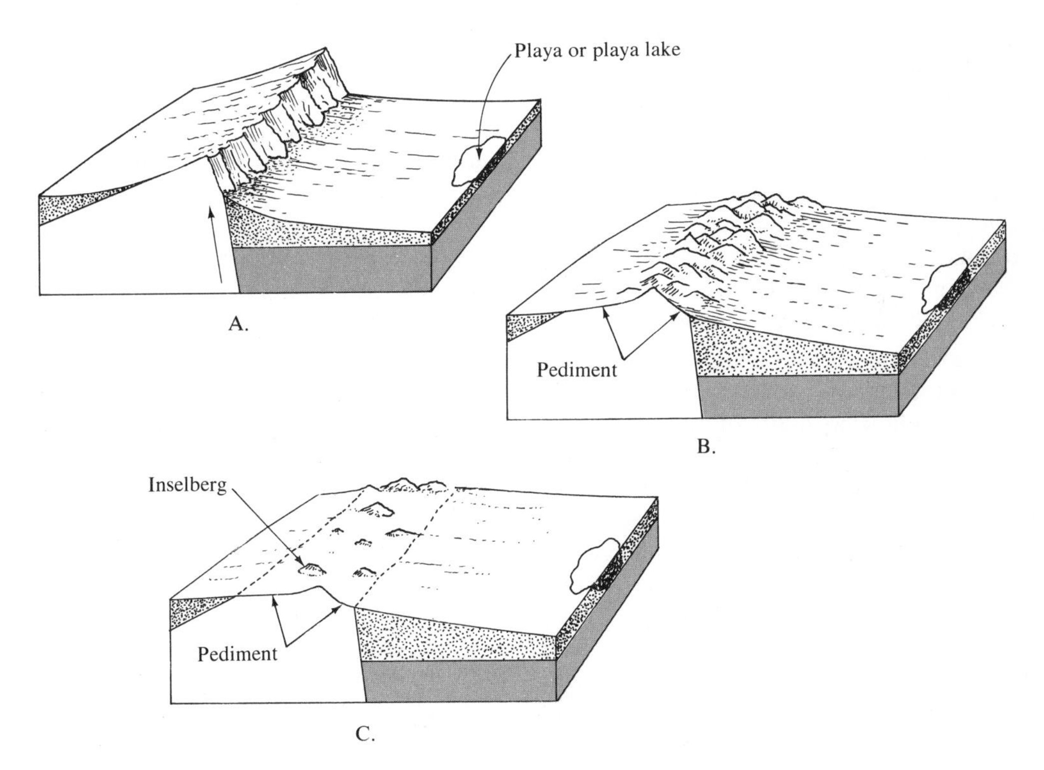

Figure 4.21 *Stages of landscape evolution in a mountainous desert.* **A.** *Early.* **B.** *Middle.* **C.** *Late. (From Foster,* Physical Geology, *3rd ed., Columbus, Ohio: Charles E. Merrill, 1979)*

Figure 4.22 *Dust storms like this one in Colorado were relatively common during the 1930s in the Dust Bowl region of the Great Plains. (Courtesy of U.S. Department of Agriculture)*

as well as southern Arizona and New Mexico. The name *Basin and Range* is an apt description for this 780,000-square-kilometer (300,000-square-mile) area, since it is characterized by more than 200 small mountain ranges which rise 900–1500 meters (3000–5000 feet) above the basins that separate them. In this region, as in others like it around the world, erosion is carried out for the most part without reference to the ocean (ultimate base level) because drainage is in the form of local interior systems. Even in areas where permanent streams flow to the ocean, few tributaries exist, and thus only a relatively narrow strip of land adjacent to the stream has sea level as the ultimate level of land reduction.

The models shown in Figure 4.21 illustrate the stages of landscape evolution in a mountainous desert. Following uplift of the mountains, running water begins carving the mass and depositing large quantities of debris in the basin. Alluvial fans form along the base of the mountains, where the gradient abruptly decreases. Further out in the basin is a flat surface called a **playa,** underlain by fine sediments. Following heavy rains, the playa becomes a lake. Soon afterward, however, the water evaporates, leaving salt deposits and a mud-cracked surface. It is during this early stage that relief is greatest, for as erosion lowers the mountains and sediment fills the basins, relief diminishes. As erosion continues, the mountain front retreats, producing a sloping bedrock surface, or **pediment,** which fringes the base of the mountains. The mountain mass is now dissected into an intricate series of valleys and divides, and the local relief has lessened. In the late stages of erosion there is continued filling of the basins with debris. Relief is at a minimum. The mountains have been reduced to isolated remnants called **inselbergs,** and the pediment has grown more extensive.

Figure 4.23 *Desert pavement composed of angular rock fragments, San Bernadino County, California. (Photo by C.S. Denny, U.S. Geological Survey)*

Each of the stages just described can be observed in the Basin and Range region. Recently uplifted mountains in an early stage of erosion are found in southern Oregon and northern Nevada. Death Valley, California, and southern Nevada fit into the more advanced middle stage, while the late stage, with its inselbergs and extensive pediments, can be seen in southern Arizona.

Wind Erosion

Moving air, like moving water, is capable of picking up loose debris and moving it to another location. *Although wind erosion is not restricted to arid and semiarid regions, it does its most effective work in these areas.* In humid places moisture binds particles together and vegetation anchors the soil so that wind erosion is negligible. For wind to be effective, dryness and scanty vegetation are important prerequisites. When such circumstances exist, wind may pick up, transport, and deposit great quantities of fine sediment. During the 1930s parts of the Great Plains experienced great dust storms (Figure 4.22). The plowing under of the natural vegetative cover for farming, followed by severe drought, made the land ripe for wind erosion, and led to the area being labeled the Dust Bowl.

Wind erosion differs from stream erosion in two significant ways. First, wind has a low density when compared to water; thus it is not capable of picking up and transporting coarse materials. Second, because wind is not confined to channels, it can spread over large areas, as well as high into the atmosphere.

One way that winds erode is by **deflation,** the lifting and removal of loose material. Since the competence of moving air is low, it can only suspend fine sediment such as clay and silt. Sand grains are rolled or skipped along the surface and comprise the bed load. Coarser particles are usually not transported by the wind. Although the effects of deflation are sometimes difficult to notice because the entire surface is being lowered at the same time, they can be significant. In portions of the Dust Bowl the land was lowered by as much as 1 meter.

The most noticeable result of deflation in some places are shallow depressions called **blowouts,** or **deflation hollows.** In the Great Plains region, from Texas north to Montana, thousands of blowouts can be seen. They range in size from small dimples less than 1 meter deep and 3 meters wide to depressions that are over 45 meters deep and several kilometers across. In the past, several explanations for the depressions were put forth, including one which suggested they were sites of buffalo wallows. Today it is believed that the vast majority were created by deflation.

In portions of many deserts the surface is characterized by a layer of coarse pebbles and gravel. Such a layer, called **desert pavement,** is created as the wind removes fine material, leaving the coarse particles behind (Figure 4.23). Once desert pavement becomes established, a process which may take hundreds of years, the surface is effectively protected from further deflation.

Wind armed with sand can abrade rock near the surface. Since sand is seldom lifted more than a meter above the surface, this *sandblasting* effect is limited in vertical extent. In areas prone

Figure 4.24 *Ventifacts. (Photo by M.R. Campbell, U.S. Geological Survey)*

Figure 4.25 *Barchan dunes. The gentle slope is on the side from which the prevailing wind is coming. (Photo by G.K. Gilbert, U.S. Geological Survey)*

to such activity telephone poles have actually been cut through near their bases. For this reason, collars are often fitted on the poles to protect them from being "sawed" down. Sandblasting may also produce interestingly shaped pebbles known as **ventifacts** (Figure 4.24). The side of the stone exposed to the prevailing wind is abraded, leaving it polished, pitted, and with sharp edges. If the wind is not consistently from one direction, or if the pebble becomes reoriented, it will have several faceted surfaces.

Wind Deposits

Wind deposits are of two distinct types: (1) accumulations of sand and (2) deposits of silt. The wind generally deposits sand in mounds or ridges called **dunes.** When moving air which is carrying sand encounters an obstruction such as a bush or rock, some of the sand is deposited in the "wind shadow" on the lee side. As the accumulation of sand grows, it becomes a more efficient trap for even more sand. In this manner dunes are created.

Characteristically, sand dunes have a gentle slope on the windward side and a steeper slope on the lee side (Figure 4.25). Unless the dune is anchored by vegetation, it will not remain stationary. Sand grains are rolled up the gentle slope and, upon reaching the crest of the dune, tumble down the steep lee side. Slowly the dune migrates along the surface in response to the force of the wind (Figure 4.26), in some cases moving as much as 20 meters per year. Several types of

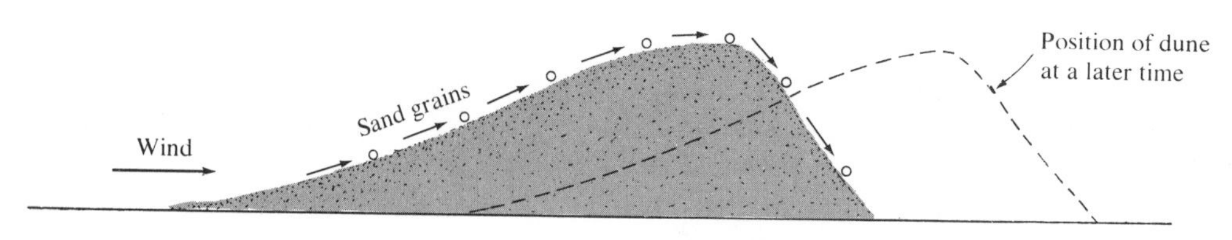

Figure 4.26 *Sand dunes migrate along the surface as sand is rolled up the windward slope and tumbles down the lee side.*

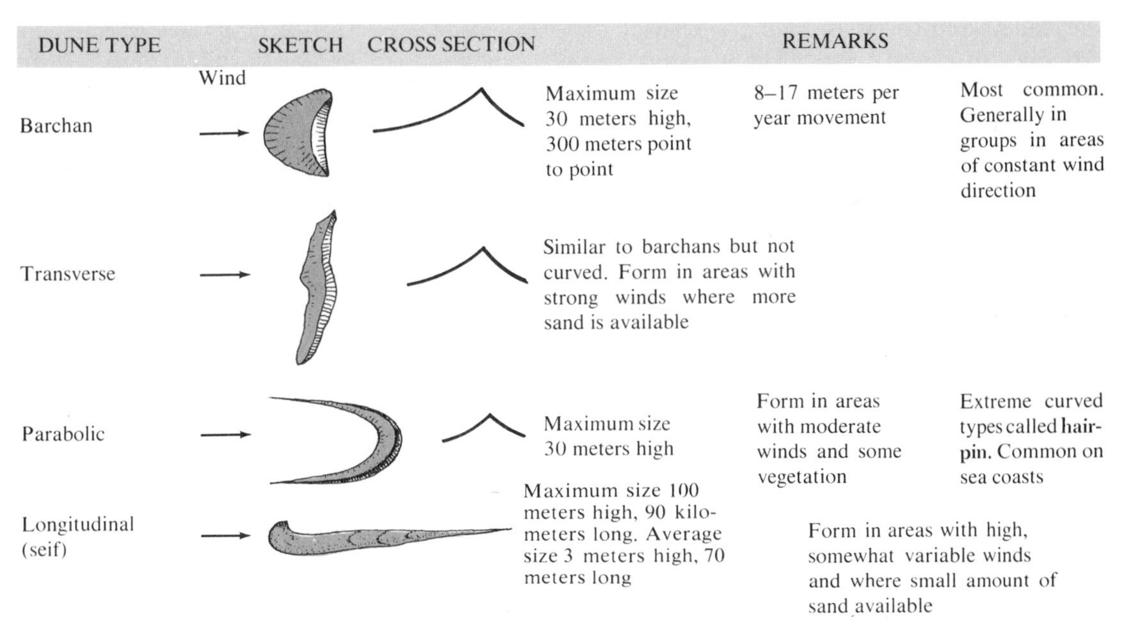

Figure 4.27 *Types of sand dunes. (From Foster,* Physical Geology, *3rd ed., Columbus, Ohio: Charles E. Merrill, 1979)*

sand dunes are recognized; their characteristics are summarized in Figure 4.27.

Figure 4.28 *A nearly vertical loess bluff.*

In some parts of the world the surface is mantled with deposits of wind-blown silt. Over a period of thousands of years dust storms deposited this material, which is called **loess** (pronounced *less* or *lus*). As you can see in Figure 4.28, when loess is breached by streams or road cuts, it tends to maintain vertical cliffs and lacks any visible layers. The thickest and most extensive loess deposits in the world occur in western and northern China, where accumulations of 30 meters are not uncommon and thicknesses of more than 60 meters have been measured. It is this buff-colored sediment which gives the Yellow River (Hwang Ho) its distinctive color.

The sources of China's loess are the arid interior regions of Asia. In the United States deposits of loess are significant in many areas, including South Dakota, Nebraska, Kansas, Iowa, Missouri, and Illinois. Unlike the deposits in China, the loess in the United States originates in glacially eroded materials. During the retreat of the glacial ice many river valleys were choked with sediment that was provided by glacial meltwater. Silt carried by the prevailing winds from these valleys settled out, forming a blanket of loess over extensive areas. The deposits are thickest and coarsest on the lee side of the valleys and rapidly thin with increased distance from the source, a fact which substantiates that loess was derived from the valleys. Today loess is being deposited along many glacial channels in Alaska.

REVIEW

1. What is a glacier? Under what circumstances does glacial ice form?
2. Where are glaciers found today? How extensive are they? How does the distribution of glaciers today compare to the extent of glacial ice during the Pleistocene epoch?
3. How many glacial advances have been recognized for North America? List them in order from first to last (Figure 4.3).
4. Describe glacial flow. At what rates do glaciers move? In an alpine glacier does all of the ice move at the same rate? Explain.
5. Why do crevasses form in the upper portion of a glacier but not below 30–60 meters (100–200 feet)?
6. Under what circumstances will the front of a glacier advance? Retreat? Remain stationary?
7. Describe the processes of glacial erosion.
8. How does a glaciated mountain valley differ from a mountain valley that was not glaciated?
9. List and describe the erosional features you might expect to see in an area where alpine glaciers exist or have recently existed.
10. What is glacial drift? What is the difference between till and outwash? What general effect do glacial deposits have on the landscape?
11. List the five basic moraine types. What do all moraines have in common? What is the significance of terminal and recessional moraines?
12. List and discuss depositional features other than moraines.
13. How does a kettle hole form?
14. How extensive are the desert and steppe regions of the earth? What are some common misconceptions about such areas?
15. What is a wash? An exotic stream?
16. Why is sea level (ultimate base level) not a significant factor influencing erosion in desert regions?
17. Describe the features and characteristics associated with each of the stages in the evolution of a mountainous desert. Where in the United States can these stages be observed?
18. Why is wind erosion relatively more important in arid regions than in humid areas?
19. List two types of wind erosion and the features which may result from each.
20. How do sand dunes migrate?
21. Although sand dunes are the best-known wind deposits, accumulations of loess are very significant in some parts of the world. What is loess? Where are such deposits found? What are the origins of this sediment?

KEY TERMS

glacier
snowfield
alpine (valley) glacier
continental glacier
Pleistocene epoch (Ice Age)
interglacial periods
zone of flow
crevasses
zone of fracture
plucking
abrasion
glacial striations
grooves
rock flour
glacial trough
hanging valley
cirque
tarn

arête
horn
col
fiord
drift
till
outwash
glacial erratics
lateral moraine
medial moraine
end moraine
terminal moraine
recessional moraine
ground moraine
outwash plain
kettle hole
drumlin
esker
kame
washes
playa
pediment
inselberg
deflation
blowout (deflation hollow)
desert pavement
ventifacts
dune
loess

Destruction in downtown Anchorage following the devastating 1964 Alaskan earthquake. (Courtesy of U.S. Geological Survey)

Earthquakes and the Earth's Interior

What is an earthquake? How does a seismograph record the location of a "quake"? Can earthquakes ever be controlled, or at least predicted? If we could view the earth's interior, what would it look like? In this chapter we shall try to answer these questions, as well as many others. The study of earthquakes is important, not only because of the devastating effect that some have on us but also because they furnish clues about the structure of the earth's interior.

QUAKE KILLS HUNDREDS

MANAGUA, Nicaragua (AP)—A disastrous earthquake rolled through this Central American city of 300,000 early yesterday, leaving a heavy toll of death and destruction. Unofficial estimates of the dead ranged as high as 18,000 but that figure appeared to be exaggerated.

"It's like standing on jelly down here," radioed U.S. communications satellite technician Ray Hashberger from a station two miles outside the city.

There were confirmed reports of at least 200 killed and thousands injured and homeless. Bodies and debris littered the streets, buildings were afire, and it appeared the death toll might come to as many as 2,000 dead.

Many of those who were not injured sat on the curbstones in a daze, surrounded by what few possessions they could save from the rubble. Many others fled the city.

Half the downtown section of the city lay in ruins as night fell. The quake devastated 36 blocks in the central area.

Fires burned out of control through the afternoon. The quake, which measured between 6 and 7 on the Richter scale of magnitude, struck at 12:40 am yesterday following a series of lesser jolts.

All normal communications, water and electrical services were out. The U.S. Embassy was among the buildings destroyed.

Many bodies lay unclaimed and unidentified in the streets. No medical facilities were functional until help began to arrive later in the day from other American countries, including the United States.

Jack Burton, an information officer in the U.S. Embassy, was in Lima, Peru, for that city's major earthquake May 31, 1970.

"The Lima quake was more gradual though it was about of the same intensity," he said. "But this quake came on like gangbusters. It knocked us on our knees. We had no warning at all."

One survivor said, "It felt like the end of the world." Thousands roamed the streets as if dazed, and other thousands fled to the countryside as smoke billowed from the rubble.

By last night, the widespread fires were said to be under control.

Smaller tremors continued to hit the city throughout the day and into the evening, loosening debris from already wrecked buildings.*

It is estimated that millions of earthquakes occur each year. Fortunately few are as devastating as the 1972 Managua earthquake just described. Generally only a few destructive earthquakes occur worldwide each year; but when they do, they merit the title "most destructive natural force." The shaking of the ground coupled with the "liquification" of some soils wreaks havoc on buildings (Figure 5.1). In addition, power and gas lines often rupture, causing numerous fires. In the 1906 San Francisco earthquake most of the damage was caused by fires which ran unchecked when broken water mains left firefighters with only trickles of water (Figure 5.2).

What Is an Earthquake?

An **earthquake** is the vibration of the earth produced by the rapid release of energy. This energy radiates in all directions from its source, or **focus,** in the form of waves analogous to those produced when a church bell is struck, vibrating the air around it. During an earthquake, and for many hours following, the earth could be described as "ringing." Even though the energy dissipates rapidly with increased distance from the focus, instruments located throughout the world record the event.

The tremendous energy released by atomic explosions or by volcanic eruptions can produce an earthquake, but these events are weak and infrequent. What mechanism then does produce a destructive earthquake? With the recent developments in the field of plate tectonics, we now realize that great sections of the earth's crust are sliding over, under, or past other sections. The last movement is shown quite nicely by the once-straight fence pictured in Figure 5.3. It is

*Courtesy of The Associated Press.

Figure 5.1 *These leaning apartment houses rest on unconsolidated soil, which imitated quicksand during the 1964 earthquake in Niigata, Japan. Although some of the buildings were hardly damaged, their new orientation left something to be desired. (Courtesy of NOAA)*

movement of this type that produces an earthquake. The actual displacement takes place along a break in the crust called a **fault.** Along most faults the movement is a slow and gradual creep. It is only when large sections become "locked" and resist movement that the potential for a major earthquake exists. Under these conditions, the rocks will bend elastically and store up energy, much like a wooden stick would if bent (Figure 5.4). Eventually, however, slippage

Figure 5.2 *San Francisco in flames after the 1906 earthquake. (Reproduced from the collection of the Library of Congress)*

Figure 5.3 *This fence was offset 2.5 meters (8.5 feet) during the 1906 San Francisco earthquake. (Photo by G.K. Gilbert, U.S. Geological Survey)*

will occur, and the rock will "snap" back to its original shape. *It is the rock returning to its original shape that produces the vibrations we know as an earthquake.* The "springing" back of the rock is called **elastic rebound,** since the rock behaves elastically, much like a stretched rubber band does when it is released.

The direction of movement between the sections of earth on either side of the fault can be vertical, horizontal, or both. An example of a fault which has primarily horizontal movement is the San Andreas fault (Figure 6.12). Its very large length, almost 960 kilometers (600 miles), and the fact that it is an ever-present threat to the city of San Francisco make it one of the most studied of all faults. During the San Francisco earthquake of 1906 the fault moved as much as 6 meters (20 feet) in some places, releasing tremendous amounts of energy. A destructive earthquake is likely to occur there again someday

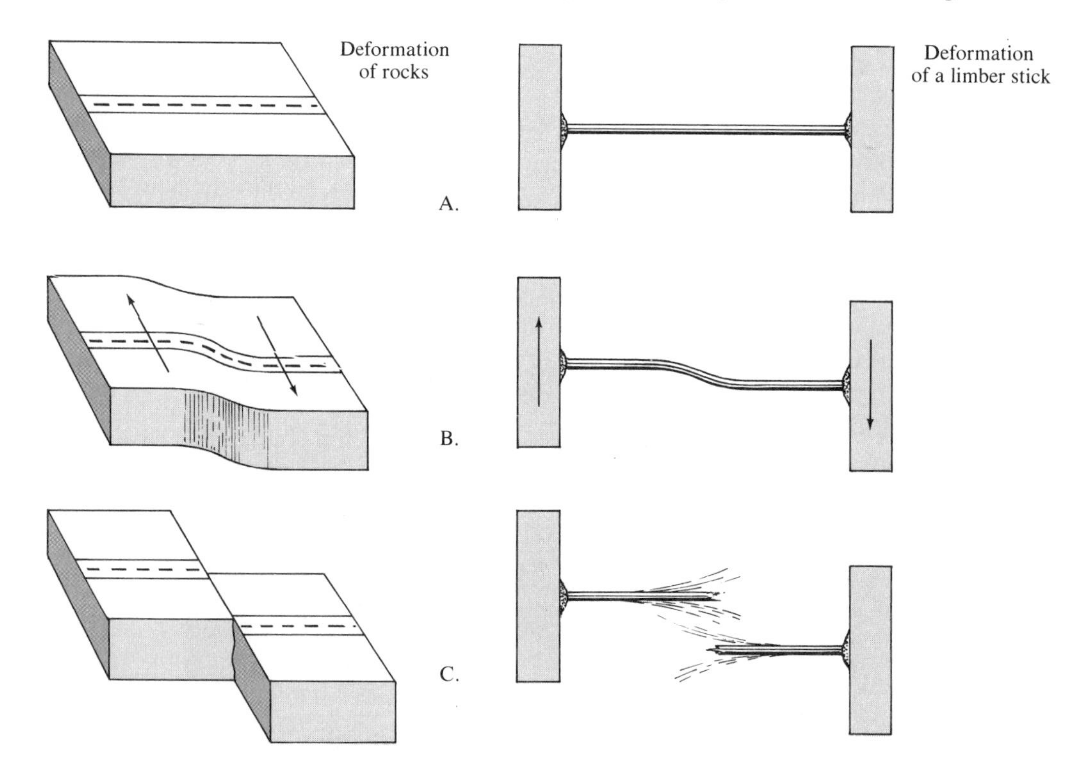

Figure 5.4 *Storage of elastic energy and its subsequent release.* **A.** *No stress applied.* **B.** *Elastic strain stored under stress.* **C.** *Release of elastic strain.*

when forces along the fault build to an appropriate level. The nature of the horizontal movement along the fault is shown by the displacement of a stream channel in Figure 5.5.

Not all movement along faults is rapid. On the contrary, much of the activity may be characterized as a slow, continuous creep.

Vertical displacement along faults, where one side is lifted higher in relation to the other, is very common. Figure 5.6 shows a scarp (cliff) produced by vertical movement. In the same manner, the 1964 Good Friday earthquake in Alaska produced almost 15 meters (50 feet) of displacement at one location.

Figure 5.5 *The displacement of a stream channel shows the horizontal movement along the San Andreas fault. (Photo by R.E. Wallace, U.S. Geological Survey)*

Earthquake Waves

The study of earthquake waves, **seismology,** dates back to attempts made by the Chinese almost 2000 years ago to determine the direction from which the earthquake waves originate. The principle used in modern **seismographs,** instruments which record earthquake waves, is rather simple. A weight is freely suspended from a support that is attached to bedrock (Figure 5.7). When waves from a distant earthquake reach the instrument, the inertia of the weight keeps it stationary, while the earth and the support vibrate. The movement of the earth in relation to the stationary weight is recorded on a rotating drum. Records obtained in this manner have provided information of great value to earth scientists (Figure 5.8). From these seismic records, they have been able to construct a picture of the

Figure 5.6 *Scarp resulting from vertical movement along a fault zone. (Photo by J.R. Stacy, U.S. Geological Survey)*

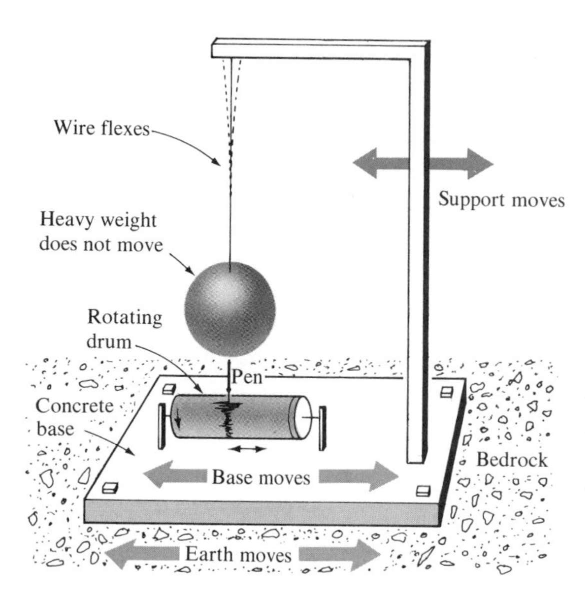

Figure 5.7 *Principle of the seismograph.*

earth's interior and gain insight and evidence which have helped to build the model for plate tectonics.

Three main types of waves are produced during an earthquake: (1) surface waves; (2) primary or P waves; and (3) secondary or S waves (Figure 5.8). *Surface waves are responsible for most of the destruction associated with an earthquake, while the main significance of P and S waves is in the study of the earth's interior.* When surface waves travel through the ground, they cause it and the buildings on it to move, similar to the way ocean swells toss a ship about (Figure 5.9A). Unlike surface waves, P and S waves travel through the interior of the earth and are bent or reflected as they enter a layer of different material, much like light is bent or reflected when it enters water. The actual motion associated with these waves as they travel through the earth is illustrated in Figures 5.9B and C. Because they travel by different modes, two important differences may be noted. *First, P waves travel fastest; therefore they reach a seismic station first. Second, P waves travel through solids, liquids, and gasses, while S waves travel only in solids.* As you will see, these differences prove important in efforts to determine the precise location of an earthquake.

Location of Earthquakes

Recall that the focus is the place where the earthquake originates, usually below the ground. The **epicenter** is the location on the surface directly above the focus (Figure 5.10). The difference in velocities of P and S waves provides a method for determining the epicenter. The principle used is analogous to a race between two autos, one faster than the other. The greater the distance of the race, the greater will be the difference in the arrival times at the finish line. Therefore the greater the interval between the arrival of the first P wave and the first S wave, the greater the distance to the earthquake. By

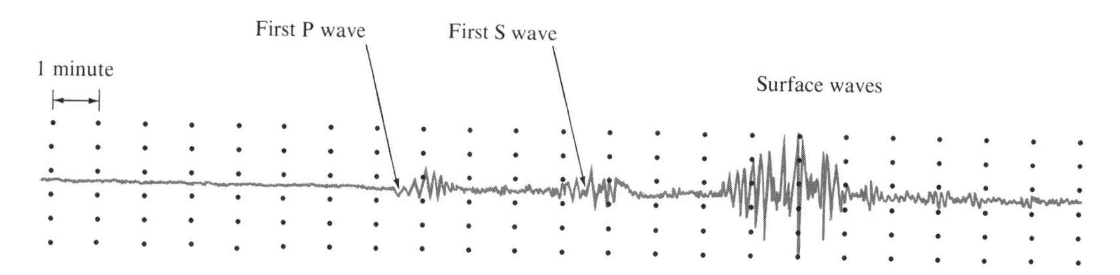

Figure 5.8 *Typical seismic record. Note the time interval between the arrival of each wave type.*

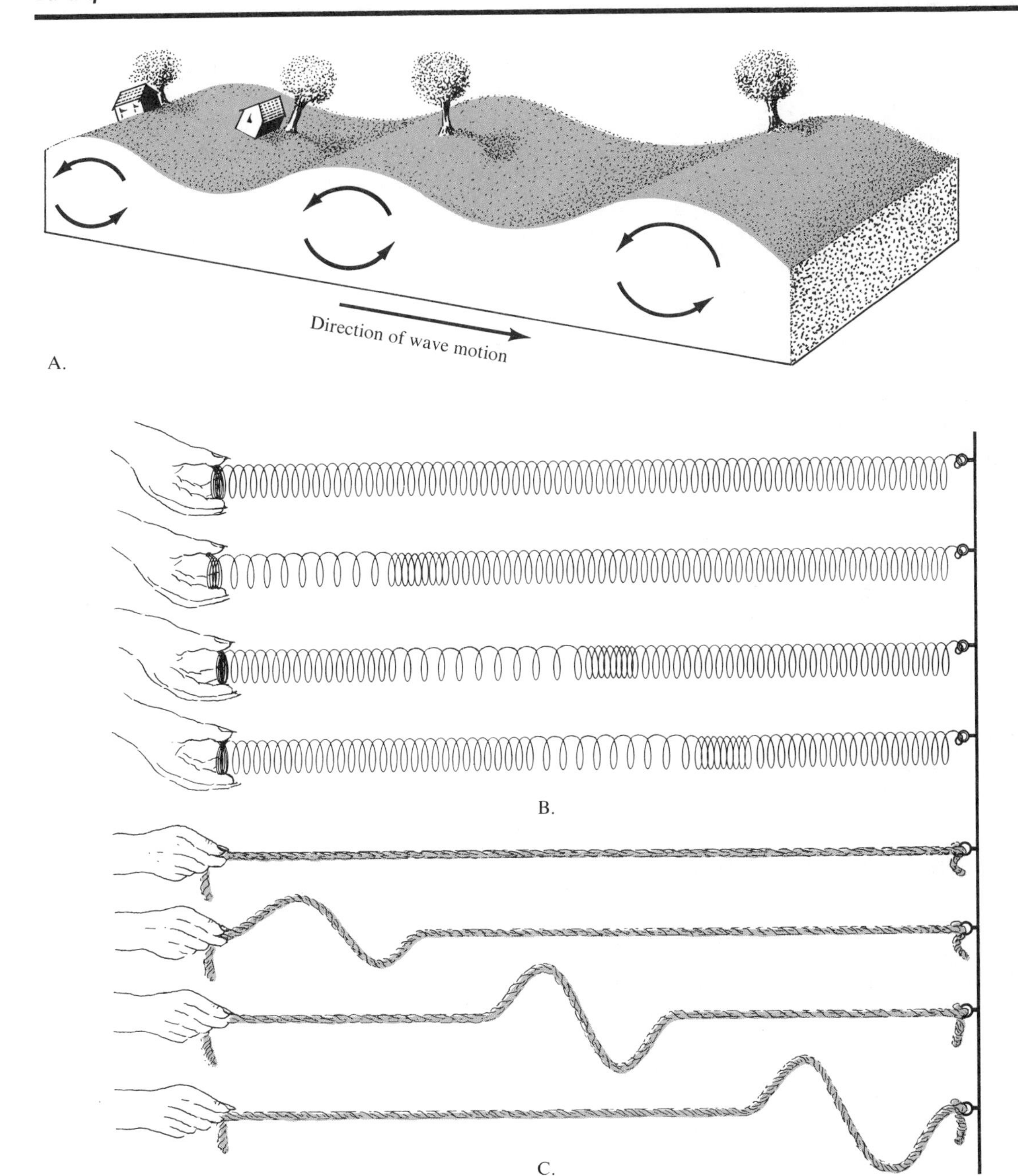

Figure 5.9 *Types of seismic waves and their characteristic motion.* **A.** *Surface waves move the land in a circular path similar to the motion of water in ocean swells.* **B.** *P waves cause the particles in the material to vibrate back and forth in the same direction as the waves move.* **C.** *S waves cause particles to oscillate at right angles to the direction of wave motion.*

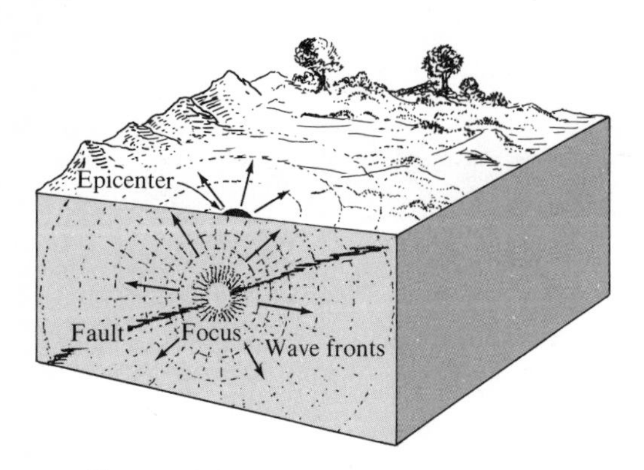

Figure 5.10 *The focus of most earthquakes is located at depth. The surface location directly above it is called the epicenter.*

using a travel-time graph as shown in Figure 5.11, the distance from the seismic station to the epicenter can be determined. If three or more seismic stations are used, a good approximation of the position of the epicenter can be made. Figure 5.12 shows the worldwide distribution of earthquake epicenters. By comparing this figure with Figure 6.3, we can see a close correlation between the location of earthquake epicenters and plate boundaries, a phenomenon which will be explored in the next chapter.

Earthquake Intensity and Magnitude

Early attempts to establish the intensity of an earthquake relied heavily on descriptions of the event. There was an obvious problem related to this method—people's accounts varied widely, making an accurate classification of the quake's intensity difficult. Then in 1902 a fairly reliable scale based on the amount of damage caused to various types of structures was developed by Giuseppe Mercalli. A modified form of this tool is presently used by the U.S. Coast and Geodetic Survey (Table 5.1). However the destruction wrought by earthquakes is not an adequate means for comparison. Many factors, including distance from the epicenter, nature of the surface materials, and building design, cause variations in the amount of damage. Consequently, methods were devised which determine the total amount of energy released during an earthquake, a measurement referred to as **magnitude.**

The **Richter scale** is widely used to describe the magnitude of a quake. On this scale magnitude is determined from the maximum motion recorded by seismic instruments. Hence if two earthquakes occur at the same distance from a seismograph, the one with the greater magnitude will cause the recording pen to move farther. By making adjustments for the weakening of seismic waves as they move from the focus, seismic stations located around the globe obtain nearly the same magnitude for an earthquake, regardless of their distance from it. The largest earthquakes recorded have had magnitudes near 8.6, while those with magnitudes less than 2.5 are usually

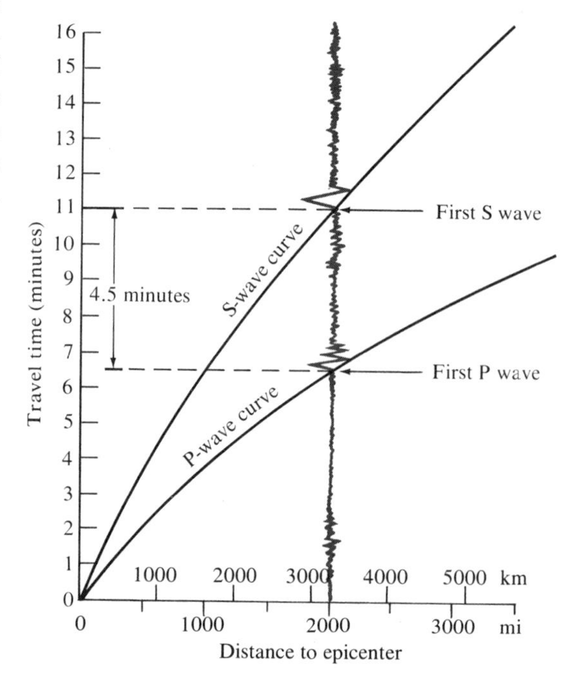

Figure 5.11 *A travel-time graph is used to determine the distance to the epicenter. The difference in arrival times of the first P and S waves in the example is 4.5 minutes. Thus the epicenter is roughly 3200 kilometers (2000 miles) away.*

Table 5.1 ***Modified Mercalli Intensity Scale***

I. Not felt except by a very few under specially favorable circumstances.

II. Felt only by a few persons at rest, especially on upper floors of buildings.

III. Felt quite noticeably indoors, especially on upper floors of buildings, but many people do not recognize as an earthquake.

IV. During the day felt indoors by many, outdoors by few. Sensation like heavy truck striking building.

V. Felt by nearly everyone, many awakened. Disturbances of trees, poles, and other tall objects sometimes noticed.

VI. Felt by all; many frightened and run outdoors. Some heavy furniture moved; few instances of fallen plaster or damaged chimneys. Damage slight.

VII. Everybody runs outdoors. Damage negligible in buildings of good design and construction; slight to moderate in well-built ordinary structures; considerable in poorly built or badly designed structures.

VIII. Damage slight in specially designed structures; considerable in ordinary substantial buildings with partial collapse; great in poorly built structures. (Fall of chimneys, factory stacks, columns, monuments, walls.)

IX. Damage considerable in specially designed structures. Buildings shifted off foundations. Ground cracked conspicuously.

X. Some well-built wooden structures destroyed. Most masonry and frame structures destroyed with foundations. Ground badly cracked.

XI. Few, if any (masonry) structures remain standing. Bridges destroyed. Broad fissures in ground.

XII. Damage total. Waves seen on ground surfaces. Objects thrown upward into air.

* SOURCE: U.S. Coast and Geodetic Survey.

not felt by humans. Table 5.2 shows how earthquake magnitudes and their effects are related.

The Richter scale is logarithmic, so that an increase of 1 magnitude signifies 10 times the ground motion and the release of roughly 30 times the energy. Thus an earthquake with a magnitude of 6.5 releases 30 times more energy than one with a magnitude of 5.5, and roughly 900 times that of a 4.5-magnitude quake. A major earthquake with a magnitude of 8.5 releases almost a thousand million times more energy than the smallest earthquakes felt by humans. This fact should dispel the notion that a moderate earthquake decreases the chances for the occurrence of a major quake in the same region. It takes thousands of moderate tremors to release the amount of energy equal to one "great" earthquake. Some of the world's major earthquakes this century include San Francisco, 1906;

Table 5.2 ***Earthquake Magnitudes and Effects***

Richter Magnitudes	Earthquake Effects
2.5	Generally not felt, but recorded
4.5	Local damage
6.0	Can be destructive in populous region
7.0	Major earthquakes. Inflict serious damage. Roughly, ten occur each year
> 8.0	Great earthquakes. Occur once every 5–10 years; produce total destruction to nearby communities

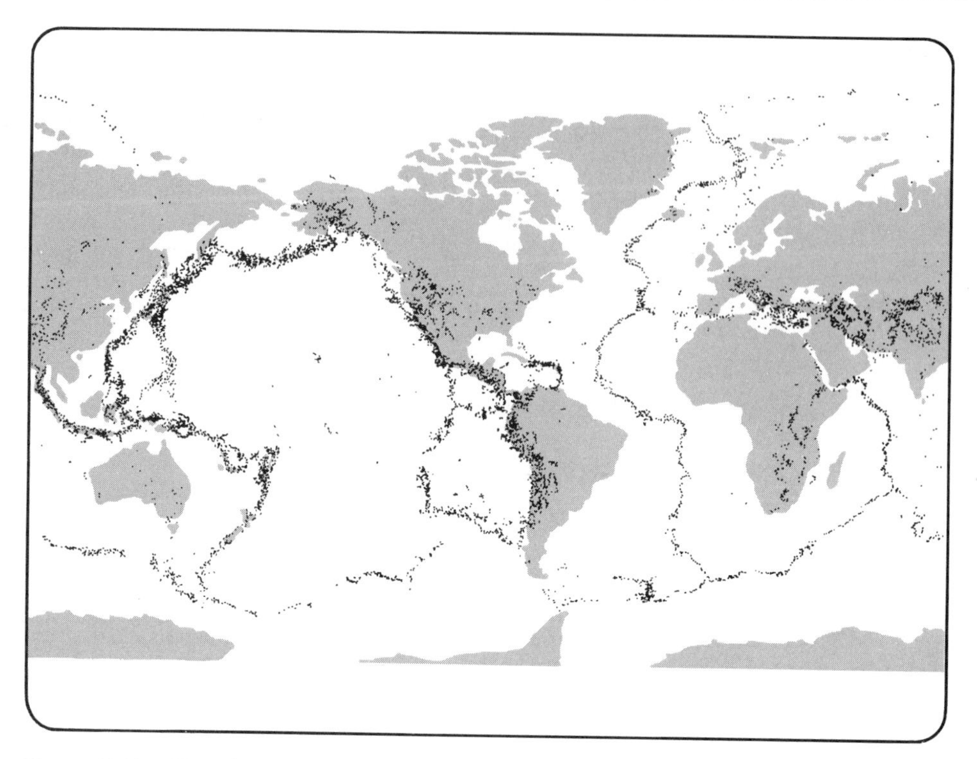

Figure 5.12 *World distribution of earthquakes for 1961–1969. (Data from NOAA)*

Japan, 1923; Chile, 1960; and Alaska, 1964. All had magnitudes in excess of 8.2.

Earthquake Destruction: The 1964 Alaskan Earthquake

The most violent earthquake to jar North America this century—the Good Friday Alaskan Earthquake—occurred at 5:36 P.M. on March 27, 1964. Felt throughout that state, the earthquake had a magnitude of 8.4–8.6 on the Richter scale and reportedly lasted 3–4 minutes. In that short period of time 114 persons lost their lives, thousands were left homeless, and the economy of the state was badly disrupted. Had the schools and business districts been open, the toll surely would have been higher. The location of the epicenter and the towns which were hardest hit by the quake are shown in Figure 5.13.

As the energy released by an earthquake travels along the earth's surface, it causes the ground to vibrate in a complex manner by moving up and down as well as from side to side. The amount of destruction attributable to the vibrations depends on several factors: (1) the intensity and duration of the vibrations; (2) the nature of the material on which the structure rests; and (3) the design of the structure. The 1964 Alaskan earthquake gave geologists new insights into the role of these factors. All of the multistory structures in Anchorage were damaged by the vibrations; the more flexible wooden-frame residential buildings fared best. A striking example of how construction variations affect earthquake damage is shown in Figure 5.14. Even structures that had been built to conform to the earthquake provisions of the Uniform Building Code of California were badly damaged. Because the California code was based upon evidence gathered from earthquakes that lasted no more than a minute, some of the damage here was the result of this quake's unusually long duration.

An earthquake often triggers events that are much greater hazards to us than the vibrations themselves. Occurrences such as landslides, fires, and seismic sea waves all may take a heavy toll. In the 1964 Alaskan earthquake landslides probably did most of the damage to artificial structures. In addition, at Valdez and Seward the violent shaking caused deltaic materials to liquify; the subsequent slumping carried both waterfronts away. Because of the threat of recurrence, these villages were relocated on more stable ground. By contrast, the town of Whittier, although located nearer the epicenter than Seward, rests on a firm foundation of granite and hence suffered little damage from the seismic vibrations. It was however damaged by a seismic sea wave.

Fires, although of little consequence in this particular earthquake, are often the most destructive result. An earthquake which rocked Japan in 1923 triggered an estimated 250 fires, devastating the city of Yokohama and destroying more than half the homes in Tokyo.

Most of the deaths in the 1964 Alaskan quake were caused by **seismic sea waves.** These destructive waves, named **tsunami** by the Japanese, who are often affected by them, are generated by

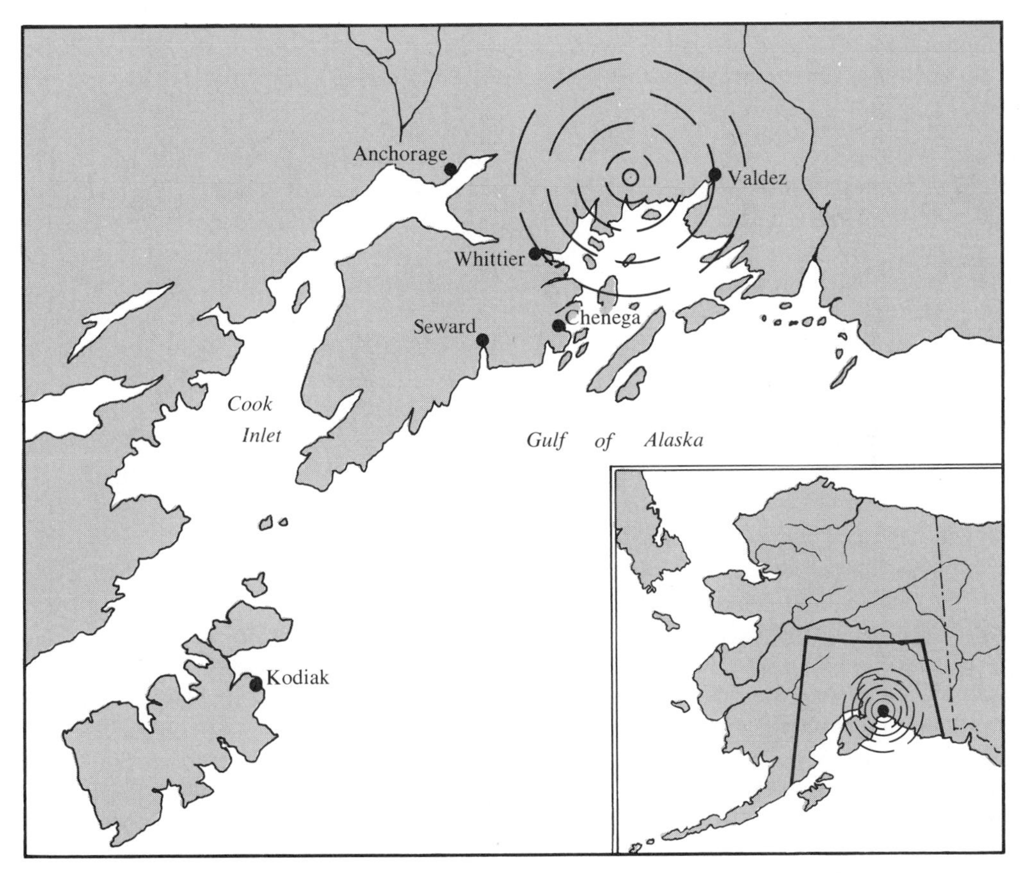

Figure 5.13 *Region most affected by the Good Friday earthquake of 1964. Note the location of the epicenter. (After U.S. Geological Survey)*

Figure 5.14 *Damage caused to the five-story J.C. Penney Co. building, Anchorage, Alaska. Very little structural damage was incurred by the adjacent building. (Courtesy of NOAA)*

movement of the ocean floor. A tsunami resembles the ripples formed when a pebble is dropped into a pond. In the open ocean its crests are usually less than 1 meter (3 feet) high, and the distance between crests often exceeds 160 kilometers (100 miles). Because of the great distance between crests, a tsunami has a very gentle slope and goes undetected by ships. However, tsunami move at speeds approaching 960 kilometers per hour, and upon entering shallow coastal waters they have been known to grow into a wall of water 30 meters (100 feet) high. The tsunami generated in the 1964 Alaskan earthquake inflicted heavy damage to communities in the vicinity of the Gulf of Alaska, completely destroying the town of Chenega. Kodiak was also heavily damaged and most of its fishing fleet destroyed when a seismic wave carried many vessels into the business district (Figure 5.15). The deaths of 107 persons have been attributed to this tsunami. By contrast, only 9 persons died in Anchorage as a direct result of the quake. Tsunami damage following the Alaskan earthquake extended along much of the west coast of North America, and in spite of a one-hour warning, 12 persons perished in Crescent City, California, where all of the deaths and most of the destruction was caused by the fifth wave. The first wave crested about 4 meters (14 feet) above low tide and was followed by three progressively smaller waves. People, believing that the tsunami had ceased, returned to the shore, only to be met by the fifth and most devastating wave, which, superimposed on high tide, crested about 6 meters (20 feet) above low tide.

Earthquake Prediction and Control

The vibrations that shook the San Fernando, California, area on the morning of February 9, 1971, inflicted 64 deaths and almost a thousand million dollars in damages (Figure 5.16)—all of this from an earthquake that lasted 60 seconds and had a moderate rating of 6.6 on the Richter scale. Fortunately, because of the early hour, freeways, businesses, and schools were sparsely occupied, reducing the possible toll. Also, had the Lower Van Norman Lake Dam, badly damaged during the earthquake, actually broken, 80,000 additional lives might have been lost and would have made it the most catastrophic event ever in the United States. This quake in populous Southern

Drumlins, such as this one in upstate New York, are depositional features associated with continental glaciers. (Courtesy of Ward's Natural Science Establishment, Inc., Rochester, N.Y.)

The Matterhorn, a glacially eroded peak in the Swiss Alps. (Courtesy of Ward's Natural Science Establishment, Inc., Rochester, N.Y.)

A hanging valley and an arête dominate this alpine scene. Glacier du Geant, France. (Courtesy of Ward's Natural Science Establishment, Inc., Rochester, N.Y.)

Figure 5.15 *A tsunami washed this fishing fleet into the heart of the village of Kodiak, Alaska. (Photo by W.R. Hansen, U.S. Geological Survey)*

California reemphasized the need for reliable methods of earthquake prediction and control.

Because of Japan's location in an earthquake-prone region, people there are also interested in earthquake prediction. They have concentrated their study on microearthquakes (fore shocks), which precede the main earthquake. It is hoped that by monitoring these seismic activities some pattern will emerge which can be used to accurately predict forthcoming tremors. In California, uplift or subsidence of the land, and changes in movement of a fault zone from a slow creep to a locked position, have been found to precede moderate earthquakes. It therefore seems reasonable that the prediction of earthquakes may be feasible by continually monitoring ground tilt, fault movement, and seismic activity. Some monitoring networks are already operational in the earthquake-prone regions of the United States; others have been proposed. Although no reliable method of short-range prediction has yet been devised, a few successful predictions have been made. In 1975 an earthquake in northeast China was predicted only hours before it occurred. By evacuating millions of people, it is believed that tens of thousands of lives were spared since many of the villages were subsequently destroyed. Unfortunately, the Chinese were able to predict, but not pin down, the exact date of the great T'ang-shan earthquake of 1976. Their long-range warning of the upcoming earthquake was not precise enough to save the more than 500,000 persons estimated to have lost their lives.

The actual control of earthquakes is another matter altogether. The discovery that humans have inadvertantly triggered many earthquakes has given earth scientists some encouragement. The most convincing evidence that people can initiate earthquakes came between 1962 and 1966, when studies on the seismic activity at the Rocky Mountain Arsenal near Denver were conducted. For a period of 80 years prior to 1962 the U.S. Coast and Geodetic Survey reported that no earthquake epicenters were located in the Denver region. In 1962 the arsenal started to dispose of wastes from its chemical warfare production into a well over 3600 meters deep. During the period of fluid waste injection, from April 1962 to September 1965, about 700 microearthquakes were reported, 75 intense enough to be felt. It is believed that the injection of water under pressure "lubricated" the fault, which had been building up strain over the years.

Other earthquakes caused by human activity have occurred in regions adjacent to large reservoirs like Lake Mead on the Arizona-Nevada line. Ever since Lake Mead was filled in 1936, hundreds of small tremors have been recorded. They are thought to have been caused by the added weight of the lake, and perhaps aided by the "lubricating" effect of the impounded water. Another large reservoir in India is believed responsible for triggering a disastrous earthquake in which 200 persons lost their lives. Underground nuclear explosions have also been responsible for initiating numerous small aftershocks, although none has been as great as the explosion itself.

The hope is that we may be able reduce the threat of earthquakes by triggering numerous small earthquakes using fluid injection or nuclear explosions. Such methods would slowly and continually release the elastic strain that

Figure 5.16 *Collapsed overpass to Golden State Freeway, San Fernando earthquake, 1971. (Photo by R.W. Wallace, U.S. Geological Survey)*

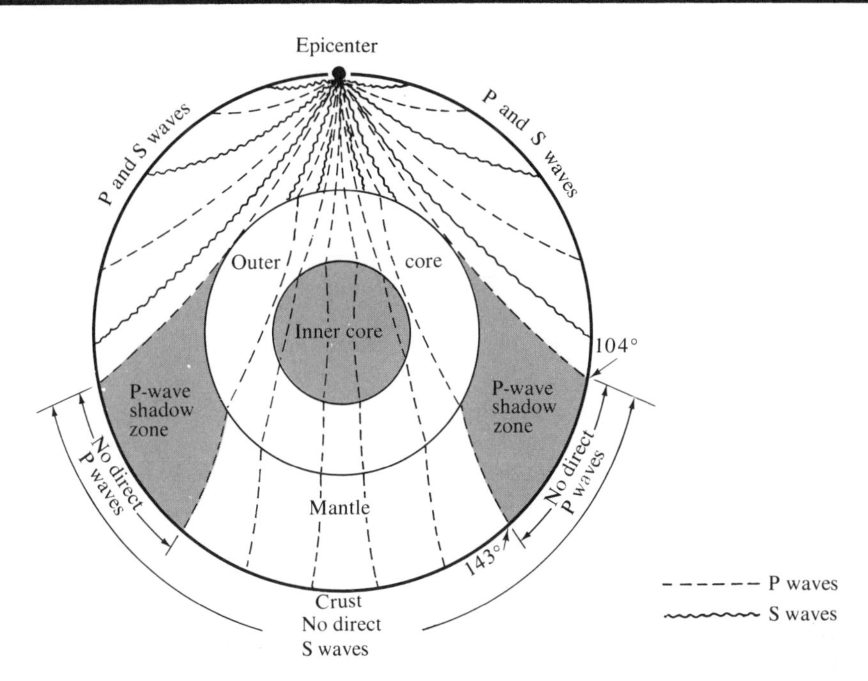

Figure 5.17 *Cross-sectional view of the earth showing internal structure and paths of P and S waves. Any place located more than 104 degrees from the epicenter will not receive direct S waves, since the outer core will not transmit them. A small shadow zone exists (from 104 to 143 degrees) for P waves. The P-wave shadow zone is caused by the bending of these waves as they pass from more rigid mantle material to less rigid core material. The seismic waves that pass through the center of the earth increase in velocity, revealing the existence of the solid inner core.*

might otherwise build up and be released as a high-magnitude earthquake. A plus for earthquake control in California is the shallow depth of earthquake foci, making drilling operations feasible. Needless to say, many tests will have to be made in remote regions before we dare risk such a venture along a fault system like the San Andreas, at least in those portions which are located in populous regions.

The Earth's Interior

The structure of the earth's interior is known almost exclusively from the behavior of P and S waves. These seismic waves travel at different velocities, depending on the nature of the layer in which they are traveling, thus indicating the position of each layer, as well as providing clues as to its composition. On this basis, the earth has been divided into four major layers: (1) the **crust,** the very thin outermost layer; (2) the **mantle,** located below the crust and having a thickness of 2900 kilometers; (3) the **outer core,** a layer about 2200 kilometers thick which has the properties of a liquid; and (4) the **inner core,** the solid innermost layer, about 1300 kilometers in radius (Figure 5.17).

In 1909 a pioneering Yugoslavian seismologist, Andrija Mohorovičić, was the first to discover variation in the earth's internal structure. By studying seismic records, he found that the velocity of waves increases abruptly below a depth of 50 kilometers. This boundary that he discovered separates the crust from the underlying mantle and is known as the **Mohorovičić discontinuity,** or **Moho,** in his honor (Figure 5.18).

The boundaries, and reasons for separating the earth into the mantle, outer core, and inner

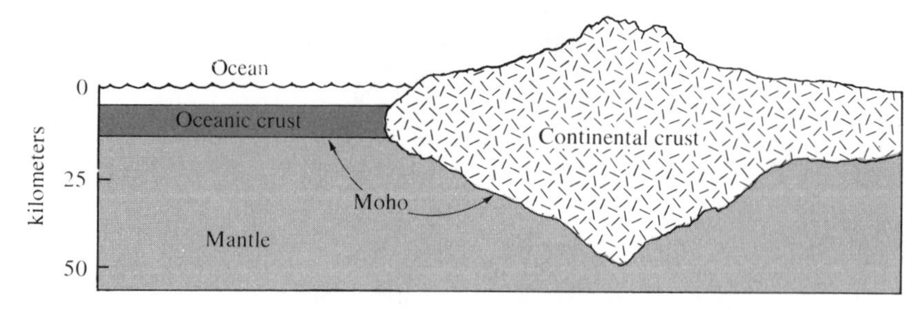

Figure 5.18 *Moho, the boundary between the crust and mantle.*

core, are illustrated in Figure 5.17. The outer core was discovered when it was found that P waves were bent and slowed upon entering it, thereby producing a **shadow zone** at the surface (Figure 5.17). Further, since the outer core does not transmit S waves, geologists conclude that it is in the molten state. The inner core, although possibly similar in composition to the outer core, transmits P waves at a higher velocity, which indicates that it is in the solid state.

A most important zone exists within the upper mantle and deserves special mention. This region, called the **asthenosphere,** or **low-velocity zone,** is located between a depth of 70 kilometers and 700 kilometers (Figure 5.19). In the asthenosphere the velocity of S waves decreases, indicating to seismologists that this zone consists partly of melted rock (approximately 10 percent). Situated above the asthenosphere is the outer solid portion of the earth, called the **lithosphere,** which includes part of the upper mantle and the crust. It is believed that the "plastic" material of the asthenosphere moves and carries along the rigid lithosphere. Thus the discovery of the asthenosphere was an important contribution to the theory which proposes that the continents drift about, which is the topic of the next chapter. *It is also from the asthenosphere that some, but not all, of the molten material for volcanic activity is thought to originate.*

Composition of the Earth

The crust of the earth varies in thickness, being greater than 70 kilometers in some mountainous regions and less than 5 kilometers in some oceanic regions (Figure 5.18). Early seismic data indicated that the continental crust, which is mostly made of felsic (granitic) rocks, is quite different in composition from the oceanic crust. Until recently, however, scientists could only speculate on the composition of oceanic crust, which lies beneath 3 kilometers of water as well as hundreds of meters of overlying sediment. With the devel-

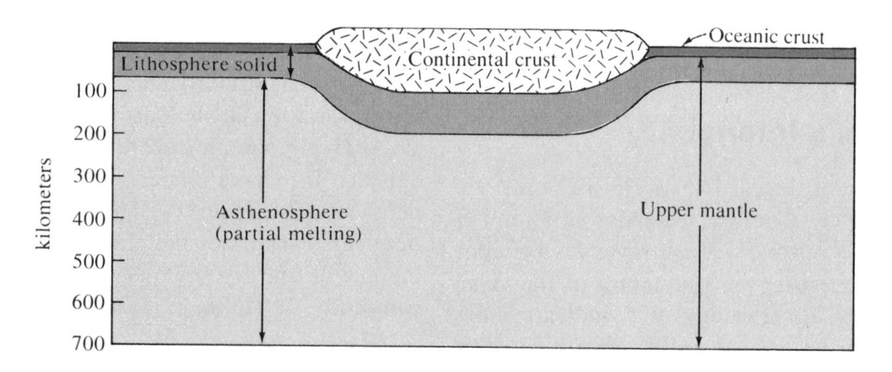

Figure 5.19 *Respective locations of the asthenosphere and lithosphere.*

opment of the deep-sea drilling ship *Glomar Challenger,* the recovery of samples from the ocean floor was made possible. The samples obtained were of mafic (basaltic) composition—indeed different from the rocks which compose the continents.

Our knowledge of the composition of the mantle and core is much more speculative. However we do have some clues. Recall that some of the lava that reaches the earth's surface originates in the partially melted asthenosphere located within the mantle. In the laboratory, experiments have shown that partial melting of a rock called peridotite results in a melt that has a basaltic composition similar to those associated with the volcanic activity of oceanic islands. Ultramafic rocks like peridotite are thought to make up the mantle and provide the lava for oceanic eruptions.

Surprisingly enough, meterorites—"shooting stars"—which fall to the earth from space, are considered evidence of the earth's inner composition. Since meteorites are part of the solar system, they are assumed to be representative samples. Their composition ranges from metallic types made of iron and nickel to stony meteorites composed of ultramafic rock similar to peridotite. Because the earth's crust contains a much smaller percentage of iron than meteorites do, geologists believe that the heavy minerals sank during the early history of the earth. By the same token, the lighter minerals may have floated to the top, creating the crust. *Thus the core of the earth is thought to be mainly iron and nickel, similar to metallic meteorites, while the surrounding mantle is believed to be composed of ultramafic rocks, similar to stony meteorites.*

The concept of a molten iron core is further supported by the existence of the earth's magnetic field, which acts as a large bar magnet. The most widely accepted mechanism explaining the magnetic field requires that the earth's core be made of a material which conducts electricity, such as iron, and which is mobile so that circulation can occur. Both of these conditions are met by the model of the earth's core that was established on the basis of seismic data.

Not only does an iron core explain the earth's magnetic field, but it also explains the high density of the inner earth, about 12 times that of water. Even under the extreme pressure at those depths, average crustal rocks with densities 2.8 times that of water would not have the density calculated for the core. But iron, which is 3 times more dense than crustal rocks, has the required density.

REVIEW

1. What is an earthquake? Under what circumstances do earthquakes occur?
2. How are faults, foci, and epicenters associated?
3. Earthquakes occur only in the rigid lithosphere, not in the plastic asthenosphere. Using the elastic rebound idea, explain this phenomenon.
4. Faults which are experiencing no active creep may be considered "safe." Rebut or defend this statement.
5. Describe the principle of a seismograph.
6. Using Figure 5.11, determine the distance between an earthquake and a seismic station if the first S wave arrives 4 minutes after the first P wave.
7. What evidence do we have that the earth's outer core is molten?
8. Contrast the physical makeup of the asthenosphere and the lithosphere.
9. The asthenosphere is important for two reasons. What are they?
10. Distinguish between the Mercalli scale and the Richter scale.
11. What is a tsunami? How is one generated?
12. Cite some reasons why an earthquake with a moderate magnitude might cause more extensive damage than a quake with a high magnitude.
13. How might earthquakes be controlled in the future?

KEY TERMS

earthquake
focus
fault
elastic rebound
seismology
seismograph
epicenter
magnitude
Richter scale
seismic sea wave (tsunami)
crust
mantle
outer core
inner core
Mohorovičić discontinuity (Moho)
shadow zone
asthenosphere (low-velocity zone)
lithosphere

The Red Sea (lower right) was created along with the Gulf of Suez and the Gulf of Aqaba as the Arabian Peninsula and Africa separated. (Courtesy of NASA)

Plate Tectonics

Will California eventually "slide" into the ocean, as some predict? Have continents really "drifted" apart over the centuries? Answers to these questions and many others which have plagued geologists for decades are now being provided by a new and exciting theory on large-scale movements taking place within the earth. This theory, called plate tectonics, represents the real frontier of the earth sciences, and its implications are so far-reaching that it can be considered the framework from which most other geologic processes should be viewed.

Continental Drift: Historical Setting

The idea that continents, particularly South America and Africa, fit together like pieces of a puzzle has been around as long as world maps. Little significance was given this idea until 1912, when Alfred Wegener, a German meteorologist, set forth his then-radical theory of **continental drift.** In his proposal Wegener suggested that a supercontinent, **Pangaea** (meaning "all land"), once existed (Figure 6.1). It was theorized that about 200 million years ago this supercontinent began breaking into smaller continents, which then "drifted" to their present positions. Wegener and others who advocated this position collected substantial evidence, including the fit of South America and Africa, to support their claims (Figure 6.2). Ancient climatic similarities, fossil evidence, and rock structures all seemed to bridge together these now separate landmasses. However, strong opposition to Wegener's theory mounted rapidly. First, Wegener stated that the larger and sturdier continents broke through the oceanic crust, much like an icebreaker cuts through ice. Tidal energy was proposed as the driving force, but was easily proved to be grossly inadequate for such a feat. Second, and probably

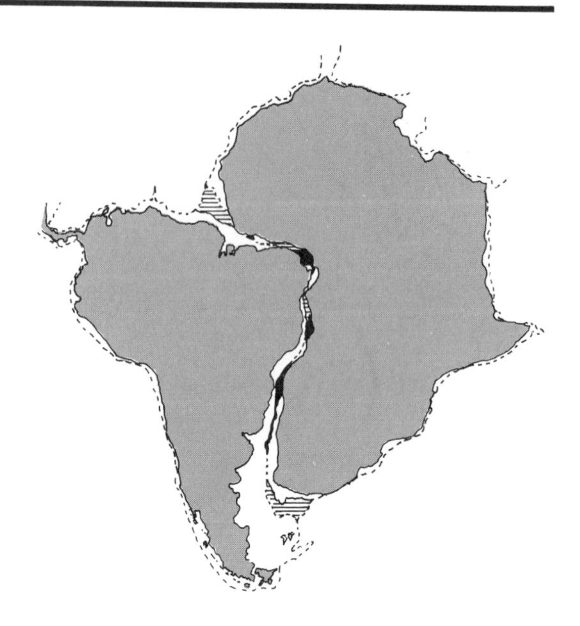

Figure 6.2 *The best fit of South America and Africa along the continental slope at a depth of 500 fathoms. (After A.G. Smith. "Continental Drift." In* Understanding the Earth, *edited by I.G. Gass. Courtesy of Artemis Press)*

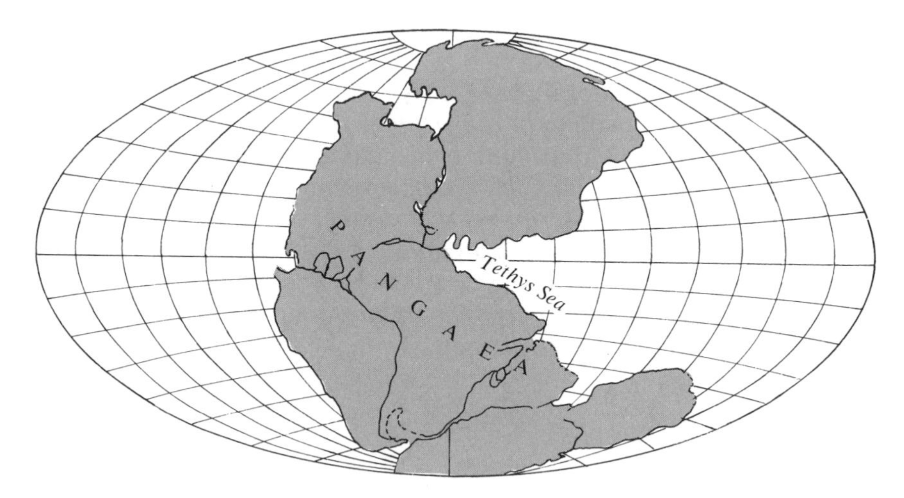

Figure 6.1 *Reconstruction of Pangaea as it is thought to have appeared 200 million years ago. (After R.S. Dietz and J.C. Holden.* Journal of Geophysical Research *75: 4943. Copyright © by American Geophysical Union)*

of greater significance, was the lack of that bit of conclusive evidence needed to jar the scientific community from its traditional position. Wegener perished on the ice sheets of Greenland in search of just that bit of evidence. Fortunately, however, his intriguing idea did not die with him.

For the few geologists who continued the search, the concept of continents in motion provided enough excitement to hold their interest. Others undoubtedly saw Pangaea as a key to explaining heretofore unexplainable observations. For example, sandstone formations found near London, England, consist of large relic sand dunes. When these sand dunes were "alive," could England have been situated nearer the equator, as the Sahara Desert is today? Coal beds located in the United States and Siberia are found to contain similar types of fossils. Is it possible that in the distant past the ancient swamps that generated these now-distant coal fields lay on adjoining landmasses?

During the years that followed Wegener's proposal great strides in technology permitted mapping of the ocean floor, and extensive data on seismic activity and the earth's magnetic field became available. By 1968 these developments led to the unfolding of a far more encompassing theory than continental drift, known as **plate tectonics.** The implications of plate tectonics are so far-reaching that it can be considered the framework from which most other geologic processes should be viewed. Since this concept is

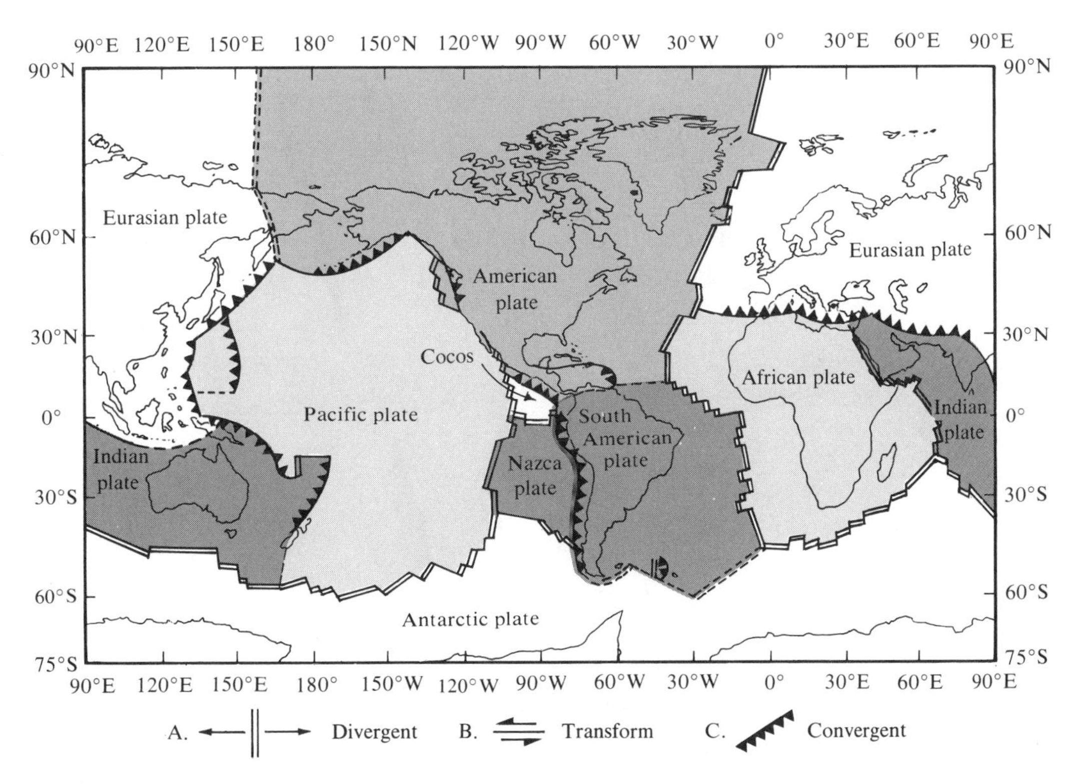

Figure 6.3 *Mosaic of rigid plates that constitute the earth's outer shell. A. is a divergent boundary, B. is a transform fault boundary, and C. is a convergent boundary. (After F.J. Sawkins, Clement G. Chase, David G. Darby, and George Rapp, Jr. The Evolving Earth, © 1974, Macmillan Publishing Co.)*

quite new, it most surely will be modified as additional information becomes available; however, the main tenets are sound and are presented here in their current state of refinement.

Plate Tectonics

The concept of plate tectonics states that the outer, solid lithosphere is made up of several individual *plates* as shown in Figure 6.3. These plates move in relation to one another upon a weaker zone in the upper mantle known as the asthenosphere (Figure 5.19). The somewhat "plastic" behavior of the asthenosphere is caused by the greater temperatures and pressures existing at greater depths. It has been demonstrated that the temperature is high enough in this zone to produce a trace of melting which adds to the mobility of this layer. Notice from Figure 6.3 that the larger plates, with the exception of the Pacific plate, are composed of both continental *and* oceanic crust—a major departure from the original drift theory, which supposed that the rigid continents moved through, not with, the ocean floor. The distance between two cities on the same plate, New York and Denver for example, remains constant, while the distance between New York and London, which are located on different plates, is changing. Since each plate moves as a distinct unit, *all major changes occur along plate boundaries.* One similarity between the plate tectonics theory and its predecessor, the continental drift theory, is the assumption of the supercontinent Pangaea. The relative positions of the individual continents that combined to make up Pangaea are shown in Figure 6.1. When the

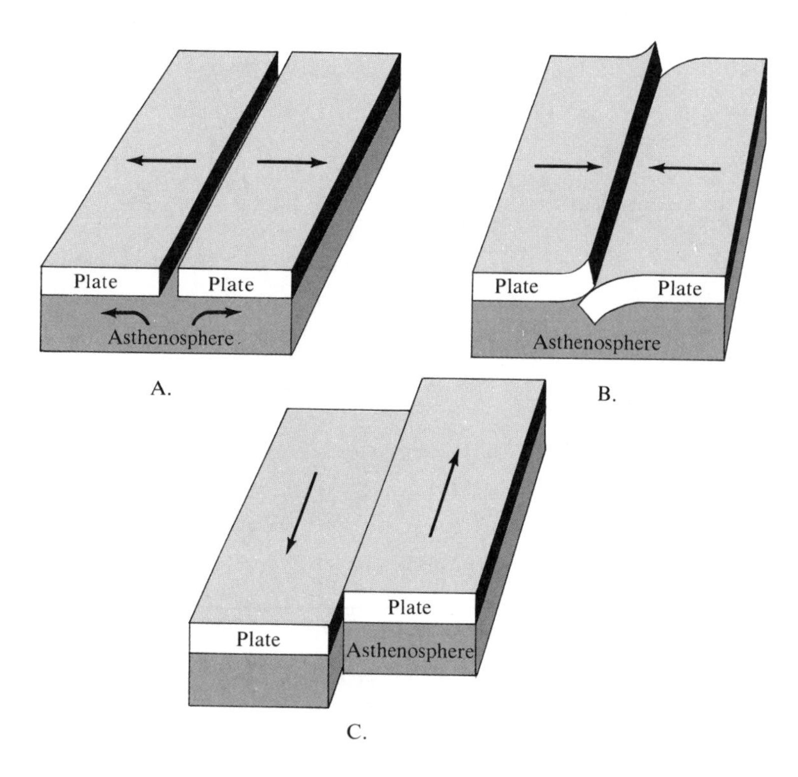

Figure 6.4 *Schematic of plate boundaries.* **A.** *Divergent boundary.* **B.** *Convergent boundary.* **C.** *Transform fault boundary.*

continental drift concept was being formed, it became popular to use the term *Laurasia* to name the northern landmasses (North America and Eurasia). The name *Gondwanaland* was used for the southern landmass which consisted of South America, Africa, Australia, Antarctica, and India. We will follow this practice when we describe the breakup of Pangaea later in the chapter.

Plate Boundaries

The first approximations of plate boundaries were made on the basis of earthquake and volcanic activity. Later work indicated the existence of three distinct types of plate boundaries, which are differentiated by the movement they exhibit (Figure 6.4). These are:

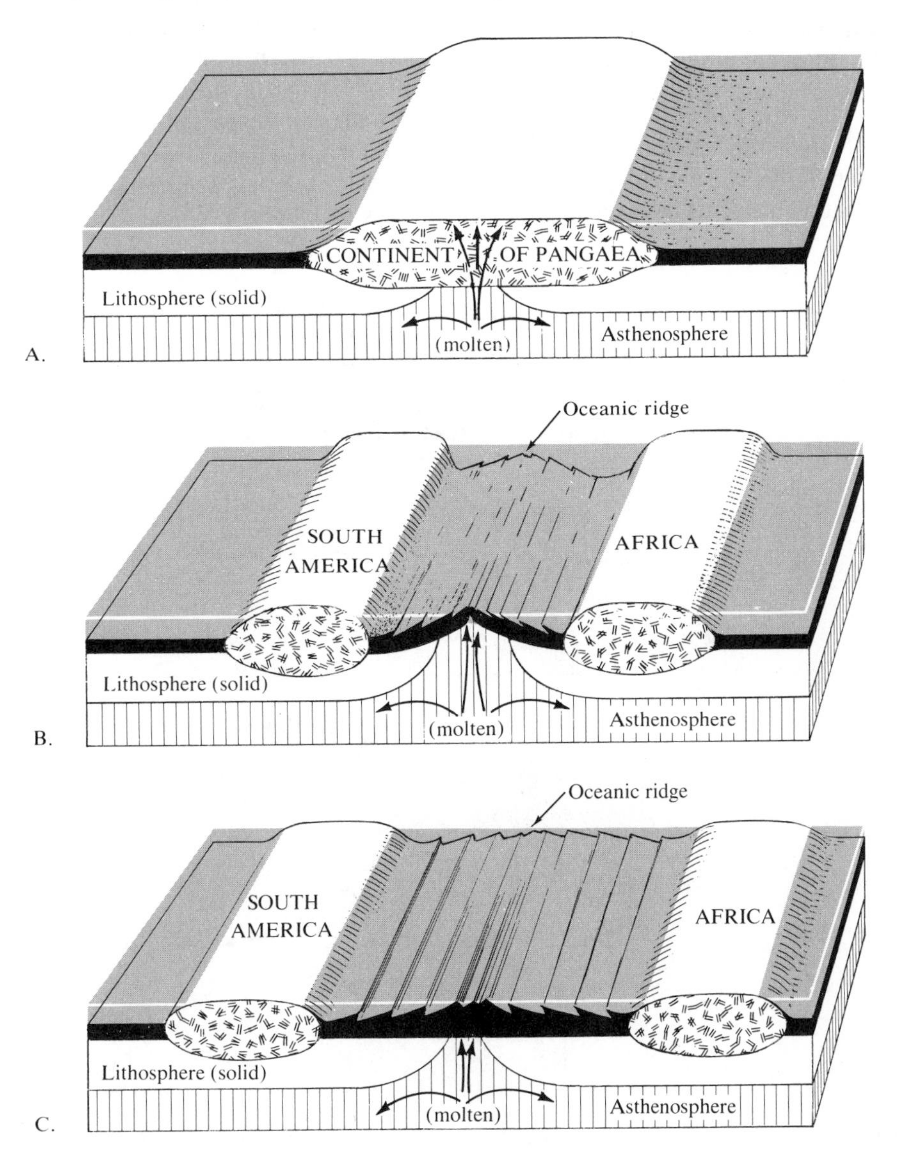

Figure 6.5 *Formation of the Atlantic Ocean by the process of sea-floor spreading.*

(1) **Divergent boundaries**—where plates move apart, leaving a gap between them.

(2) **Convergent boundaries**—where plates move together, causing one to go under the other, as happens with oceanic crust; or where plates collide, which occurs when the leading edges are made of continental crust.

(3) **Transform fault boundaries**—where plates slide past each other, scraping and deforming as they pass.

Each plate is bounded by a combination of these zones (Figure 6.3). Movement along one boundary requires that adjustments be made at another.

Divergent Boundaries

The divergent boundaries, where plate spreading occurs, are situated at the mid-oceanic ridges. As the plates separate the gap created is quickly filled with molten rock that oozes up from the hot asthenosphere (Figure 6.5). This material slowly cools to produce a new sliver of sea floor. Successive separations and fillings continue to add new oceanic crust (lithosphere) between diverging plates. This mechanism, which has produced the floor of the Atlantic Ocean during the past 200 million years, is called **sea-floor spreading.** The typical rate of spreading at these ridges has been estimated to be 5 centimeters (2 inches) per year. This seemingly slow

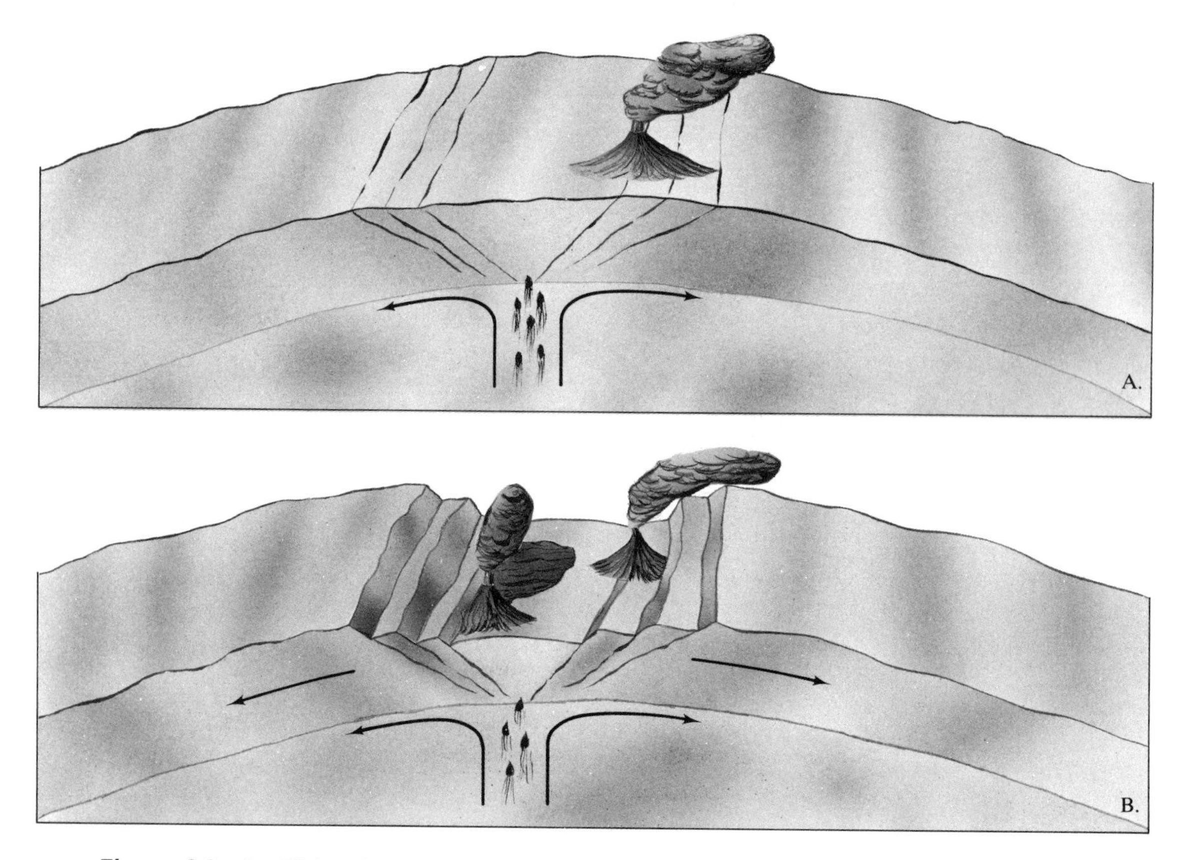

Figure 6.6 **A.** *Rising hot rock upwarps the crust, causing numerous cracks in the rigid lithosphere.* **B.** *As the crust is pulled apart, large slabs of rock sink, generating a rift zone. Volcanic activity is seen associated with the rifting.*

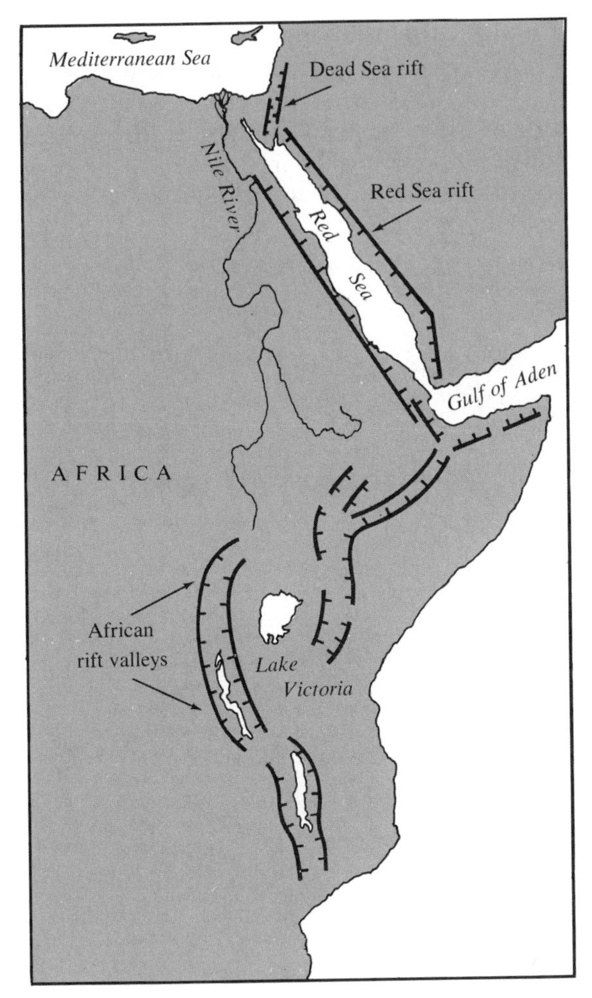

Figure 6.7 *East African rift valleys and associated features.*

rate is rapid enough to have opened and reclosed the Atlantic Ocean more than ten times during the 5000-million-year history of our planet, although it probably did not. All spreading centers are not ancient. The Red Sea is thought to be the site of a recently formed spreading center. It is here that the Arabian Peninsula separated from Africa and moved toward the northwest (see chapter-opening photo).

The actual breakup of a continent is thought to be initiated by upwelling of hot rock from below. The effect of this activity is upwarping of the crust in the region directly above this active zone. The crustal stretching associated with the upwarping generates numerous cracks as shown in Figure 6.6A. As the hot rock spreads laterally away from the region of upwelling, the broken lithosphere is pulled apart. Many of the broken slabs slide downward into the gaps created by the separating plates (Figure 6.6B). During this phase of the continental breakup, volcanic activity will be prevalent.

The rather large down-faulted valleys that are created by the process just described are called **rifts,** or **rift valleys.** The Great Rift Valley of East Africa is such a feature (Figure 6.7). What we see in East Africa appears to be the initial stage in the breakup of that continent. If the process should continue, in a short time, geologically speaking, this region of East Africa will part from the mainland in much the same way as did the Arabian Peninsula just a few million years ago.

As the spreading process continues, the rift valley will lengthen and deepen, eventually extending out into the ocean. At this point the valley will have become a linear sea with an outlet into the ocean (see chapter-opening photo). The region of initial splitting will be the site of volcanic activity which continually fills the cracks formed in the sea floor, thereby creating a larger sea. This active zone is also the site where we would expect the development of a new oceanic ridge.

The location of active spreading centers, the mid-oceanic ridges, are characterized by an elevated position and numerous volcanic structures which have grown upon the newly formed crust. In a few places such as Iceland, a part of the Mid-Atlantic Ridge, these spreading centers have actually grown above sea level. The volcanic island of Surtsey, which first emerged from the ocean depths in 1963, is such a feature (Figure 7.19). Although volcanic activity does contribute to the height of a ridge, the high temperature of the intruding magma and the slow rate at which it cools is the primary reason for its elevated position. As the newly formed lithosphere travels away from the spreading center it gradually cools and contracts. This effect accounts for the increase in ocean depth that exists away from the ridge. It takes almost 100 million years before cooling and contraction ceases completely and the crust reaches a normal level.

Convergent Boundaries

At oceanic ridges new lithosphere is being generated; however since the total surface area of the earth remains constant, lithosphere must also be destroyed. The *zone of plate convergence* is the site of destruction. When a sec-

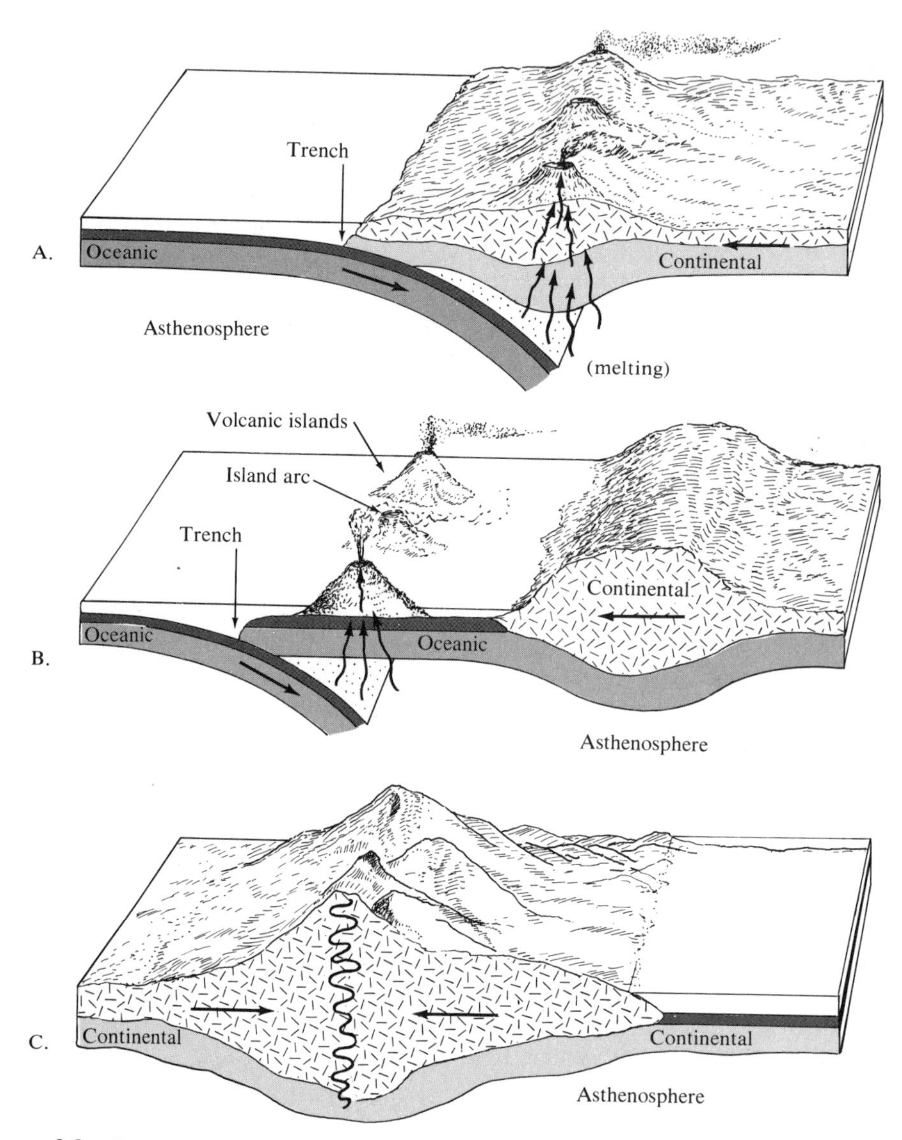

Figure 6.8 *Zones of plate convergence.* **A.** *Oceanic-continental.* **B.** *Oceanic-oceanic.* **C.** *Continental-continental.*

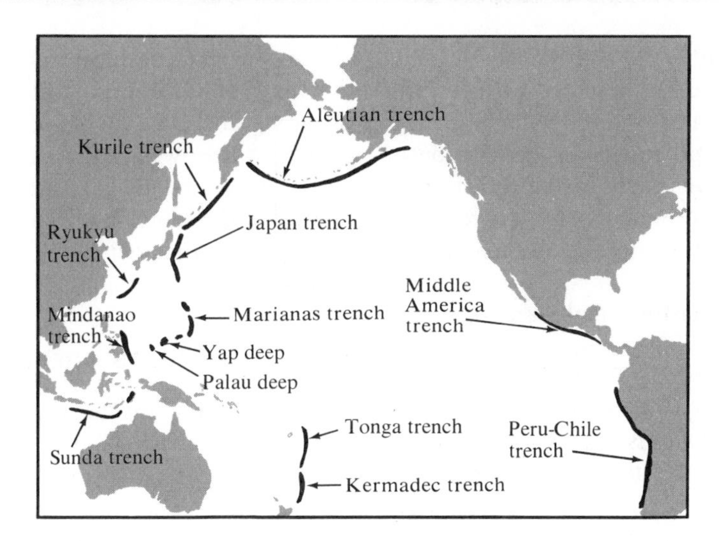

Figure 6.9 *Distribution of the world's major oceanic trenches. (From Foster,* Physical Geology, *3rd ed., Columbus, Ohio: Charles E. Merrill, 1979)*

tion of continental crust converges with oceanic crust, the less dense continental material remains "floating" while the more dense oceanic crust sinks into the asthenosphere (Figure 6.8A). Upon entering this hot region the solid plate and any sediments carried with it begin to melt. The result is the production and upward movement of molten magma, which upon reaching the surface gives rise to volcanic eruptions (Figure 6.8A). The volcanic Andes Mountains of South America are believed to have been produced by such activity when the Nazca plate melted as it plunged beneath the continent of South America (Figure 6.3). The numerous earthquakes generated within the Andes testify to the activity beyond our view.

The region where an oceanic plate is sliding into the asthenosphere because of convergence is called a **subduction zone.** As the oceanic plate is thrust beneath the overriding plate, it bends, thereby producing a linear **oceanic trench** adjacent to the zone of subduction (Figure 6.8A). Some trenches, like the Japanese and the Aleutian, are hundreds of kilometers long and 8–10 kilometers deep (Figure 6.9).

When two oceanic plates converge, and one is subducted, volcanic activity is also initiated. However, in this case, the volcanoes form on the ocean floor (Figure 6.8B). If the activity is sustained, new land will emerge from the ocean depths. Japan typifies such a volcanic system, which is called an **island arc.** Island arcs are located adjacent to an ocean trench where active subduction and melting of the lithosphere are taking place. In the early stage of its development, an island arc is a linear belt of numerous small volcanic islands. The Aleutian, Mariana, and Tonga islands are such features.

If two plates carrying continental crust move together, they collide rather than slide one beneath the other (Figure 6.8C). This occurs because of the lighter composition, and thus greater buoyancy, of continental crust. Such a collision occurred when the once-separate continent of India "rammed" into Asia, producing one of the most spectacular mountain ranges on earth, the Himalayas (Figure 6.10). During the collision the continental crust buckled, fractured, and generally shortened. The collision was accompanied by volcanic activity which "welded" together the once-separate landmasses.

Transform Fault Boundaries

Some boundaries, represented by transform faults, are located where plates slip past each other without production or destruction of the crust. These faults form in the direction of plate movement and were first discovered in association with offsets in the oceanic ridges (Figure 6.11).

Although most transform faults are located in oceanic crust, a few are situated within continents. The San Andreas fault of California is such a phenomenon (Figure 6.12). Along this fault the Pacific plate is moving toward the northwest, past the North American plate. If this movement continues for millions of years, that part of California west of the fault zone, including the Baja Peninsula, will become an island off the west coast of the United States and Canada, and could eventually reach Alaska. However, the more immediate concerns are the earthquakes triggered by movements along this fault system.

Plate Tectonics and Earthquakes

In the preceding chapter we noted the close association of plate boundaries and earthquakes. In trench regions, where subduction zones ex-

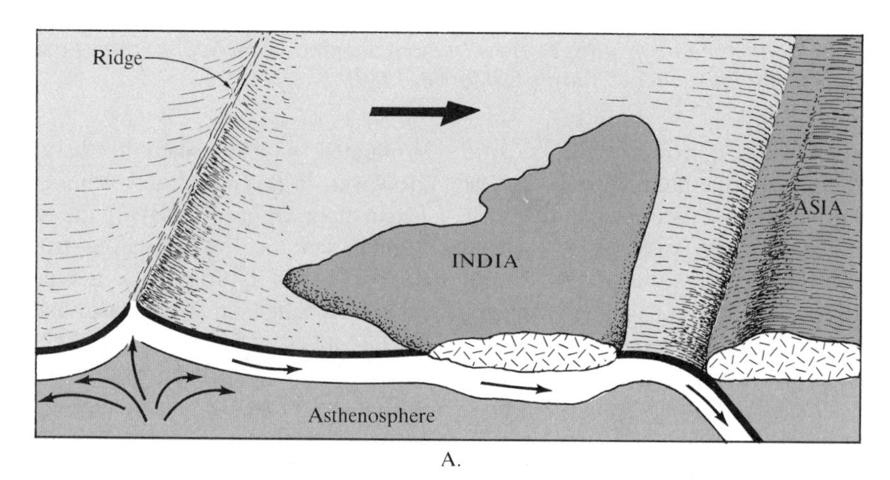

A.

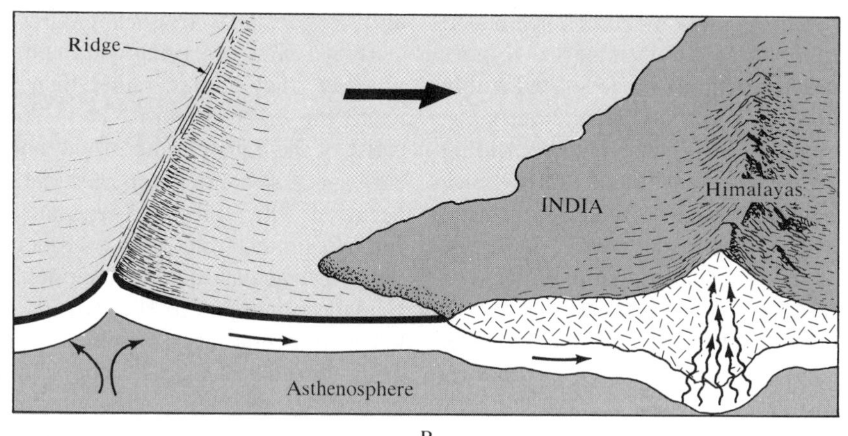

B.

Figure 6.10 *India "rammed" Asia about 45 million years ago, producing the majestic Himalayas.*

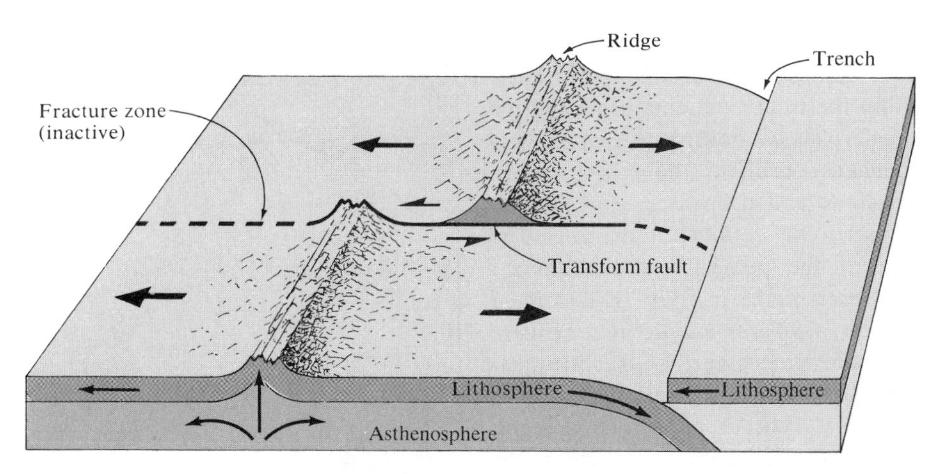

Figure 6.11 *Transform faults often separate offset ridges. Note that the lithosphere is moving in opposite directions in the region between the two ridges, while elsewhere along the fracture zone both plates move in the same direction.*

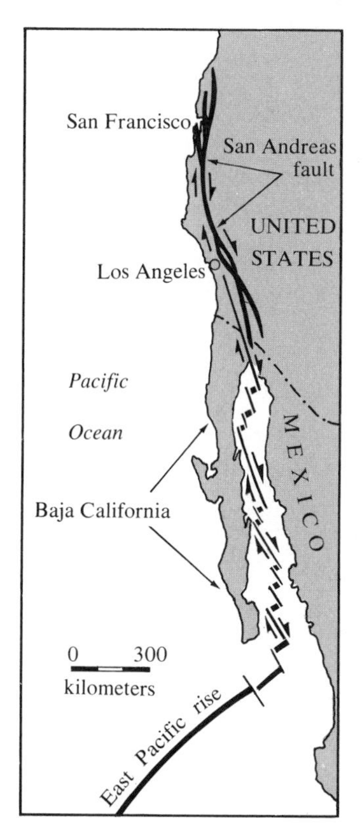

Figure 6.12 *Along the San Andreas fault the Pacific plate is moving northwestward in relation to the American plate. (Drawing after W.A. Elders et al. "Crustal Spreading in Southern California."* Science *178: 15-24. Copyright © 1972 by the American Association for the Advancement of Science; photo by R.E. Wallace, U.S. Geological Survey)*

ist, this association is most spectacular. Studies comparing the depths of earthquake foci to their location within the trench systems produced the interesting pattern shown in Figure 6.13. Shallow focus earthquakes occur in the trench areas, while intermediate and deep focus earthquakes take place closer to the continental margins. The interpretation of the pattern is quite simple: *The depth of earthquake activity is associated with the position of the descending plate* (Figure 6.14). Very shallow focus earthquakes are generated at the outer edge of the oceanic trenches where the plate is bent to permit its descent. Earthquakes like these, which occur within the oceanic crust, can generate the destructive sea waves called tsunami. Shallow focus earthquakes are also produced by friction as the descending plate scrapes beneath the overriding slab. Because of the rigid nature and close proximity to the surface of the material involved, shallow focus earthquakes trigger the greatest destruction, as exemplified by the 1964 Alaskan quake.

As the slab continues its descent into the asthenosphere, deeper focus earthquakes are produced. These earthquakes appear to be formed within the descending plate and are caused by resistance to its downward movement. Since the

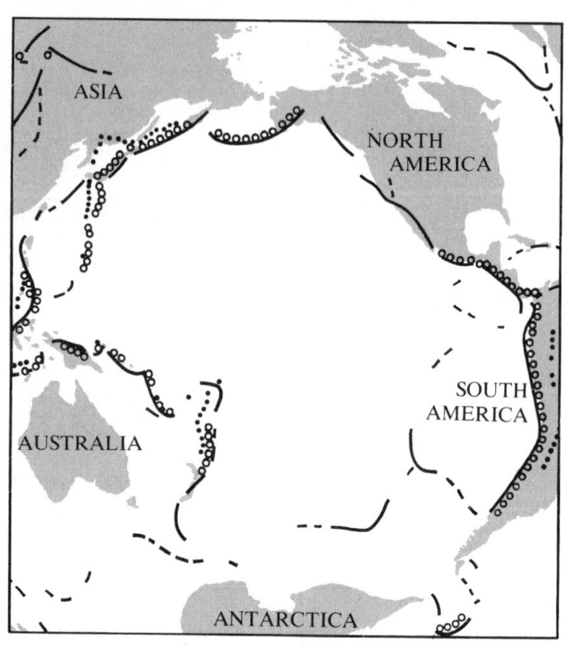

Figure 6.13 *Distribution of shallow, intermediate, and deep focus earthquakes. (After B. Gutenberg and C.F. Richter.* Seismicity of the Earth and Associated Phenomena. *Copyright 1949, 1954 by Princeton University Press, Figure 33, p. 90. Reprinted by permission of Princeton University Press)*

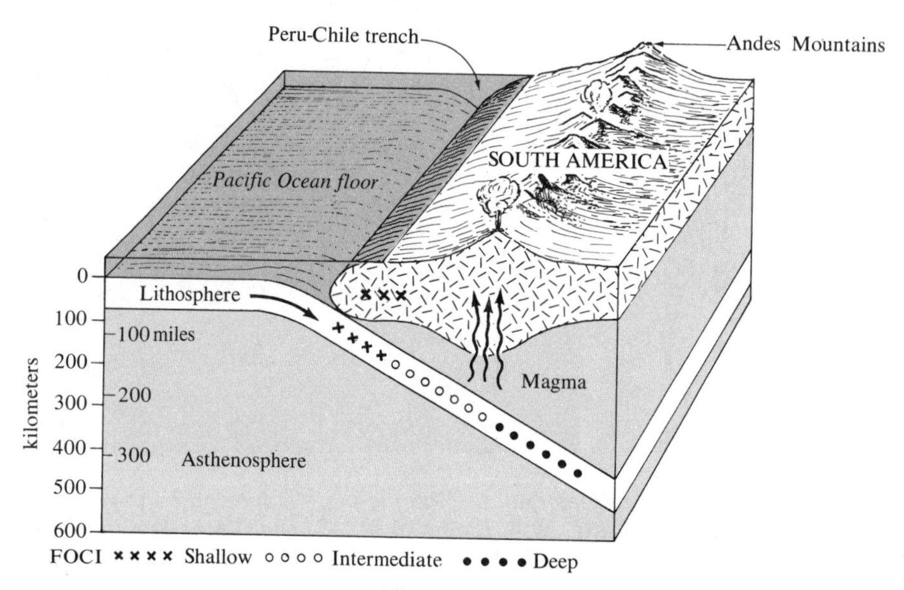

Figure 6.14 *Relationship between the descending plate and depth of earthquake foci.*

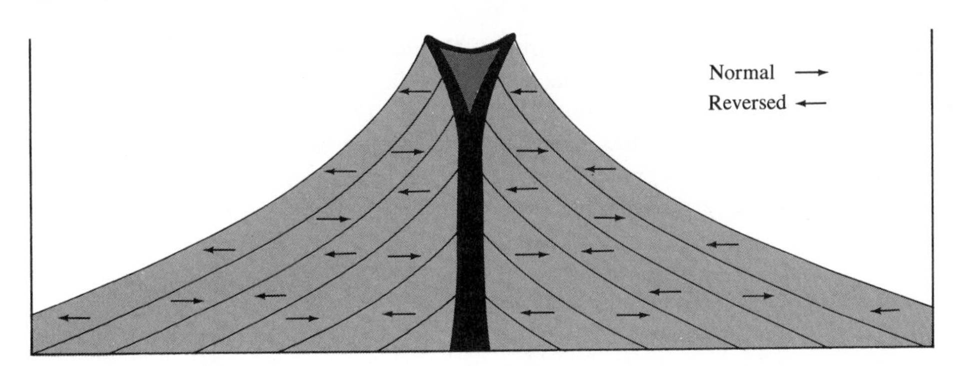

Figure 6.15 *Idealized diagram illustrating how lava flows record reversals in the earth's magnetic field.*

earthquakes occur within the rigid lithosphere rather than within the "plastic" asthenosphere, they provide a method for tracking the plate's descent into the mantle (Figure 6.14). Extremely few earthquakes have been recorded below 700 kilometers (450 miles), possibly because the lithosphere has been completely incorporated into the mantle by the time it reaches that depth.

Deep focus earthquakes take place at depths below 300 kilometers (200 miles) and are associated exclusively with subduction zones, while the earthquakes which occur along divergent zones and transform faults have shallow foci. How does this fact fit the model of plate tectonics? Recall that earthquakes result from the rapid release of strain which can only be built up in the rigid material of the lithosphere, not in the "plastic" asthenosphere. According to the plate tectonic theory, the subduction zones are the only regions where rigid material is forced to great depths, and therefore should be the only sites of deep focus earthquake activity. *The absence of deep focus earthquakes along divergent and transform fault boundaries adds much credibility to the plate tectonic theory.*

Plate Tectonics and Fossil Magnetism

Much of our knowledge of sea-floor spreading and continental drift comes from the study of the earth's magnetic field. Anyone who has used a compass to find direction knows that the earth's magnetic field has a north and a south pole which align closely, but not exactly, with the respective geographic poles. The compass needle, which is simply a magnet free to move about, will line up with the earth's magnetic field and point toward the magnetic poles. Some minerals found in rock are magnetic; hence they behave much like a compass needle. Above a certain temperature called the *Curie point* these magnetic materials lose their magnetism. However, when these magnetic materials again cool below their Curie point (about 500° C) they become magnetized in the direction of the earth's magnetic field. Once solidified, these minerals will remain "frozen" in this position. Should the rock be moved or the magnetic poles change position, the minerals will retain their original alignment. Rocks which were solidified millions of years ago and "remember" the location of the poles at the time of their formation are said to possess **fossil,** or **remnant, magnetism.**

A study of fossil magnetism in a region where several volcanic eruptions had occurred on widely separate occasions led to an interesting discovery. The orientation of minerals in each of the separate lava flows was found to be different. This indicated that between volcanic eruptions the magnetic poles had moved to a new location. Consequently, the concept of **polar wandering** was born. Our knowledge of the magnetic field indicates that its poles must stay relatively near the geographic poles because the rotation of the earth aids in generating them. If the geographic poles do not wander, which we believe is true, neither can the magnetic poles. A more acceptable explanation for polar wandering is provided by the theory of plate tectonics. If the magnetic poles remained stationary, their apparent movement could be produced by moving the continents. Using fossil magnetism, scientists are working to determine the paths taken by the drifting continents.

Another milestone in the field of magnetism came when scientists learned that the earth's magnetic field periodically reverses polarity;

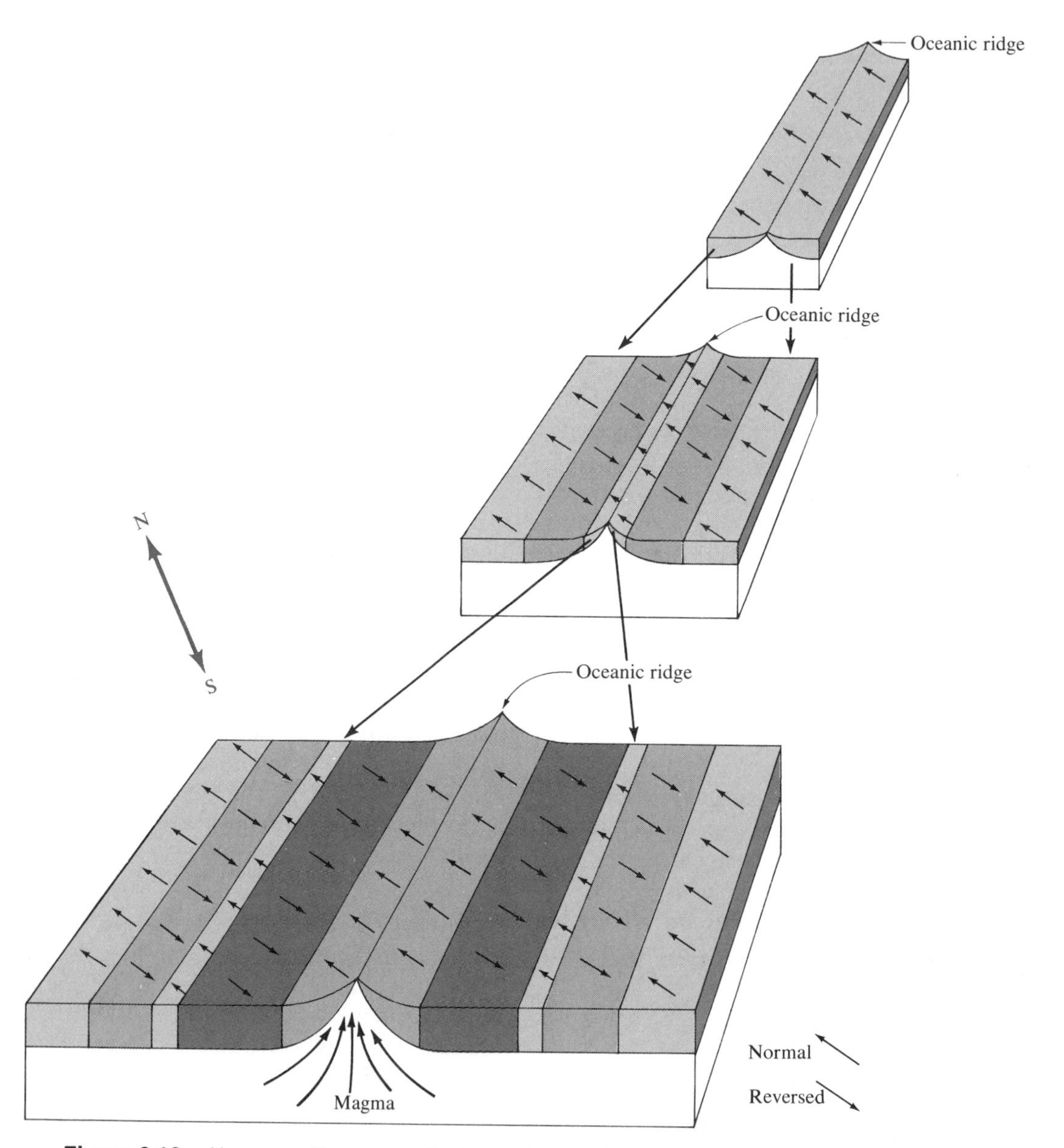

Figure 6.16 *New sea floor records the polarity of the magnetic field at that time. Hence it behaves much like a tape recorder, as it continually keeps count of changes in the earth's magnetic field.*

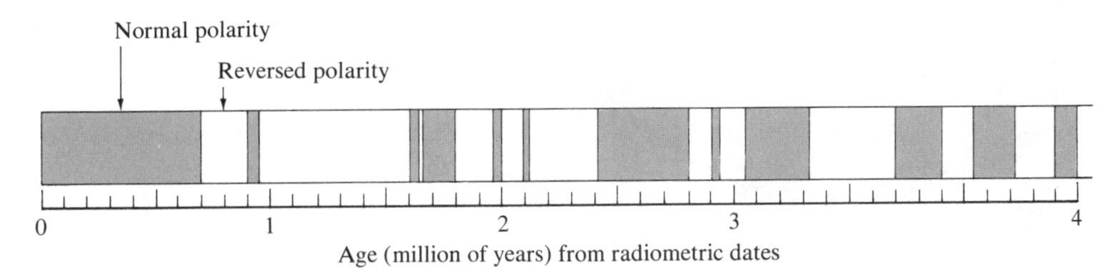

Figure 6.17 *Time scale of main magnetic reversals in the more recent past. (Data from A. Cox; drawing from Foster,* Physical Geology, *3rd ed., Columbus, Ohio: Charles E. Merrill, 1979)*

that is, the north pole becomes the south pole. Someone holding a compass during such a reversal would see the north end of the needle spin toward the south pole. Fossil magnetism, which indicates the direction of the poles, also tells whether the existing field has *normal* polarity like today's field or the opposite, *reverse* polarity (Figure 6.15). Magnetic surveys taken across the oceanic ridges disclosed symmetrically distributed strips of normally and reversed magnetized rocks (Figure 6.16). To explain this unexpected discovery, Professor Hess of Princeton University proposed the theory of sea-floor spreading. At the oceanic ridges, and in a mostly continuous manner, new sea floor is being produced, and in so doing, recording the reversals in the magnetic field. The time lapse between the reversals ranges from thousands to hundreds of thousands of years (Figure 6.17). By determining the amount of new sea floor produced during a magnetic reversal of known length, the rate of spreading can be calculated. Rates of a few centimeters per year are most typical of the divergent oceanic ridges.

Testing the Model

From every area of geology new data are being compiled and used to test the concept of plate tectonics. The model just presented will surely undergo changes; however, the basic premises appear to be sound and able to withstand the test of time. Although most geologists have accepted this theory energetically, there remains an ever-diminishing number who reject it in part or in total. Some of the evidence supporting plate tectonics and sea-floor spreading has already been presented in this chapter. In addition, the following testimonies are submitted.

The existence of numerous submerged volcanoes, or **seamounts,** on the floor of the ocean was discovered after World War II. Some of these volcanic structures are located hundreds of meters below sea level and have flat tops which in-

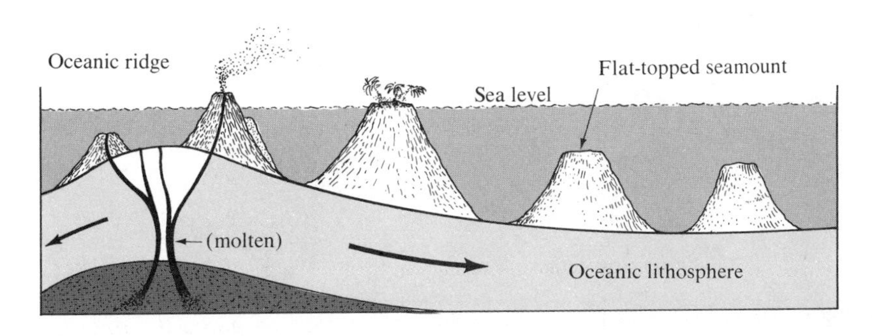

Figure 6.18 *Formation of flat-topped seamounts.*

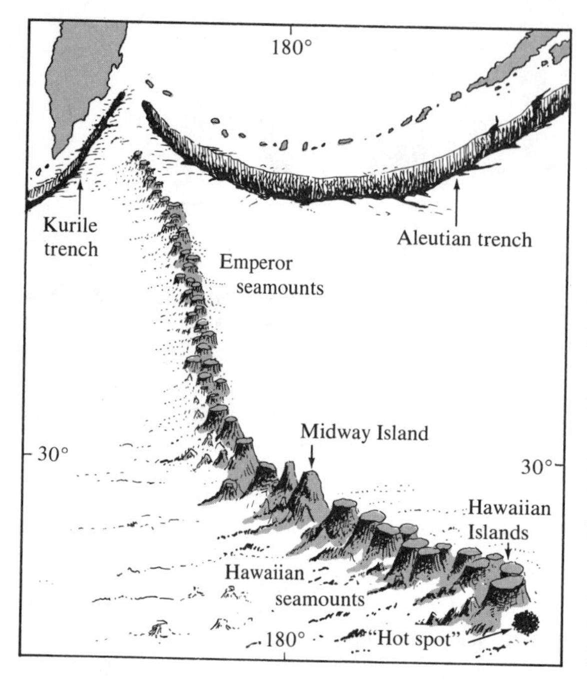

Figure 6.19 *Chain of islands and seamounts that extends from Hawaii to the Aleutian Trench.*

dicate erosion by wave action. Oceanographers were, at the time of discovery, puzzled by this phenomenon. How can seamounts acquire flat tops if they were always below sea level? Now, using the plate tectonic theory, we realize that many of these volcanoes were produced near the oceanic ridges. The larger ones emerged from the sea, where their tops were exposed to erosion. Because of sea-floor spreading, many of these have since been submerged as they were carried away from the ridge area (Figure 6.18).

Mapping of the seamounts also brought into focus the chain of volcanic structures which extends between the Hawaiian Islands and the Midway Islands and then continues northward toward the Aleutian Trench (Figure 6.19). Radiometric dating of these seamounts revealed that those closest to the trench were formed nearly 70 million years ago, while the last structure in this long chain, the island of Hawaii, rose from the sea less than 1 million years ago (Figure 6.20). Researchers have proposed that a "hot spot" exists within the mantle and emits magma onto the overlying sea floor. Presumably, as the Pacific plate moved over the "hot spot," successive volcanic structures emerged. The age of each volcano indicates the time when it was situated over the "hot spot." Kauai is the oldest island in the Hawaiian chain. Five million years ago, when it was positioned over the "hot spot," it was the only Hawaiian island in existence (Figure 6.20).

Aboard the *Glomar Challenger,* an oceanographic research vessel, deep-sea sampling in the

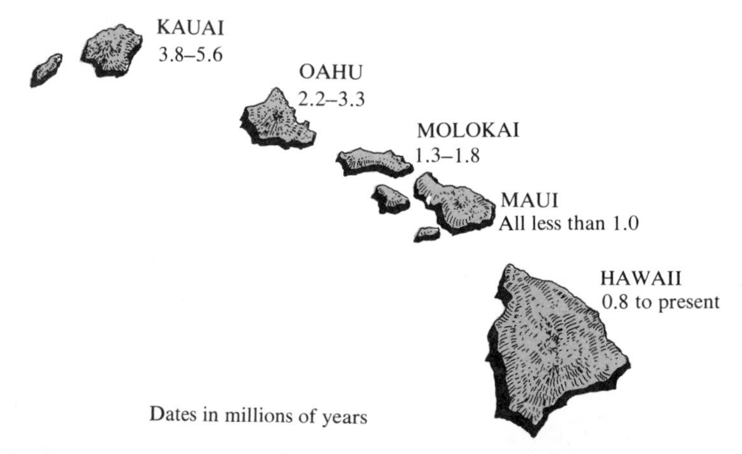

Figure 6.20 *Radiometric dating of the Hawaiian Islands revealed the decreasing age of the volcanic activity toward Hawaii. (Data from I. McDougall)*

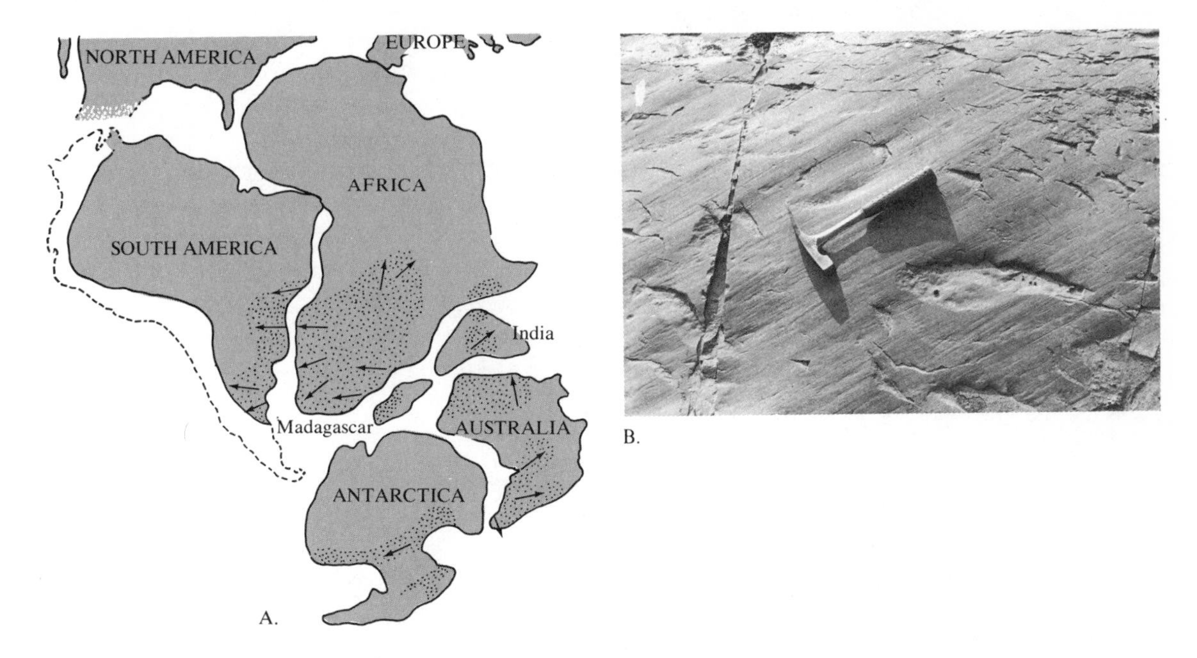

Figure 6.21 **A.** *Direction of ice movement in the southern supercontinent which was called Gondwanaland by the founders of the continental drift concept.* **B.** *Glacial striation in the bedrock of Hallet Cove, South Australia, indicates direction of ice movement. (Photo by W.B. Hamilton, U.S. Geological Survey)*

early 1970s clearly supported the concept of sea-floor spreading. As we might expect, the youngest sea-floor samples were obtained from the mid-oceanic ridge areas, while progressively older samples were acquired at increasingly greater distances from the ridge. To date, no oceanic crust with an age in excess of 200 million years has been located. By comparison, some continental crust has been dated at more than 3.7 billion years. It is unlikely that oceanic crust older than 200 million years will ever be found in place, since it is continually being recycled into the mantle. The lack of any appreciable amount of sediments mantling the ridge areas also testifies to their recent formation.

We do not want to give the impression that all of the support for the continental drift concept came from space-age technology. On the contrary, Wegener and others who took up the search found some very persuasive evidence. They discovered the existence of a peculiar fossil flora, genus name *Glossopteris,* in India, Africa, Australia, and South America. Much later *Glossopteris* was also discovered in Antarctica. How could this fossil flora be so similar in places separated by thousands of kilometers of open ocean? In these same regions, and in sequence with the strata containing the flora, glacial deposits exist. Some of the bedrock contains glacial grooves and striations which clearly indicate that the ice came from the sea (Figure 6.21). How do glaciers move from water to land? How does a glacier develop in the hot, arid regions of Australia? As compelling as their evidence may have been, it was 50 years before most accepted it and the logical conclusions to which it led.

Pangaea: Before and After

Although some of the information is incomplete, some of the gross details of the migrations of individual continents over the past 500 million

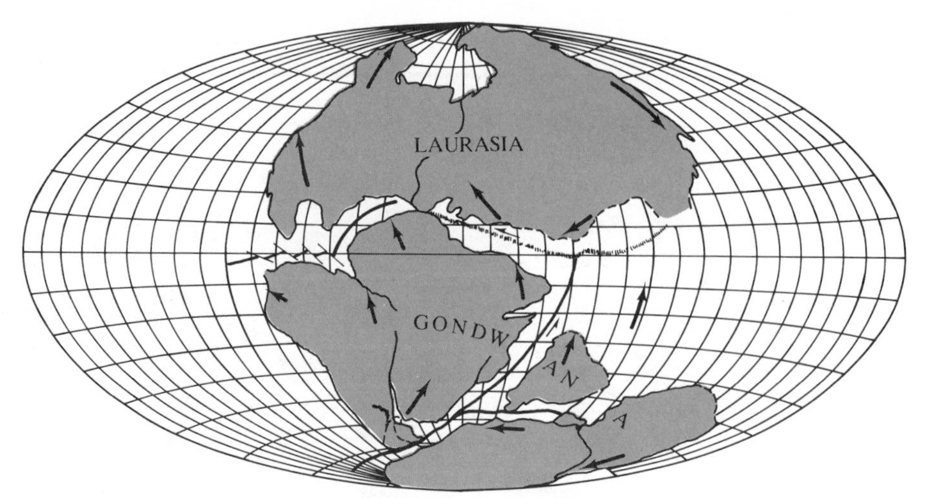

A. 180 million years ago (Triassic period).

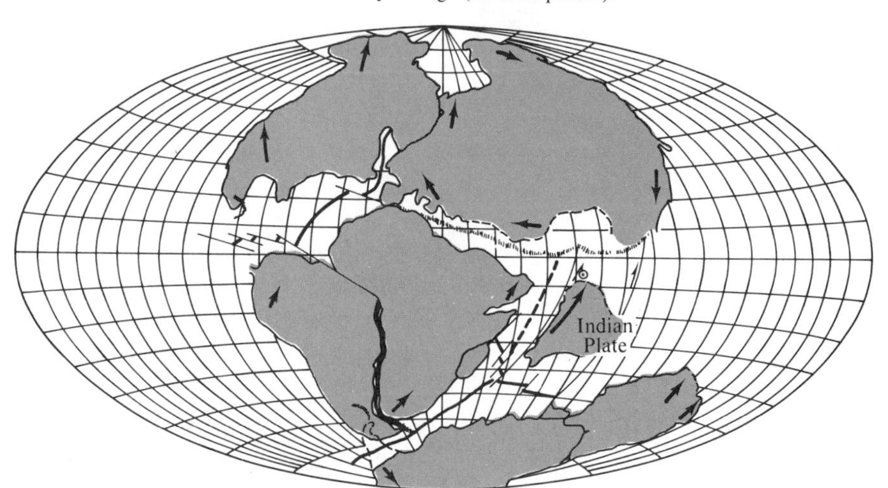

B. 135 million years ago (Jurassic period).

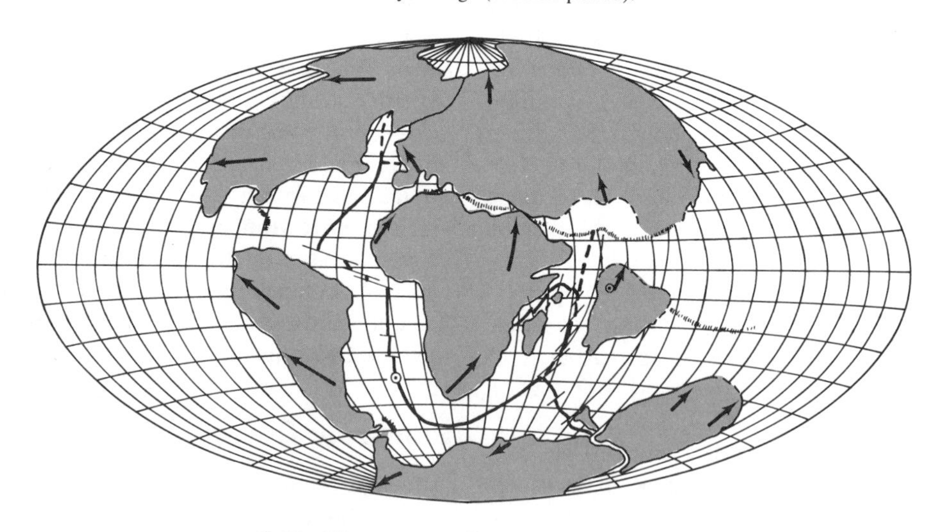

C. 65 million years ago (Cretaceous period).

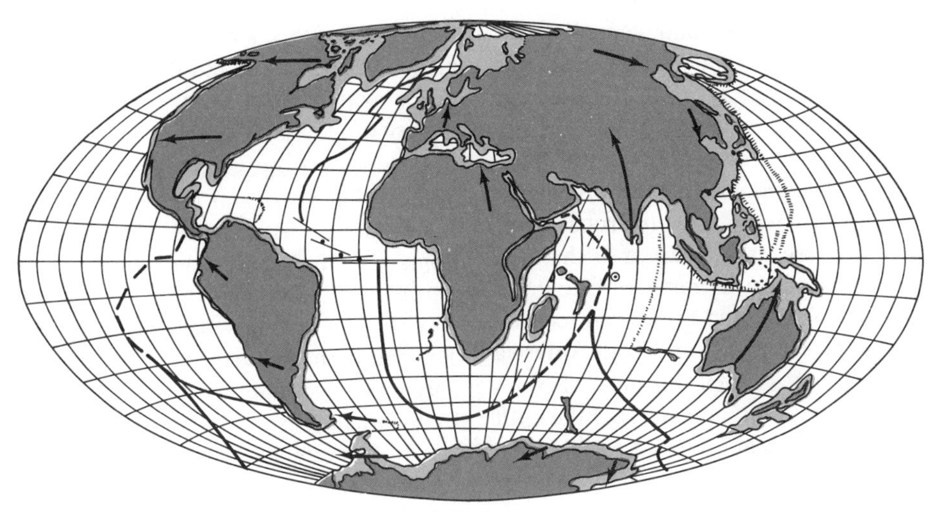

D. Present.

Figure 6.22 *Several views of the breakup of Pangaea over a period of 200 million years according to Dietz and Holden. (Robert S. Dietz and John C. Holden,* Journal of Geophysical Research *75: 4939–56, 1970. Copyright by American Geophysical Union)*

years has been roughly projected by Robert Dietz, John Holden, and others. By extrapolating plate motion back in time using such evidence as the orientation of volcanic chains, the distribution and movements along transform faults, and paleomagnetism, Dietz and Holden were able to reconstruct Pangaea as shown in Figure 6.1. Further, the use of radioactive dating helped to organize the time frame for the formation and eventual breakup of Pangaea, and the relatively stationary position of "hot spots" through time helped to fix the geographic locations of the continents.

Breakup of Pangaea

The initial fragmentation of Pangaea began about 200 million years ago. Figures 6.22A–D illustrate the breakup and subsequent paths taken by the landmasses involved. As we can readily see in Figure 6.22A, two major rifts accompanied the initial breakup. The rift zone created between North America and Africa generated numerous outpourings of Triassic age basalts which are presently visible along the eastern seaboard of the United States. Radiometric dating of these basalts indicates that the split occurred nearly 200 million years ago; we can use this date as the birth date of the North Atlantic. The rift in southern Pangaea (Gondwanaland) developed a "Y" shape which sent India on its northward journey and simultaneously separated South America-Africa from Australia-Antarctica.

Figure 6.22B illustrates the position of the continents about 135 million years ago, about the time Africa and South America began drifting apart. India can be seen halfway into its journey to Asia, and the North Atlantic has widened considerably. By the end of the Cretaceous period 65 million years ago, Madagascar had separated from Africa and the South Atlantic had opened (Figure 6.22C). At this juncture India had drifted over a "hot spot" which generated numerous fluid basalt flows across India in a region now called the Deccan Plateau. These lava flows are very similar to those that make up the Columbia Plateau of our Northwest (Figure 7.13).

The current map (Figure 6.22D) shows India in contact with Asia, an event that occurred about 40 million years ago and generated the highest mountains on earth along with the Tibetan Highlands. It is interesting that the average height of Tibet is 5000 meters, higher than any spot in the forty-eight contiguous states. It is thought that India's continued northward migration is the source of energy for numerous earthquakes which have recently occurred in the area of collision and as far away as northern China. By comparing Figures 6.22C and D, we can see that the separation of Greenland from North America was a recent event in geologic history. Also notice the recent separation that generated the Gulf of California, an event that occurred less than 10 million years ago.

Before Pangaea

Prior to the formation of Pangaea, the landmasses had probably gone through several episodes of fragmentation similar to what we see happening today. Also like today, these ancient continents moved away from each other only to collide again on the opposite side of the earth. During the period between 500 and 225 million years ago the fragments of an earlier dispersal began to collect to form the continent of Pangaea. Remnants of this activity are the Ural Mountains of the Soviet Union and the Appalachian Mountains, which flank the east coast of North America.

It appears from available evidence that about 500 million years ago the northern continent of Laurasia was fragmented into three major sections—North America, northern Europe (southern Europe was part of Africa), and Siberia —with each section separated by a sizable ocean. The southern continent of Gondwanaland probably was intact and lay near the south pole. The first collision is believed to have occurred as North America and Europe closed the pre-North Atlantic ocean. This activity resulted in the formation of the northern Appalachians. Parts of the floor of the former ocean can be seen high above sea level in Nova Scotia. The sliver of eastern Canada seaward of the zone of this collision is truly a gift from Europe. It is also thought that before North America and Europe collided, part of Scotland and Ireland were attached to the North American plate. During the joining of North America and Europe, Siberia was closing the gap between itself and Europe, which lay farther to the west. This closing cul-

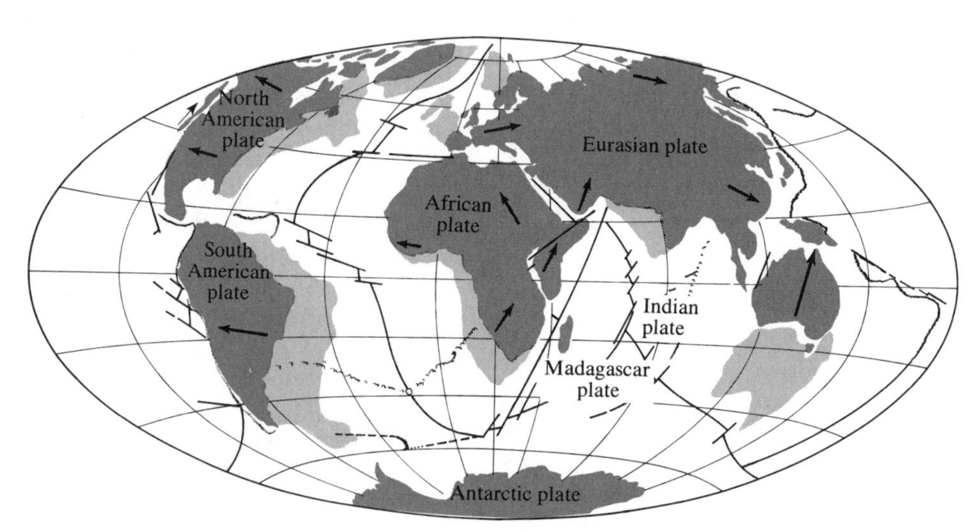

Figure 6.23 *The world as it may look 50 million years from now. (From "The Breakup of Pangaea," Robert S. Dietz and John C. Holden. Copyright © 1970 by Scientific American, Inc. All rights reserved)*

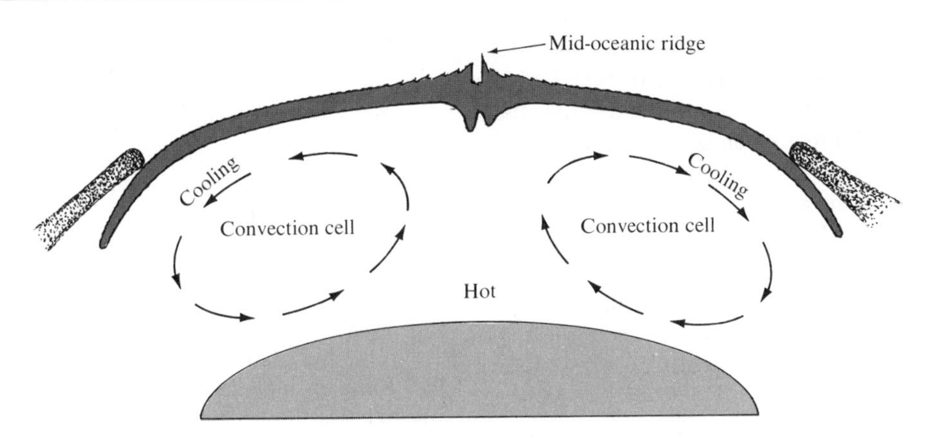

Figure 6.24 *Convection currents within the earth, thought to be a possible mechanism responsible for plate motion.*

minated about 300–350 million years ago in the formation of the Ural Mountains. The consolidation of these landmasses completed the northern continent of Laurasia.

During the next 50 million years or so the northern and southern landmasses converged, producing the supercontinent of Pangaea. It was at this time (about 250–300 million years ago) that Africa and North America collided to produce the southern Appalachians.

A Look into the Future

After Dietz and Holden drew together the events that over the past 225 million years resulted in the current configuration of the continents, they went one step further and extrapolated plate motion into the future. Figure 6.23 illustrates what they envision the earth's landmasses will look like 50 million years from now. Important changes are seen in Africa. A new sea emerges where East Africa parts company with the mainland in a manner similar to the formation of the Red Sea during the separation of the Arabian Peninsula. In North America we see that the Baja Peninsula and the portion of Southern California that lies west of the San Andreas fault has slid past the North American plate. Should this northward migration take place as predicted, Los Angeles and San Francisco will pass by each other.

In another part of the world Africa has moved slowly toward Europe, initiating perhaps the next major mountain-building stage on our dynamic planet. Australia is seen on a collison course with New Guinea and Asia, and North and South America are once again separating. These projections into the future, although interesting, must be viewed with caution since many assumptions must be correct for these events to unfold as just described. Nevertheless, similar types of changes in the shapes and positions of the continents will undoubtedly occur for many years to come.

The Driving Mechanism

Plate tectonics describes plate motion and the effects of this motion rather than its cause. Consequently, its acceptance does not rely on a knowledge of the driving force. This is fortunate, since none of the mechanisms yet proposed can account for all of the major facets of plate motion. Nevertheless, it is generally believed that the unequal distribution of heat deep within the earth is the underlying cause of this movement. This in turn is thought by many geologists to generate convection cells within the mantle (Figure 6.24). The warmer, less dense material of the lower mantle rises very slowly in the regions of the oceanic ridges. As it spreads laterally it cools, becomes more dense, and begins to sink back into the mantle, where it is reheated. It should be noted that rocks need not be molten to flow. Just as a hot, yet solid metal can be pounded into different shapes, so too can rock move when it is subjected to heat and extreme pressure. Heat measurements indicate a higher rate of heat flow at the oceanic ridges than at the trenches, a good indication that convection cells do exist. However, many of the details of their motions remain unclear. How many cells exist? At what depth do they originate?

Although unequal distribution of heat is generally accepted as the underlying driving force of plates, many geologists are not convinced that convection currents exist. Thus many other mechanisms which may play an important role in plate motion have been suggested. It is now accepted that the oceanic lithosphere has a greater density than the asthenosphere which supports it from below. This being the case, it has been proposed that once a plate begins to descend (in the vicinity of a trench), the heavy sinking slab might pull the trailing lithosphere along. This slab pull theory, as it is known, is similar to another model which suggests that the weight of the lithosphere sliding down an ocean ridge could push the plates along. Another model proposes that rather narrow, hot plumes of rock could move up from below the asthenosphere. Upon reaching the lithosphere these plumes would spread laterally and carry the plates along. These hot plumes ("hot spots") reveal themselves on the surface as volcanoes in places such as Hawaii and Iceland. We must conclude that although many mechanisms have been proposed, no clear answer to the question of what drives the plates exists at this time.

REVIEW

1. What was probably the first bit of evidence which led scientists to believe that the continents were once connected?
2. What was Pangaea?
3. On what basis were plate boundaries first established?
4. Where and how is new lithosphere being formed? Destroyed? Why must the production and destruction of the lithosphere be going on at about the same rate?
5. Why is the oceanic portion of a lithospheric plate destroyed while the continental portion is not?
6. Relate the formation of the Andes Mountains to the movement of plates.
7. Discuss the formation of an island arc.
8. In what ways may the origin of the Japanese Islands be considered similar to the formation of the Andes Mountains? How do they differ?
9. Differentiate between transform faults and the other two types of plate boundaries.
10. Some people predict that California will sink into the ocean. Is this idea consistent with the concept of plate tectonics?
11. Describe the distribution of earthquake epicenters and foci depths as they relate to oceanic trench systems.
12. Explain why flat-topped seamounts are considered evidence for sea-floor spreading.
13. If the "hot spot" concept proves correct, what direction was the Pacific plate moving while the Emperor seamounts were being produced (see Figure 6.19)? While the Hawaiian seamounts were being produced?
14. With what type of plate boundary are the following places or features associated (be as specific as possible): Himalayas, Aleutian Islands, Red Sea, Andes Mountains, San Andreas fault, Iceland, Japan?

KEY TERMS

continental drift
Pangaea
plate tectonics
divergent boundaries
convergent boundaries
transform fault boundaries
sea-floor spreading
rift (rift valley)
subduction zone
oceanic trench
island arc
fossil (remnant) magnetism
polar wandering
seamounts

The volcano Parícutin a few months after its inception. (Photo by Tad Nichols)

Volcanic Eruptions
Volcanoes
Fissure Eruptions
Intrusive Igneous Activity
Igneous Activity and Plate Tectonics

7

Igneous Activity

The significance of igneous activity may not be obvious at first glance. However since volcanoes extrude molten rock which formed at great depth, they provide the only windows we have for direct observation of the processes which occur many kilometers below the earth's surface. Furthermore, the gasses emitted during volcanic eruptions are thought to be the material from which the atmosphere and oceans evolved. Either of these facts is reason enough for igneous activity to warrant our attention. All volcanic eruptions are spectacular, but why are some destructive and others quiescent? Is the island of Hawaii a volcano as high as Mt. Everest resting upon the ocean floor? Do the large volcanoes of Washington and Oregon present a threat to human lives? This chapter considers these and other questions as we explore the formation and movement of magma.

In 1943, about 200 miles west of Mexico City, the volcano Parícutin was born (see chapter-opening photo). The site of the eruption was a cornfield owned by Dionisio Pulido, who with his wife, Paula, witnessed the event as they were preparing the field for planting. For two weeks prior to the first eruption, numerous earth tremors caused apprehension in the village of Parícutin about 3½ kilometers away. Then around 4:00 P.M. February 20 smoke with a sulfurous odor rose from a small hole which had been in the cornfield for as long as Dionisio could remember. During the night hot, glowing rock fragments produced a spectacular fireworks display. The next day the cone grew to a height of 40 meters and by the fifth day it was over 100 meters high. By that time explosive eruptions were throwing hot fragments 1000 meters above the crater rim. The larger fragments fell near the crater, some remaining incandescent as they rolled down the slope. These fragments built an aesthetically pleasing cone, while finer ash fell over a larger area, burning and eventually covering the village of Parícutin. Within two years the cone had reached 400 meters and would rise only a few tens of meters more.

The first lava flow came from a fissure that had opened just north of the cone; but after a few months of activity, flows began to emerge from the base of the cone itself. In June of 1944 a clinkery flow 10 meters (30 feet) thick moved over the village of San Juan Parangaricutiro, leaving only the church steeple exposed (Figure 7.1). After nine years the activity ceased almost as quickly as it began. Now Parícutin is just another one of the numerous volcanoes which dot the landscape in this region of Mexico. Like the others, it will probably not erupt again.

Parícutin is the only volcano whose formation has been observed by geologists from beginning to end and is special for that reason alone. However, it should be noted that Parícutin represents only one example of the many diverse processes which collectively fall under the heading igneous activity. As we saw in Chapter 1, igneous activity is the movement of molten rock called magma inside as well as outside the crust. The former process is called intrusive, or plutonic, activity. Although we can never see intrusive activity, it is nonetheless important. Because most of what we know about igneous activity comes from the study of volcanic eruptions, we will begin our discussion there.

Figure 7.1 *The village of San Juan Parangaricutiro engulfed by lava from Parícutin, shown in the background. Only the church towers remain. (Photo by Tad Nichols)*

Volcanic Eruptions

Volcanic activity is generally thought to produce a picturesque, cone-shaped structure which periodically erupts in a violent manner. Although eruptions are sometimes violent, many are very quiescent. The factors which determine the nature of the eruptions include the magma's composition, its temperature, and the amount of dissolved gasses it contains. These first two factors, temperature and composition, primarily affect the magma's mobility, or **viscosity.** The more viscous a material, the greater its resistance to flow. For example, molasses is more viscous than water. The effect of temperatures on viscosity is easily visualized. Just as heating molasses makes it more fluid, hot magma produces fluid lava. So, as a lava flow cools and crystallizes, it loses its mobility and eventually comes to a halt.

As molten rock begins to crystallize in a magma chamber, the dissolved gasses tend to separate from the minerals and migrate upward. These gasses play a major role in volcanic eruptions, for they provide the force which propels the magma from the cone. Highly viscous magmas often plug the vent, trapping the gasses below. Under this condition there is a buildup of pressure until the gasses explosively eject the semimolten rock from the volcano. Very fluid magmas, on the other hand, allow the gasses to escape, producing rather quiet eruptions. *Thus the quantity of dissolved gasses in the magma, as well as the ease with which they can escape, determines to a large degree the nature of the eruption.*

The mineral composition of magmas was discussed in Chapter 1 when igneous rocks were classified. One of the major differences between the various igneous rocks and the magmas from which they originate is their silica (SiO_2) content. Magmas which produce mafic rocks like basalt contain about 50 percent silica, while rocks of felsic composition (granite and its extrusive equivalent, rhyolite) contain over 70 percent silica. The intermediate rock types, andesite and diorite, contain around 60 percent silica. Because the silica molecules link together, they impede the flow of magma. Consequently, the viscosity of magma is directly related to its silica content as well as its temperature. *The lower the percentage of silica in a magma, the greater the fluidity.*

Basaltic lavas, because of their low silica content, are usually very fluid. In the Hawaiian Is-

A.

B.

Figure 7.2 **A.** *Typical pahoehoe (ropy) lava flow, Kilauea, Hawaii. The upper flow is covering an earlier flow (smooth surface). (Photo by W. Cross, U.S. Geological Survey)* **B.** *Sketch of pahoehoe lava. (From Foster,* Physical Geology, *3rd ed., Columbus, Ohio: Charles E. Merrill, 1979)*

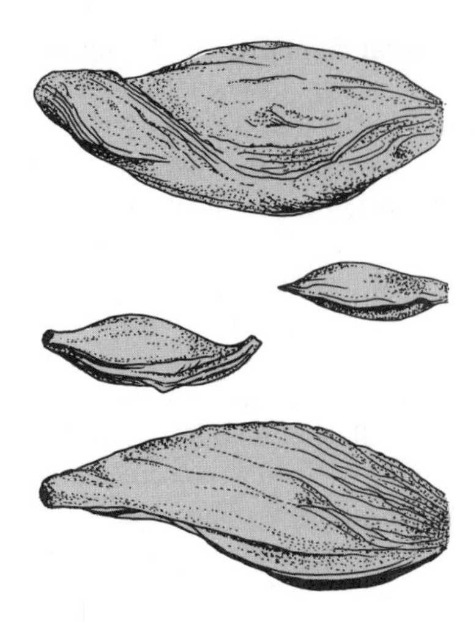

Figure 7.3 *Volcanic bombs. Ejected lava fragments take on a streamlined shape as they sail through the air. (From Foster,* Physical Geology, *3rd ed., Columbus, Ohio: Charles E. Merrill, 1979)*

lands such lavas have been clocked at speeds of tens of kilometers per hour on steep slopes. However, this rate is rare. In contrast, the movement of rhyolitic lava is often too slow to be perceived. When fluid basaltic lavas of the Hawaiian type cool and crystallize, they form a rather smooth skin on top which wrinkles as the still-molten subsurface lava continues to advance (Figure 7.2). Their ropy appearances led the natives to call them **pahoehoe** ("ropy") **flows.** Slower-moving lavas generally have a sharp, jagged surface and are called **aa** (pronounced "ah ah") **flows.** The name reflects what a barefooted Hawaiian would say while walking across it! The lava which flowed from Parícutin and buried San Juan Parangaricutiro was of the aa type (Figure 7.1).

Besides minerals, magma contains varied amounts of dissolved gasses which are held in the molten rock by confining pressure, just as carbon dioxide is held in soft drinks. As with soft drinks, as soon as the pressure in the magma is reduced, the gasses begin to escape. Because of this, it is difficult to determine the amount of gas originally contained in the molten rock.

It is believed that the gaseous portion of most magmas makes up 1–5 percent of the total volume. Although the percentage may be small, the actual quantity of gas emitted can exceed thousands of tons per day. The composition of the gasses is also of interest to scientists, since the atmosphere and the oceans were produced from them. Evidence from Hawaiian eruptions indicates that the gasses are composed of 70 percent water vapor, 15 percent carbon dioxide, 5 percent each of nitrogen and sulfur compounds, and lesser amounts of chlorine, hydrogen, and argon. Sulfur is the most easily recognized because of its pungent odor.

As fluid basaltic magma nears the surface the dissolved gasses are allowed to escape freely and continually. Hence lava flows are the rule. The tops of these flows often retain the holes

Figure 7.4 *Fiery clouds (nuées ardentes) race down the slope of Mt. Mayon, Philippines, 1968. (Courtesy of Smithsonian Center for Short-Lived Phenomena)*

left by the escaping gasses. Hardened lava containing these voids is called **scoria** (Figure 1.13). With viscous magmas, particularly those having a rhyolitic or andesitic composition, the gasses are less able to escape and may build up an internal pressure capable of producing a violent eruption. Upon release these superheated gasses expand a thousandfold as they blow pulverized rock and lava from the vent. The particles ejected by the escaping gasses are called **pyroclastics.** These fragments range in size from very fine *dust,* to sand-sized volcanic *ash,* and to the larger *volcanic bombs* and *blocks.* When the hot ash, usually glassy shards, falls, it sticks together to form a **welded tuff.** Vast sheets of this material cover parts of the western United States. Volcanic bombs are semimolten upon ejection and often take on a streamlined shape (Figure 7.3).

Volcanic debris can be scattered great distances from its source, particularly the fine dust, which may be blasted into the upper atmosphere, where it remains for months. There it produces brilliant sunsets and has on occasion slightly lowered the earth's average temperature.

The "year without a summer," 1815, was caused by one of the greatest eruptions on record when Tambora, a volcano in Indonesia, blew its top. The amount of material extruded in a brief moment was a hundred times greater than the quantity of debris extruded from Parícutin in nine years.

The most devastating eruptions occur when hot incandescent ash and gasses are ejected, producing a fiery cloud called a **nuée ardente** which moves at speeds up to 160 kilometers (100 miles) per hour down steep volcanic slopes (Figure 7.4). The expanding gasses buoy the solid particles, carrying them in an avalanche fashion.

In 1902 a nuée ardente from Mt. Pelée, a small volcano in the West Indies, destroyed the

Figure 7.5 *St. Pierre as it appeared shortly after the eruption of Mt. Pelée, 1902. (Reproduced from the collection of the Library of Congress)*

town of St. Pierre, killing its total population of 28,000 except for a prisoner protected in a dungeon, a shoemaker, and a few persons who were on ships in the harbor (Figure 7.5). Satis N. Coleman, in *Volcanoes, New and Old,* relates a vivid account of this event, which lasted less than five minutes:

> I saw St. Pierre destroyed. The city was blotted out by one great flash of fire. Nearly 40,000 people were killed at once. Of eighteen vessels lying in the roads, only one, the British steamship *Roddam* escaped and she, I hear, lost more than half of those on board. It was a dying crew that took her out. Our boat, the *Roraima,* arrived at St. Pierre early Thursday morning. For hours before entering the roadstead we could see flames and smoke rising from Mt. Pelée. No one on board had any idea of danger. Captain G.T. Muggah was on the bridge, all hands got on deck to see the show. The spectacle was magnificent. As we approached St. Pierre we could distinguish the rolling and leaping of red flames that belched from the mountain in huge volumes and gushed into the sky. Enormous clouds of black smoke hung over the volcano. There was a constant muffled roar. It was like the biggest oil refinery in the world burning up on the mountain top. There was a tremendous explosion about 7:45, soon after we got in. The mountain was blown to pieces. There was no warning. The side of the volcano was ripped out and there was hurled straight toward us a solid wall of flame. It sounded like a thousand cannons.
>
> The wave of fire was on us and over us like a flash of lightning. It was like a hurricane of fire. I saw it strike the cable steamship *Grappler* broadside on, and capsize her. From end to end she burst into flames and then sank. The fire rolled in mass straight down upon St. Pierre and the shipping. The town vanished before our eyes.
>
> The air grew stifling hot and we were in the thick of it. Wherever the mass of fire struck the sea, the water boiled and sent up vast columns of steam. The sea was torn into huge whirlpools that careened toward the open sea. One of these horrible, hot whirlpools swung under the *Roraima* and pulled her down on her beam end with the suction. She careened way over to port, and then the fire hurricane from the volcano smashed her, and over she went on the opposite side. The fire wave swept off the masts and smokestacks as if they were cut by a knife. . . . The blast of fire from the volcano lasted only a few minutes. It shrivelled and set fire to everything it touched. Thousands of casks of rum were stored in St. Pierre, and these were exploded by the terrific heat. The burning rum ran in streams down every street and out into the sea. This blazing rum set fire to the *Roraima* several times . . . Before the volcano burst, the landings of St. Pierre were covered with people. After the explosion, not one living soul was seen on land.*

Volcanoes

Successive eruptions from a single vent build a mountainous accumulation of material known as a **volcano.** Three types of volcanoes are generally recognized: shield volcanoes, cinder cones, and composite cones (stratovolcanoes). Each has a characteristic eruptive style and, consequently, a characteristic form.

When highly fluid lava is extruded, the volcano takes the shape of a broad, slightly domed structure called a **shield volcano** (Figure 7.6). Shield volcanoes are built primarily of basaltic lava flows and contain only a small percentage of pyroclastic material. They have a slope of a few degrees at their flanks and no more than 10 degrees near their summit, as exemplified by the volcanoes of the Hawaiian Islands. Mauna Loa, the largest volcano on earth, is one of five shield volcanoes that together make up the island of Hawaii (Figure 7.6). It rests on the ocean floor 5000 meters below sea level and its summit stands over 4000 meters above the water. Nearly one million years and numerous eruptive cycles were required to build this truly gigantic volcanic pile. Many other volcanic islands have been built in a similar manner from the ocean depths. Because of the fluid nature of this lava, Hawaiian-type eruptions are very quiescent.

As the name suggests, **cinder cones** are built of ejected lava fragments. Because unconsoli-

*New York: John Day, 1946, pp. 80–81.

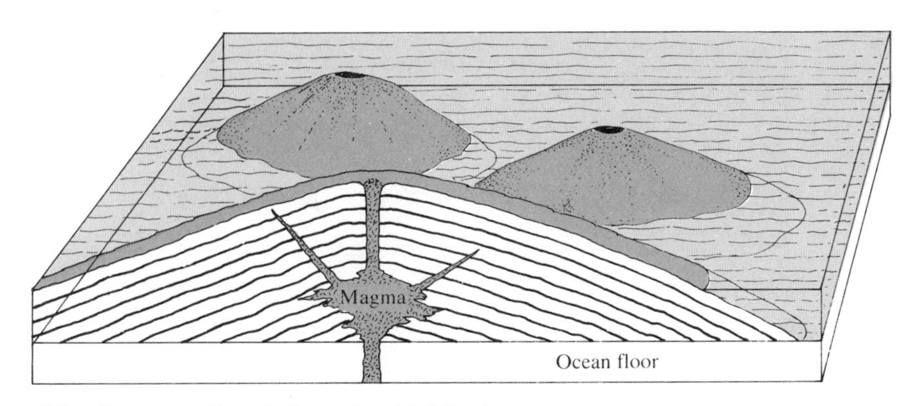

Figure 7.6 *Cross-sectional view of a shield volcano.*

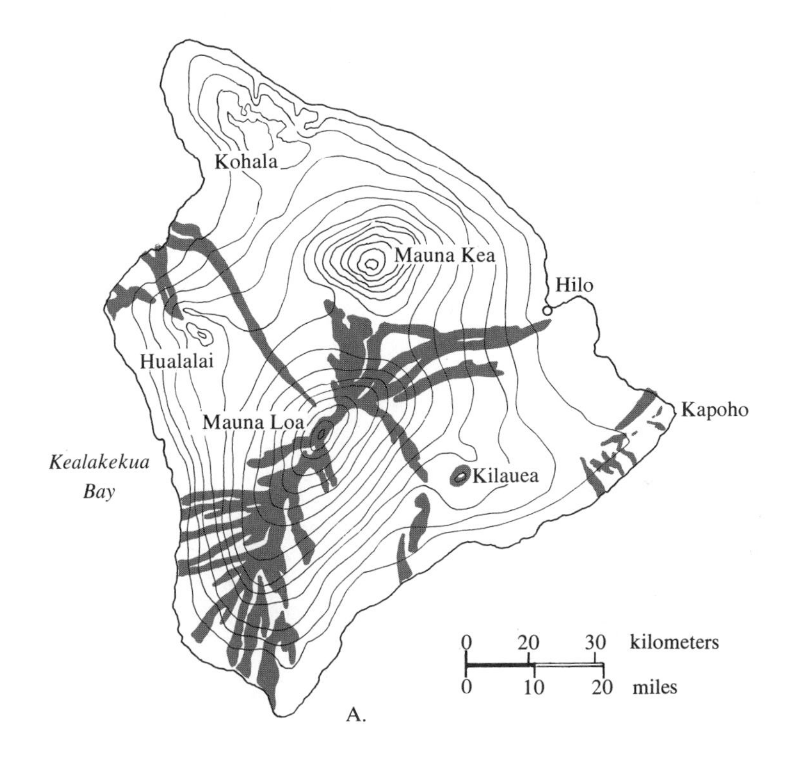

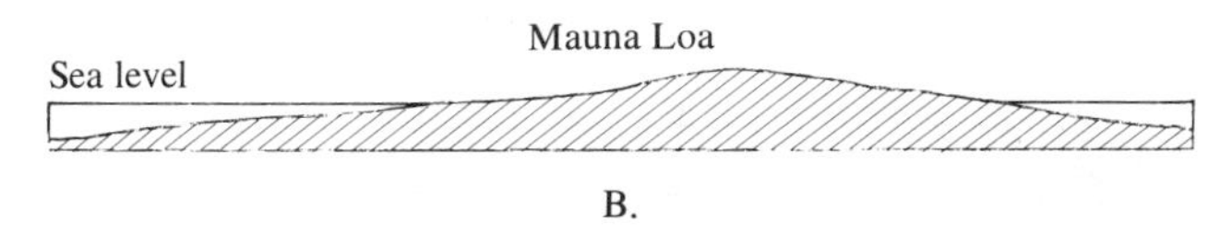

Figure 7.7 *Map of the island of Hawaii.* **A.** *Five volcanoes collectively make up the island.* **B.** *Illustration of the very gentle slope which is characteristic of a shield volcano (no vertical exaggeration). (A. and B. after U.S. Geological Survey and Foster,* Physical Geology, *3rd ed., Columbus, Ohio: Charles E. Merrill, 1979)*

Figure 7.8 *Fresh-looking cinder cone and lava flow among numerous older cones on the Colorado Plateau north of Flagstaff, Arizona. (Photo by John S. Shelton)*

dated pyroclastic material has a high angle of repose (between 30 and 40 degrees), these volcanoes have the steepest slopes. Cinder cones are small in size, usually less than 300 meters (1000 feet) high, and often form as parasitic cones on larger volcanoes. They also occur in

Figure 7.9 *Mt. Fuji, Japan, is a majestic composite cone. (Courtesy of Japan National Tourist Organization)*

groups where they appear to represent the last phase of activity in a region of older basaltic flows (Figure 7.8). This may result because the contributing magma has cooled and become more viscous. Parícutin, which was described at the beginning of this chapter, is a good example of the formation and structure of a cinder cone.

Composite cones, or **stratovolcanoes,** consist of alternating layers of lava flows and pyroclastic materials. The contributing magma has an andesitic-to-rhyolitic composition and hence is rather viscous. For long periods a composite cone may extrude lava; then it may suddenly revert to periods of violent pyroclastic ejections. Some volcanoes have gone through stages in which both activities occurred during the same eruptive phase. The resulting structure is a prominent cone with a steep summit and rather gently sloping flanks. Two of the earth's most picturesque cones, Mt. Mayon in the Philippines and Fujiyama in Japan (Figure 7.9) are of this type.

Although composite cones are the most picturesque, they also represent the most devastating type of volcanic activity. Their eruption can be violent and unexpected, as was the eruption of the Italian volcano Vesuvius in 79 A.D. Prior to the eruption the volcano had been dormant for centuries. On August 24, however, tranquility ceased, and the city of Pompeii (near Naples) and an unknown number of its residents were buried. They remained so for nearly seventeen centuries, until the city was rediscovered and excavated. The large volcanoes that make up the Cascade Range of the western United States, including Mt. Rainier, Mt. Hood, and Mt. Shasta, are also composite cones. Like Vesuvius prior to 79 A.D., they show no signs of activity.

Located at the summit of most volcanoes is a steep-walled depression called a **crater.** The crater is connected to the magma chamber via a pipelike **conduit,** or **vent.** When fluid magma moves up the pipe, it is stored in the crater, until it fills and overflows. During periods of inactivity back flow can drain the lava lake. Lava that is very viscous often forms a plug in the crater which slowly rises or is blown out, enlarging the crater.

Some volcanoes have unusually large craters termed **calderas.** Most calderas form when the summit of a volcano collapses into the partially emptied magma chamber below. Located in such

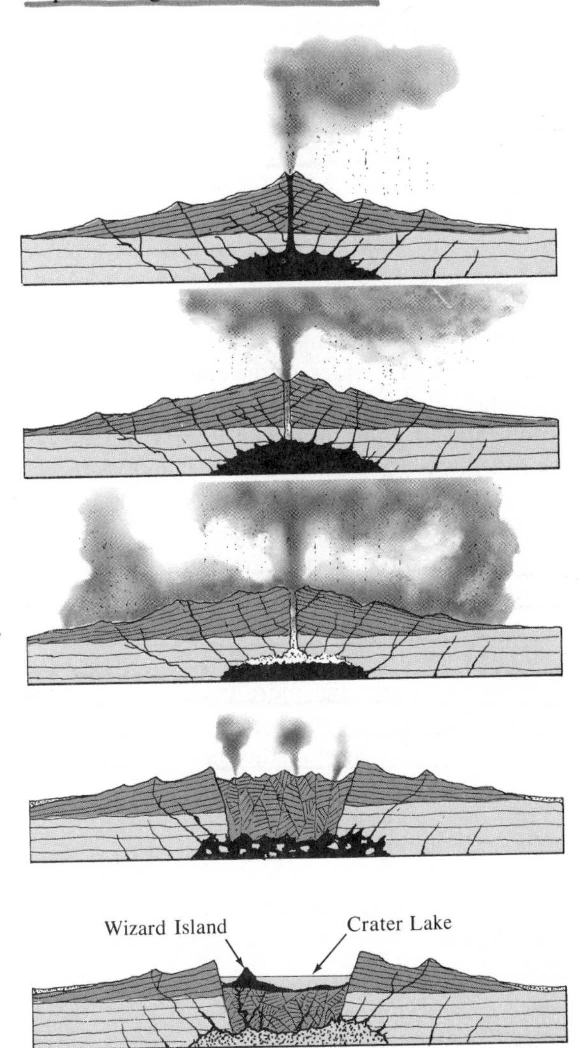

Figure 7.10 *Sequence of events that formed Crater Lake, Oregon. About 6000 years ago, the summit of former Mount Mazama collapsed following a violent eruption which partly emptied the magma chamber. Subsequent eruptions produced the cinder cone called Wizard Island. Rainfall and groundwater contributed to form the lake. (After H. Williams.* The Ancient Volcanoes of Oregon, *p. 47. Courtesy of the University of Oregon)*

a depression, Crater Lake in Oregon formed when the upper 1500 meters of a once-prominent 3600-meter composite cone collapsed (Figure 7.10). The lake is 10 kilometers across and over 600 meters deep (Figure 7.11). After the collapse, a small cinder cone formed Wizard Island in the lake, providing a reminder of a once-active stage. On rare occasions calderas are produced when a volcano decapitates itself. The eruption of the Indonesian volcano Krakatoa in 1883 is an example of such a situation. This explosion was heard nearly 5000 kilometers away, as pumice and ash rose 15,000 meters into the air. After almost 24 hours of total darkness the sky cleared somewhat, revealing that two-thirds of the island had disappeared. Much of the once-lofty volcano probably fell into the space left by the 18 cubic kilometers (4 cubic miles) of debris which was extruded. A 30-meter (100-foot) high tsunami generated by the explosion took its toll of 36,000 lives in nearby Java and Sumatra.

Volcanoes, like all land areas, are continually being lowered by the forces of erosion and mass wasting. Cinder cones, because of their unconsolidated nature, are the easiest to remove. As erosion progresses the more resistant rock that occupies the pipe may remain standing above the terrain long after most of the cone has vanished.

Figure 7.11 *Crater Lake occupies a caldera about 10 kilometers (6 miles) wide. (Courtesy of U.S. Geological Survey)*

Figure 7.12 *Shiprock, New Mexico, is the remnant neck of a volcano which erosion has almost completely removed (Photo by John S. Shelton)*

Shiprock, New Mexico, is thought to be such a feature, called a **volcanic neck** (Figure 7.12).

Fissure Eruptions

Although volcanic eruptions from a central vent are the most familiar, by far the largest amounts of volcanic material are extruded from cracks in the crust called **fissures.** Rather than building a cone, these long narrow cracks distribute volcanic materials over a wide area. An extensive area in the northwestern United States known as the Columbia Plateau was formed in this manner. Here, numerous fissure eruptions extruded very fluid basaltic lava. Successive flows, some 100 meters thick, buried the old landscape as they built a lava plain, which in some places is nearly a mile thick (Figure 7.13). The fluidity is evident, since some lava remained molten long enough to flow 160 kilometers from its source. The term **flood basalts** appropriately describes these water-like flows.

Although vast continental areas are covered with basalt flows, the greatest activity of this type occurs hidden from view on the ocean bottom. Fissure eruptions along mid-oceanic ridges are creating new sea floor, and at a rather rapid rate, since all of the sea floor is less than 200 million years old. Consequently, in the estimated lifespan of the earth this process could have created 25 times as much ocean floor as presently exists; and it may indeed have done so.

When magma with a high silica content is emitted from a fissure, ash and pumice deposits are the rule. Hot shards of glass are carried by the wind and deposited in layers which upon hardening resemble lava flows. Much of the igneous activity in Yellowstone National Park area is of this type. In the northeastern part of the park, 27 fossil forests, one upon the other, can be seen. During periods of inactivity a forest developed upon the newly formed volcanic surface, only to be covered by ash from the next eruptive stage.

Figure 7.13 *Basalt flows of the Columbia Plateau. (Photo by E.T. Jones, U.S. Geological Survey)*

Intrusive Igneous Activity

Although we have separated the discussions of intrusive and extrusive igneous activity, remember that they occur simultaneously. Figure 7.14 shows several types of intrusive igneous bodies

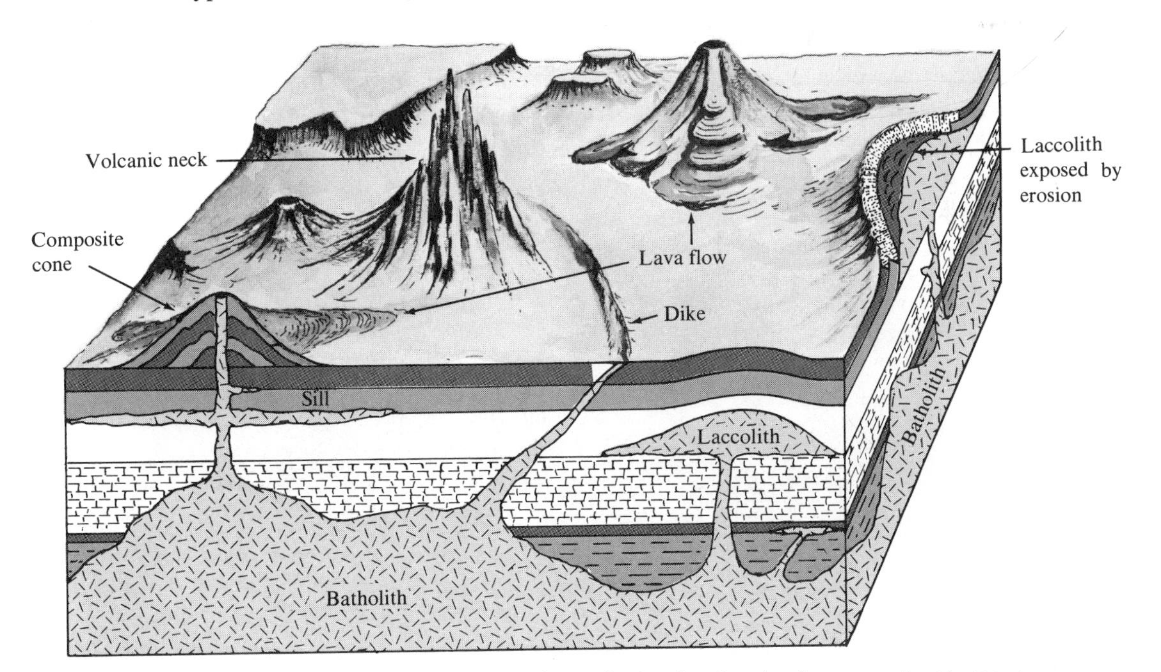

Figure 7.14 *Block diagram illustrating intrusive and extrusive igneous structures.*

Figure 7.15 *Dikes in the Spanish Peak region of Colorado. The igneous dikes are more resistant to erosion than the rock into which they were intruded. (Photo by G.W. Stose, U.S. Geological Survey)*

which form when magma solidifies within the earth's crust. They are generally grouped according to their shape (tabular or massive) and their relationship to the rocks into which they have intruded. **Discordant** bodies cut across preexisting structures, while **concordant** bodies intrude parallel to preexisting surfaces such as bedding planes. **Tabular masses** have a sheetlike appearance, and when discordant, they are known as **dikes.** Dikes are much longer than they are wide, and because they are often more resistant than the rock into which they intrude, they form ridges when exposed by erosion (Figure 7.15). Tabular masses which are concordant are called **sills** (Figure 7.16). When igneous rocks within a sill cool, they contract, producing fractures. In these and other igneous bodies, particularly lava flows, the fractures can develop into patterns known as

Figure 7.16 *Taylor Glacier region, Antarctica. The dark horizontal band is a sill intruded into flat-lying sandstone. (Photo by W.B. Hamilton, U.S. Geological Survey)*

Figure 7.17 *Devil's Post Pile National Monument, California, exhibits columnar joints and the columns that result. (Courtesy of National Park Service, U.S. Department of the Interior)*

columnar joints, which are shown in Figure 7.17. **Batholiths,** the largest intrusive bodies, are massive in shape, and when they occupy the core of a mountainous area, as does the Sierra Nevada batholith, they extend for hundreds of kilometers. Intrusive structures which resemble sills except for their lens shape are called **laccoliths.** The upwarping of overlying rocks caused by the intrusion of laccoliths often produces a dome-shaped structure at the surface.

Igneous Activity and Plate Tectonics

The origin of magma has been a controversial topic in geology almost from the very beginning of the science. How do magmas of different compositions arise? Why do volcanoes in the deep oceanic basins extrude only basaltic lava, while those adjacent to oceanic trenches extrude mainly andesitic and rhyolitic lava? Why does an area of igneous activity commonly called the "ring of fire" surround the Pacific Ocean? New insight gained from the plate tectonic theory is providing some answers to these old questions. According to this theory, three principle zones of volcanic activity exist. These are (1) spreading centers; (2) subduction zones; and (3) regions within the plates themselves (Figure 7.18).

We will first examine the origin of magma as it relates to plate tectonics and then look at the activity that occurs at these zones.

Origin of Magma

Magma is produced when rock is heated to its melting point. Rocks of granitic composition melt at temperatures considerably below 1000° C (1850° F), while basaltic rocks require temperatures closer to 1200° C (2200° F) before they will melt. However individual minerals in a rock, no matter what its composition, have different melting points. *As a rock is heated, the minerals with the lowest melting points liquify first.*

If the molten portion oozes from the mixture before melting is complete, it will have a composition different from that of the original rock. This process, known as **partial melting,** produces a magma with a higher silica content than the parent material. Consequently, a melt generated in this fashion is nearer in composition to the granitic (felsic) end of the spectrum (Table 1.3). *The process of partial melting is thought to produce most, if not all, magmas.*

What are the sources of heat for melting rock? Part of it is believed to come from decay of radioactive materials which are thought to be concentrated in the upper mantle and crust. Workers in underground mines have long recognized that temperature increases with depth. The rate of temperature rise varies from place to place but is thought to average about 1° C per 30 meters (2° F per 100 feet) in the upper mantle and crust. This figure is known as the **thermal gradient.** If temperature were the only factor

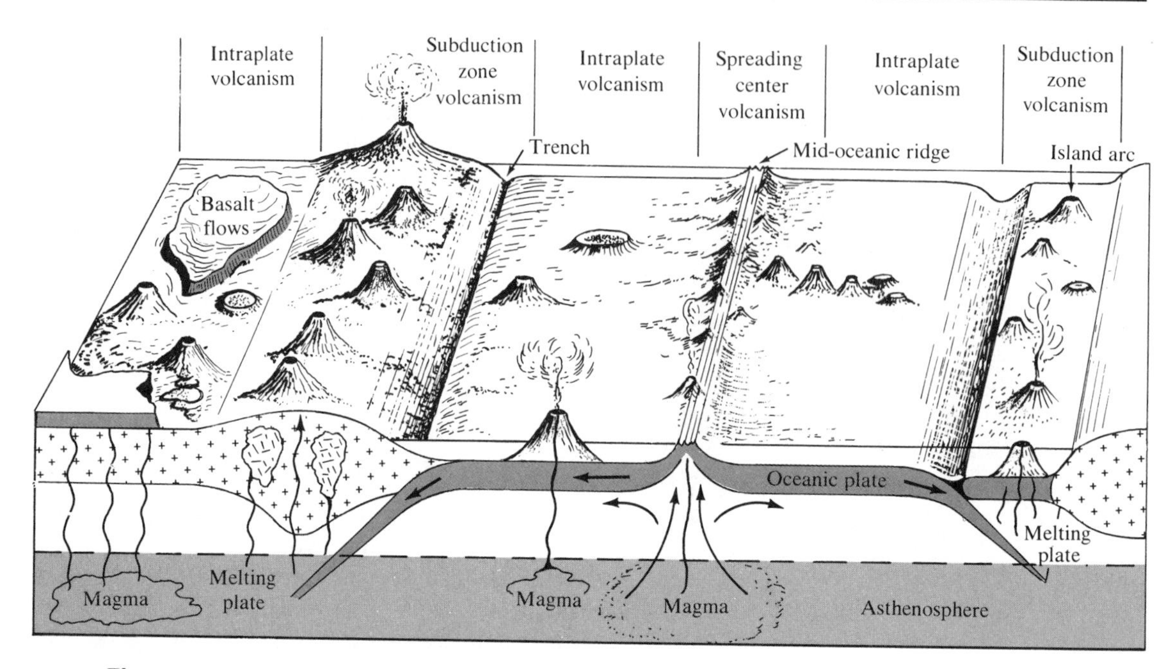

Figure 7.18 *The three zones of volcanism. Two of these zones are plate boundaries, and the third includes the areas within the plates themselves.*

affecting the melting of rocks, the earth would be a molten ball covered with only a thin solid outer shell. However pressure also increases with depth, and changes in pressure alter the rock's melting point. In general, *an increase in the confining pressure causes an increase in the rock's melting point.* For example, one of the feldspar minerals melts at 1000° C on the surface but requires a temperature over 1400° C to melt under the pressure found at a depth of 100 kilometers. Consequently, the increase in temperature with depth is partially offset by the increase in pressure. Another important factor affecting the melting point of rock is its water content. *The more water present, the lower the melting point.* The effect of water on the melting point is magnified by increased pressure. Thus "wet" rock under pressure has a much lower melting point than dry rock of the same composition.

Basaltic Magma. Most basaltic magma is believed to originate in the partially molten asthenosphere at depths exceeding 200 kilometers (120 miles). Laboratory experiments indicate that at the temperatures and pressures found at this depth partial melting of the ultramafic rock peridotite will produce a basaltic magma with a low water content. Since the magma is lighter than the surrounding rock, it has a tendency to rise. Because basaltic magma forms at great

Figure 7.19 *Surtsey emerged from the ocean just south of Iceland in 1963. (Courtesy of Icelandic Photo and Press Service)*

depths, we might expect that most of it would crystallize before it reached the surface. However as magma moves upward the confining pressure diminishes, thus reducing the rocks melting point. As long as the magma migrates fast enough, the drop in temperature can be offset by the drop in melting point.

Volcanologists studying the Hawaiian Islands have obtained seismic data which indicate that the basaltic lava composing these islands does indeed originate in the asthenosphere. Prior to the eruptive stage basaltic lava migrates upward and enters smaller chambers within the volcano. This influx causes the volcano to swell and often generates small earth tremors. Because the semi-molten asthenosphere probably encircles the earth, it is possible for any location to experience volcanic activity of the Hawaiian type. Why such activity is initiated in some locations and not in others remains an unanswered question.

Andesitic Magma. Andesitic lava, which is between basaltic and rhyolitic lava in composition, is thought to originate in several ways. One method is believed to be the fractional crystallization of a basaltic magma (see Chapter 1). When a magma of basaltic composition crystallizes slowly in a large magma chamber, minerals which are higher in iron and magnesium crystallize first. These minerals settle out, leaving the liquid higher in potassium, sodium, and silicon. Then the less dense liquid portion moves upward as does oil in water. If the magma reaches the surface, the resulting lava flows have an andesitic composition. However, the close association of andesitic lava with subduction zones indicates the existence of yet another mode of formation. It is believed that as cold oceanic lithosphere is thrust into the asthenosphere it begins to melt. Recall that the oceanic crust has a basaltic composition. The partial melting of these wet, sediment-laden basalts could yield a magma of andesitic composition. They would buoy upward after a sufficient quantity melted to overcome the resistant material above them. The volcanic Andes Mountains, from which

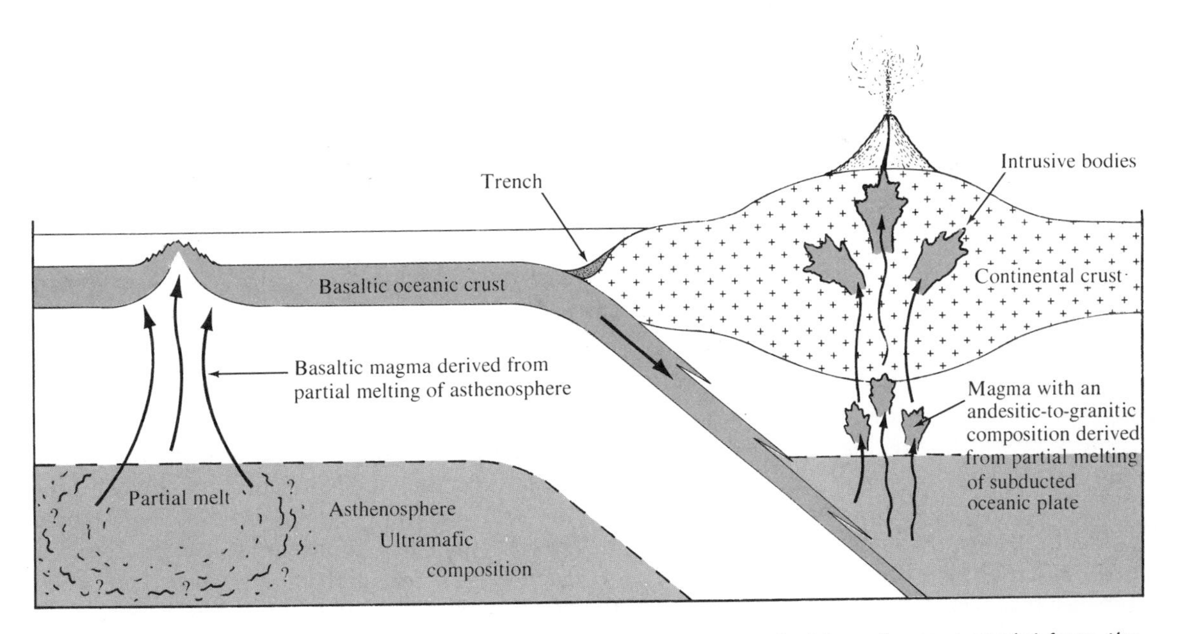

Figure 7.20 *Schematic drawing of the two-stage process that transforms material from the asthenosphere into continental crust. Once continental crust is generated, its low density appears to keep it afloat forever.*

Figure 7.21 *Mt. Shasta, California, one of the most picturesque volcanoes of the Cascade Range. (Photo by John S. Shelton)*

andesite obtains its name, are thought to be an example of this mechanism at work.

Rhyolitic Magma. Rhyolitic lavas are known to be extruded only from volcanoes located on continental crust. This suggests that remelting of continental crust might be a mechanism for their formation. Because the continental crust is higher in silica than the oceanic crust, its melting point is much lower. In addition, the continental crust often contains water, which would further lower its melting point. Burial of "wet" continental crust to depths of 50 kilometers (30 miles) is thought to be sufficient to trigger melting. Another possible source of rhyolitic magmas is the melting of sediments weathered from continental materials. The sediments carried along with subducting oceanic plates must surely melt and generate some rhyolitic magma. Rhyolitic magma formed in this manner has a high water content. Recall that as magma rises the confining pressure decreases, and this in turn reduces the effect that water has on a rock's melting point. Consequently, most rhyolitic magmas lose their mobility before reaching the surface and produce the intrusive rock granite. Erosion eventually exposes these large intrusive bodies of granitic composition known as batholiths.

Volcanic Activity and Plate Boundaries

Spreading Center Volcanism. As stated earlier, the greatest volcanic extrusions are believed to occur at mid-oceanic ridges where sea-floor spreading is active (Figure 7.18). As the lithosphere spreads apart, basaltic magma derived from the partially melted asthenosphere fills the gap. Some reaches the ocean floor and spreads laterally away from the ridge area. On occasion numerous eruptions produce a volcanic cone which may rise above sea level as did Surtsey in 1963 (Figure 7.19).

Submerged flat-topped volcanic cones are believed to be remnants of these volcanic islands which formed at the oceanic ridges. First, erosion lowered the islands to sea level; as they were carried along on the cooling and contracting ocean floor, they sank.

Iceland provides geologists with a platform to observe sea-floor spreading in action. Like Surtsey, Iceland is a part of the Mid-Atlantic ridge system that protrudes above sea level. It represents a situation where lava is supplied faster than the sea-floor spreading is occurring. Here, pockets of lava collect below and erupt at rather frequent intervals. These reservoirs allow for some fractional crystallization, so that later eruptions have somewhat different compositions than earlier ones.

Subduction Zone Volcanism. The volcanic activity associated with convergent zones where oceanic crust is being subducted and melted is of great interest to geologists. Sites of this activity are where new continental crust is being generated and represent an initial stage in the formation of a complex mountain system (see Chapter 8). The magma generated by the melting of the subducted oceanic plate moves upward and builds a volcanic arc adjacent and nearly parallel to the trench. In its early stages the island arc is made up of a chain of volcanoes, some of which rise slightly above sea level, like the Mariana Islands of the South Pacific. Later stages in the development of an island arc are exemplified by Japan and the Alaskan Peninsula.

The volcanic and associated intrusive rocks which make up present-day island arc systems are mostly andesitic-to-granitic in composition. Magma of this composition often produces explosive eruptions; hence some of the most violent volcanic activity, as at Mt. Pelée, has been associated with the island arc systems. Furthermore, their composition very closely approximates the average composition of continental crust. It has been common practice for geologists to call intrusive rocks of intermediate-to-felsic composition granite. However "true" granite makes up only a small portion of the continental crust.

Many geologists believe that a developing island arc represents the second stage in the formation of continental crust from the material of the upper mantle. First, partial melting of the ultramafic rock in the asthenosphere yields the basaltic magma, which after solidifying becomes new oceanic crust at the mid-oceanic ridges (Figure 7.20). This newly formed crust is continually wedged away from the ridges as new sea floor is produced. Eventually it reaches a trench, where it is subducted and partially melted, generating the intermediate and felsic magma which migrates upward to produce the island-arc systems themselves. Since the continental rocks are less dense than the underlying material, they seem to remain afloat forever. That being the case, *continental crust appears to be growing larger at the expense of oceanic crust.* Even the sediment derived from the erosion of continental crust is carried downward into the asthenosphere with a subducting oceanic plate. Here it melts and returns to the continent. Although the continental crust, once formed, remains afloat, it does break apart and is carried along in a conveyor-belt fashion until it collides with another landmass. Australia, which broke from Antarctica and is being rafted northward, will probably join Asia in much the same manner as did India.

Not all volcanic activity in subduction zones occurs in association with island arcs. As stated in Chapter 6, the volcanic peaks in the Andes Mountains were generated by the subduction of the Nazca plate under the South American continent. Similarly, the chain of composite cones which runs from northern California through Washington State are thought to represent volcanic activity in a subduction zone which has since been abandoned (Figure 7.21).

Intraplate Volcanism. The factors which trigger volcanic activity within the plate proper are more difficult to establish. Volcanic islands that form upon oceanic crust have a basaltic composition, which points to the upper mantle as the probable source for the magma. Recall that seismic data indicate that the Hawaiian Islands do in fact tap the asthenosphere. But just what initiates this intraplate activity is unknown.

Volcanism that occurs on continents away from plate boundaries is the most difficult to explain. Activity like that in the Yellowstone area produced lava and ash flows of rhyolitic composition, while extensive basaltic flows covered vast portions of our Northwest, producing the Columbia Plateau. Yet these greatly different areas lie only a short distance apart.

Some of the volcanic activity that has occurred in the interior of a continent may be associated with an attempt to produce a spreading center. The rift zone in Africa is considered to be the beginning stage of such a breakup. If a spreading center develops, Africa will split into two sections in much the same way as North America separated from Europe or, more recently, as the Arabian Peninsula separated from Africa, producing the Red Sea.

REVIEW

1. What is the difference between magma and lava?
2. What three factors determine the nature of a volcanic eruption? What role does each play?
3. Describe pahoehoe and aa lava.
4. List the main gasses released during a volcanic eruption.
5. Describe scoria and explain how it is produced.
6. Describe each type of pyroclastic material.
7. Compare and contrast the main types of volcanoes (size, shape, eruptive style, and so forth).
8. Name an example of each of the three types of volcanoes.
9. Compare the formation of Hawaii with that of Parícutin.
10. How is a caldera different from a crater?
11. Describe the formation of Crater Lake. Compare it to the caldera which formed during the eruption of Krakatoa.
12. What is Shiprock, New Mexico, and how did it form?
13. Describe each of the four intrusive features discussed in the text.
14. Explain how most magma is thought to originate.
15. What do the hot springs and geysers of the Yellowstone National Park indicate about the thermal gradient in that region?
16. Where and by what process is basaltic magma thought to originate? Andesitic magma? Rhyolitic magma?
17. Do the rocks of Iceland have a composition similar to that of continental crust or oceanic crust? Explain.
18. Describe how flat-topped submarine volcanoes are thought to originate.
19. What similarity exists between the mountains of the Cascade Range of Washington and Fujiyama of Japan?
20. Explain the steps by which continental crust is thought to originate from material in the asthenosphere.
21. Why do geologists believe that the continental crust grows larger at the expense of oceanic crust?

KEY TERMS

viscosity
pahoehoe
aa
scoria
pyroclastics
welded tuff
nuée ardente
volcano
shield volcano
cinder cone
composite cone (stratovolcano)
crater
conduit (vent)
caldera
fissure
flood basalt
discordant
concordant
tabular
dike
sill
columnar joints
batholith
laccolith
partial melting
thermal gradient

Intensely folded rock strata provide evidence of the forces altering the earth's crust. (Photo by W.B. Hamilton, U.S. Geological Survey)

Orogenesis
Crustal Uplift
Rock Deformation
Mountain Types
Mountain Building and Plate Tectonics

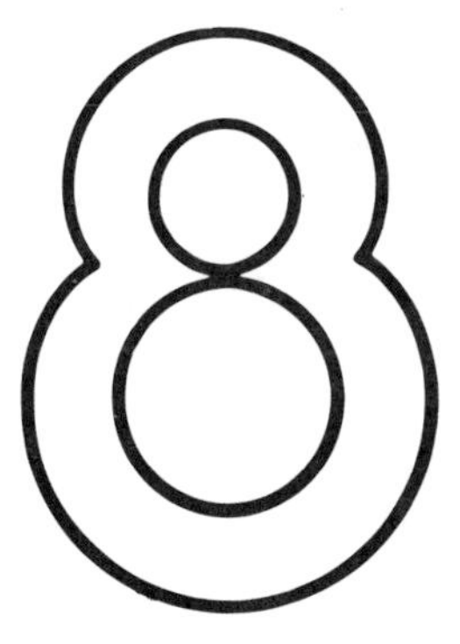

Mountain Building

Mountains provide some of the most spectacular scenery on our planet. This splendor has been captured by poets, painters, and songwriters alike. Geologists believe that at some time all continental regions were mountainous masses and have concluded that the continents grow by the addition of mountains to their flanks. Consequently, by unraveling the secrets of mountain formation, geologists will have taken a major step in determining the evolution of the earth. If continents do indeed grow by adding mountains to their flanks, how do geologists explain the existence of mountains (the Urals for example) that are located in the interior of a landmass? In order to answer this and other related questions, this chapter attempts to piece together the sequence of events which is believed to generate these lofty structures.

Mountains are often spectacular features which rise several hundred meters or more above the surrounding terrain. Some occur as single isolated masses; the volcanic cone Kilimanjaro, for example, stands almost 6000 meters (20,000 feet) above sea level overlooking the grasslands of East Africa. Others make up a portion of an extensive mountainous chain, such as the American Cordillera, which runs continuously from the tip of South America through Alaska. Some chains, like the Himalayas, are youthful, gigantic mountains that are still rising, while others are very old and nearly worn down, as exemplified by the Appalachian Mountains in the eastern United States.

Although every mountain system is unique, the similarities of systems are generally greater than the differences. Mountain systems show evidence of enormous forces which have folded, faulted, and generally deformed large sections of the earth's crust (see chapter-opening photo). The processes of folding and faulting have contributed to their majestic appearance, but much of the credit for their beauty must be given to the work of water and glacial ice which sculpture these uplifted masses in an unending effort to lower them to sea level.

The name given to the group of processes which collectively produce a mountain system is **orogenesis,** from the Greek *oros* ("mountain") and *genesis* ("to come into being").

Orogenesis

It has long been recognized that complex mountain systems have many features in common—enough, in fact, that geologists have concluded that they must have comparable orogenic histories. Many young mountains parallel the coasts of continents. They are made up of thick sequences of sedimentary rock, occasionally totaling more than 15,000 meters, which have been folded, faulted, and intruded by igneous bodies. Until the last decade it had generally been accepted that these sediments accumulated in a slowly subsiding trough called a **geosyncline.** After great thicknesses of sediment had built up, horizontal forces from the seaward side of the geosyncline began to squeeze the sediments, shortening and thickening the crust, producing a high-standing mountain system while simultaneously pushing much of the sediment deeper into the earth. It was believed that the melting of these deeply buried sediments generated magma which moved upward, intruding the overlying unmelted sediments. Thus a complex mountain chain containing folded and faulted sedimentary rocks surrounding a core of igneous intrusions and metamorphic rocks was formed.

Although the geosynclinal concept of mountain building just discussed has many merits, it fails to account for the underlying cause of orogenesis. What produced the subsidence in the geosyncline? Why did the force come from the sea to squeeze the sediments? Why does a mountain range run nearly continuously along the west coasts of North and South America? These unsolved questions forced geologists to continue to evaluate the question of mountain building.

The first encompassing explanation of orogenesis came only a decade ago as part of the plate tectonic theory. As noted before, the idea of plates colliding has opened many new and exciting avenues to geologists. Before examining mountain building according to the plate tectonic model however, it will be advantageous first to view the processes of crustal uplifting and deformation.

Crustal Uplift

If you were to climb a mountain, it would not be unusual to find evidence, like a fish fossil, which would indicate that the sedimentary rock composing the mountain was once below sea level. This find would be rather convincing proof that some drastic changes occurred between the time the animal died and when you found its fossilized remains. Evidence like this fills the geologic record and is even present in the historical record. For example, Figure 8.1 shows three columns that are the ruins of a Roman-built temple. These remains have clam borings to a height of about 6 meters (20 feet), indicating that the land upon which the temple was built submerged and was later partially uplifted. Those clam borings might also be explained by changes in sea level; however a change in sea level is not recorded at any other location for that same time period. A good bit of evidence for crustal uplift can be found along coastlines. When a coastal area remains stationary for a period of time, wave action cuts a gently sloping bench. In parts of California old wave-cut benches can be found hundreds of meters above sea level (Figure 8.2). Each represents a period when that area was at sea level.

Unfortunately the reasons for uplift are not always as easy to determine as the evidence for the movements. We do know that the force of gravity must play an important role in determining the elevation of the land. It is believed that the lithosphere floats on top of the partially molten asthenosphere, much like logs float on

Figure 8.1 *Remaining columns of the ancient Roman temple of Jupiter Serapis, Pozzuoli, Italy. Clam borings 6 meters above sea level indicate former submergence. (Charles Lyell.* Principiples of Geology, *12th ed., 1875. Courtesy of Flint and Skinner.* Physical Geology. *John Wiley & Sons)*

Figure 8.2 *Wave-cut terraces on the Palos Verdes Hills south of Los Angeles, California. Once at sea level, the highest terraces are now about 400 meters above it. (Photo by John S. Shelton)*

water. By carrying this analogy a little farther we can see why mountains stand higher than the surrounding areas. Imagine two logs, one much thicker than the other, floating in water. The larger log will float higher than the smaller log. In the same manner, *mountainous regions are believed to represent unusually thick sections of the earth's crust, while areas of low elevation have crust of normal thickness.* The mountains, like the thick log, not only stand high above the surface but also have a "root" which extends farther into the supporting material below (Figure 5.18). The existence of these roots has been confirmed by seismic and gravitational data. Carrying this idea one more step, the crust beneath the oceans must be thinner than that of the continents because it sits so much lower. This is true, but the greater density of the oceanic crust is another factor that contributes to its lower position.

The concept of a floating lithosphere in gravitational balance is called **isostasy.** The theory of isostatic balance can account for some uplifting. Remember that the crust is floating on the partially molten asthenosphere. As erosion strips off the tops of mountains, the mountains themselves will "bob" up. (Visualize what happens to a ship as its cargo is being unloaded.) The processes of uplifting and erosion will continue until the "root" of the mountain has reached the same height as the surrounding crust (Figure 8.3), at which time the crust in that area will have attained a normal thickness. As the mountain wears down, the weight of the eroded sediment dumped on the adjacent coastal region will cause it to subside.

If the isostatic theory is correct, we would expect that if weight is added to the crust it will respond by subsiding, and when weight is removed there will be uplifting. Evidence of this type exists, strongly supporting the concept of *isostatic adjustment.* When Hoover Dam was built in the 1930s, the impounded water of Lake Mead and the millions of tons of sediment it collected caused regional subsidence. Another classic example is provided by nature. When the glaciers occupied sections of North America during the last ice age, the added weight of the 3-kilometer thick masses caused downwarping of the earth's crust. Since the ice has melted, uplifting of as much as 330 meters has occurred in the Hudson Bay region, where the thickest ice existed.

In summary, *mountains are unusually thick sections of the earth's crust which stand above the surrounding region and because of isostatic adjustment tend to remain so until erosion and uplifting bring their roots up to the depth of the surrounding crust*. But how do these thick sections of the earth's crust come into existence?

Rock Deformation

When rocks are subjected to stresses which are greater than their strength, they begin to deform, usually by folding and faulting. It is easy to visualize how rock breaks, but for a long time geologists were puzzled by the ability of seemingly rigid material to be bent into intricate folds without being appreciably broken during the process (Figure 8.4). In an attempt to find an answer to this puzzle, geologists turned to the laboratory, where they subjected rocks to stresses while simulating the conditions believed to be found at different depths within the crust. Although all rock types deform somewhat differently, general characteristics of rock deformation were determined by these experiments. It was found that when stress is applied slowly, rocks first respond by deforming elastically. Elastical changes are reversible; that is, like a rubber band, the rock will return to its original shape

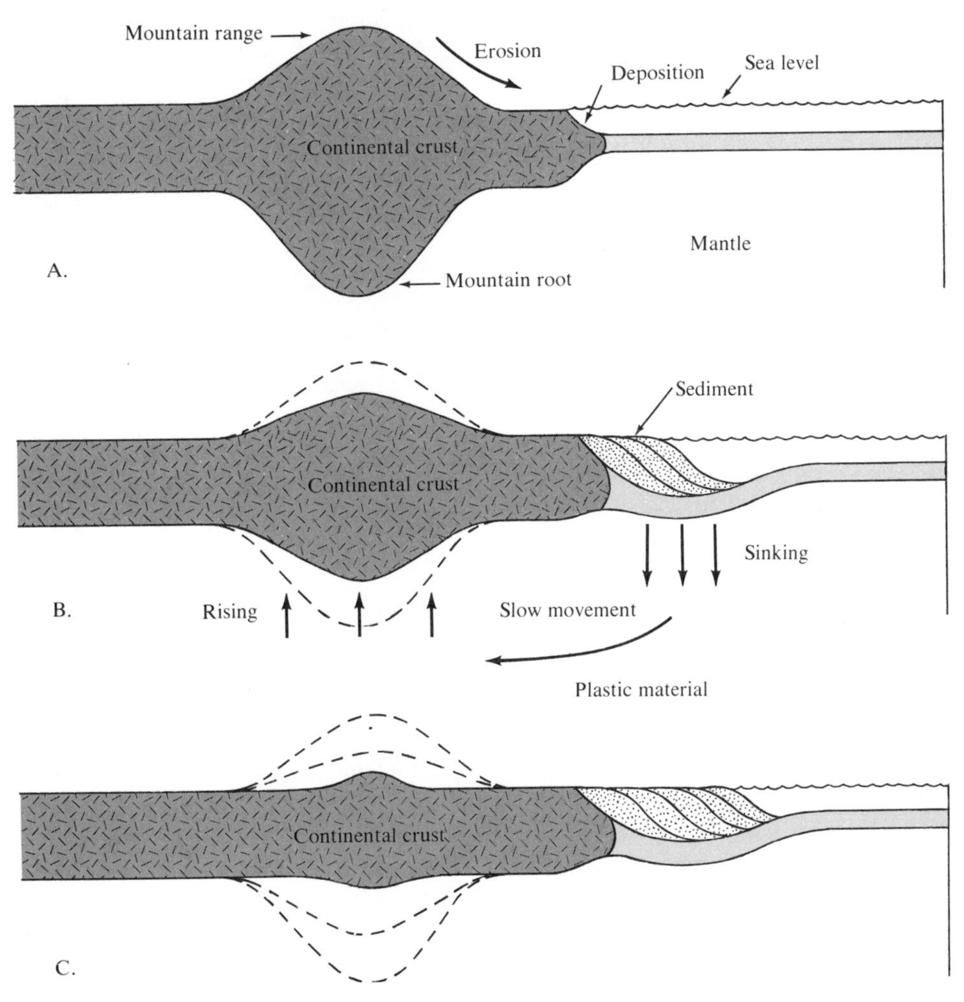

Figure 8.3 *Sequence of sketches illustrating how the combined effect of erosion and isostatic adjustment results in a thinning of the crust in mountainous regions.*

Figure 8.4 *Intricately folded rock of Cabbage Island, Maine. (Photo by W.B. Hamilton, U.S. Geological Survey)*

when the stress is removed. Thus elastic changes are temporary. However, once the elastic limit is surpassed, rock either ruptures or deforms plastically. *Plastic changes are permanent and result in a change in shape by folding and flowing* (Figure 8.4). Laboratory experiments confirm that under the high temperatures and pressures found at great depth, most rocks deform plastically by folding rather than breaking. Rocks tested under surface conditions also deform elastically, but once they exceed their elastic limit, most rupture. Recall that the energy for an earthquake is believed to come from the stored elastic energy which is released as rock ruptures and returns to its original shape.

One factor that cannot be duplicated in the laboratory is geologic time. We know that if stress is applied quickly, as with a hammer, rocks tend to break. On the other hand, these same materials may deform plastically if the stress is applied over extended periods. For example, marble benches have been known to sag under their own weight over a period of a hundred years. In nature, small forces applied over long periods of time surely play an important role in deforming rock strata.

Figure 8.5 *Green River formation, Colorado. Faulting caused the 3-meter (10-foot) vertical displacement of these sandstone beds. (Photo by D.E. Winchester, U.S. Geological Survey)*

Faulting

Faults are fractures along which movement has taken place. Several categories of faults are recognized on the basis of the relative movement between the material on both sides of the fault plane (Figure 8.5). The movement can be horizontal, vertical, or both. Faults having primarily vertical movement are called **normal faults** when the rock above the fault plane moves *down* relative to the rock below (Figure 8.6A); **reverse faults** appear when the rock above the fault plane moves *up* relative to the rock below

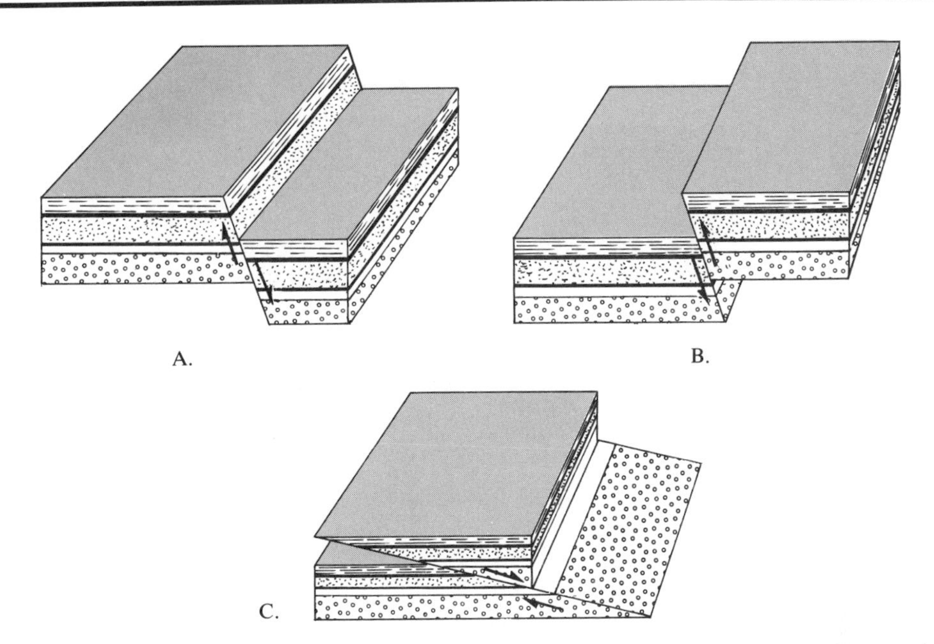

Figure 8.6 *Block diagrams of the principal fault types that exhibit primarily vertical movement.* **A.** *Normal fault.* **B.** *Reverse fault.* **C.** *Thrust fault.*

the fault plane (Figure 8.6B). Reverse faults having a very low angle are referred to as **thrust faults** (Figure 8.6C). In mountains such as the Alps thrust faults have displaced rock for many kilometers over adjacent strata. Faults having mainly horizontal movement are called **strike-slip faults.** The movement is said to be *right lateral* if the land on the opposite side of the fault moves to your right as you face the fault, and *left lateral* if it moves to your left (Figure 8.7). Many large strike-slip faults are plate boundaries, in which case they are called **transform faults.** The San Andreas fault is a good example of such a feature. When faults have both vertical and horizontal movement they are called **oblique faults.**

Figure 8.7 *Imperial Valley, California. Movement along this strike-slip fault displaced the trees of this orange grove about 4 meters. (Photo by John S. Shelton)*

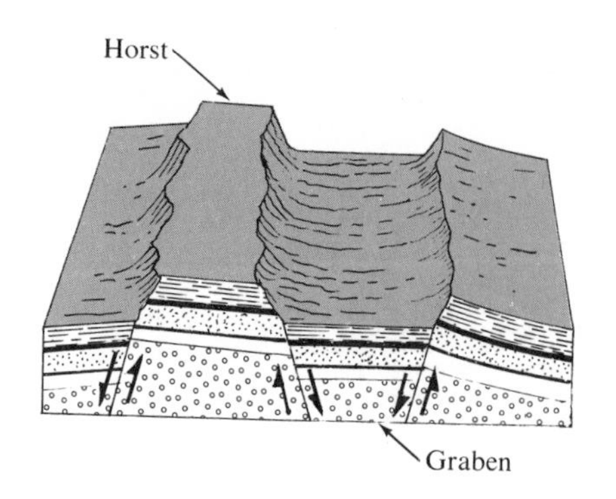

Figure 8.8 *Diagrammatic sketch of downfaulted block (graben) and upfaulted block (horst).*

Fault motion provides the geologist with a method of determining the nature of the forces at work within the earth. *Normal faults indicate the existence of tensional forces which pull the crust apart.* This pulling apart can be accomplished either by forces from below that cause the surface to stretch and break or by horizontal forces that actually rip the crust apart. The latter situation is thought to occur in spreading centers where plate divergence is prevalent. Here, a central block which is bounded by normal faults, called a **graben,** drops as the plates separate (Figure 8.8). These grabens produce an elongated trenchlike structure which is bounded by upfaulted structures called **horsts.** The Great Rift Valley of East Africa is made up of several gigantic grabens, above which tilted horsts produce a linear mountainous topography. This valley, nearly 6000 kilometers (3600 miles) long, is the site where the excavation of the earliest fossil human took place. Other rift valleys include the Rhine Valley of Germany and the valley of the Dead Sea in Israel. Even larger rift valleys occur in mid-oceanic ridges where seafloor spreading is in progress.

The blocks involved in reverse and thrust faulting are moving together; therefore geologists conclude that compressional forces are at work. As you may have suspected, the primary regions of this activity are thought to be the convergent zones, where plates are colliding. Compressional forces generally produce folds as well as faults and result in a general thickening and shortening of the material involved.

Folding

During mountain building, flat-lying sedimentary rocks are often bent into a series of broad folds, much like those formed when you hold the ends of a sheet of paper and push them together. The result of folding is a shortening and thickening of the crust. When the stresses are great, the sedimentary strata can be pushed into very tight folds, one upon the other. Figure 8.9 shows some of the typical structures formed during folding. The linear upfolded forms are

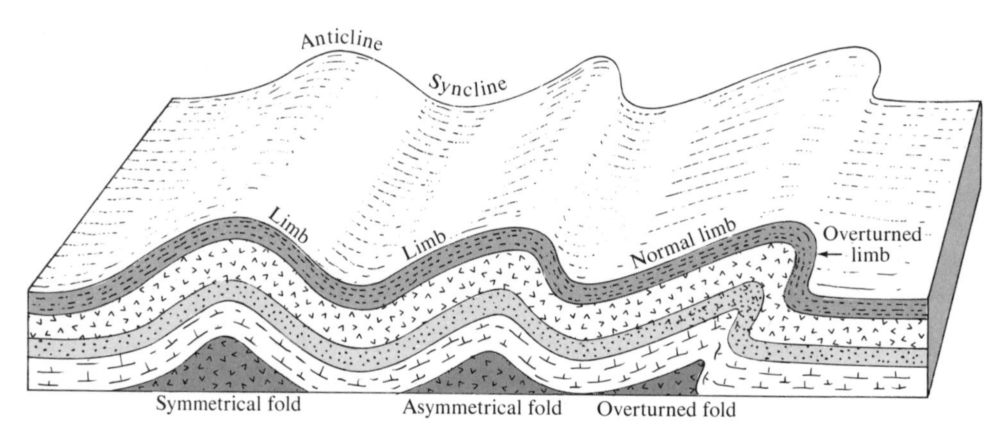

Figure 8.9 *Simplified block diagram of principal types of folded strata.*

Figure 8.10 *A simple anticline exposed along the banks of the Potomac River. (Photo by I.C. Russell, U.S. Geological Survey)*

called **anticlines** (Figure 8.10), and the down-folded troughs are referred to as **synclines.**

Depending on their orientation, anticlines and synclines are said to be symmetrical, asymmetrical, or overturned if one limb has been tilted beyond the vertical (Figure 8.9). These folds do not continue forever; their ends die out much like the wrinkles in cloth do. The ends of these folds are said to be plunging. Figure 8.11 shows the pattern produced when erosion strips the top from folded structures and exposes their interior. Note that the outcrop pattern of an anticline points in the direction it is plunging. A good example of the kind of topography which results when erosional forces attact folded sedimentary strata is found in the Ridge and Valley Province of the Appalachians (Figure 8.12). In this region resistant sandstone beds remain as ridges separated by valleys cut into the more easily eroded shale or limestone beds.

Some folds, like some thrust faults, are believed to be the result of compressional forces associated with continental collisions that squeeze the entrapped rock, much like a giant vise. This deformation leaves the strata intensely folded and faulted.

Although plate collision is an important mechanism of deformation, evidence points to

Figure 8.11 *Sheep Mountain, a doubly plunging anticline. Note that erosion has cut the flanking sedimentary beds into low ridges that make a "V" pointing in the direction of plunge. (Photo by John S. Shelton)*

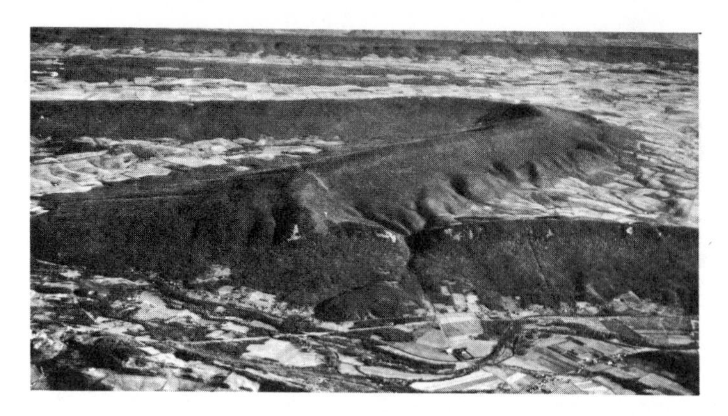

Figure 8.12 *Ridge and Valley Province near Hollidaysburg, Pennsylvania. The zigzag (Z) fold shown is typical of plunging folds. (Photo by John S. Shelton)*

the existence of yet another process—**gravity sliding.** Simply stated, gravity causes large sheets of sediment to slide downslope as a unit, folding and breaking them in the process (Figure 8.13). This mechanism seems to be particularly significant in the deformation of thick units of unconsolidated sediment that are folded while the underlying bedrock over which they move remains unchanged. It has been shown that gravity sliding can take place on a very gentle slope which develops during that stage in the mountain-building process when the igneous intrusions are implaced. These igneous bodies thicken the crust and apparently cause some isostatic uplifting. Some of the folding in the Appalachian Mountains is thought to have occurred in this manner. In this case, the region of igneous implacement and uplifting occurred mainly to the east of our present coastline, where it has since been eroded and covered by younger sedimentary rocks of the coastal plain. The strata nearest the intrusive bodies are the most highly deformed and metamorphosed. As we move westward, the strata change from tight, overturned folds to broad, open folds that eventually become nearly flat lying. The direction of movement was definitely from east to west. Although much of the deformation was caused by gravity sliding, some must also be attributed to the implacement of the igneous bodies which pushed the sediment aside.

Gentle upwarping of the crust forms a **dome,** and downwarping creates a **basin.** The Black Hills of South Dakota exemplify domal structures where erosion has stripped away the upwarped sedimentary beds, exposing older igneous and metamorphic rocks in the center. Several examples of basins also exist in the United States. The basins of Michigan and Illinois have very

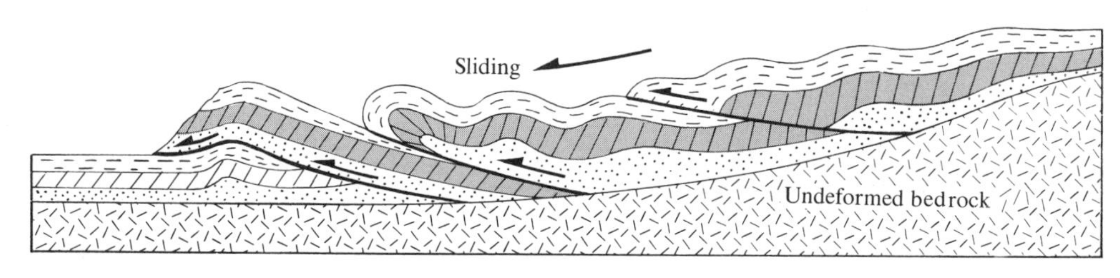

Figure 8.13 *Idealized cross section illustrating gravity sliding. Note the thrust faults which permit older strata to override younger strata.*

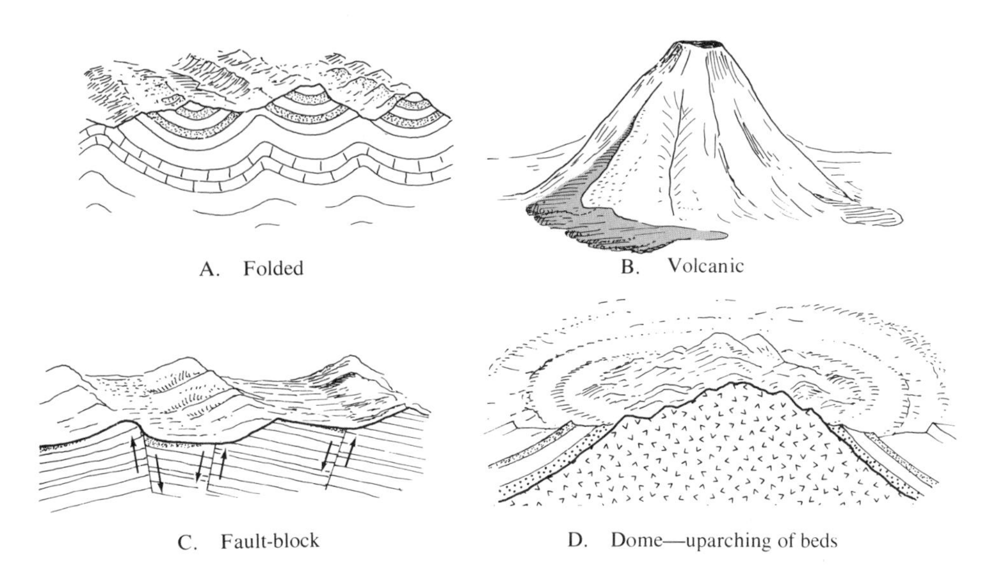

Figure 8.14 *Classification of mountains. (From Foster,* Physical Geology, *3rd ed., Columbus, Ohio: Charles E. Merrill, 1979)*

gently sloping beds similar to saucers. Because their beds slope at such a low angle, large basins are usually identified by the age of the rocks that compose them. The youngest rocks are found near the center, while the oldest rocks are found at the flanks, just the opposite of what we find for a domal structure.

Mountain Types

Even though no two mountain ranges are the same, they can be classified if we select their *most* dominant characteristics. Using this approach, four main categories of mountains emerge: (1) folded mountains; (2) volcanic mountains; (3) fault-block mountains; and (4) upwarped (domed) mountains (Figure 8.14).

Folded Mountains

Folded mountains comprise the largest and most complex mountain systems. Although folding is the dominant characteristic, faulting and igneous activity are always present in varying degrees. The Alps, Urals, Himalayas, Rockies, and Appalachians are all of this type. Since folded mountains represent the world's major mountain systems, the process of mountain building is usually described in terms of their formation. Thus a separate section on mountain building and plate tectonics is devoted primarily to the formation of folded mountains.

Volcanic Mountains

Volcanic mountains are the easiest to visualize. They form from the extrusion of lavas and pyroclastic materials, which if continued long enough, produces gigantic volcanic piles. Most volcanic activity occurs within the ocean basins, particularly at the mid-oceanic ridges, which are extensive volcanic mountain ranges that tower high above the ocean floor. Although mid-oceanic ridges seldom extend above sea level, a recent exception was the 1963 emergence of the volcanic island of Surtsey just south of Iceland. Other examples of the volcanic process at work are the volcanic arc systems. Volcanic mountain ranges also occur on land. The Cascade Range of Washington and Oregon is made up of numerous composite cones, including Mounts Rainier, Hood, and Shasta. Many of the peaks in the western Andes of South America are volcanic in origin as well.

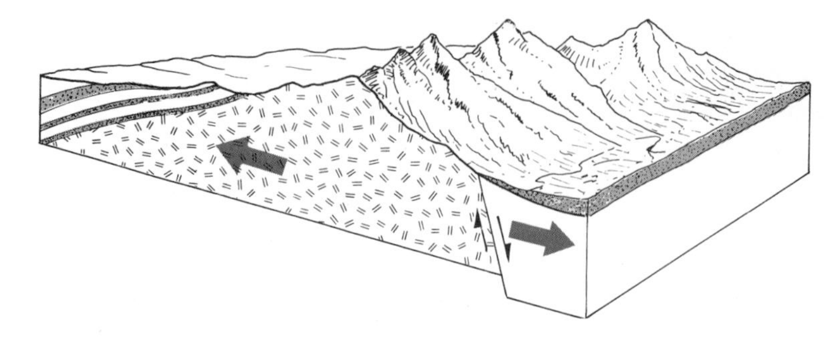

Figure 8.15 *Drawing of the fault block that constitutes the Teton Range. (After U.S. Geological Survey)*

Fault-block Mountains

Fault-block mountains are bounded by high-angle normal faults. Some are associated with rift valleys such as those in East Africa, while others appear to be formed by vertical uplifting. A notable example of fault-block mountains is found in the Basin and Range Province of the southwestern United States. Here the crust is literally broken into hundreds of pieces which have been tilted, giving rise to nearly parallel mountain ranges, about 80 kilometers in length, which rise precipitously above the basins. Since in this case the breakup was accompanied by volcanic activity, it has been suggested that the tilting occurred because of the loss of support from below when the magma moved to the surface. Other examples of fault-block mountains in the United States include the Teton Range of Wyoming and the Sierra Nevada of California. Both are faulted along their eastern flanks, which were uplifted as the block tilted downward to the west (Figure 8.15). Looking from Jackson Hole and Owens Valley respectively, the eastern fronts of these ranges rise over 2 kilometers, making them two of the most spectacular mountain fronts in the United States (Figure 8.16).

Upwarped (Domed) Mountains

Upwarped mountains possibly represent the most diverse group. Some, like the Black Hills of South Dakota and the Adirondack Mountains of New York, are made up of older igneous and metamorphic bedrocks which were once eroded flat and mantled with sediment. As they were upwarped erosion stripped away the

Figure 8.16 *Imposing east face of the Sierra Nevada as viewed from Owens Valley. (Photo by W.C. Mendenhall, U.S. Geological Survey)*

Lone Star Geyser in Yellowstone National Park. (Courtesy of Ward's Natural Science Establishment, Inc., Rochester, N.Y.)

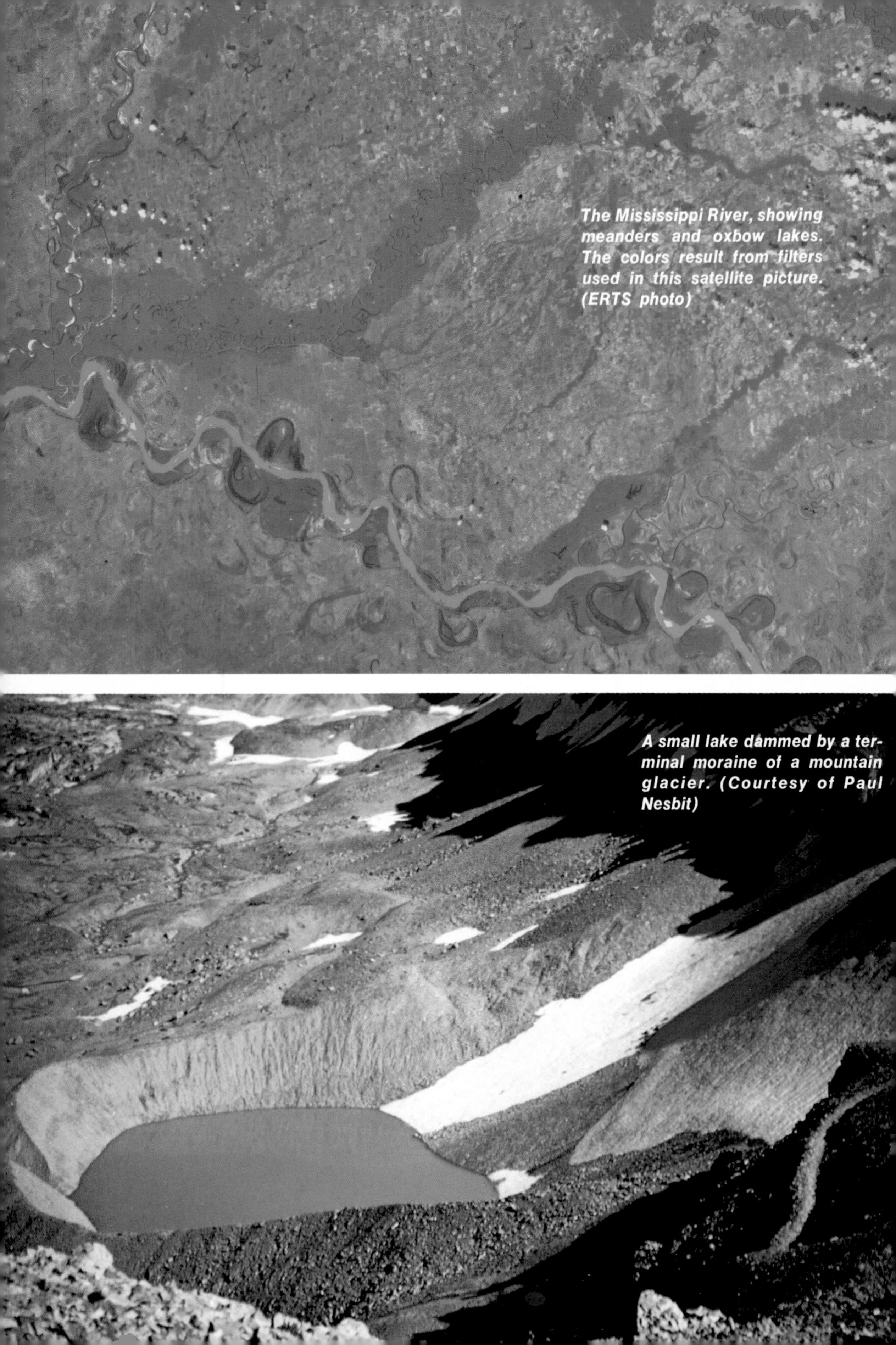

The Mississippi River, showing meanders and oxbow lakes. The colors result from filters used in this satellite picture. (ERTS photo)

A small lake dammed by a terminal moraine of a mountain glacier. (Courtesy of Paul Nesbit)

veneer of sedimentary strata, leaving a core of igneous and metamorphic rocks standing above the surrounding terrain (Figure 8.17). Others have been formed by the implacement of an igneous body such as a batholith or laccolith (Figure 7.14). As the intrusion is implaced, upwarping of the overlying material may occur. Because the intrusive rocks are more resistant to erosion, these structures will remain elevated long after the upwarped sedimentary beds have been eroded away.

Although the structure of the Appalachian Mountains was produced by folding and faulting, the present mountainous topography is the result of recent upwarping (Figure 8.18). Prior to this upwarping the Appalachians had been eroded to a nearly flat plain (peneplain). Uplifting rejuvenated the forces of erosion, which then quickly cut valleys into the softer sedimentary rock and left the harder beds standing as ridges, producing the ridge-and-valley topography so well exemplified in Pennsylvania.

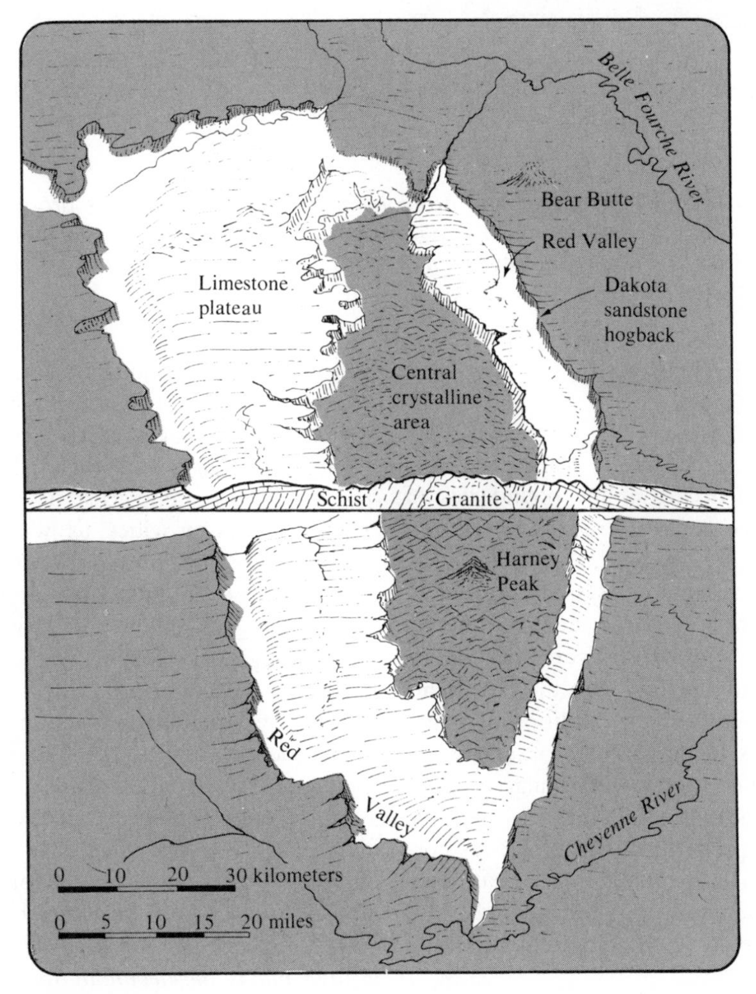

Figure 8.17 *The Black Hills of South Dakota, an example of an upwarped mountain whose resistant igneous and metamorphic central core has been exposed by erosion. (After Arthur N. Strahler.* Introduction to Physical Geography, *3rd ed. Copyright © 1973, by John Wiley & Sons, Inc. Reprinted by permission)*

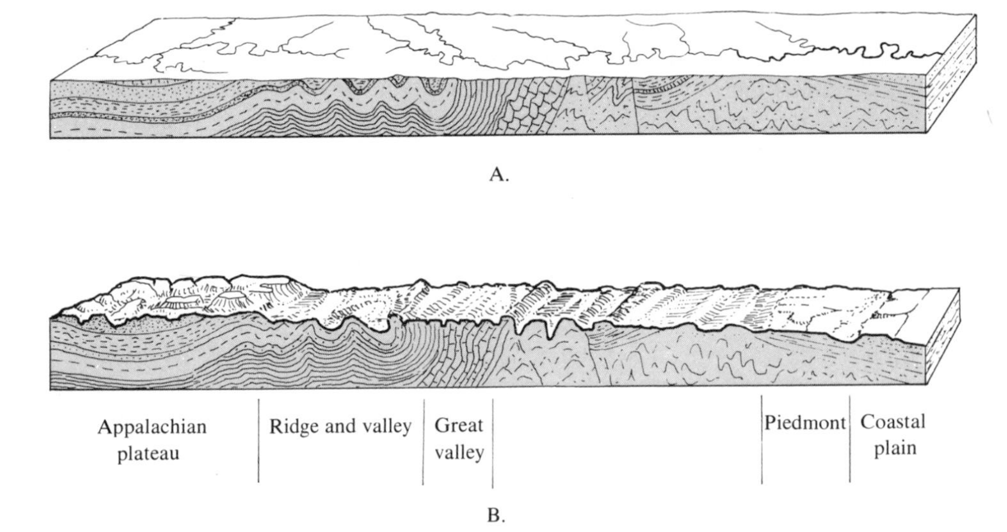

Figure 8.18 *Recent upwarping and erosion in the Appalachians has produced the present topography.* **A.** *Eroded Appalachians.* **B.** *Uplifting and rejuvination of the Appalachians. (Drawn by Erwin Raisz, from "Stream Sculpture on the Atlantic Slope," by Douglas Johnson, © 1931, used with permission of Columbia University Press)*

Mountain Building and Plate Tectonics

According to the theory of plate tectonics, mountain building occurs at the boundaries of colliding plates and is closely related to the processes which are at work producing island arcs. Just as happens in island arc systems, mountain building is thought to generate new continental crust. Although this idea is still very speculative, many geologists now believe that the continents grow sliver by sliver with the addition of linear mountainous belts to their flanks. Examples of such growth include the Coastal Range of California, the Andes of South America, and the Pyrenees of Europe. If this concept proves correct, it means that all continental areas once stood as high mountainous regions, which were subsequently lowered to their present elevation by erosion.

Orogenesis at Subduction Zones

The first stage in the development of a mountain system according to the plate tectonic model is thought to occur prior to the formation of the subduction zone. At that time the continental margin is inactive; that is, it is not a plate boundary, but part of the same plate as the adjoining oceanic crust. The east coast of the United States provides us with a present-day example of an inactive continental margin. Here, like inactive continental margins worldwide, the process of sedimentation is producing the thick wedge of shallow-water sediments which make up the continental shelf (Figure 8.19A). In the deep water beyond the shelf turbidity currents (dense slurries of mud and water) are depositing sediment upon the continental rise. Most recently geologists have learned that these deposits are synonymous with the geosynclinal sediments that undergo deformation in the process of mountain building. Hence, *geologists have concluded that the geoyscline is not a linear trough but is the accumulation of sediments on the continental shelf and continental rise.*

At some point in time the continental margin becomes an active subduction zone and the process of deformation is initiated (Figure 8.19B).

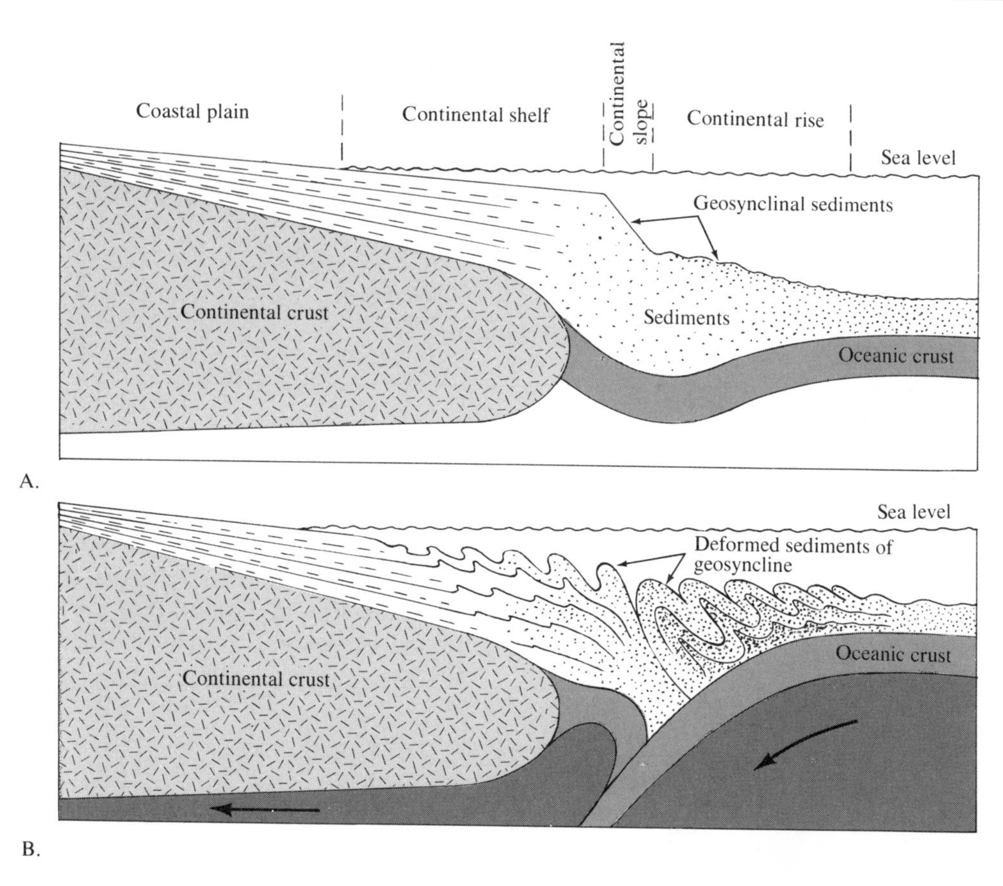

Figure 8.19 *Cross section showing the extent of sedimentation at* **A.** *the inactive continental margin and* **B.** *the active continental margin.*

A good place to examine an active continental margin is the west coast of South America. Here the Nazca plate is being subducted beneath the South American plate in the zone of the Peru-Chile trench (Figure 6.3). This subduction zone was probably formed in conjunction with the breakup of the supercontinent Pangaea (Figure 6.1). As South America was torn from Africa and carried westward, the oceanic crust adjacent to the west coast of South America was bent and thrust under the continental plate. However the oceanic crust does not give way without some effect on the overriding plate. In the case of South America, the Nazca plate caused deformation of the geosynclinal sediments which flanked the continental margin, producing the original folded and faulted part of the complex mountain system we now call the Eastern Andes. Subsequent subduction and partial melting of the Nazca plate initiated yet another stage—the development of the volcanic arc. In any arc system the volcanic activity is the most noticeable, but far greater quantities of magma are implaced below the surface, forming numerous batholiths. The effect of this is to thicken the crust. In response to isostatic adjustment, uplifting will follow. Thus the arc, because of uplifting and volcanic activity, may rise many kilometers above the height of the oceanic trench. The first volcanic mountains of the Andean arc formed seaward of the original folded mountains. As the submerging plate extended itself

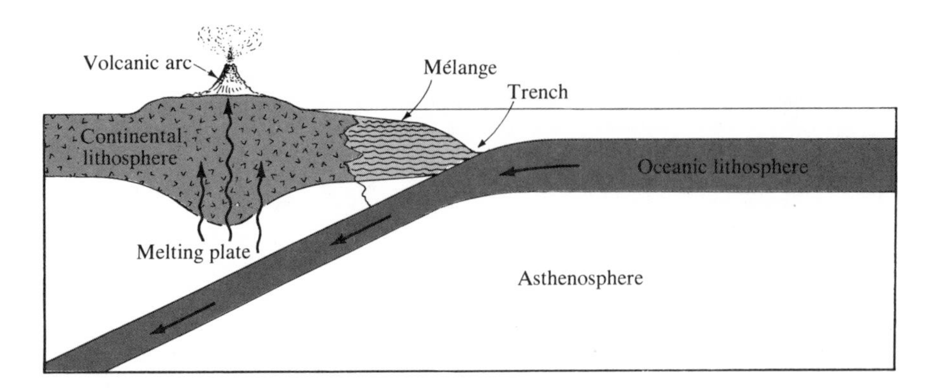

Figure 8.20 *Idealized cross section showing the formation of a mélange. These highly folded and faulted structures are characterized by high-pressure, low-temperature metamorphic rocks.*

eastward, it initiated volcanic activity in the same direction. Hence the present-day volcanic activity of the Andes is found eastward of the initial volcanoes, which have for the most part been eroded away. Remnants of the original volcanic arc are the exposed batholiths which make up the western flank of the Andes.

Some of the sediments from the eroding arc are carried seaward, where they are deposited between the volcanic arc and trench or on oc-

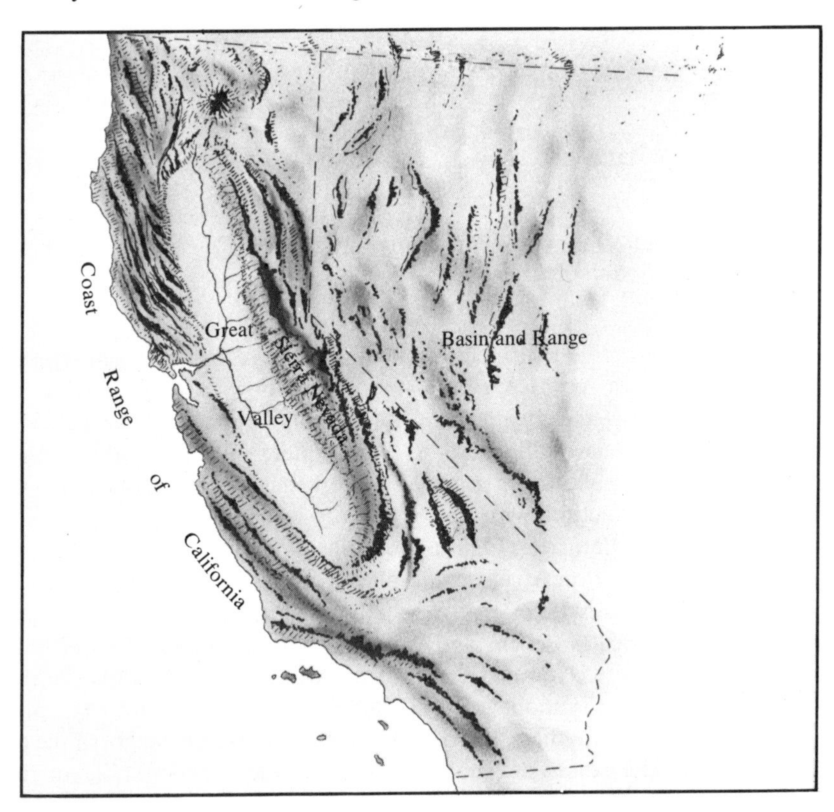

Figure 8.21 *Map of the California Coastal Range and the Sierra Nevada.*

casion in the trench itself. These sediments are also deformed by the converging plates (Figure 8.20), producing highly folded and faulted metamorphic rocks which contain slivers of oceanic crust that have been sheared off the descending plate. This aggregate of rock is termed a **mélange.** These complex structures are also found associated with the subduction zones of island arc systems like Japan. The metamorphic rocks of a mélange are formed under great pressure from the converging plates, but at rather low temperatures. Consequently, they can be distinguished from the metamorphic rocks that form at much higher temperatures in association with intrusive igneous bodies. When a mélange is found in the interior of a continent, it represents the relic of a former submergent zone. Such a circumstance provides an important clue for the interpretation of the geologic history of that region.

The California Coast Range and the Sierra Nevada are thought to be the remnants of a volcanic arc-and-trench system mélange as just described (Figure 8.21). These parallel mountainous belts were produced by the subduction of the Pacific plate under the North American plate. The Sierra Nevada batholith is the remnant volcanic arc and was produced by many separate surges of magma derived from the melting of the Pacific plate. Subsequent uplifting and erosion removed most of the evidence of the volcanic activity and exposed the batholith. In the trench region preexisting sediments and those provided by the eroding volcanic arc were strongly folded and faulted into the complex mélange which presently constitutes the Coast Range of California. The uplift of the Coast Range took place only in the recent geologic past, as evidenced by sediment which still covers part of it. The subduction of the Pacific plate ceased about 30 million years ago when the North American plate collided with a spreading ridge in the Pacific. At that time the San Andreas fault became the boundary between the plates, changing a former convergent plate boundary into a transform fault boundary where the plates slide past each other rather than under one another.

Continental Collisions

Up to this point we have discussed the formation of mountain systems associated with subduction zones where the leading edge of only one of the two converging plates contained continental crust. However it is possible for both of the colliding plates to be carrying continental crust. Because continental lithosphere is too bouyant to undergo subduction, a collision results. An excellent example occurred about 45 million years ago when India collided with Asia (Figure 6.10). India, previously a part of Antarctica, was rafted nearly 5000 kilometers due north before the collision occurred. The result was the formation of the spectacular Himalaya Mountains and the Tibetan Highlands. Although most of the oceanic crust which separated these landmasses prior to the collision was subducted, some was caught up in the squeeze along with sediment that lay offshore and can now be found elevated high above sea level.

The spreading center that separated India from Antarctica and moved it northward is still active; hence India continues to move into Asia at an estimated rate of a few centimeters per year. Now, however, numerous earthquakes have been recorded off the south coast of India, and geologists believe a new subduction zone is in the making. If formed, it would provide a disposal site for the crust of the Indian Ocean, which is continually being produced at the spreading center to the southwest. If this is the case, India's northward journey may soon come to an end. A very similar collision is believed to have occurred when the European continent collided with the Siberian continent to produce the Ural Mountains which extend in a north-south direction through the Soviet Union. Prior to the advent of plate tectonics geologists had difficulty explaining how thousands of meters of sediment were deposited and then became highly deformed while in the middle of a landmass like Asia.

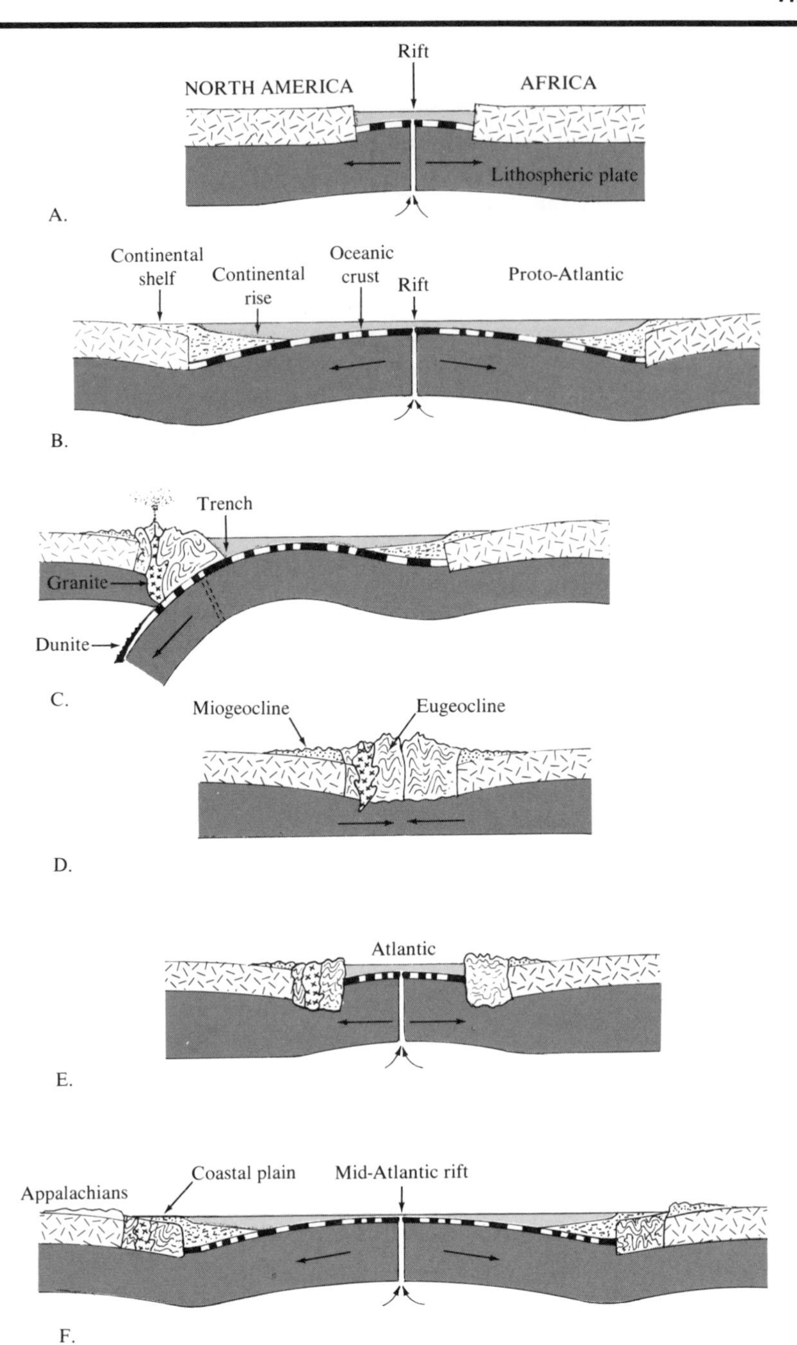

Figure 8.22 *Evolution of the Appalachians. According to the plate tectonic model, North America and Africa split some 600 million years ago to produce the ancestral Atlantic Ocean. Sediments that collected on the flank of North America were crumpled into the Appalachians as these two continents merged 350 million years ago. They again separated about 180 million years ago to produce the North Atlantic Ocean. (After "Geosynclines, Mountains, and Continent-Building," R.S. Dietz. Copyright © 1972 by Scientific American, Inc. All rights reserved)*

Other mountain ranges which show evidence of continental collisions are the Alps and the Appalachians. The Appalachian system was produced by the collision of North America and northern Africa (Figure 8.22). Although they have since separated, less than 200 million years ago these two continents were juxtaposed as part of the supercontinent Pangaea. It was this joining, geologists believe, that provided the impetus which generated this once-spectacular mountain system bordering the east coast of the United States. Prior to this collision this area was probably very similar to what we find today. Extensive continental shelf sediments flanked its margins, except that the sediments were deposited on a continental shelf which was located in the position of the present Ridge and Valley Province. Then, for reasons which are not yet fully explainable, the pre-Atlantic basin began to close, and North America and Africa converged. Studies of the sedimentary rocks in the Appalachians indicate that volcanic activity was initiated during the period of plate convergence. Some batholiths can be found in New England, but much of region to the south, which experienced concurrent igneous activity, is buried below the present-day continental shelf sediments.

Some geologists believe that this was not the first closing of the Atlantic, nor will it be the last. Should the Atlantic again close, the sediments which have accumulated along the east coast of the United States would surely be deformed into a spectacular mountain system. In this manner another sliver of land would be added to the continent.

In summary, *the plate tectonics model provides the most viable explanation for orogenesis.* The basic steps in this process, in an admittedly oversimplified form, include the following:

(1) Shallow- and deep-water sediments are deposited at an inactive continental margin.

(2) The continental margin becomes active. Plate convergence causes subduction of the oceanic plate, producing a trench system. Deformation of the deep-water sediments of the continental rise begins.

(3) Partial melting of the submerging oceanic plate initiates the period of igneous activity which produces the volcanic arc. The implaced intrusive bodies metamorphose the intruded sediments, forming the igneous-metamorphic core of the embryo mountain. Uplifting of the arc causes gravity sliding and deformation of the shelf sediments.

(4) Erosion of the volcanic arc carries sediments to the trench area. These sediments are deformed by the plunging plate or in some instances by continental collisions.

(5) Continued uplift of the thickened and deformed crust produces the lofty folded mountain system.

The evolution of each mountain system will have to be reevaluated in terms of the plate tectonics model. This process will shed new light on their evolutionary history and also be used to evaluate the model itself. In this way new insights into our dynamic planet will emerge.

REVIEW

1. State the three lines of evidence which support the concept of crustal uplift. Can you think of another?
2. What happens to a floating object when weight is added? Subtracted? How do these principles apply to changes in the elevation of mountains? What term is applied to the adjustment that causes crustal uplift of this type?
3. List two lines of evidence which support the idea of a floating crust.
4. What conditions favor rock deformation by folding? By faulting?
5. Using the concept of elastic rebound, explain why earthquakes cannot occur at great depths.
6. Compare the movement of normal and reverse faults. What type of force produces each?
7. At which of the three types of plate boundaries does normal faulting predominate? Reverse faulting? Strike-slip faulting?
8. Describe a horst and a graben. Explain how a graben valley forms and name one.

9. Compare and contrast anticlines and synclines. Domes and basins. Anticlines and domes.
10. Although we classify many mountains as folded, why might this description be misleading?
11. Name the site where sediments are deposited and have a good chance of being squeezed into a mountain range.
12. Which type of plate boundary is most directly associated with mountain building?
13. Describe a mélange and explain its formation.
14. In what way are the islands of Japan, the Sierra Nevada, and the Western Andes similar?
15. Would the discovery of a sliver of oceanic crust in the interior of a continent tend to support or refute the theory of plate tectonics? Why?
16. Why might it have been difficult for geologists to conclude that the Appalachian Mountains were formed by plate collision if examples like the Himalayas did not exist?
17. Why do geologists believe that the northward journey of India is near an end?
18. How does the plate tectonic theory explain the existence of fossil marine life on top of the Ural Mountains?
19. In your own words, briefly enumerate the steps involved in the formation of a mountain system according to the plate tectonic model.

KEY TERMS

orogenesis
geosyncline
isostasy
normal fault
reverse fault
thrust fault
strike-slip fault
transform fault
oblique fault
graben
horst
anticline
syncline
gravity sliding
dome
basin
folded mountains
volcanic mountains
fault-block mountains
upwarped mountains
mélange

The strata exposed in the Grand Canyon contain clues to millions of years of earth history. (Photo by N.W. Carkhuff, U.S. Geological Survey)

Geologic Time and Earth History

Almost 200 years ago James Hutton recognized that the earth was very old. But how old? Scientists tried to date the earth for many years, but their attempts were not very fruitful. Instead they had to rely on techniques which helped them place events in their proper order without knowing how long ago they occurred. How was this done? What part did fossils play? With the discovery of radioactivity and the rather recent development of radiometric dating techniques, geologists have been able to assign fairly accurate specific dates to the events of earth history. What is radioactivity? Why is it a good "clock" for dating the geologic past? In this chapter we shall attempt to answer the questions raised here, as well as others.

In 1869 John Wesley Powell, who was later to head the U.S. Geological Survey, led a pioneering expedition down the Colorado River and through the Grand Canyon (Figure 9.1). Writing about the strata that were exposed by the downcutting of the river, Powell said that, ". . . the canyons of this region would be a Book of Revelations in the rock-leaved Bible of geology." Powell was undoubtedly impressed with the countless millions of years of earth history exposed along the walls of the Grand Canyon (Figure 9.2). Interpreting earth history is a prime goal of the science of geology. Like a modern-day sleuth, the geologist must interpret the clues found preserved in the rocks. By studying rocks, especially sedimentary rocks, and the features they contain, geologists can often unravel the complexities of the past.

Events by themselves, however, have little meaning until they are put into a time perspective. Studying history, whether it be the Civil War or the Age of Dinosaurs, requires a calendar. Among the major contributions that geology has made to the knowledge of humankind is the geologic calendar and the concept that earth history is exceedingly long. Over many years geologists

Figure 9.1 **A.** *Start of the expedition from Green River Station. A drawing from Powell's 1875 book.* **B.** *Major John Wesley Powell, pioneering geologist and the second director of the U.S. Geological Survey. (A. and B. courtesy of U.S. Geological Survey)*

Figure 9.2 *Grand Canyon as depicted in an extraordinary drawing by W.H. Holmes. (Courtesy of U.S. Geological Survey)*

have devised a time scale of earth history—a calendar where geologic events can be put in their proper place. Geologists, recognizing that earth history has spanned an immense amount of time, worked at finding out just how old the earth is.

Early Methods of Dating the Earth

Current methods of radiometric dating put the age of the earth between 4500 and 5000 million years. However, this great age for the earth is a relatively recent discovery. Although James Hutton and others who accepted the principle of uniformitarianism believed the earth was very old, they had no way of knowing its exact age. Solutions to this dating problem were sought, and several methods were subsequently devised.

One method involved the rate at which sediment is deposited. Some geologists reasoned that if they could determine the rate that sediment accumulates, and could further ascertain the total thickness of sedimentary rock that had been deposited during earth history, they could accurately estimate the length of geologic time. All that was necessary was to divide the rate of sediment accumulation into the total thickness of sedimentary rock. Unfortunately this method was riddled with difficulties, some of which are as follows:

(1) Different sediments accumulate at different rates under varying conditions. Thus determining an overall rate of sediment accumulation is extremely difficult. Further, if such a rate is determined, it does not necessarily mean that the same rate can be applied to the past.

(2) Since no single locality has a complete geologic column, estimates of the total thickness of sedimentary rocks had to be compiled by adding together the maximum known thickness of rocks of each age. These estimates had to be

revised each time a thicker section was discovered.

(3) Sediment compacts when it is lithified; thus a correction for compaction had to be made.

Needless to say, estimates of the earth's age varied considerably as different scientists attempted this method. The figures representing the maximum thickness of sedimentary rock ranged from 9600 meters (32,000 feet) to over 100,500 meters (335,000 feet). The amount of time for 0.3 meter (1 foot) of sediment to accumulate varied from 100 years to over 8600 years. The age of the earth as calculated by this method therefore ranged from 3 million to 1500 million years!

Another method for dating the earth involved the salinity of the oceans, which were assumed to originally have been fresh water. Scientists felt that if they could accurately estimate the quantity of salt being carried to the ocean each year by rivers and the total amount of salt currently in the oceans, they could determine the length of geologic time by dividing the latter figure by the former. Near the turn of the twentieth century John Joly calculated the age of the earth at about 90 million years using this method. Joly, however, had no accurate notion of the amount of salt lost from the oceans because of deposition, evaporation, and winds blowing salt inland. It is also probable that the rate of salt accumulation has not always been constant. Thus Joly's estimate for the age of the earth was not accurate. However, both of the methods for dating the earth that have just been described indicated that the earth was considerably older than the 6000 years given it by Archbishop Usher (Chapter 1).

Perhaps the most influential estimates of the age of the earth were compiled by the well-known and highly respected physicist Lord Kelvin in the latter part of the nineteenth century. Since Kelvin's estimates required few assumptions and were based on precise measurements, they were widely accepted for a time. One of Kelvin's methods was founded on the widely held assumption that the earth had originally been molten and had cooled to its present condition. Although his data and calculations were limited, Kelvin still made it quite obvious that the earth could not be more than 100 million years old, and likely much less. The second of Kelvin's estimates was based on the fact that the source of the sun's tremendous output of energy was of a conventional nature (nuclear fusion and radioactivity had not yet been discovered). His calculations indicated that the sun could only have illuminated the earth for a few tens of millions of years. Furthermore, he said that in the past it had been much hotter and in the future it would become much cooler. He believed the earth was inhabitable for organisms for a period of only 20–40 million years. Kelvin's apparently irrefutable estimates had a rather profound impact:

> Evolutionists found it virtually impossible to accept these figures, but all they had were educated guesses in the face of Kelvin's potent mathematics. Darwin and others compromised their original theories in their later years in an effort to reconcile evolution and uniformitarianism with the physicists' estimates. Eventually, however, they were vindicated.*

Radioactivity and Radiometric Dating

Most chemical elements are stable and do not change. However some are unstable, constantly releasing heat as their nuclei break apart or decay. This is the heat that keeps the interior of the earth so very hot and is the source of the heat which Kelvin was measuring when he thought he was measuring the "cooling" earth.

In Chapter 1 we learned that an atom is composed of electrons, protons, and neutrons. Electrons have a negative charge and protons have a positive charge. Since a neutron is actually a proton and electron combined, it has no charge. Protons and neutrons are found in the center, or nucleus, of the atom, and electrons spin around the nucleus in definite paths, or

*Leigh W. Mintz. *Historical Geology: The Science of a Dynamic Earth,* 2nd ed. (Columbus, Ohio: Merrill, 1977), pp. 84–85.

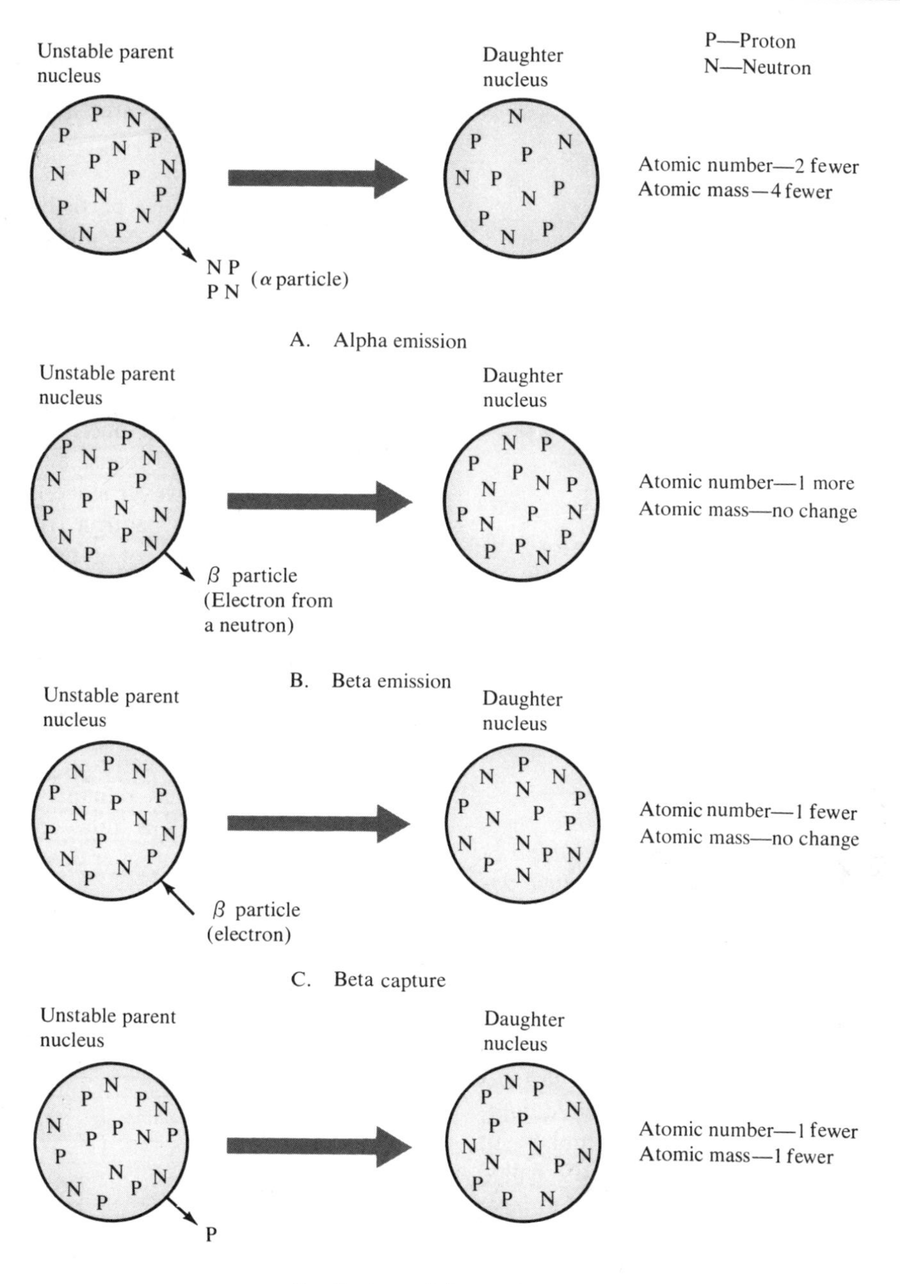

Figure 9.3 *Common types of radioactive decay. Notice that in each case the number of protons (atomic number) in the nucleus changes, thus yielding a different element.*

orbits. Practically all (99.9 percent) of the mass of an atom is found in the nucleus, indicating that electrons have practically no mass at all. By adding together the number of protons and neutrons in the nucleus, the mass of the atom is determined. The atomic number (the atom's identify-

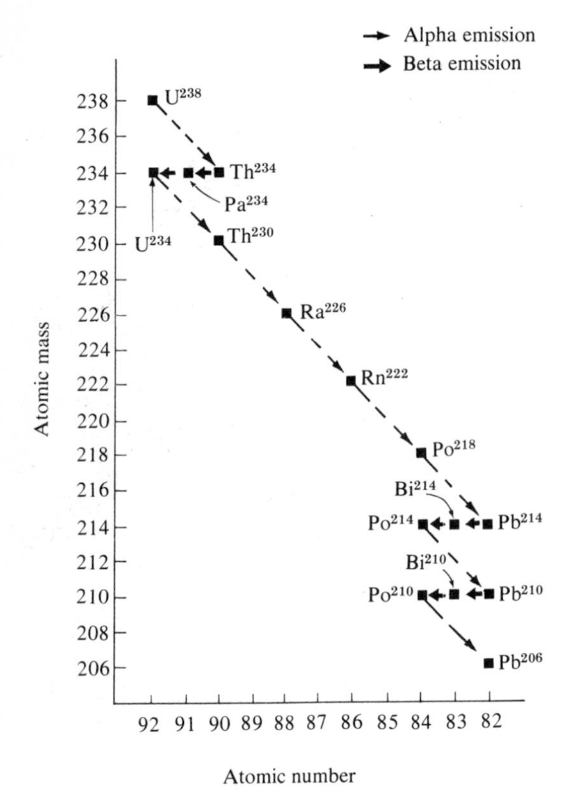

Figure 9.4 *The most common isotope of uranium (U^{92}_{238}) is an example of a radioactive decay series. Before the stable end product (Pb^{82}_{206}) is reached, many different isotopes are produced as intermediate steps.*

ing number) is equal to the number of protons. *Every element has a different number of protons in the nucleus, and thus a different atomic number.* Many elements have different numbers of neutrons in the nucleus. Such elements, called isotopes, have different atomic weights, but have the same atomic number.

The forces which bind protons and neutrons together in the nucleus are very strong; however the nature of these forces is still poorly understood. In some isotopes the nuclei are unstable; that is, the forces which bind the protons and neutrons together are not strong enough. As a result, the nuclei spontaneously break apart or decay, a process called **radioactivity.** What happens when unstable nuclei break apart? Four of the common types of radioactive decay are illustrated in Figure 9.3 and are summarized as follows:

(1) Alpha particles (α particles) may be emitted from the nucleus. An alpha particle is composed of 2 protons and 2 neutrons. Thus the emission of an alpha particle means the atomic mass of the isotope is reduced by 4 and that the atomic number is lowered by 2.

(2) When a beta particle (β particle), or electron, is given off from a nucleus, the atomic mass remains unchanged, because electrons have practically no mass. However since the electron must have come from a neutron (remember, a neutron is a combination of a proton and an electron), the nucleus contains one more proton than before. Therefore, the atomic number increases by 1.

(3) Sometimes a beta particle (electron) is captured by the nucleus. The electron combines with a proton and forms a neutron. As in the last example, the atomic mass remains unchanged. However since the nucleus now contains one fewer proton, the atomic number drops by 1.

(4) When a proton is ejected from the nucleus, the atomic number drops by 1, as does the atomic mass.

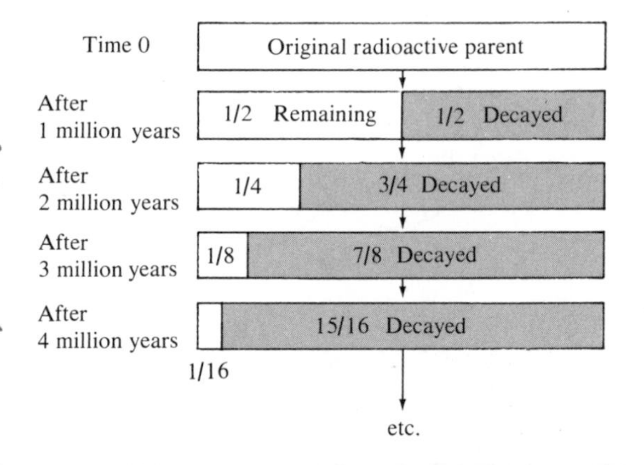

Figure 9.5 *Decay of a hypothetical radioactive isotope with a half-life of 1 million years.*

Erosion of a thick sandstone has formed Monument Valley, Utah. (Photo by Emil Muench)

The Sinai Peninsula is bordered by rifts believed to be caused by sea-floor spreading. (Courtesy of NASA)

Paricutin Volcano in eruption at night, 1943. (Photo by K. Segerstrom, U.S. Geological Survey)

Rainbow Bridge, a spectacular arch in Rainbow Bridge National Monument, Utah. (Photo by W.B. Hamilton, U.S. Geological Survey)

The radioactive isotope is often referred to as the **parent,** and the elements resulting from the decay of the parent are termed the **daughter products.** Figure 9.4 provides an example of radioactive decay. Here it may be seen that when the radioactive parent, uranium-238 (atomic number 92, atomic mass 238), decays, it emits 8 alpha particles and 6 beta particles before becoming the stable daughter product lead-206 (atomic number 82, atomic mass 206).

Certainly among the most important results of the discovery of radioactivity is that it provided a reliable means of calculating the ages of rocks and minerals which contain radioactive isotopes, a procedure referred to as **radiometric dating.** Why is radiometric dating reliable? The answer lies in the fact that *the rate at which radioactive isotopes decay is constant and unaffected by any chemical or physical agents.*

The amount of time it takes for one-half of a radioactive substance to decay, called its **half-life,** is a common way of expressing the rate of radioactive disintegration. If we began with a pound of radioactive material, half a pound would decay after one half-life, half the remaining amount would break down after another half-life, and so on.

Figure 9.5 illustrates the principle of radiometric dating using a hypothetical radioactive parent that decays directly into the stable daughter product. Its half-life is 1 million years. By calculating the percentages of radioactive parent and stable daughter product, the age of the specimen can be determined. In this example, when the quantities of parent and daughter are equal (ratio 1:1), we know that one half-life has transpired and that the specimen is 1 million years old. When the ratio of parent to daughter reaches 1:15, we know the sample is 4 million years old.

Of the many radioactive isotopes that exist in nature, only four have proved significant in providing radiometric ages for ancient rocks. Table 9.1 summarizes these most frequently used isotopes. Others are either very rare or have half-lives that are too short or much too long to be useful. Rubidium-87 and the two isotopes of uranium are used only for dating rocks that are millions of years old, but potassium-40 is more versatile. Although the half-life of potassium-40 is 1300 million years, recent analytic techniques have made it possible to detect the tiny amounts of its stable daughter product, argon-40, in rocks as young as 50,000 years.

To date more recent events, carbon-14 (also called **radiocarbon**), the radioactive isotope of carbon, is used. Since it has a half-life of only 5730 years, it can be used for dating events from the historic past, as well as those from recent geologic history. Until the late 1970s radio-

Table 9.1 ***Frequently Used Radioactive Isotopes***

Radioactive Parent	Half-life (years)	Stable Daughter Product	Minerals and Rocks Commonly Dated
Uranium-238	4510 million	Lead-206	Zircon; uraninite; pitchblende
Uranium-235	713 million	Lead-207	Zircon; uraninite; pitchblende
Potassium-40	1300 million	Argon-40	Muscovite; biotite; hornblende; glauconite; sanidine; whole volcanic rock
Rubidium-87	47,000 million	Strontium-87	Muscovite; biotite; lepidolite; microcline; glauconite; whole metamorphic rock

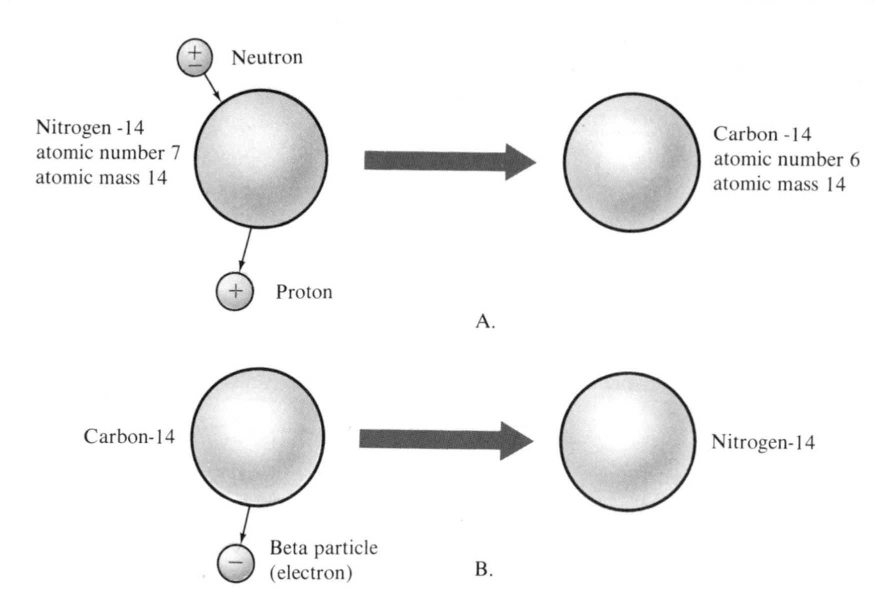

Figure 9.6 **A.** *Production and* **B.** *decay of carbon-14.*

carbon was useful in dating events only as far back as 40,000–50,000 years. However, as was the case with potassium-40, the development of more sophisticated analytical techniques has increased the usefulness of this "clock." Carbon-14 can now be used to date events as far back as 75,000 years. This is a significant accomplishment because it means that geologists can now date many Ice Age phenomena that previously could not be dated accurately.

Carbon-14 is continuously produced in the upper atmosphere as a consequence of cosmic ray bombardment, in which cosmic rays (high-energy nuclear particles) shatter the nuclei of gasses, releasing neutrons. The neutrons are absorbed by nitrogen (atomic number 7, atomic mass 14), causing its nucleus to emit a proton. Thus the atomic number drops by 1 (to 6), and a new element, carbon-14, is created (Figure 9.6A). This isotope of carbon is quickly incorporated into carbon dioxide, circulates in the atmosphere, and is absorbed by living matter. As a result, all organisms contain a small amount of carbon-14.

While an organism is alive the decaying radiocarbon is continually replaced, and the proportions of carbon-14 and carbon-12 (the normal isotope of carbon) remain constant. However when the plant or animal dies, the amount of carbon-14 gradually decreases as it decays to nitrogen-14 by beta emission (Figure 9.6B). Therefore by comparing the proportions of carbon-14 and carbon-12 in a sample, radiocarbon dates can be determined. Although carbon-14 is only useful in dating the last small fraction of geologic time, it has become a very valuable tool for anthropologists, archeologists, and historians, as well as for geologists who study very recent earth history. In fact, the development of radiocarbon dating was considered so important that the chemist who discovered this application, Willard F. Libby, received a Nobel Prize.

Bear in mind that although the basic principle of radiometric dating is rather simple, the actual procedure is quite complex, for the chemical analysis which determines the quantities of parent and daughter that are present must be painstakingly precise. In addition, some radioactive materials do not decay directly into the stable daughter product as was the case with our hypothetical example, a fact which may further

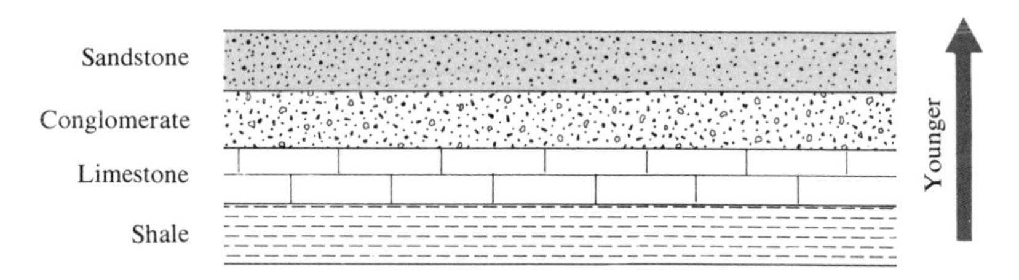

Figure 9.7 *Applying the law of superposition to this cross section, the shale bed is oldest, and the sandstone is youngest.*

complicate the analysis. In the case of uranium-238, there are thirteen intermediate unstable daughter products formed before the fourteenth, and last daughter product, the stable isotope lead-206, is produced (Figure 9.4).

Radiometric dating methods have produced literally thousands of dates for events in earth history. Rocks from several localities have been dated at more than 3000 million years, and geologists realize that still-older rocks exist. For example, a granite from South Africa which has been dated at 3200 million years contains inclusions of quartzite. Quartzite is a metamorphic rock which originally was the sedimentary rock sandstone. Since sandstone is the product of the lithification of sediments produced by the weathering of preexisting rocks, we have a positive indication that older rocks existed.

Radiometric dating has vindicated the ideas of Hutton, Darwin, and others who over 150 years ago assumed that geologic time must be immense. Indeed, it has proven that there has been enough time for the slow processes we can observe to have accomplished tremendous tasks.

The Magnitude of Geologic Time

The magnitude of geologic time is a most difficult concept to grasp, because we must learn to think in spans of time that far exceed our common experience. Earth features, which seem to be everlasting and unchanging to us and in fact to generations of people, are indeed slowly changing. Thus over millions of years, mountains rise and are eroded to hills, and rivers excavate deep canyons. How long is 5000 million years? If you were to begin counting to 5000 million at the rate of one number per second and continued 24 hours a day, 7 days a week, and never stopped, it would take about two lifetimes (150 years) to reach 5000 million! Don L. Eicher gives us another basis for comparison:

> Compress for example, the entire 4.5 billion [4500 million] years of geologic time into a single year. On that scale, the oldest rocks we know date from about mid-March. Living things first appeared in the sea in May. Land plants and animals emerged in late November and the widespread swamps that formed the Pennsylvanian coal deposits flourished for about four days in early December. Dinosaurs became dominant in mid-December, but disappeared on the 26th, at about the time the Rocky Mountains were first

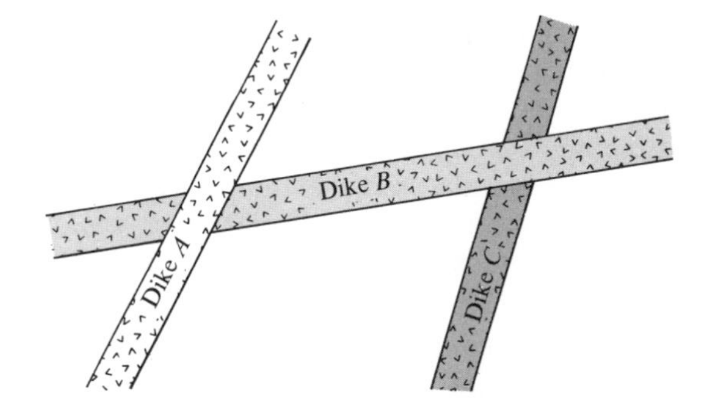

Figure 9.8 *Cross-cutting relationships. All of the dikes are younger than the rock into which they were intruded. Since Dike* B *cuts through Dike* C *and Dike* A *cuts through Dike* B*, the order of intrusion, from oldest to youngest, is Dike* C*, Dike* B*, Dike* A.

uplifted. Manlike creatures appeared sometime during the evening of December 31st, and the most recent continental ice sheets began to recede from the Great Lakes area and from northern Europe about 1 minute and 15 seconds before midnight on the 31st. Rome ruled the Western world for 5 seconds from 11:59:45 to 11:59:50. Columbus discovered America 3 seconds before midnight, and the science of geology was born with the writings of James Hutton just slightly more than one second before the end of our eventful year of years.*

Relative Dating

Radiometric dating results in specific dates for rock units which represent various events in the earth's distant past. We can now state with some confidence that particular geologic events took place a certain number of years ago. Such dates are referred to as **absolute dates,** for they pinpoint the time in history when something took place. Prior to the discovery of radioactivity and the development of the technology of radiometric dating, geologists had no precise method of absolute dating and had to rely solely on relative dating. **Relative dating** means that rocks are placed in their proper sequence or order. Relative dating will not tell us how long ago something took place, only that it followed one event and preceded another. The relative dating techniques which were developed are still widely used. Absolute dating methods did not replace these techniques; they simply supplemented them. To establish a relative time scale, a few simple principles or rules had to be discovered and applied. Although they may seem rather obvious to us today, their discovery was a very important scientific achievement.

The most basic of all the principles of relative dating is the **law of superposition,** which simply states that *in an undeformed sequence of sedimentary rocks, each bed is older than the one above it and younger than the one below.* This rule also applies to other surface-deposited materials such as lava flows and beds of ash from volcanic eruptions. Applying this law to the beds shown in Figure 9.7, we can easily place each of the layers in its proper place, with the sandstone being the youngest and the shale the oldest. In reality, there are many cases in which the deformation of mountain building has overturned the strata, making the sequence of beds difficult to determine. When such circumstances occur, they must be realized and corrected for by the geologist.

When igneous intrusions or faults cut through other rocks, they are assumed to be younger than the things they cut. For example, when two dikes intersect, the older one must have been opened up in order to allow the younger one to cut through it. The younger dike would therefore be continuous, while the older dike would be interrupted at the point of their intersection. Figure 9.8 illustrates this principle of **cross-cutting.**

Layers of rock are said to be **conformable** when they are found to have been deposited without interruption. However there is no place on earth that contains a complete set of conformable strata. Even for a particular span of time, many locations do not have a complete sequence of rocks representing the entire period. All such breaks in the rock record are termed **unconformities.** Figure 9.9 illustrates some of the ways in which unconformities may develop. Perhaps the most easily recognized type of unconformity consists of tilted or folded sedimentary rocks that are overlain by other, more flat-lying strata. These are called **angular unconformities** and indicate that the period of distortion (folding or tilting) and erosion is not represented by sedimentary rocks (Figure 9.10). **Disconformities** are often much more difficult to recognize because the strata on either side of these unconformities are essentially parallel. Disconformities may represent either a period of nondeposition or a period of erosion.

By applying the principles of relative dating to the hypothetical geologic cross section shown

* Don L. Eicher. *Geologic Time* (Englewood Cliffs, New Jersey: Prentice-Hall, 1968), p. 19.

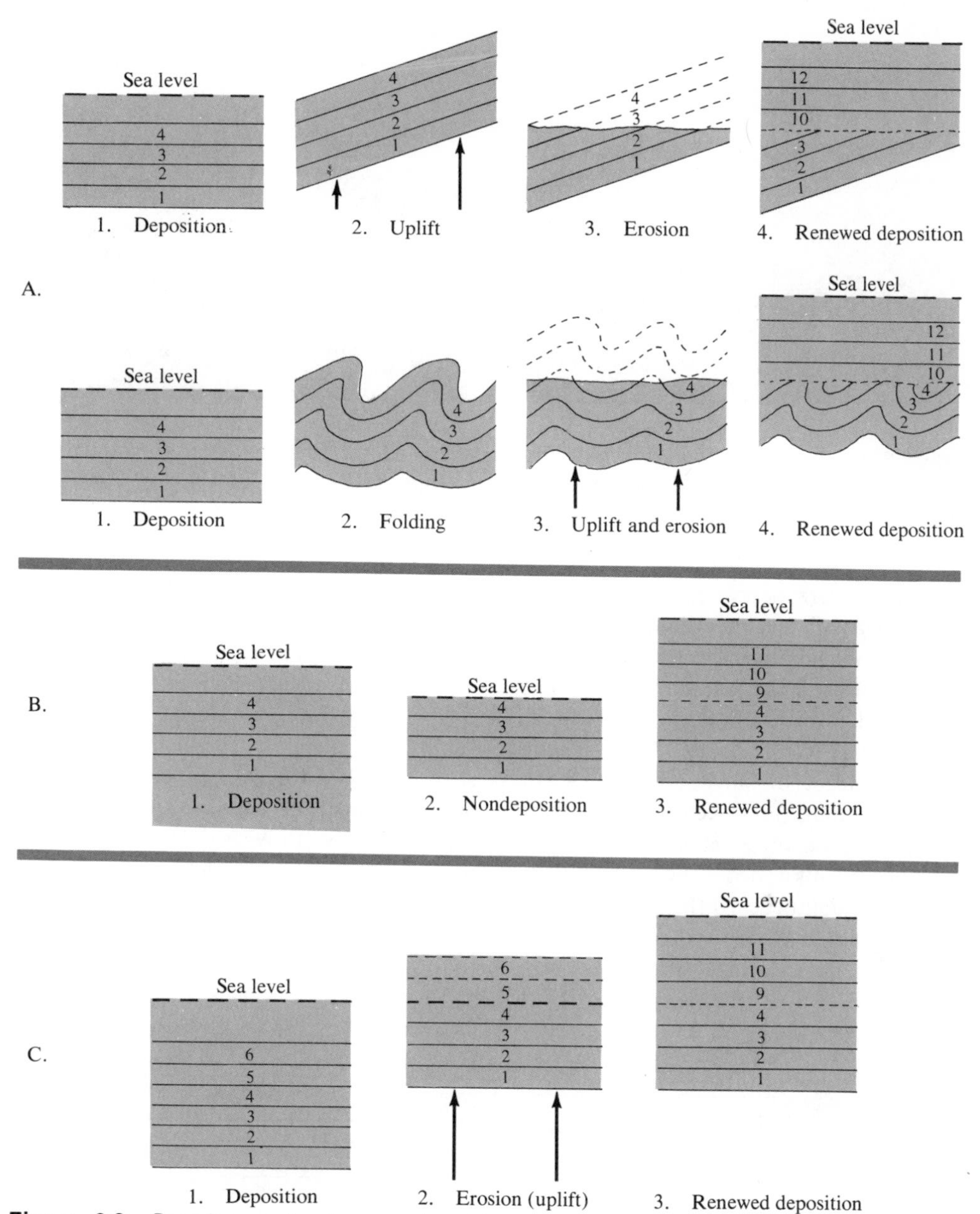

Figure 9.9 *Development of unconformities.* **A.** *Two examples of angular unconformity.* **B.** *Nondeposition resulting in a disconformity.* **C.** *Erosion resulting in a discomformity. In each case a gap in the rock record has been created. Notice that angular unconformities are easier to identify than disconformities. (From Foster,* Physical Geology, *3rd ed., Columbus, Ohio: Charles E. Merrill, 1979)*

in Figure 9.11, the rocks and the events in earth history they represent may be placed in their proper sequence. The following statements summarize the logic used to interpret the cross section:

(1) Applying the law of superposition, beds

Figure 9.10 *Angular unconformity. The tilted beds of Laramie Sandstone are overlain by the horizontal beds of the Wasatch Conglomerate, Park County, Wyoming. (Photo by C.A. Fisher, U.S. Geological Survey)*

10, 9, 8, and 6 were deposited, in that order. Since bed 7 is a sill (a concordant igneous intrusion), it is younger than the rocks that were intruded. Further evidence that the sill is younger than beds 8 and 6 are the inclusions in the sill of fragments from these beds. If the igneous mass contains pieces of surrounding rock, the surrounding rock must have been there first.

(2) Following the intrusion of the sill (bed 7), the intrusion of the dike (bed 11) occurred. Since the dike cuts through beds 10 through 6, it must be younger than all of them.

(3) Next, the rocks were tilted and then eroded. We know the tilting happened first because the upturned ends of the strata have been eroded. The tilting and erosion, followed by further deposition, produced an angular unconformity.

(4) Beds 5 through 1 were deposited in that order, again using the law of superposition. Although the lava flow (bed 4) is not a sedimentary rock layer, it is a surface-deposited layer, and thus superposition may be applied.

(5) Finally, the irregular surface and the

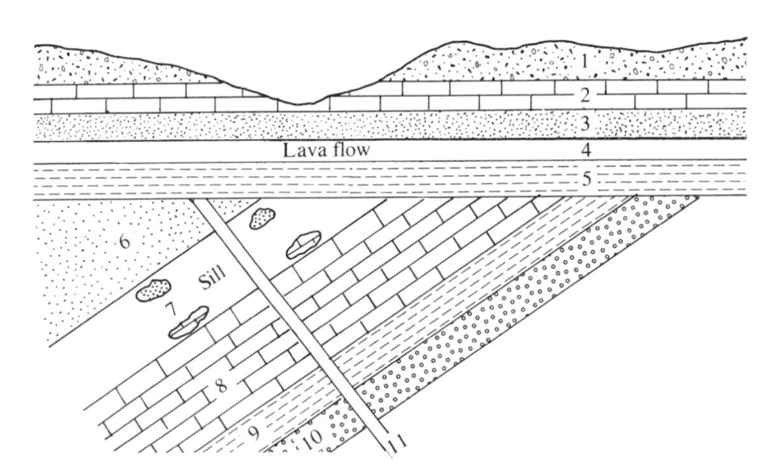

Figure 9.11 *Geologic cross section of a hypothetical region.*

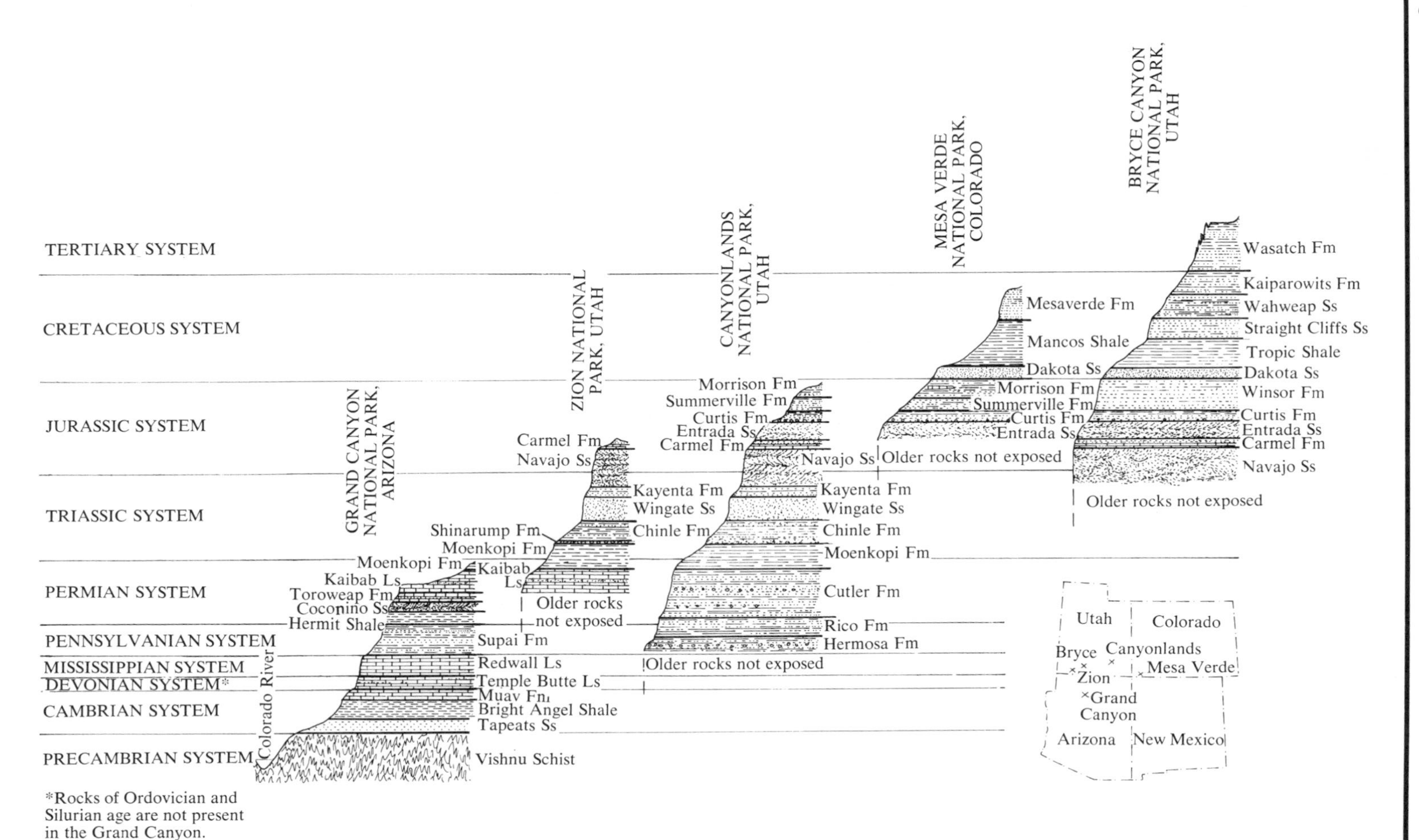

Figure 9.12 *Correlation of strata at several locations on the Colorado Plateau reveals the total extent of sedimentary rocks in the region. (After U.S. Geological Survey and Foster,* Physical Geology, *3rd ed., Columbus, Ohio: Charles E. Merrill, 1979)*

stream valley indicate that another gap in the rock record is being produced by erosion.

In the foregoing example our goal was to establish a relative time scale for the rocks and events in the area of the cross section. Remember, we do not have any idea how many years of earth history are represented, nor do we know how this area compares to any other.

Correlation

In order to develop a geologic calendar that is applicable to the whole earth, rocks of similar age in different regions must be matched up. Such a task is referred to as **correlation.** Within a limited area there are several methods of correlating the rocks of one locality with those of another. A bed or series of beds may be traced simply by walking along the outcropping edges. However this may not be possible when the continuity of the bed is interrupted. Correlation over short distances is often achieved by noting the place of a bed in a sequence of strata, or a bed may be identified in another location if it is composed of very distinctive or uncommon minerals. By correlating the rocks from one place to another, a more comprehensive view of the geologic history of a region is possible. Figure 9.12, for example, shows the correlation of strata at several places on the Colorado Plateau. No single locale contains a complete sequence, but correlation reveals the total extent of sedimentary rocks.

Most geologic studies involve rather small areas. Although they are usually important in their own right, their full value is realized only when they are correlated with other regions. Although the methods just described may be sufficient to trace a rock formation over relatively short distances, they are not adequate for matching up rocks at great distance. When correlation between widely separated areas or between continents is the objective, the geologist must rely upon fossils.

Fossils

Paleontology, the branch of geology devoted to the study of ancient life, is based on the study of fossils. Originally the term **fossil** referred to any curious object dug from the ground, but it is now used to mean the remains or traces of organisms preserved from the geologic past. Although geologists studying past life must have a firm background in the biological sciences, they are often quick to point out the differences between the two fields: "If the remains stink they belong to zoology, but if not, to paleontology."

Fossils are of many types. The remains of relatively recent organisms may not have been altered at all. Such objects as teeth, bones, and shells are common examples (Figure 9.13). Far less common are entire animals, flesh included, that have been preserved because of rather unusual circumstances. Remains of prehistoric elephants called mammoths that were frozen in the Arctic tundra of Siberia and Alaska are examples (Figure 9.14A), as are insects that were entrapped in tree resin (amber), which then hardened (Figure 9.14B). More commonly, fossils are petrified (literally, "turned into stone"), meaning that the original substance, like wood or shells, has been replaced by mineral matter, or that the open spaces have been filled by minerals (Figure 9.15). In addition, many fossils, rather than being the remains (altered or not) of prehistoric life, are just traces of past life. Shell casts and molds, footprints, trails, and worm burrows are all examples (Figure 9.16).

Sometimes objects which people thick are fossils are merely rocks with an accidental resemblance. George Gaylord Simpson, a noted paleontologist, related this experience:

> Only the other day I was offered for sale at a large price 'the petrified leg of a woman.' I was called a liar and a cheat when I explained that it was only a piece of volcanic rock with an accidental (and very slight) resemblance to the vision in the mind of the owner.

Figure 9.13 *Fossil shark teeth, an example of unaltered remains.*

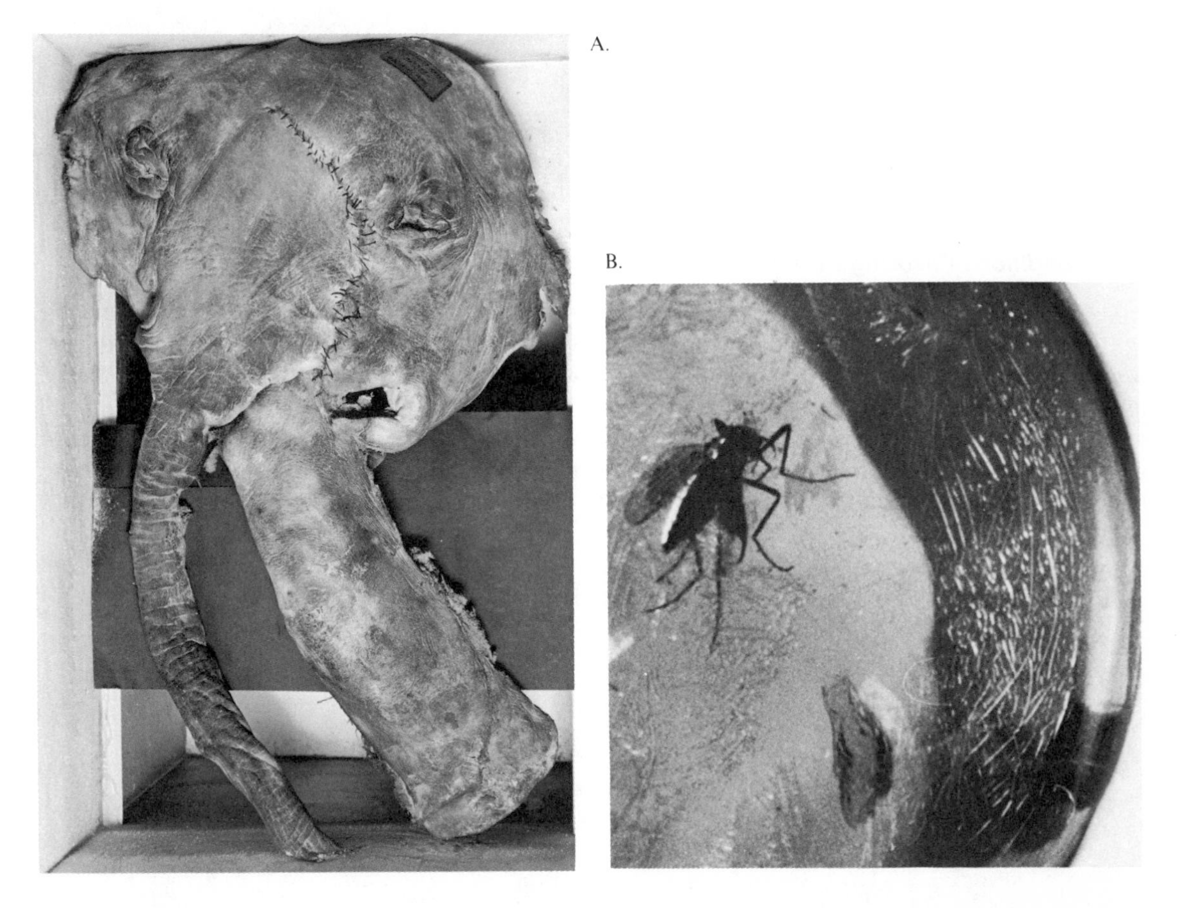

Figure 9.14 **A.** *Well-preserved young wooly mammoth dug out of frozen ground in Alaska. (Courtesy of the American Museum of Natural History)* **B.** *Insect in amber. (Courtesy of Ward's Natural Science Establishment, Inc., Rochester, N.Y.)*

A.

B.

Figure 9.15 **A.** *Petrified wood in Petrified Forest National Park, Arizona. (Photo by N.H. Darton, U.S. Geological Survey)* **B.** *Petrified bones of the large dinosaur Brontosaurus. (Photo by H.B. Robinson, National Park Service, U.S. Department of the Interior)*

Conditions Favoring Preservation

Only a tiny fraction of the organisms that have lived during the geologic past have been preserved as fossils. Normally the remains of an animal or plant are totally destroyed. Under what circumstances are they preserved? Two special conditions appear to be necessary: rapid burial and the possession of hard parts.

Usually when an organism perishes, its soft parts are quickly eaten by scavengers or decomposed by bacteria. The remaining hard parts are then weathered, eventually crumbling into dust. Occasionally, however, the remains are buried by sediment. In this situation, scavengers and weathering cannot disturb them because they have been removed from the environment where these destructive forces operate. *Rapid burial*

A.

B.

Figure 9.16 **A.** *Natural casts of several different shelled invertebrates. (Photo by E.B. Hardin, U.S. Geological Survey)* **B.** *Dinosaur footprint in sandstone, Jefferson County, Colorado. (Photo by J.R. Stacy, U.S. Geological Survey)*

following death therefore is an important condition favoring preservation.

In addition, *animals and plants have a much better chance of being preserved as part of the fossil record if they have hard parts.* Although traces and imprints of soft-bodied animals such as jellyfish, worms, and insects exist, they are, to say the least, rare. Flesh usually decays so rapidly that preservation is exceedingly remote. Hard parts like shells, bones, teeth, and wood predominate in the record of past life.

Because preservation is contingent on special conditions, the record of life in the geologic past is slanted. While the record of those organisms with hard parts that lived in areas of sedimentation is quite complete, we only get an occasional glimpse at the rest (Figure 9.17).

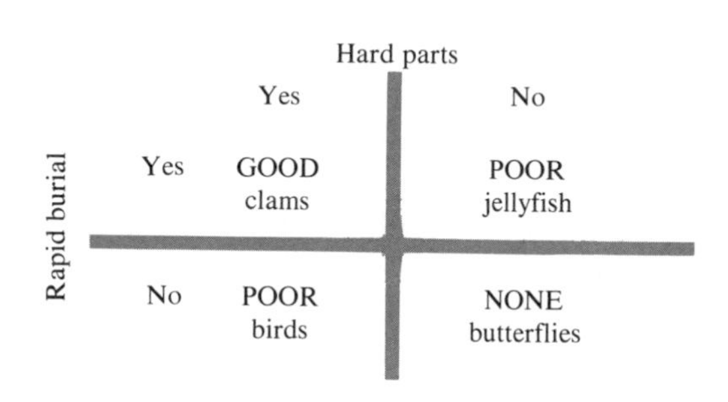

Figure 9.17 *Organisms which live in areas of sedimentation and possess hard parts are the most likely to be preserved. If hard parts are lacking or rapid burial does not occur, the liklihood of preservation diminishes. If neither condition exists, as with butterflies, there is practically no chance of fossilization. (From Leo F. Laporte,* Ancient Environments, *Englewood Cliffs, New Jersey: Prentice-Hall, Inc, 1968, p. 75)*

Fossils and Correlation

Although the existence of fossils had been known for centuries, it was not until the late 1700s and early 1800s that their significance as geologic tools was made evident. During this period an English engineer and canal builder, William Smith, discovered that each rock formation in the canals contained fossils unlike those in the beds either above or below. Further, he noted that sedimentary strata in widely separated areas could be identified by their distinctive fossil content. Based upon Smith's classic observations and the findings of many geologists who followed, one of the most important and basic principles in historical geology was formulated: *Fossil organisms succeed one another in a definite and determinable order, and therefore any time period can be recognized by its fossil content.* This has come to be known as the **principle of faunal succession.** In other words, when fossils are arranged according to their age, they do not present a random or haphazard picture. To the contrary, fossils show progressive changes from simple to complex and reveal the advancement of life through time. For example, an Age of Trilobites is recognized quite early in the fossil

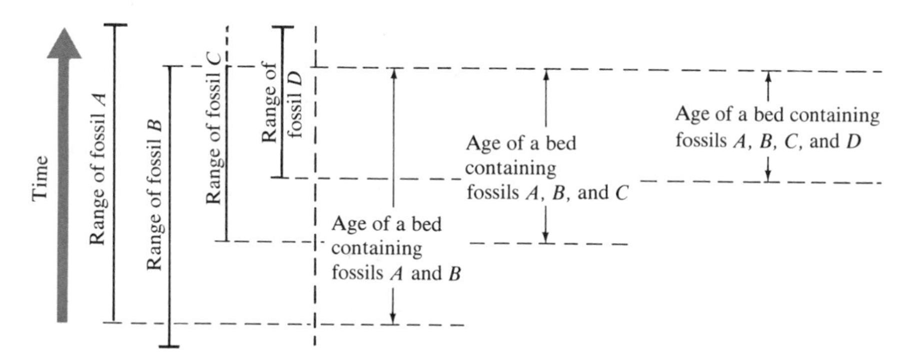

Figure 9.18 *Use of overlapping ranges of fossils helps date rocks more exactly than using a single fossil. (From Foster,* Physical Geology, *3rd ed., Columbus, Ohio: Charles E. Merrill, 1979)*

record. Then, in succession, paleontologists recognize an Age of Fishes, an Age of Coal Swamps, an Age of Reptiles, and an Age of Mammals. These "ages" pertain to groups that were especially plentiful and characteristic during particular time periods. Within each of the "ages" there are many subdivisions based, for example, on certain species of trilobites, and certain types of fish, reptiles, and so on. This same succession of dominant organisms, never out of order, is found on every major landmass.

Since fossils were found to be time indicators, they became the most useful means of correlating rocks of similar age in different regions. Geologists pay particular attention to certain fossils called **index,** or **guide, fossils.** Since these fossils are widespread geographically and are limited to a short span of geologic time, their presence provides an important method of matching rocks of the same age. Rock formations, however, do not always contain a specific index fossil. In such situations, groups of fossils are used to establish the age of the bed. Figure 9.18 illustrates how a group of fossils may be used to date rocks more precisely than could be accomplished by the use of any one of the fossils.

In addition to being important and often essential tools for correlation, fossils are also important environmental indicators. Although much can be deduced about past environments by studying the nature and characteristics of sedimentary rocks, a close examination of the fossils present can usually provide a great deal more information. For example, when the remains of certain clam shells are found in limestone, the geologist can deduce that the region was once covered by a shallow sea. Also, by using what we know of living organisms, we can conclude that fossil animals with thick shells capable of withstanding pounding and surging waves inhabited shorelines. On the other hand, animals with thin, delicate shells probably indicate deep, calm offshore waters. Hence by taking a closer look at the types of fossils, the approximate shoreline may be identified. Further, fossils can be used to indicate the former temperature of the water. Certain kinds of present-day corals must live in warm and shallow tropical seas like those around Florida and the Bahamas. When similar types of coral are found in ancient limestones, they give a good estimate of the marine environment that must have existed when they were alive. The preceding are just a few brief examples of how fossils can help unravel the complex story of earth history.

The Geologic Calendar

The whole of geologic history has been subdivided into units of varying magnitude which together comprise the calendar of earth history (Figure 9.19). The major units of the calendar were delineated during the nineteenth century, principally by workers in western Europe and Great Britain. Since absolute dating was not a reality during this time, the entire calendar was created using methods of relative dating. It has only been recently that absolute dates have been added to the calendar.

By examining Figure 9.19, you can see that the largest of the subdivisions of the geologic calendar are called **eras.** Three eras are currently recognized: the **Paleozoic** ("ancient life"), the **Mesozoic** ("middle life"), and the **Cenozoic** ("recent life"). As the names imply, the eras are bounded by quite profound worldwide changes in life forms. Each of the eras is subdivided into time units known as **periods.** The Paleozoic has seven, the Mesozoic three, and the Cenozoic two. Since we are currently living in the Cenozoic era, there may be more periods yet to come. Each period is characterized by a somewhat less profound change in life forms as compared with the eras. Finally, each of the twelve periods are further divided into still smaller units called **epochs.** Except for the seven epochs which have been named for the periods of the Cenozoic era, those of other periods are not named.

Notice that the detail of the geologic calendar does not begin until about 600 million years ago, the date for the beginning of the first period of the Paleozoic era, the Cambrian period. The more than 4000 million years prior to the Cambrian is simply referred to as the **Precambrian.** Why is the Precambrian not subdivided into numerous eras, periods, and epochs? The reason is that Precambrian history is not known in great enough detail. The quantity of information geologists have deciphered about the earth's past is somewhat analogous to the detail of human history. The farther back we go, the less that is known. Certainly more data and information exist about the past ten years than for the first decade of the twentieth century; the events of the nineteenth century have been documented much better than the events of the first century A.D.; and so on. So it is with earth history. The more recent past has the freshest, least disturbed, and most observable record. The farther back in time the geologist goes, the more fragmented the

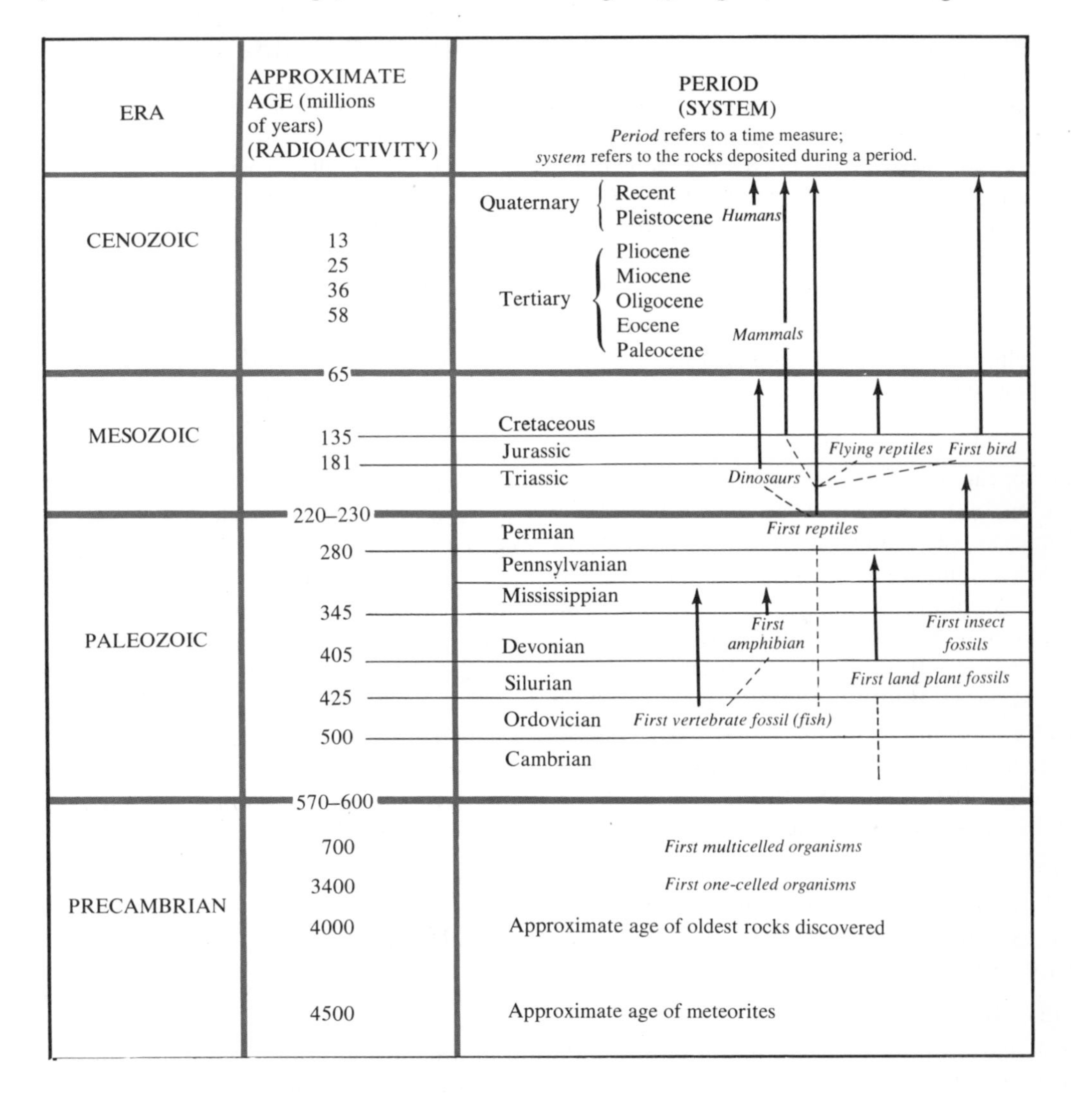

Figure 9.19 *The geologic calendar. The absolute dates were added quite recently, long after the calendar had been established using relative dating techniques. (From Foster,* Physical Geology, *3rd ed., Columbus, Ohio: Charles E. Merrill, 1979)*

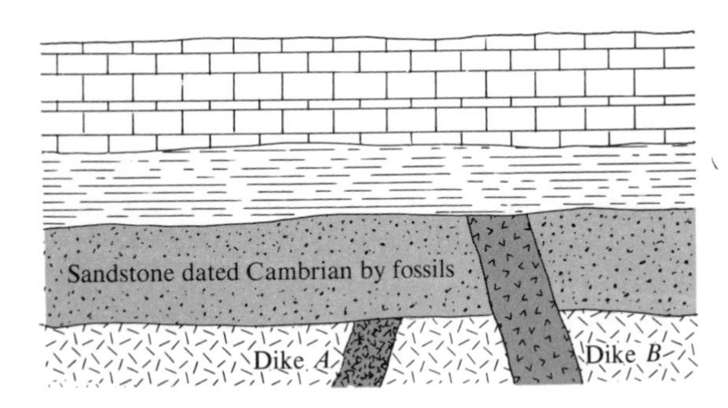

Figure 9.20 *Since radiometric dating is generally confined to igneous rocks, an absolute age for this bed of Cambrian sandstone must be determined by its relationships to Dikes* A *and* B. *In this example, the sandstone is older than Dike* B *and younger than Dike* A. *(From Foster,* Physical Geology, *3rd ed., Columbus, Ohio: Charles E. Merrill, 1979)*

record and clues become. There are other reasons to explain our lack of a detailed time scale for this vast segment of earth history:

(1) The first abundant fossil evidence does not appear in the geologic record until the beginning of the Cambrian period. Prior to the Cambrian, very simple life forms such as algae, bacteria, fungi, worms, and sponges predominated. All of these organisms lack hard parts, an important prerequisite for fossilization. Therefore there is only a meager Precambrian fossil record. Many exposures of Precambrian rocks have been studied in some detail, but correlation is exceedingly difficult without fossils. Consequently, there is no general agreement on the subdivisions of the Precambrian.

(2) Most Precambrian rocks are covered by rocks of a more recent age. Since exposures are few and far between, the study of Precambrian history is very difficult.

(3) Because Precambrian rocks are very old, most have been subjected to a great many changes. The bulk of the Precambrian rock record is composed of highly distorted metamorphic rocks. This makes the interpretation of past environments very hard, because many of the clues found in sedimentary rocks have been destroyed.

With the development of radiometric dating methods, a solution to the troublesome task of dating and correlating Precambrian rocks is at hand. Untangling the complex Precambrian record, however, is still many years away.

Difficulties in Dating the Geologic Calendar

Although reasonably accurate absolute dates have been worked out for the periods of the geologic calendar (see Figure 9.19), the task is not without its difficulties. The primary difficulty in assigning absolute dates to units of time lies in the fact that radioactive elements are typically restricted to igneous rocks. Even if a clastic rock included sediment which contained a radioactive mineral, the rock could not be dated, for the grains in a clastic sedimentary rock are not the same age as the rock in which they occur. The sediments composing such a rock may have been weathered from rocks of diverse ages. Thus the age of a mineral in a sedimentary rock only tells us that the rock can be no older.

On the other hand, in igneous rocks the minerals and rock form simultaneously; the age of the mineral containing a radioactive isotope is the same age as the rock. Therefore, in order to date sedimentary strata, the geologist must relate them to igneous masses, as in Figure 9.20. The problem of dating the periods illustrates the necessity of combining laboratory dating methods with geological information.

REVIEW

1. Describe two early methods for dating the earth. How old was the earth thought to be according to these estimates? List some weaknesses of each method.
2. If a radioactive isotope of thorium (atomic number 90, atomic mass 232) emits 6 alpha particles and 4 beta particles during the

course of radioactive decay, what is the atomic number and atomic mass of the stable daughter product?

3. Why is radiometric dating the most reliable method of dating the geologic past?
4. A hypothetical radioactive isotope has a half-life of 10,000 years. If the ratio of radioactive parent to stable daughter product is 1:3, how old is the rock containing the radioactive material?
5. Assume that the age of the earth is 5000 million years.
   a. What fraction of geologic time is represented by recorded history (assume 5000 years for the length of recorded history)?
   b. The first abundant fossil evidence does not appear until the beginning of the Cambrian period (600 million years ago). What percent of geologic time is represented by abundant fossil evidence?
6. Distinguish between absolute and relative dating.
7. What is the law of superposition? How are cross-cutting relationships used in relative dating?
8. Draw a geologic cross section of an area that has experienced the following sequence of events. After completing the cross section, label all the unconformities.
   a. Deposition of four sedimentary beds.
   b. Emergence above sea level and the erosion of two sedimentary layers.
   c. Submergence, followed by the deposition of three more sedimentary beds.
   d. Folding.
   e. Intrusion of a dike.
   f. Erosion to a flat surface.
   g. Deposition of four more sedimentary beds.
   h. Tilting of the entire region.
   i. Erosion of the entire region to a flat surface.
   j. Deposition of two more sedimentary layers.
9. What is meant by the term *correlation?*
10. Describe some different types of fossils. What organisms have the best chance of being preserved as fossils?
11. Describe William Smith's important contribution to the science of geology.
12. Why are fossils such useful tools in correlation?
13. In addition to being important aids in dating and correlating rocks, how else are fossils helpful in geologic investigations?
14. What subdivisions make up the geologic calendar? What is the primary basis for differentiating the eras?
15. Why is the Precambrian not subdivided into smaller units?
16. Briefly describe the difficulties in assigning absolute dates to layers of sedimentary rock.

KEY TERMS

radioactivity
parent
daughter product
radiometric dating
half-life
radiocarbon
absolute dates
relative dating
law of superposition
cross-cutting
conformable
unconformity
angular unconformity
disconformity
correlation
paleontology
fossil
principle of faunal succession
index (guide) fossil
era
Paleozoic
Mesozoic
Cenozoic
period
epoch
Precambrian

part 2

the oceans

The H.M.S. Challenger. *(From C.W. Thomson and Sir John Murray,* Report on the Scientific Results of the Voyage of the H.M.S. *Challenger, Vol. 1. Great Britain: Challenger Office, 1895, Plate 1)*

10 Extent, Topography, and Sediments

How deep is the ocean? How much of the earth is covered by the global sea? These are basic questions, but they went unanswered for thousands of years. Although people's interest in the oceans undoubtedly dates back to ancient times, it was not until rather recently that these seemingly simple questions began to be answered. The beginning of this chapter deals with these answers. Then, we shall take a brief glimpse at the history of oceanography. The remainder of the chapter provides a look at the earth beneath the sea. Suppose that all of the water were drained from the ocean. What would we see? Plains? Mountains? Canyons? Plateaus? Indeed, the ocean conceals all of these features, and more. In fact, the topography of the ocean floor is as varied as that of any continent. And what about the carpet of sediment that mantles the sea floor? Where did it come from and what can we learn by examining it?

Calling the earth the "water planet" is indeed appropriate, because nearly 71 percent of its surface is covered by the global ocean. Although the ocean comprises a much greater percentage of the earth's surface than the continents, it was only in the relatively recent past that the ocean became an important focus of study. Recently, there has been a virtual explosion of data about the oceans, and with it, oceanography has grown dramatically.

Oceanography is actually not a science in itself; rather it involves the application of all the sciences in a comprehensive and interrelated study of the oceans in all of their aspects and relationships. A brief discussion of some of the major subdivisions of oceanography follows. *Physical oceanography* is primarily concerned with energy transmission through ocean water, specifically with such items as wave formation and propagation, currents, tides, energy exchange between ocean and atmosphere, and penetration of light and sound. *Chemical oceanography* is a study of the chemical properties of seawater, of the cause and effect of variation of these properties with time and from place to place, and of the means of measuring these properties. *Biological oceanography* is the study of the interrelationship of marine life with its oceanic environment. The study includes the distribution, life cycles, and population fluctuations of marine organisms. Finally, *geological oceanography* deals with the floor and shore of the oceans and embraces such subjects as submarine topography, geological structure, erosion, and sedimentation. *The interrelationship of all the sciences is a chief characteristic of oceanography.**

The Seven Seas?

The term *sea* is used in two different ways. It is often used interchangeably with *ocean* in referring to bodies of salt water. On the other hand, a sea is also considered to be part of an ocean, or to be a body of water that is substantially smaller than an ocean.

The familiar expression *seven seas* dates back to ancient times and refers to the seas known to the Mohammedans prior to the fifteenth century. These included the Mediterranean Sea, the Red Sea, the East African Sea, the West African Sea, the China Sea, the Persian Gulf, and the Indian Ocean. In more recent times Rudyard Kipling popularized the term *seven seas* by using it as the title for a book of poetry. In fact, since Kipling's time there has been a tendency to divide the world ocean into seven parts to retain this legendary number. The popular division is the North Atlantic, South Atlantic, North Pacific, South Pacific, Indian, Arctic, and Antarctic. Although some speak of an Antarctic Ocean, most (including the International Hydrographic Bureau) regard these waters of the South Polar region merely as extensions of the Atlantic, Pacific, and Indian oceans. *Actually, of course, all limits of oceans are arbitrary, as there is only one global ocean.*

Extent of the Oceans

The area of the earth is about 510 million square kilometers (197 million square miles). Of this total, approximately 360 million square kilometers (140 million square miles), or 71 percent, are represented by the oceans and marginal seas. The remaining 29 percent, 150 million square kilometers (57 million square miles), is represented by the continents, which protrude from the water like enormous islands.

By studying a globe or world map, it is readily apparent that the continents and oceans are not evenly divided between the Northern and Southern hemispheres. When we compute the percentages of land and water in the Northern Hemisphere, we find that nearly 61 percent of the surface is water, while about 39 percent is

*Adapted from H. Dubach and R. Tabor, *Questions About the Oceans,* National Oceanographic Data Center Publication G-13, p. 67.

land. In the Southern Hemisphere, on the other hand, almost 81 percent of the surface is water, and only 19 percent is land. It is no wonder then that the Northern Hemisphere is called the *land hemisphere,* and the Southern Hemisphere the *water hemisphere* (Figure 10.1).

Figure 10.2 shows the distribution of land and water for each 5-degree zone in the Northern and Southern hemispheres. *Between latitudes 45 degrees north and 70 degrees north there is actually more land than water, while between 40 degrees south and 65 degrees south there is almost no land to interrupt the oceanic and atmospheric circulation.*

The volume of the ocean basins is many times greater than the volume of the continents above sea level. In fact, the volume of all land above sea level is only 1/18 that of the ocean. Figure 10.3 helps to illustrate this point. Note that the mean elevation of the land surface is 840 meters (2755 feet), while the average depth of the ocean is more than 4.5 times this figure—3800 meters (12,465 feet). Therefore, if the solid earth was perfectly smooth (level) and round, the oceans would cover it to a depth of more than 2000 meters.

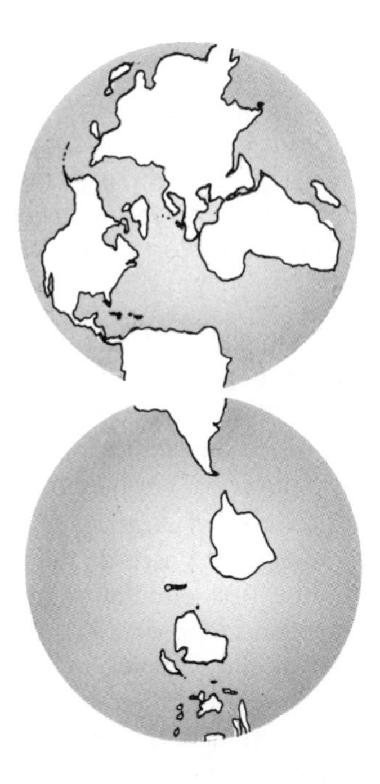

Figure 10.1 *Two views of the earth from space, showing the uneven distribution of land and water. Most of the land occurs in the Northern Hemisphere; the Southern Hemisphere may properly be termed the water hemisphere.*

A comparison of the three major oceans reveals that the Pacific is by far the largest; it is nearly as large as the Atlantic and Indian oceans combined (Figure 10.2). The Pacific contains slightly more than half of the water in the world ocean, and because it includes few shallow seas along its margins, it has the greatest average depth—3940 meters (12,900 feet).

The Atlantic, bounded by almost parallel continental margins, is a relatively narrow ocean when compared to the Pacific. When the Arctic Ocean is included, the Atlantic has the greatest north-south extent and connects the two polar regions. Because the Atlantic has many shallow adjacent seas, including the Caribbean, the Gulf of Mexico, the Baltic, and the Mediterranean, as well as wide continental shelves along its borders, it is the shallowest of the three oceans, with an average depth of 3310 meters (10,850 feet).

The Indian Ocean, lying south of Asia and east of Africa, ranks third in area. Its average depth, however, is between those of the Pacific and Atlantic—3840 meters (12,600 feet).

A Brief History of Oceanography

There is no precise date or event which marks the beginning of oceanography; in a sense it began whenever people started using the oceans as a means of travel. A history of oceanography often begins with the Phoenicians, who were great sailors primarily interested in trade rather than exploration. For more than 2000 years (from approximately 2700 B.C. to 600 B.C.) they sailed the coasts of Europe, Africa, and Asia. They are said to have been the first who dared to

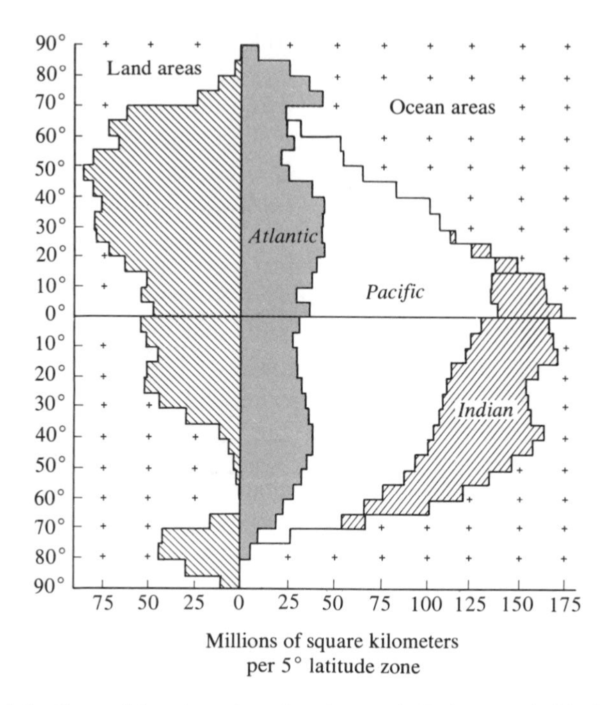

Figure 10.2 *Distribution of land and water in each 5-degree latitude belt. (From M. Grant Gross,* Oceanography: A View of the Earth, *2nd ed., Englewood Cliffs, N.J.: Prentice-Hall, 1977; and M. Grant Gross,* Oceanography, *3rd ed., Columbus, Ohio: Charles E. Merrill, 1976. Data from G. Wüst, W. Brogmus, and E. Noodt, "Die zonale Verteilung von Salzgehalt, Niederschlag, Verdunstung, Temperatur und Dichte an der Oberfläche der Ozeane,"* Kieler Meeresforschungen, *Band X, 1954:139)*

sail at night, guided by the North Star. Although it is believed that they may have ventured far out into the Atlantic, they generally kept their vessels close to shore. To protect their trade, the Phoenicians kept their voyages secret and left no written record of their journeys.

During the Middle Ages, which followed the fall of Rome, mapmaking and travel nearly ceased in the countries bordering the Mediterranean; exploration, however, continued in the north. During the ninth and tenth centuries the Vikings established settlements in Iceland and Greenland. The notable voyages of Eric the Red and Leif Ericson took them to Canada and Newfoundland. However the knowledge gained by these bold Norse explorers had little significance, because it never came to the attention of those in Europe who wrote the books and made the geopolitical decisions.

At the beginning of the Renaissance ocean exploration again began to flourish with notable expeditions like that of Diaz, who rounded the Cape of Good Hope and returned to Portugal in 1497. Other pioneering voyages of this period include the discovery of America by Columbus in 1492, the voyage of Vasco da Gama to India in 1499, and the first circumnavigation of the earth by Magellan's expedition between 1519 and 1522. All of these travels were motivated by a desire for trade and discovery. The oceans were a means of transportation, not an object of study, although gaining some knowledge of the oceans was indeed a necessary and unavoidable part of the job.

There is little argument today that modern geology began with the writings of James Hutton, and that he, without question, deserves the title Founder of Modern Geology. The title Founder of Oceanography, however, is not as easily bestowed. It is sometimes given to the Englishman Edward Forbes (1815–1854), and sometimes conferred upon an American navel officer, Matthew Fontaine Maury (1806–1873).

The contributions of Edward Forbes, who late in his life held the distinguished Chair of Natural History at the University of Edinburgh, centered primarily on life in the sea. Forbes began the analysis of ocean water and was a pioneer in the use of the dredge and in the study of oceanic life zones. About 1850 he prepared a map showing the distribution of marine life, probably the first attempt to divide the oceans into provinces on scientific grounds. Perhaps his most famous theory was that an **azoic** ("without life") **zone** existed within the ocean beyond a depth of 550 meters (1800 feet). Although he is often best remembered for his erroneous theory of the azoic zone, his pioneering work was instrumental in developing the modern science of marine biology and led the way for many who followed.

The modern science of the oceans is often dated from the American navel officer Matthew Fontaine Maury, the Pathfinder of the Sea. In 1855 Maury published what many consider the first textbook on oceanography, entitled *The Physical Geography of the Sea.* The volume includes chapters on the Gulf Stream, the atmosphere, currents, the depths of the ocean, winds, climates, drifts, storms, and more. The book was the end result of a great deal of data gathering and indicated that its author had a profound grasp of the concept that the sea is a single dynamic mechanism. After an injury kept Maury from sailing, he began accumulating and compiling data from ships' logs on currents and winds. In his pursuit of information, Maury enlisted the aid of mariners from all types of ships and of many nationalities. Up to this time there had

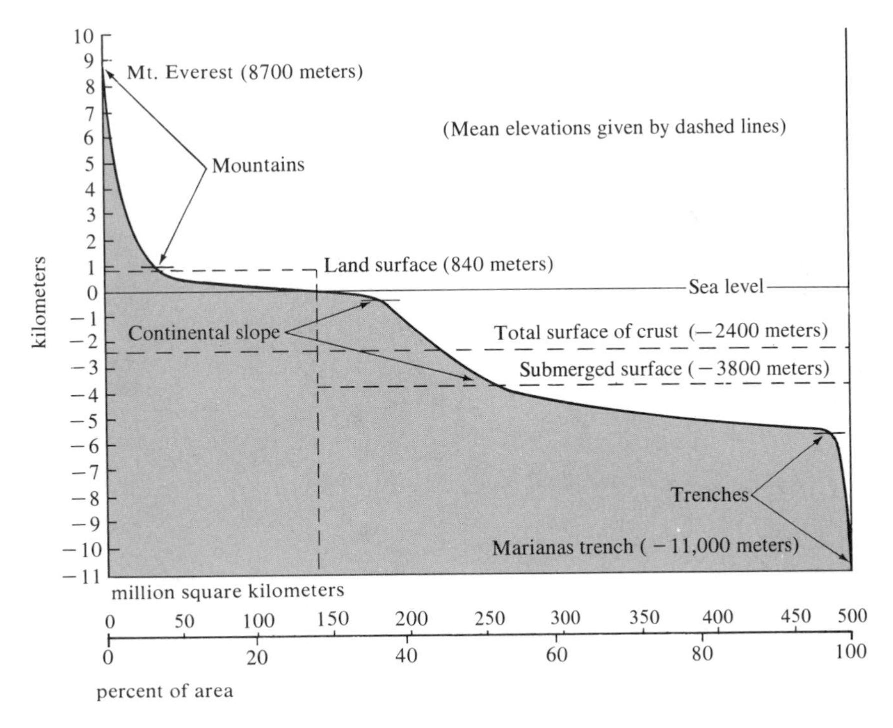

Figure 10.3 *Elevations of the earth's crust. (From Foster,* Physical Geology, *3rd ed., Columbus, Ohio: Charles E. Merrill, 1979)*

been little correlated knowledge of wind, weather, tides, and currents; each sailor learned the hard way. His organization of these data revealed previously unknown patterns of ocean currents, and his charts helped reduce sailing times considerably. For example, the time required to sail from the east coast of the United States to Rio de Janeiro, Brazil, was reduced by as much as 10 days, and the trip to California by way of Cape Horn was shortened by 30 days. He also produced the first bathometric map of the North Atlantic, which showed ocean depths at 1000-fathom intervals (1 fathom equals 1.8 meters or 6 feet, the approximate distance from fingertip to fingertip of a person with outstretched arms).

Another event of great significance to modern oceanography was the historic 3½-year voyage of the H.M.S. *Challenger,* the ship pictured at the opening of the chapter. Beginning in December 1872, and ending in May 1876, the *Challenger* expedition made the first, and perhaps most comprehensive, study of the global ocean attempted by one agency. The 110,000-kilometer (69,000-mile) trip took the ship and its crew of scientists to every ocean except the Arctic (Figure 10.4). Periodically they sampled the total depth of water, temperatures at various depths, weather conditions, and the rate and direction of surface and subsurface currents. Samples of water and life were collected at various levels, and the life and sediment of the bottom were sampled with dredges. Over 700 new genera and 4000 new species of animal and plant life were discovered. The data collected clearly indicated that the oceans are teeming with undiscovered life and showed without a question that life exists at great depths, disproving Forbes's theory of an azoic zone. The scientific results of the voyage were published over a 15-year period and filled 50 large volumes. The "*Challenger* Deep-Sea Exploring Expedition," as it was officially called, opened a great descriptive era of oceanography. It is perhaps fitting that it is from this voyage that the term *oceanography* was born.

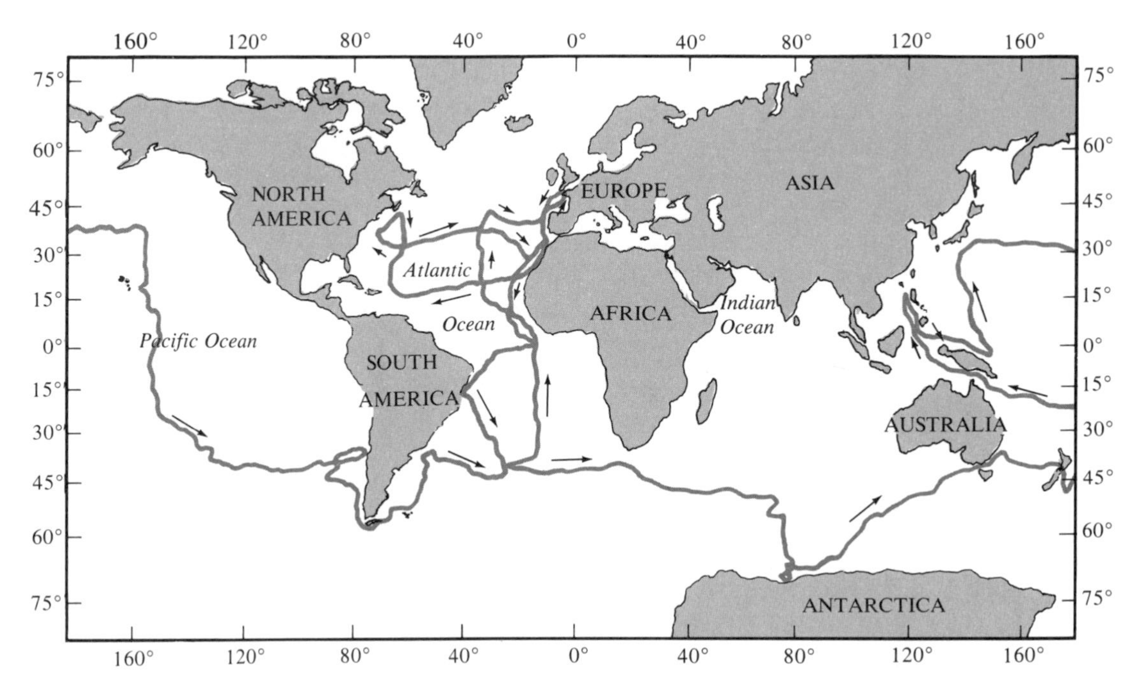

Figure 10.4 *General outline of the voyage of the* Challenger. *(From H.S. Bailey, Jr., "The Voyage of the Challenger." Copyright © 1953 by Scientific American, Inc. All rights reserved)*

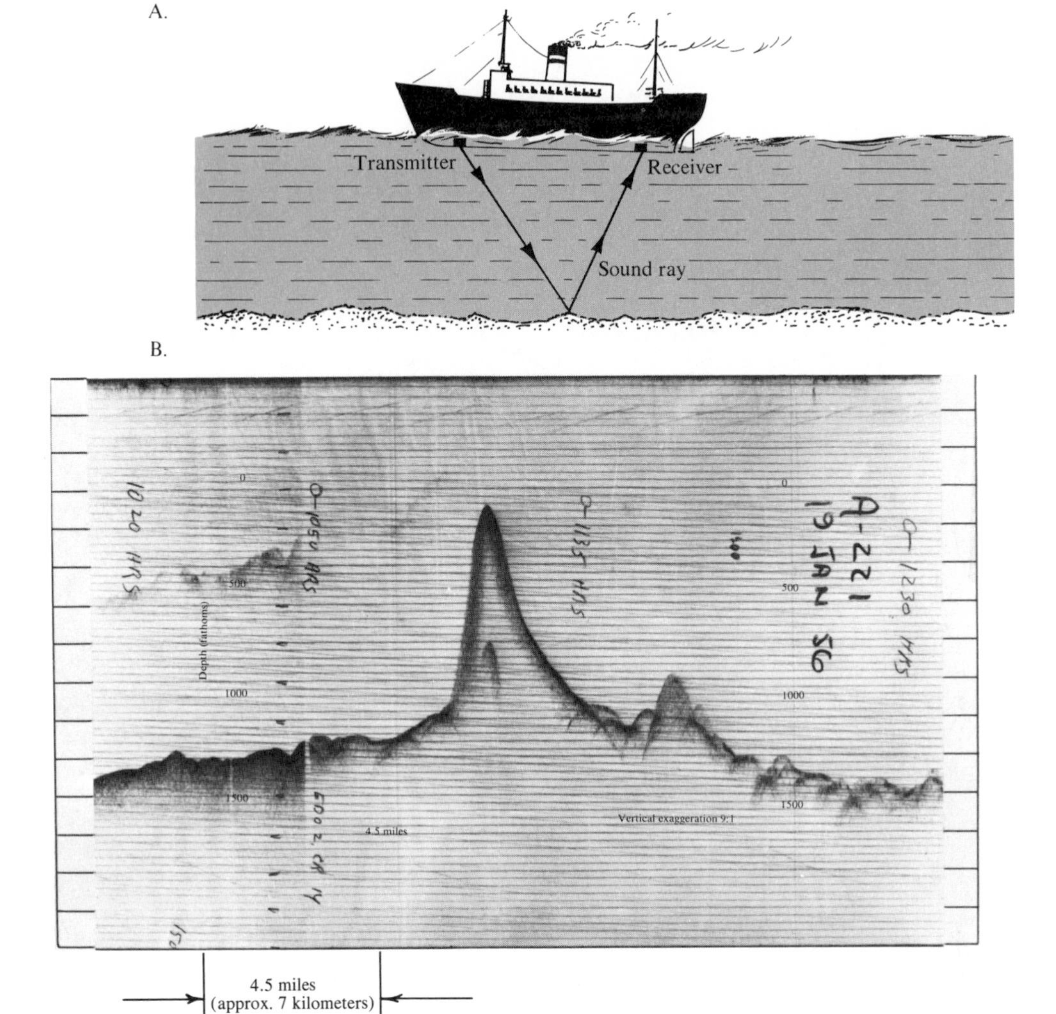

Figure 10.5 **A.** *Echo sounder. Depth = ½ speed of sound × echo time. (From Foster,* Physical Geology, *3rd ed., Columbus, Ohio: Charles E. Merrill, 1979)* **B.** *Sea-floor profile made by an echo sounder. (Courtesy of Woods Hole Oceanographic Institution)*

Certainly one of the most remarkable men who contributed to our knowledge of the oceans was the Norwegian adventurer and scientist Fridtjof Nansen. Although he is perhaps best known to oceanographers as a polar explorer, Nansen was also an artist, a zoologist, and a winner of the Nobel Peace Prize. After studying the waters of the Arctic basin, Nansen developed a theory concerning surface currents in this ice-choked water body. After convincing his government and the British Royal Geographical Society to help him test his theory, a special ship, the *Fram,* was built. The *Fram* was designed so that expanding ice would not crush it, but rather would force it up to the surface. For three years, from September 1893 to August 1896, the *Fram* and its crew drifted with the ice. During this pioneering voyage Nansen not only made impor-

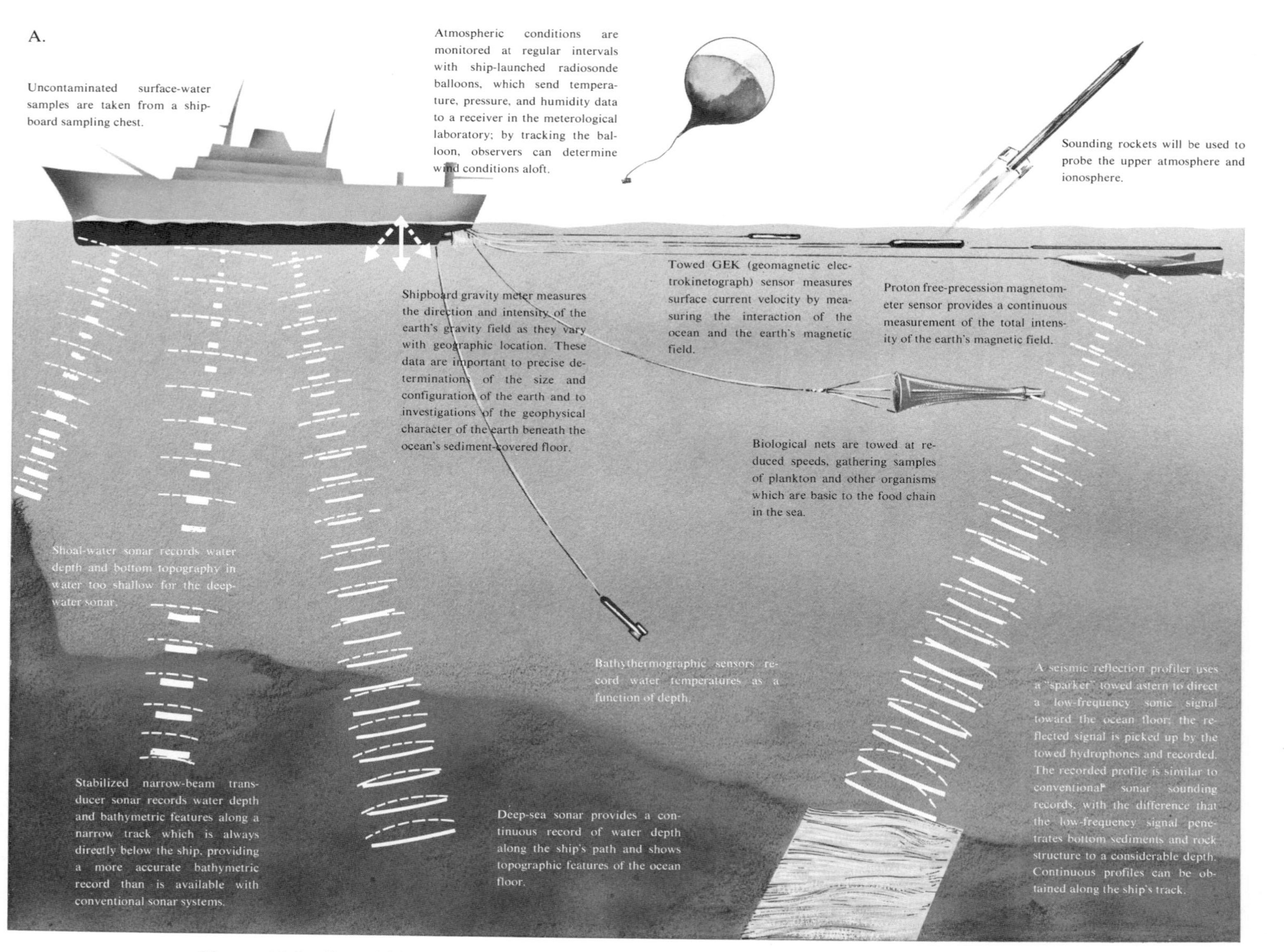

Figure 10.6 *Capabilities of a modern research vessel.* **A.** *Underway capability.* **B.** *On-station capability.*

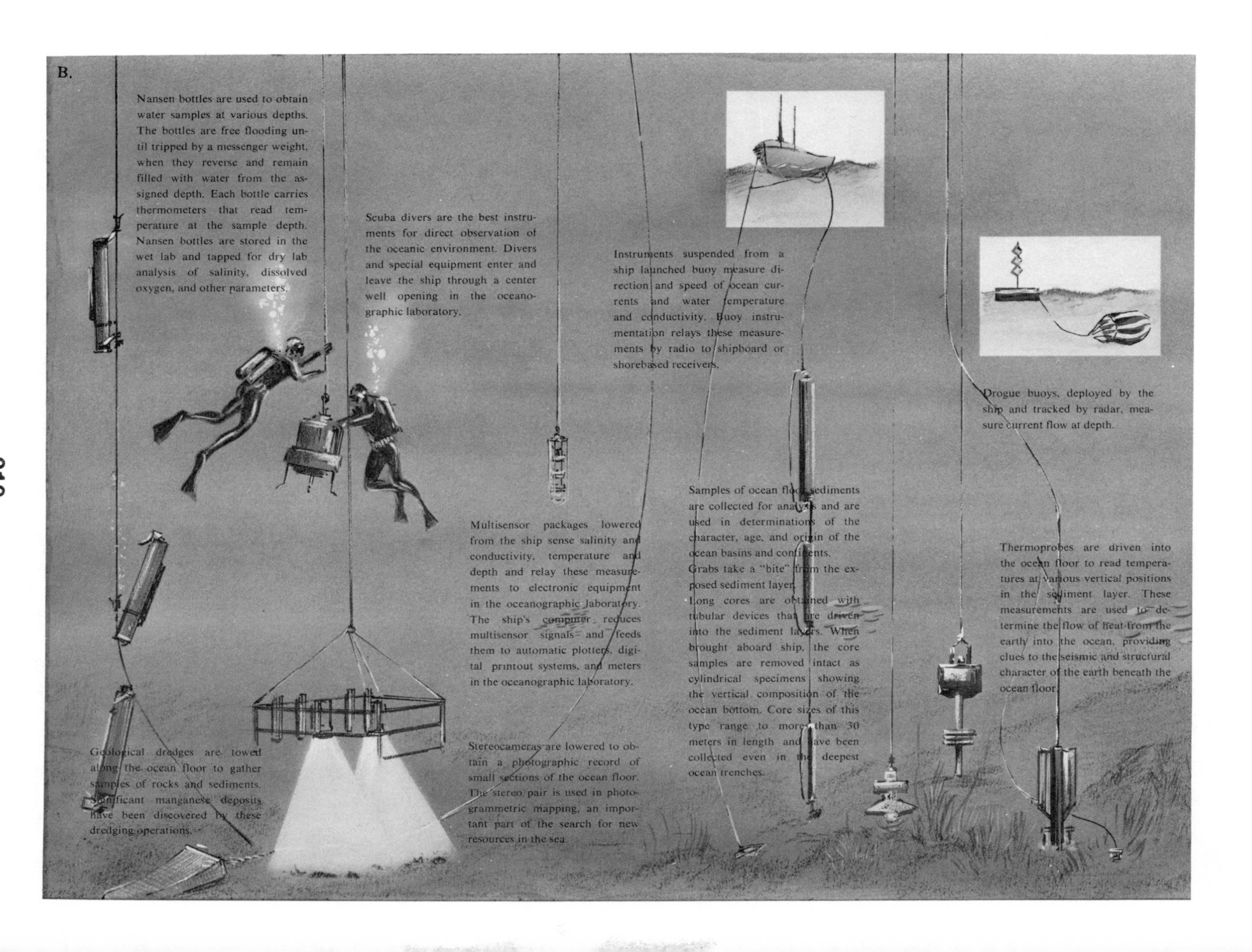
B.
Nansen bottles are used to obtain water samples at various depths. The bottles are free flooding until tripped by a messenger weight, when they reverse and remain filled with water from the assigned depth. Each bottle carries thermometers that read temperature at the sample depth. Nansen bottles are stored in the wet lab and tapped for dry lab analysis of salinity, dissolved oxygen, and other parameters.
Scuba divers are the best instruments for direct observation of the oceanic environment. Divers and special equipment enter and leave the ship through a center well opening in the oceanographic laboratory.
Instruments suspended from a ship launched buoy measure direction and speed of ocean currents and water temperature and conductivity. Buoy instrumentation relays these measurements by radio to shipboard or shorebased receivers.
Drogue buoys, deployed by the ship and tracked by radar, measure current flow at depth.
Multisensor packages lowered from the ship sense salinity and conductivity, temperature and depth and relay these measurements to electronic equipment in the oceanographic laboratory. The ship's computer reduces multisensor signals and feeds them to automatic plotters, digital printout systems, and meters in the oceanographic laboratory.
Samples of ocean floor sediments are collected for analysis and are used in determinations of the character, age, and origin of the ocean basins and continents.
Grabs take a "bite" from the exposed sediment layer.
Long cores are obtained with tubular devices that are driven into the sediment layers. When brought aboard ship, the core samples are removed intact as cylindrical specimens showing the vertical composition of the ocean bottom. Core sizes of this type range to more than 30 meters in length and have been collected even in the deepest ocean trenches.
Thermoprobes are driven into the ocean floor to read temperatures at various vertical positions in the sediment layer. These measurements are used to determine the flow of heat from the earth into the ocean, providing clues to the seismic and structural character of the earth beneath the ocean floor.
Geological dredges are towed along the ocean floor to gather samples of rocks and sediments. Significant manganese deposits have been discovered by these dredging operations.
Stereocameras are lowered to obtain a photographic record of small sections of the ocean floor. The stereo pair is used in photogrammetric mapping, an important part of the search for new resources in the sea.

tant observations about ocean currents but also found that this ice-covered sea reached to true oceanic depths. Further, the drift of the *Fram* disproved the commonly held notion that a continent existed in this northern ocean and showed that the ice covering the polar area throughout the year was not of glacial origin but was a freely moving ice pack that formed directly upon the water surface. Following the voyage, Nansen, realizing the need for more accurate measurements of salinity and temperature, designed an instrument that has come to be called the **Nansen bottle** (Figure 11.4). The Nansen bottle is still one of the standard oceanographic sampling devices.

The voyage during 1925–1927 of the German ship *Meteor* has been considered the sequel to the *Challenger* expedition. This voyage was perhaps the first modern oceanographic trip because it had the advantage of more modern equipment such as the Nansen bottle and revolutionary electronic depth-sounding equipment (echo sounder). Prior to the invention of the echo sounder, ocean depth measurements had to be taken with a weighted line. In deep water the time required for the line to reach bottom can exceed an hour and a half, while the task of hauling in the line is even more time-consuming. It is obvious why only a limited number of such soundings were practicable, and therefore why our knowledge of the sea floor remained skimpy. With the development of the echo sounder, depths could be determined in a matter of seconds or minutes, while the vessel was in motion (Figure 10.5).

The echo sounder apparatus works by transmitting sound waves from the ship toward the ocean bottom. A delicate receiver catches the echo from the bottom, and a highly accurate clock measures the time interval in small fractions of a second. By knowing the velocity of the sound waves in the water (about 1500 meters or 4900 feet per second), the depth can be calculated very precisely. Since ultrasonic waves are often used because they can be distinguished from the audible sounds made during the operation of the ship, the "sound" is not detectible by human ears.

Since World War II interest in and research concerning the ocean have grown swiftly. Today we live in a period of analytical oceanography which has replaced the period when purely descriptive studies were the rule. Although sampling and systematic observations are still important, more emphasis is given to theories and hypotheses. Data are gathered which specifically relate to certain ideas, with the goal of proving or disproving these ideas. There are now hundreds of vessels engaged in oceanographic research, some having a special purpose, others equipped to do many tasks. The National Oceanic and Atmospheric Administration's (NOAA) ships *Discoverer* and *Researcher* are examples of the new generation of research and survey vessels. These ships have the capability of taking the measure of the global oceanic system (Figure 10.6). Although there seems to have been a great emphasis on ocean study over the past century, there is still much more to learn, for every question that is answered about the oceans leads to additional questions that demand answers.

The Earth beneath the Sea

If all the water was drained from the ocean basins, what kind of surface would be revealed? It would not be the quiet, subdued topography as was once thought, but a surface characterized by a great diversity of features—towering mountain chains, deep canyons, and flat plains. In fact, the scenery would be just as varied as that on the continents (Figure 10.7).

Oceanographers studying the topography of the ocean basins have delineated three major units: continental margins, the ocean basin floor, and mid-ocean ridges. The map in Figure 10.8 outlines these provinces for the North Atlantic, and the profile at the bottom of the illustration shows the varied topography. Such profiles usually have their vertical dimension exaggerated many times—40 times in this case—to make

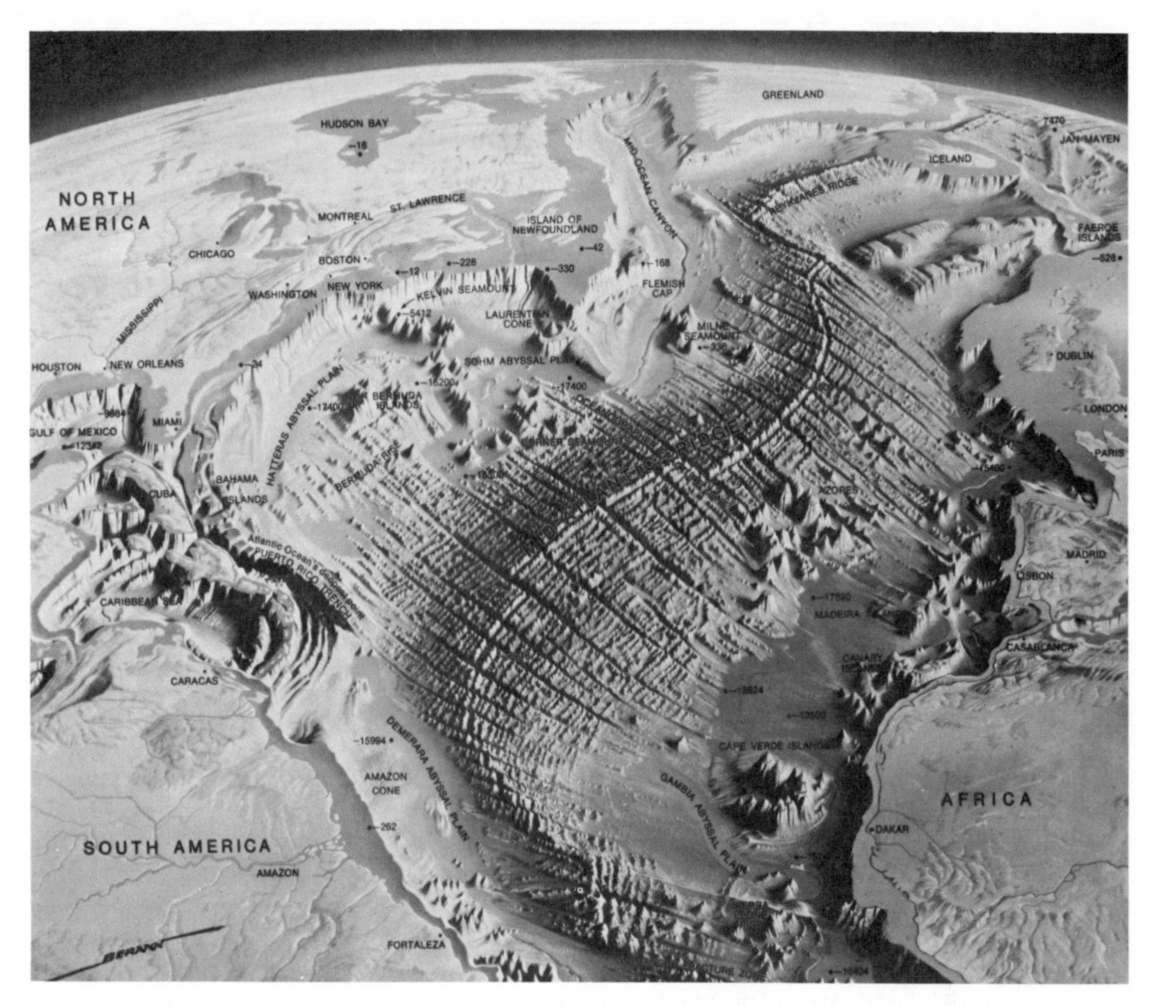

Figure 10.7 *The ocean floor is characterized by a great diversity of features. This is an artist's view of what would be seen if all of the water were removed from the Atlantic Ocean basin. (From a painting by Heinrich Berann; courtesy of Aluminum Company of America)*

topographic features more conspicuous. Because of this, the slopes shown in the profile of the sea floor in Figure 10.8 appear to be much steeper than they actually are.

The Continental Shelf

The features comprising the continental margin include the continental shelf, the continental slope, and the continental rise (Figure 10.9A). The first of these parts, the **continental shelf,** is a gently sloping submerged surface that extends from the edge of the continent towards the ocean basin. Since it is underlain by continental-type crust, it is clearly a flooded extension of the continents. Continental shelves vary in width from very narrow to broad, averaging about 65 kilometers (40 miles) wide. The general slope of the shelf is slight, averaging less than a 0.3-meter (1-foot) drop for each 150–300 meters (500–1000 feet) of horizontal distance. It would appear to the eye as a flat surface. Although the boundary is conventionally stated as the 100-

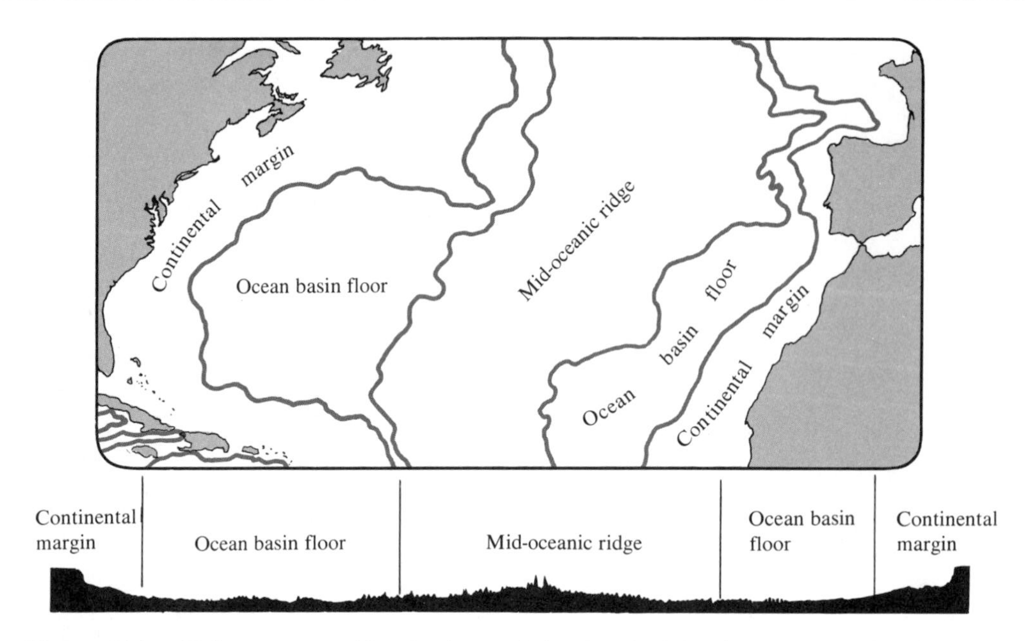

Figure 10.8 *Major topographic divisions of the North Atlantic and a profile from New England to the coast of North Africa. (After B.C. Heezen, M. Tharp, and M. Ewing, "The Floors of the Oceans,"* Geological Society of America Special Paper 65, *p. 16)*

fathom (180-meter or 600-foot) submarine contour, there are many local variations. *A better definition of the edge of the shelf is the point where a rapid steepening of the gradient occurs, marking the beginning of the continental slope* (Figure 10.9B).

The continental shelves represent 7.5 percent of the total area of the oceans, which is equivalent to about 18 percent of the earth's total land area. These areas have taken on increased economic and political significance since they have been found to be sites of important mineral deposits, including large reservoirs of petrolium and natural gas, as well as huge sand and gravel deposits. Of course, the waters of the continental shelf contain many important fishing grounds that are significant sources of food.

When compared with many parts of the deep ocean floor, the surface of the continental shelf is relatively featureless. However this is not to say that the shelves are completely smooth. The most profound features are long valleys running from the coastline into deeper waters. Many of these valleys are the seaward extensions of river valleys on the adjacent landmass. Such valleys were excavated during the Pleistocene epoch (Ice Age). During this time great quantities of water were tied up in vast ice sheets on the continents, causing sea level to drop by 90–120 meters (300–400 feet) and exposing the continental shelves. Because of this, rivers extended their courses, and land-dwelling plants and animals inhabited the newly exposed portions of the continents. Today these areas are covered by the sea and are inhabited by marine organisms. Dredging along the eastern coast of North America has produced the remains of numerous land dwellers, including mammoths, mastadons, and horses. Bottom sampling has also revealed that freshwater peat bogs existed, adding to the evidence that the continental shelves were once land areas.

Continental Slope and Rise

Marking the seaward edge of the continental shelf is the **continental slope,** a feature characterized by a steep gradient as compared with the shelf, which leads into deep water. While the slope varies from place to place, it has an average drop of about 70 meters per kilometer (370 feet per mile). The continental slope marks the boundary between the continental crust and the oceanic crust.

Along some mountainous coasts the continental slope descends abruptly into deep oceanic trenches which intervene between the continent and ocean basin. In such cases, the shelf is very narrow or does not exist at all. The side of the trench and the continental slope are essentially the same feature and grade into the adjacent mountains which tower thousands of meters above sea level. An example of such a situation occurs along the west coast of South America (Figure 10.10). Here the vertical distance from

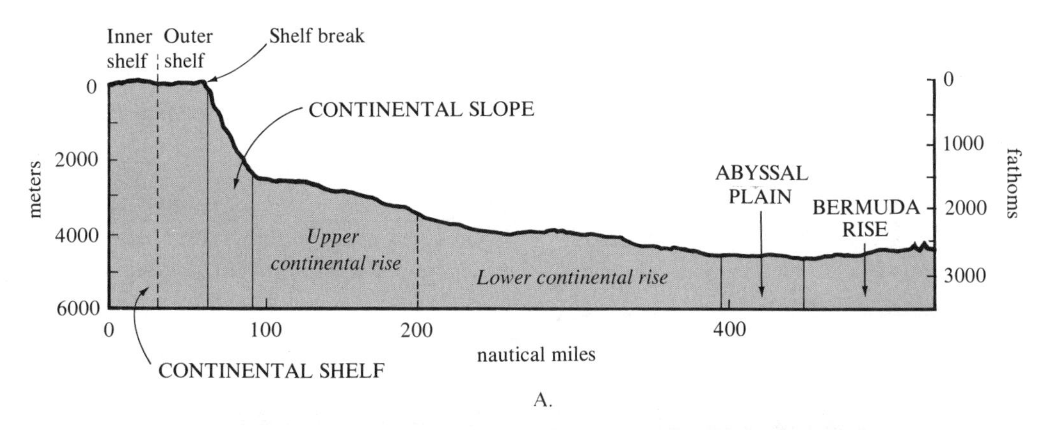

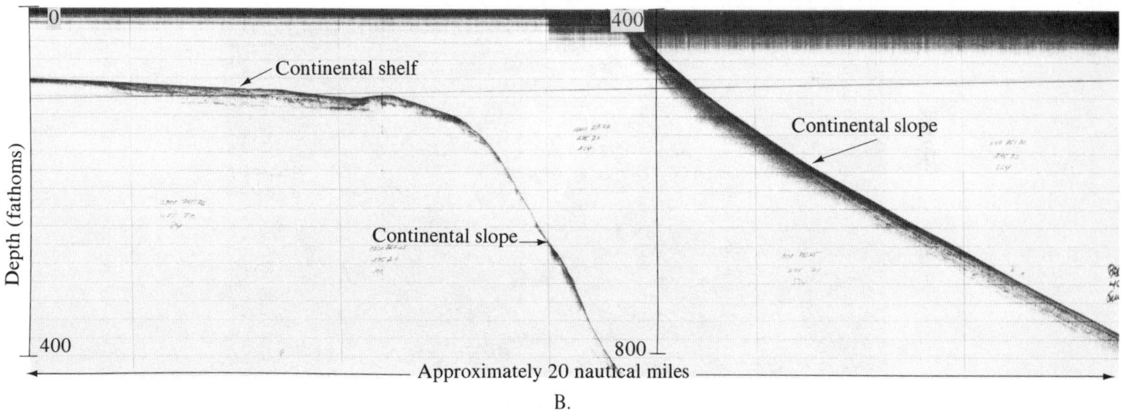

Figure 10.9 **A.** *Schematic profile showing the provinces of the continental margin. (From B.C. Heezen, M. Tharp, and M. Ewing, "The Floors of the Oceans,"* Geological Society of America Special Paper 65, *p. 26)* **B.** *Portion of a trace made by a precision depth recorder showing the outer continental shelf and continental slope off the coast of Brazil. (Courtesy of Submarine Topography Department, Lamont-Doherty Geological Observatory)*

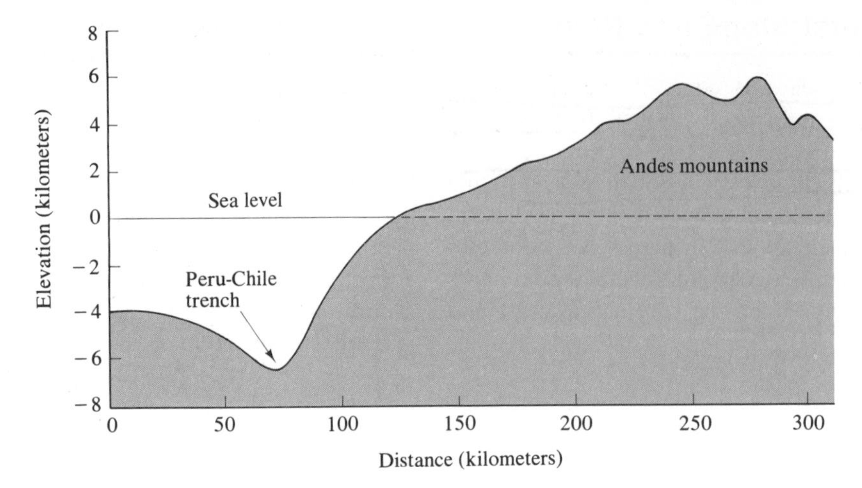

Figure 10.10 *A profile across the Peru-Chile trench and the Andes Mountains. The vertical scale is exaggerated ten times.*

the high peaks of the Andes Mountains to the floor of the deep Peru-Chile trench bordering the continent exceeds 12,200 meters (40,000 feet).

In regions where trenches do not exist the steep continental slopes merge into a more gradual incline known as the **continental rise** (Figure 10.9). Here the gradient lessens to between 4 and 8 meters per kilometer (20 and 40 feet per mile). While the width of the continental slope averages about 20 kilometers (12 miles), the

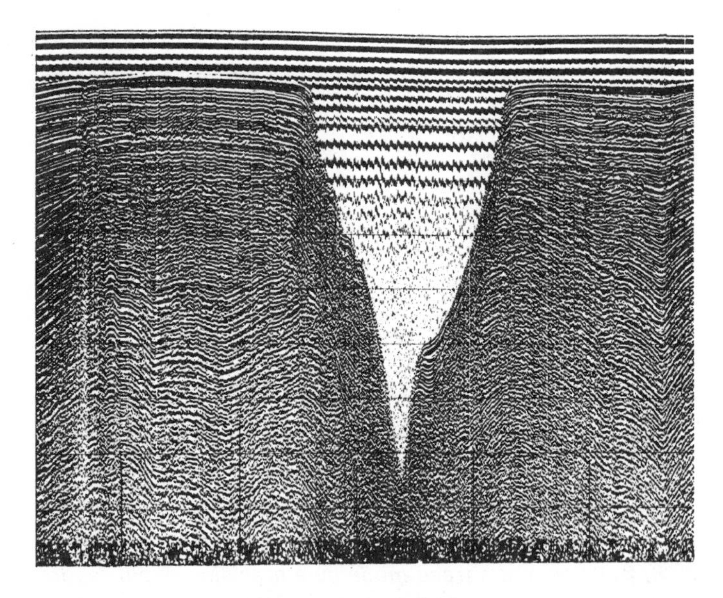

Figure 10.11 *An echo-sounding profile of the Congo submarine canyon off the west coast of Africa. The bottom of the canyon is about 3 kilometers below the level of the sea floor of the continental shelf. Across the top, the canyon is more than 10 kilometers wide. (Courtesy of K.O. Emery, Woods Hole Oceanographic Institution)*

The earth's surface is usually partially covered by clouds except in the arid regions such as North Africa, the rust-colored area, and the Mediterranean Sea. (Courtesy of NASA)

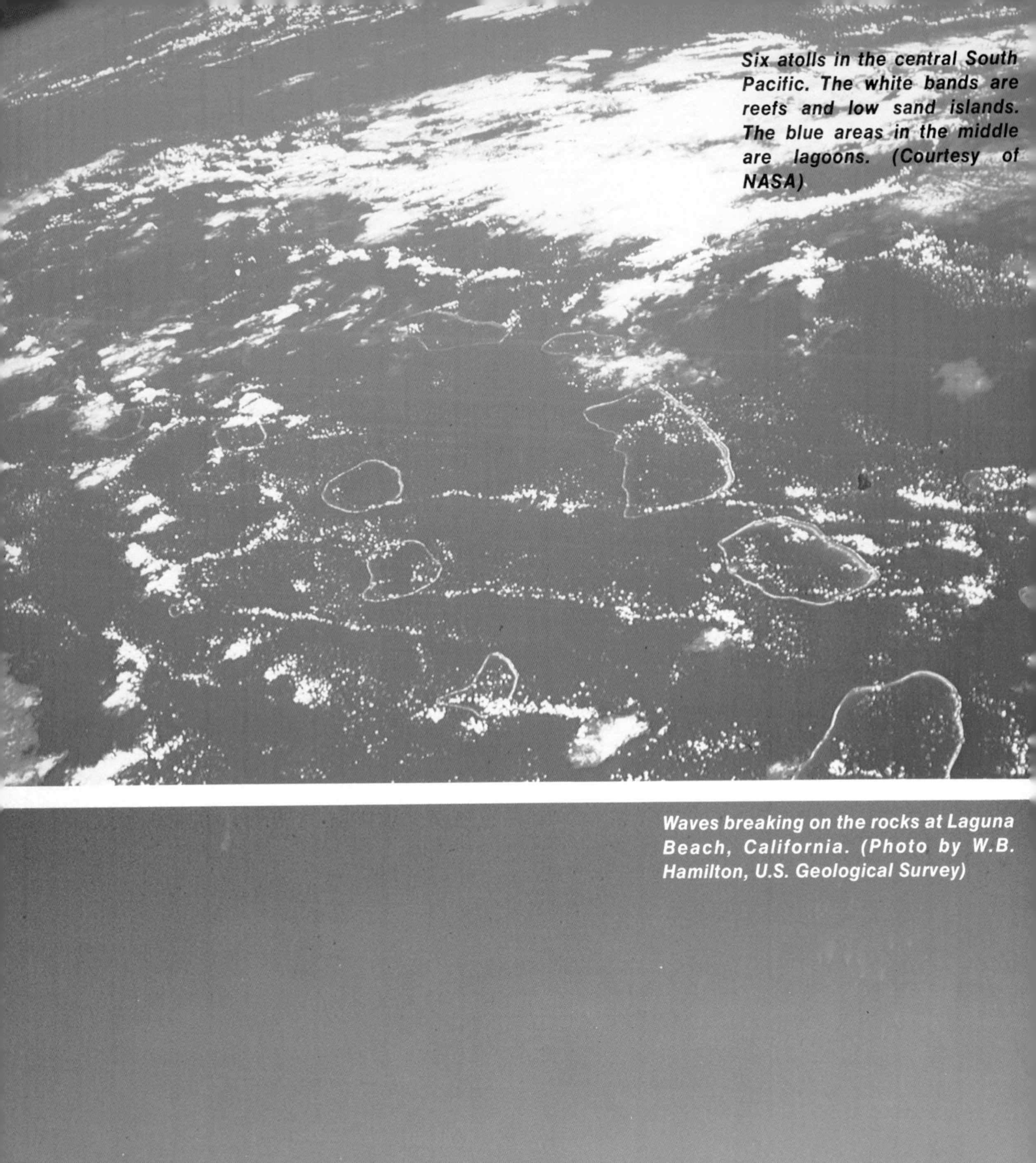

Six atolls in the central South Pacific. The white bands are reefs and low sand islands. The blue areas in the middle are lagoons. (Courtesy of NASA)

Waves breaking on the rocks at Laguna Beach, California. (Photo by W.B. Hamilton, U.S. Geological Survey)

continental rise may reach for hundreds of kilometers. This feature consists of a thick accumulation of sediment that moved downslope from the continental shelf to the deep ocean floor. Although rises are relatively featureless, their surfaces are occasionally interrupted by submarine canyons or by volcanoes that have not yet been completely buried by the sediments.

The deep, steep-sided valleys known as **submarine canyons** originate on the continental slope and may reach to water depths of 3 kilometers (2 miles). Although some of these canyons appear to be the seaward extensions of valleys that were carved on the continental shelf during the Ice Age, there are others which do not line up in this manner. Furthermore, the canyons reach depths far below the maximum lowering of sea level, which indicates that they were created by some process that operates below the ocean surface (Figure 10.11). Most of the available information seems to favor the view that submarine canyons have been excavated by turbidity currents.

Turbidity Currents

Turbidity currents are downslope movements of dense, sediment-laden water. They are created when sand and mud on the continental shelf and slope are dislodged, perhaps by an earthquake, and are thrown into suspension. Since the muddy water is denser than the clearer water above, it flows down the slope, eroding and accumulating more sediment as it continues to gain speed (Figure 10.12). The erosional work repeatedly carried on by these muddy torrents eventually results in the excavation of submarine canyons.

Often turbidity currents continue across the continental rise, still cutting channels, until they finally lose their force and come to rest along the bottom of the ocean basin. As the speed of the current slows, the suspended sediments settle out. First, the coarser sand is dropped, followed by successively finer deposits of silt and then clay. Consequently, the deposits, called **turbidites,** are characterized by a decrease in sediment grain size from bottom to top, a phenomenon known as **graded bedding.**

Although it had been known for many years that turbidity currents occur in lakes, their existence in the oceans was not proven until the 1950s. Prior to this time geologists and oceanographers thought that deposits in the deep waters of the ocean accumulated by individual grains settling slowly through the water, while the formation of submarine canyons was a controversial topic which seemed to have no widely accepted solution.

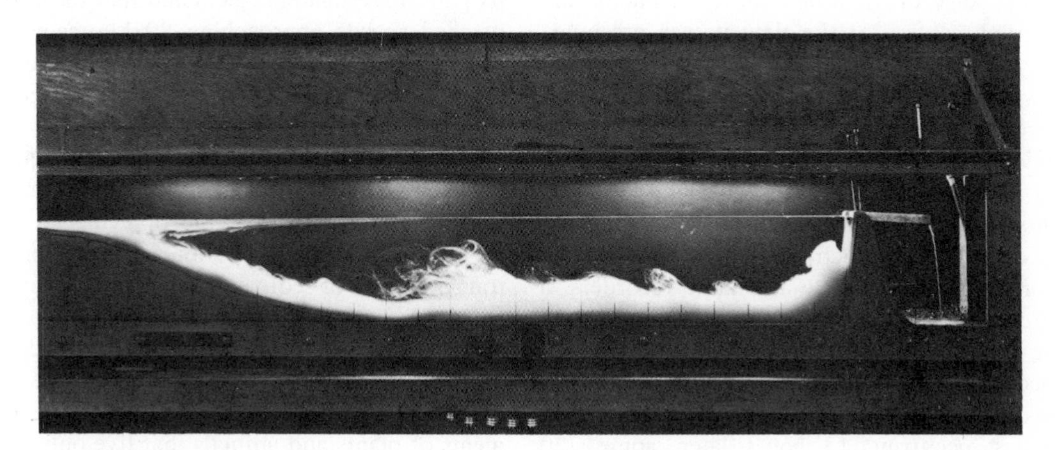

Figure 10.12 *Turbidity current produced in a water-filled laboratory tank. (Courtesy of H.S. Bell, Sedimentation Laboratory, California Institute of Technology)*

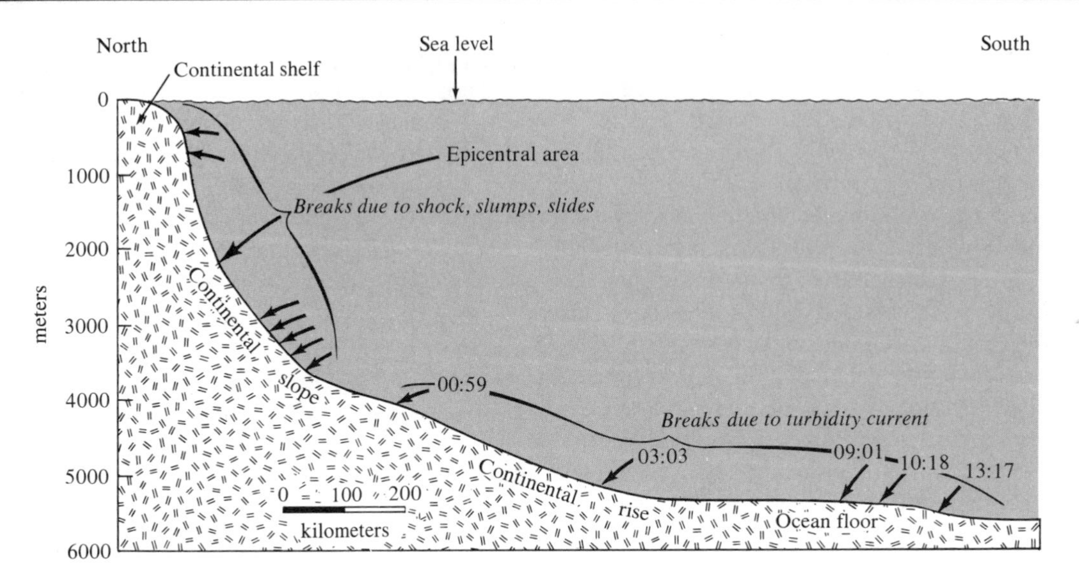

Figure 10.13 *Profile of the sea floor showing the events of the November 18, 1929, earthquake off the shores of Newfoundland. The arrows point to the cable breaks; the numbers show times of the breaks in hours and minutes after the earthquake. The vertical scale is greatly exaggerated. (After B.C. Heezen and M. Ewing, "Turbidity Currents and Submarine Slump and the 1929 Grand Banks Earthquake,"* American Journal of Science *250: 867)*

Two lines of evidence helped confirm the presence of turbidity currents in the ocean and establish them as important mechanisms of submarine erosion and sediment transportation. The first important evidence came from records of a rather severe earthquake that took place off the coast of Newfoundland in 1929 and resulted in the breakage of 13 transatlantic telephone and telegraph cables. At the time it was presumed that the tremor had caused the multiple breaks. However when the data were examined, it appeared that this was not the case. After plotting the locations of the breaks on a map, it was seen that all the breaks had occurred along the steep continental slope and the gentler continental rise. Since the time of each break was known from information provided by automatic recorders, a pattern of what had happened could be deduced. The breaks high up on the continental slope took place first, almost concurrently with the earthquake. The other breaks happened in succession, the last occurring 13 hours later, some 720 kilometers (450 miles) from the source of the quake (Figure 10.13). The breaks downslope had obviously taken place too long after the tremor to have been caused by the shock of the earthquake. The existence of a turbidity current, triggered by the quake, thus appeared as a plausible alternative. As the avalanche of sediment-choked water raced downslope it snapped the cables in its path. Investigators calculated that the current reached speeds approaching 80 kilometers (50 miles) per hour on the steep slopes and about 24 kilometers (15 miles) per hour on the gentler slopes below. Subsequent investigations of cable breaks in other areas revealed a similar sequence of events.

A second compelling line of evidence relating turbidity currents to submarine erosion and transportation of sediment came from the examination of deep-sea sediment samples. These cores show that extensive graded beds of sand, silt, and clay exist in the quiet waters of the deep ocean. Some of the samples also include fragments of plants and animals that live only in the shallower waters of the continental shelves. No mechanism other than turbidity currents could explain the existence of these deposits.

Although there is still much to be learned about the complex workings of turbidity currents, it has been well established that they are a very important mechanism of sediment transport in the ocean. *By the action of turbidity currents, submarine canyons are created and sediments are carried to the deep ocean floor where they form flat plains.*

Features of the Ocean Basin Floor

Between the continental margin and the oceanic ridge system lies the ocean basin floor (see Figure 10.8). The size of this region—almost 30 percent of the earth's surface—is roughly comparable to the percentage of the surface that projects above the sea as land. Here we find remarkably flat regions, known as abyssal plains, steep-sided volcanic peaks, called seamounts, and deep-ocean trenches, which are dramatically deep grooves in the ocean floor.

Deep-ocean Trenches

Deep-ocean trenches are long, relatively narrow features that represent the deepest parts of the ocean. Several in the western Pacific approach or exceed 10,000 meters (32,800 feet) deep, and at least a portion of one, the Challenger Deep in the Marianas Trench, is more than 11,000 meters (36,000 feet) below sea level. Table 10.1 presents some dimensional characteristics of trenches.

Although deep-ocean trenches represent only a very small portion of the area of the ocean floor, they are nevertheless very significant features geologically. Trenches are the sites where moving crustal plates are destroyed as they plunge back into the mantle (Figure 6.8). In addition to the earthquakes created as one plate descends beneath another, volcanic activity is also associated with trench regions. Trenches in the open ocean are paralleled by volcanic island arcs, while volcanic mountains, like the Andes, may be found paralleling trenches that are adjacent to continents. As was mentioned in Chapter 6, the melting of a descending plate produces the molten rock that leads to this volcanic activity.

Table 10.1 *Dimensions of Some Deep-ocean Trenches*

Trench	Depth (kilometers)	Average Width (kilometers)	Length (kilometers)
Aleutian	7.7	50	3700
Japan	8.4	100	800
Java	7.5	80	4500
Kuril-Kamchatka	10.5	120	2200
Marinas	11.0	70	2550
Middle America	6.7	40	2800
Peru-Chile	8.1	100	5900
Philippine	10.5	60	1400
Puerto Rico	8.4	120	1550
South Sandwich	8.4	90	1450
Tonga	10.8	55	1400

Abyssal Plains

Abyssal plains are extremely flat features of the deep ocean floor; in fact, *abyssal plains are likely the most level areas on the earth*. The plain found off the coast of Argentina, for example, has less than 3 meters (10 feet) of relief over a distance exceeding 1300 kilometers (800 miles).

By employing seismic profilers, instruments whose signals penetrate far below the ocean floor, researchers have shown that abyssal plains consist of thick accumulations of sediment that were deposited atop the low, rough portions of the ocean floor. The nature of the sediment indicates that these plains consist primarily of sediments transported far out to sea by turbidity currents. The turbidite deposits are interbedded with sediments composed of minute clay-sized particles that continuously settle onto the ocean floor.

Abyssal plains are found as part of the sea floor in all of the oceans. However they are more widespread where there are no deep ocean trenches adjacent to the continents. Since the Atlantic Ocean has fewer trenches to act as traps for the sediments carried down the continental slope, it has more extensive abyssal plains than the Pacific.

Seamounts

Dotting the ocean floors are isolated volcanic peaks called **seamounts** that rise at least 1000 meters (3000 feet) above the surrounding topography. Although these steep-sided conical peaks have been discovered in all of the oceans, the greatest number have been identified in the Pacific.

Many of these undersea volcanoes begin to rise near oceanic ridges, divergent plate boundaries where the plates of the lithosphere move apart (see Chapter 6). They continue to grow as they ride along on the moving plate. If the volcano rises fast enough, it emerges as an island. Examples in the Atlantic include the Azores, Ascension, Tristan da Cunha, and St. Helena. While the volcanoes exist as islands, their tops are eroded by the action of waves. After millions of years they eventually sink as the moving plate slowly carries them from the oceanic ridge area (see Figure 6.18). In other instances these eroded volcanoes are submerged because the oceanic ridge becomes inactive and subsides. After sinking, they become flat-topped seamounts called **guyots.**

Mid-ocean Ridges

Mid-ocean ridges are found in all of the major oceans and represent more than 20 percent of the earth's surface. They are certainly the most prominent topographic features in the oceans, for they form an almost continuous mountain range, varying in width from 500 to 5000 kilometers (300 to 3000 miles), which extends for some 64,000 kilometers (40,000 miles). The crests of the ridges are marked by deep clefts, or **rifts,** and are flanked by ridges and lines of peaks that extend outward for hundreds of kilometers (Figure 10.7). The axes of the ridges are marked by frequent earthquakes and are characterized by a much higher heat flow through the crust. The rifts at the center of the ridges are the sites where new magma wells up from the asthenosphere below, continually creating new oceanic crust. *The rifts therefore represent divergent plate boundaries where sea-floor spreading is taking place.*

The rate at which the sea floor spreads affects the topography of the ridge systems. Relatively fast spreading yields broad elevations and gentle slopes, like those along the East Pacific Rise. On the other hand, the steeper flanks of the Mid-Atlantic Ridge were formed because the plates are spreading apart more slowly. The slopes are steeper at a slow spreading center because the crust moves away so slowly (less than 3 centimeters, or 1.8 inches, per year) that the molten material piles up below the ridge.

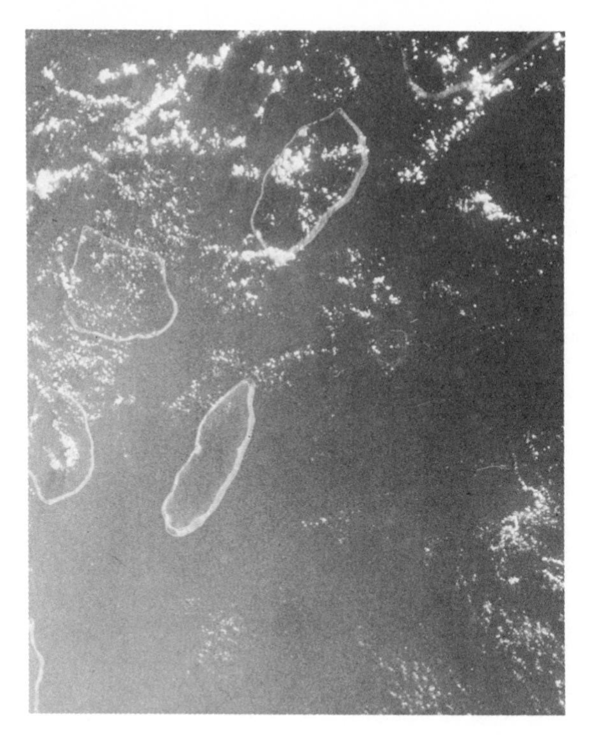

Figure 10.14 *View from space of a group of atolls in the Pacific Ocean. (Courtesy of NASA)*

Coral Reefs and Atolls

Coral reefs are among the most picturesque features found in the ocean. They are constructed primarily from the calcareous (calcite-rich) skeletal remains and secretions of corals and certain algae. The term *coral reef* is somewhat misleading in that it makes no mention of the skeletons of many small animals and plants found inside the branching framework built by the corals; nor of the fact that limy secretions of algae help bind the entire structure together.

Coral reefs are confined largely to the warm waters of the Pacific and Indian oceans, although a few occur elsewhere. Since reef-building corals grow best in waters with an average annual temperature of about 24° C (75° F), their location is in part the result of their need for warm water. They can survive neither sudden temperature changes nor prolonged exposure to temperatures below 18° C (65° F). In addition, these reef-builders require clear sunlit water. Consequently, the limiting depth of active reef growth is about 45 meters (150 feet).

From 1831 to 1836 the naturalist Charles Darwin was aboard the British ship *Beagle* on a surveying expedition that circumnavigated the globe. One outcome of Darwin's studies during the five-year voyage was a theory on the formation of coral islands, or **atolls.** As Figure 10.14 illustrates, atolls consist of a continuous or broken ring of coral reef surrounding a central lagoon. From the time that Darwin first studied them, until shortly after World War II, their manner of origin challenged people's curiosity.

Darwin's theory explained what seemed to be a paradox; that is, how can corals, which require warm, shallow, sunlit water no deeper than 45 meters (150 feet) to live, create structures that reach thousands of meters to the floor of the ocean? Commenting on this in *The Voyage of the Beagle,* Darwin stated:

> . . . from the fact of the reef-building corals not living at great depths, it is absolutely certain that throughout these vast areas, wherever there is now an atoll, a foundation must have originally existed within a depth of from 20 to 30 fathoms from the surface.

The essence of Darwin's theory was that coral reefs form on the flanks of sinking volcanic islands. Illustrated in Figure 10.15, as the island slowly sinks beneath the sea, the corals continue to build upward.

> For as mountain after mountain, and island after island, slowly sank beneath the water, fresh bases would be successively afforded for the growth of the corals.

In succeeding years there were challenges to Darwin's theory. These arguments were not finally settled until after World War II when the United States made extensive studies of two atolls (Eniwetok and Bikini) that were going to become sites for testing atomic bombs. The deep

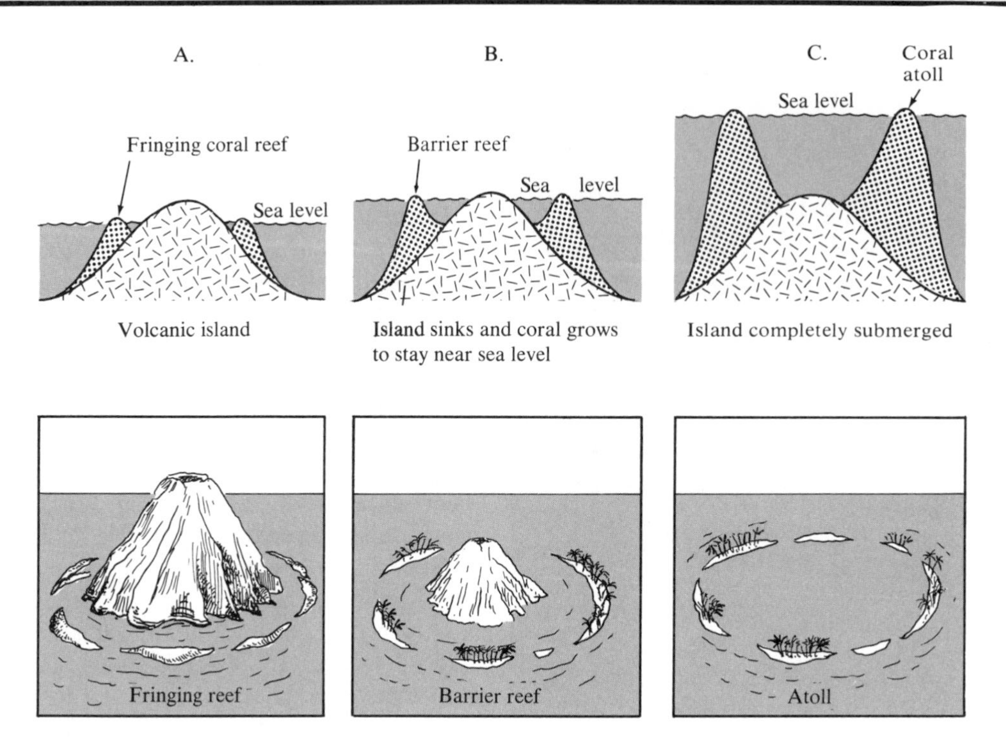

Figure 10.15 *Formation of a coral atoll. (In part from Foster,* Physical Geology, *3rd ed., Columbus, Ohio: Charles E. Merrill, 1979)*

drilling of these atolls revealed volcanic rock underlying the thick coral reef structure. This finding was a striking confirmation of Darwin's theory.

Atolls, like guyots, owe their existence to a gradual sinking of the crust. Recently the theory of plate tectonics has offered a reasonable explanation for this subsidence, as crustal plates slowly move away from the higher oceanic ridge to the lower ocean basin floor or gradually sink as the ridge becomes inactive.

Sea-floor Sediments

Except for a few areas of exposed rock, the ocean floor is mantled with sediment. Part of it has been carried down the continental slope by turbidity currents, and the rest has slowly settled to the bottom from above. The thickness of this carpet of debris varies considerably. In the Pacific Ocean uncompacted sediment measures about 600 meters (2000 feet) or less. In the Atlantic the thickness varies from 500 to 1000 meters (1600 to 3300 feet), while in some deep trenches which act as sediment traps accumulations exceed 9 kilometers (5.5 miles). Since the sediment contains about 50 percent water, squeezing the water out would yield a layer only one-half as thick.

Although accumulations of sand-sized particles are found on the deep ocean floor, mud is the most common sediment covering this region. Muds also predominate on the continental shelf and slopes, but the sediments in these areas are coarser overall because of greater quantities of sand. Sampling has shown that sands are generally deposited on the continental shelf, forming beaches along the shore. However in some cases this coarse sediment, which is expected to be found near the shore, occurs in irregular patches at greater depths near the seaward limits of the continental shelves. While some of the sand may have been deposited by local currents that are capable of moving coarse sediment far from shore, the bulk of it appears to be the result of sand deposition on ancient beaches. Such beaches formed during the Ice Age, when sea level was much lower than it is today. These patches of sand were then submerged as sea level rose again.

Types of Sea-floor Sediments

Sea-floor sediments may be classified according to their origin into three broad categories: (1) lithogenous ("derived from rocks") sediment; (2) biogenous ("derived from organisms") sediment; and (3) hydrogenous ("derived from water") sediment. Although each category is discussed separately, it should be remembered that all sea-floor sediments are mixtures. No body of sediment comes from a single source.

Lithogenous Sediment. Lithogenous sediment consists primarily of mineral grains which are products of weathering on the continents. They are transported to the oceans by rivers, glaciers, and wind. Other sources include wave erosion along coasts and volcanic eruptions both on land and in the sea.

Since the smallest particles may take many years to settle to the ocean floor, they may be transported for thousands of kilometers by ocean currents. Some fine particles are also carried great distances by winds before they are dropped on the water surface. As a consequence, virtually every part of the ocean receives some lithogenous sediment. However, the rate at which this sediment accumulates on the ocean floor away from the continents is indeed very slow. It may take from 25,000 to 250,000 years for a 3-centimeter layer to form. On the other hand, near the mouths of large rivers on the continental margins lithogenous sediment accumulates rapidly.

Since deep-sea lithogenous sediments remain suspended in the water for a very long time, there is ample opportunity for chemical reactions to occur. Because of this, the colors of these sediments are often red or brown, resulting when iron on the particle or in the water reacts with dissolved oxygen in the water and leaves a coating of iron oxide (rust). Conversely, lithogenous sediments that are accumulating rapidly, as they do near the continents, are not in contact with the seawater for a long enough time to react with the dissolved oxygen. Consequently, these sediments usually do not acquire a red or brown color.

Biogenous Sediment. Biogenous sediment consists of material of marine organic origin, that is, shells and skeletons of marine animals and plants (Figure 10.16). This debris is produced mostly by microscopic organisms living in the sunlit waters near the ocean surface. The remains then "rain" down upon the sea floor. Although sediments are called biogenous, they must contain only 30 percent (by volume) biogenous material to be so classified.

The most common of the biogenous sediments are known as **calcareous oozes,** meaning they are primarily calcium carbonate ($CaCO_3$). These sediments, which cover nearly one-half of the deep ocean bottom, are produced by organisms that live their lives in warm surface waters. Since the deep, cold water of the ocean contains much carbon dioxide (CO_2), the calcium carbonate-rich shells are completely dissolved by the time they reach great depths. Hence calcareous oozes are rare below depths of 4500 meters (14,800 feet).

Other biogenous sediments include siliceous ooze and phosphate-rich materials. The former is composed primarily of opaline skeletons of diatoms (single-celled algae) and radiolaria (single-celled animals), while the latter is derived from the bones, teeth, and scales of fish and other marine organisms.

Hydrogenous Sediment. Hydrogenous sediment consists of minerals that crystallize directly from seawater through various chemical processes. For example, some limestone is formed in this manner. Although the calcium carbonate that makes up most limestones is biogenous in origin, some limestones are formed when calcite precipitates directly from the water.

One of the principal examples of hydrogenous sediment, and one of the most important sedi-

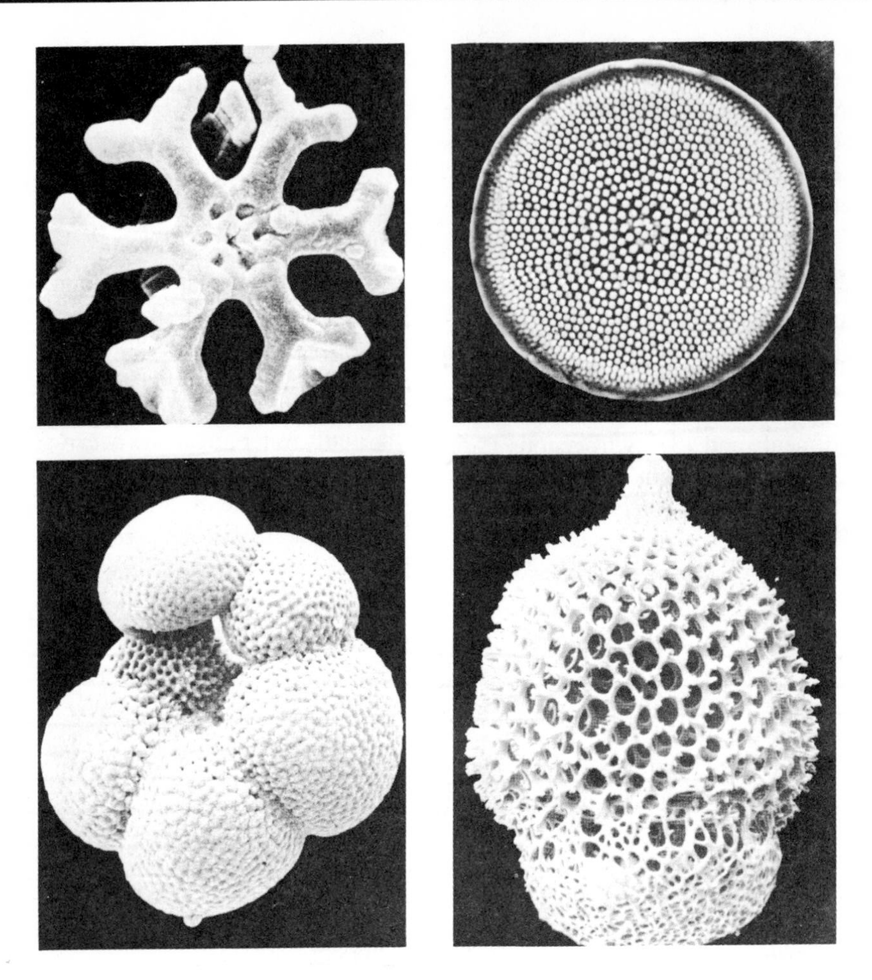

Figure 10.16 *Enlarged photomicrographs of typical calcareous and siliceous materials from minute animals and plants that compose biogenous sediments. (Courtesy of Deep Sea Drilling Project, Scripps Institution of Oceanography)*

ments on the ocean floor in terms of economic potential, are **manganese nodules.** These rounded, blackish, potato-sized nodules are composed of a complex mixture of minerals that form very slowly on the floor of the ocean (Figure 10.17). In fact, their rate of formation represents one of the slowest chemical reactions known. It may take 1000 years or more for a layer 1 millimeter thick to accumulate. Some portions of the sea floor are littered with these deposits while other areas lack manganese nodules altogether. The reason is that nodules cannot grow to any substantial size in regions where sedimentation is relatively heavy because they are buried too quickly. As a result, manganese nodules are rare on the floor of the Atlantic, which is well supplied with lithogenous sediment, but are quite common on the floor of the Pacific, where sediment accumulation is much slower.

Although manganese nodules contain approximately 23 percent manganese, the interest in them as a potential resource lies in the fact that many valuable metals are enriched in them. In addition to manganese, the average nodule contains 6 percent iron, about 1 percent each of copper and nickel, and a slightly lesser quantity

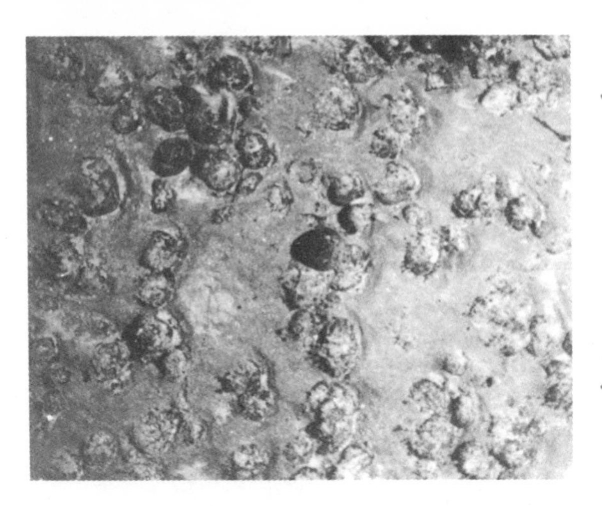

Figure 10.17 *Manganese nodules are potential sources of useful metals. (Courtesy of U.S. Geological Survey)*

of cobalt. However, before manganese nodules prove to be a valuable commercial source for these metals, a couple of serious problems must be solved. Since the greatest concentrations lie at depths in excess of 3800 meters (12,500 feet), mining these deposits will have to await an economical means of dredging or scooping them from the deep ocean floor. Further, once they are collected, the nodules will require a substantially different processing technique than is currently used for other ores.

Sea-floor Sediments and Climatic Change

Since instrumental records of climate elements go back only a couple of hundred years (at best), how do scientists find out about climates and climatic changes prior to that time? The obvious answer is that they must reconstruct past climates from indirect evidence; that is, they must examine and analyze phenomena that respond to and reflect changing atmospheric conditions. One of the most interesting and important techniques for analyzing the earth's climatic history is the study of sediments from the ocean floor.

Although sea-floor sediments are of many types, most contain the remains of organisms that once lived near the sea surface (the ocean-atmosphere interface). When such near-surface organisms die, their shells slowly settle to the floor of the ocean where they become part of the sedimentary record. One reason that sea-floor sediments are useful recorders of worldwide climatic change is that the numbers and types of organisms living near the sea surface change as the climate changes. This principle is explained by Richard Foster Flint as follows:

> . . . we would expect that in any area of the ocean/atmosphere interface the average annual temperature of the surface water of the ocean would approximate that of the contiguous atmosphere. The temperature equilibrium established between surface seawater and the air above it should mean that . . . changes in climate should be reflected in changes in organisms living near the surface of the deep sea. . . . When we recall that the sea-floor sediments in vast areas of the ocean consist mainly of shells of pelagic Foraminifers, and that these animals are sensitive to variations in water temperature, the connection between such sediments and climatic change becomes obvious.*

Thus, in seeking to understand climatic change as well as other environmental transformations, scientists have become increasingly interested in the huge reservoir of data in sea-floor sediments.

Sea-floor Sediments and Sea-floor Spreading

Specially constructed research vessels like the *Glomar Challenger* (Figure 10.18) are capable of drilling into the ocean floor to considerable depths, even in as much as 3000 meters (10,000 feet) of water. Such technology has aided researchers immeasurably in studying the sediments of the ocean floor. Drilling into the ocean floor, for example, has yielded data which help substantiate the theory of plate tectonics.

---

* *Glacial and Quaternary Geology* (New York: Wiley, 1971), p. 718.

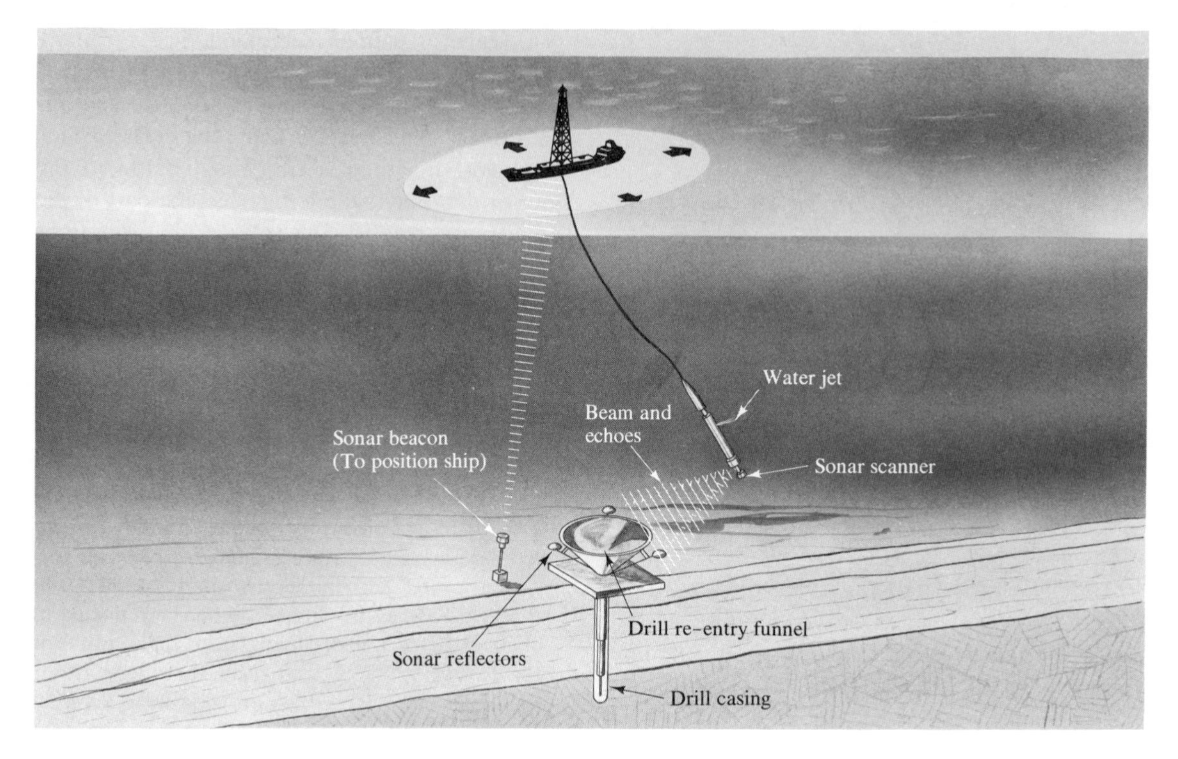

Figure 10.18 *The* Glomar Challenger *possesses remarkable operating capabilities. The ship can be positioned in water too deep for anchors and remain at a drilling site as long as desired because of a dynamic positioning system that maintains the ship's position within a radius of about 100 meters. When necessary, a previously drilled hole can be reentered. This may sound simple, but remember that the hole is only 12 centimeters wide and may be thousands of meters below the ship. At the time reentry is attempted, the drill string is relowered with a sonar scanner that emits sound signals attached. These signals are echoed back from three reflectors spaced around the funnel. Position information is relayed to the ship and a water jet is used to steer the drill bit directly over the funnel. (Based on a National Science Foundation report)*

Cores taken near oceanic ridges have shown that only relatively thin layers of young sediment overlie the oceanic crust. With an increase in distance from the ridge, the sediment generally thickens, and the age of the sediments also increases.

At one time it was thought that sediment accumulations must be very thick since oceanic sedimentation had likely been an ongoing process since the earth's very early history. When sediments on parts of the deep ocean floor were found to be less than 200 meters (650 feet) thick, scientists were puzzled. Today, however, this fact finds a reasonable explanation in the theory of sea-floor spreading and plate tectonics. *While new sea floor is created at the oceanic ridges, the sea-floor sediments and the crustal plates on which they ride are continually destroyed where they plunge back into the mantle near the deep ocean trenches.*

REVIEW

1. Define oceanography. Briefly distinguish several subdivisions of this broad discipline.

2. To what does the familiar expression "the Seven Seas" refer today? Did it always have its current meaning?
3. How does the area covered by the oceans compare with that of the continents? Describe the distribution of land and water on earth.
4. Answer the following questions about the Atlantic, Pacific, and Indian oceans:
   a. Which is the largest in area? Which is smallest?
   b. Which has the greatest north-south extent?
   c. Which is shallowest? Explain why its average depth is least.
5. How does the average depth of the ocean compare to the average elevation of the continents (see Figure 10.3)?
6. Briefly describe the contributions to oceanography of the following: Edward Forbes, Matthew Fontaine Maury, Fridtjof Nansen, the *Challenger* expedition, the *Meteor* expedition.
7. Assuming that the average speed of sound waves in water is 1500 meters per second, determine the water depth if the signal sent out by an echo sounder requires 6 seconds to strike bottom and return to the recorder (see Figure 10.5).
8. How does the continental slope differ from the continental shelf and the continental rise?
9. Defend or rebut the statement "Most submarine canyons were formed during the Ice Age when rivers extended their courses seaward."
10. What are turbidities? What is meant by the term *graded bedding?*
11. Discuss the evidence that helped confirm the existence of turbidity currents in the ocean and establish them as significant mechanisms of erosion and sediment transport.
12. The sediment that covers the abyssal plain is largely derived from the adjacent continents. Why are there only a few abyssal plains in the margins of the Pacific while the Atlantic has extensive abyssal plains?
13. What evidence could be used to prove that guyots were once at the sea surface?
14. How are mid-ocean ridges and deep-ocean trenches related to sea-floor spreading?
15. What is an atoll? Describe Darwin's theory on the origin of atolls. Was the theory ever confirmed?
16. Differentiate between the three basic types of sea-floor sediment.
17. How fast do lithogenous sediments accumulate in the open ocean?
18. If you were to examine biogenous sediment that had been taken from a depth of 4500 meters (15,000 feet), would it more likely be rich in calcareous or siliceous materials? Explain.
19. Why are manganese nodules fairly abundant on the floor of the Pacific but rare on the floor of the Atlantic?
20. Why are sea-floor sediments useful in studying climates of the past?
21. In what way has deep-sea drilling yielded data that help substantiate the theory of plate tectonics?

KEY TERMS

oceanography
azoic zone
Nansen bottle
echo sounder
continental shelf
shelf break
continental slope
continental rise
submarine canyon
turbidity current
turbidites
graded bedding
deep-ocean trench
abyssal plain
seamount
guyot
mid-ocean ridge
coral reef
atoll
lithogenous sediment
biogenous sediment
calcareous ooze
hydrogenous sediment
manganese nodules

Wave breaking on Point Lobos, near Carmel, California. (Photo by M.R. Campbell, U.S. Geological Survey)

11

Composition and Movements

The restless waters of the ocean are constantly on the move. The winds generate surface currents, the moon and sun produce the tides, and density differences create the deep ocean circulation. Further, waves carry the energy from storms to distant shores, where their impact erodes the land. This chapter examines the movements of ocean waters and their importance after taking a look at the composition of seawater and at the layered structure of the ocean.

Composition of Seawater

Seawater is a complex solution of salts. It consists of about 3.5 percent (by weight) dissolved mineral substances. Although the percentage of salts may seem small, the actual quantity is huge. If all of the water were evaporated from the oceans, a layer of salt approaching 60 meters (200 feet) thick would cover the entire ocean floor. The **salinity,** that is, the proportion of dissolved salts to pure water, is commonly expressed in parts per thousand rather than as a percentage (parts per hundred). The symbol used to express parts per thousand is ‰. Thus the average salinity of the ocean is about 35 ‰.

The principal elements that contribute to the salinity of the ocean are shown in Figure 11.1. If we attempted to make our own seawater, we could come reasonably close by following the recipe shown in Table 11.1. From this table it is evident that most of the salt is sodium chloride (common table salt), not a surprising revelation when we consider the proportions of elements shown in Figure 11.1. Sodium chloride together with the next four most abundant salts comprises 99 percent of the salt in the sea. Although only seven elements make up these five most abundant salts, seawater probably contains all of the earth's other naturally occurring elements. Despite their presence in minute quantities, many of these elements are very important in maintaining the necessary chemical environment for life in the sea.

Table 11.1 ***Recipe for Artificial Seawater***

MIX:	
Sodium chloride ($NaCl$)	23.48 grams
Magnesium chloride ($MgCl_2$)	4.98
Sodium sulfate ($NaSO_4$)	3.92
Calcium chloride ($CaCl_2$)	1.10
Potassium chloride (KCl)	0.66
Sodium bicarbonate ($NaHCO_3$)	0.192
Potassium bromide (KBr)	0.096
Hydrogen borate (H_3BO_3)	0.026
Strontium chloride ($SrCl_2$)	0.024
Sodium fluoride (NaF)	0.003
ADD:	
Water (H_2O) to form 1000 grams of solution.	

The relative abundance of the major components in sea salt are essentially constant, no matter where the ocean is sampled. Variations in salinity, therefore, are primarily a consequence of changes in the water content of the solution. As a result, *high salinities are found where evaporation is high, as is the case in the dry subtropics. Conversely, where heavy precipita-*

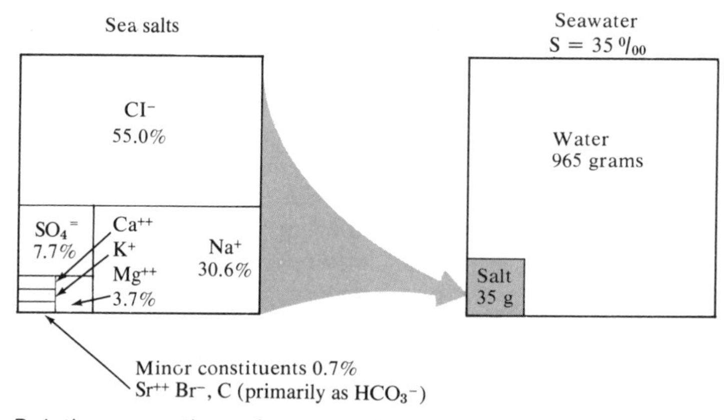

Figure 11.1 *Relative proportions of water and dissolved salts in seawater.*

tion dilutes ocean waters, as in the mid-latitudes and near the equator, lower salinities prevail (Figure 11.2). While salinity variations in the open ocean normally range from 33 $^0/_{00}$ to 37 $^0/_{00}$, some seas demonstrate extraordinary extremes. For example, in the restricted waters of the Persian Gulf and the Red Sea, where evaporation far exceeds precipitation, salinities may exceed 42 $^0/_{00}$. Conversely, very low salinities occur where large quantities of fresh water are supplied by rivers and precipitation. Such is the case for the Baltic Sea, where salinity varies from 2 $^0/_{00}$ to 7 $^0/_{00}$.

What are the sources for the vast quantities of salts in the ocean? The products of the chemical weathering of rocks on the continents constitute one source. These soluble materials are delivered to the oceans by streams at a rate that has been estimated at more than 2500 million tons annually. The second major source of the elements found in ocean water is the earth's interior. Through volcanic eruptions, large quantities of water and dissolved gasses have been emitted during much of geologic time. This process, called **outgassing,** is thought to be the principal source of water in the oceans as well as in the atmosphere. Certain elements, notably chlorine, bromine, sulfur, and boron, are much more abundant in the ocean than in the earth's crust. Since they result from outgassing, their abundance in the sea tends to confirm the hypothesis that volcanic action is largely responsible for the present oceans.

Although rivers and volcanic activity continually contribute materials to the oceans, the salinity of seawater is not increasing. In fact, many oceanographers believe that the composition of seawater has been relatively consistent for a large span of geologic time. Why doesn't the sea get saltier? Obviously, material is being removed just as rapidly as it is added. Some of the elements are withdrawn from seawater by plants and animals as they build shells and skeletons. Others are removed when chemically

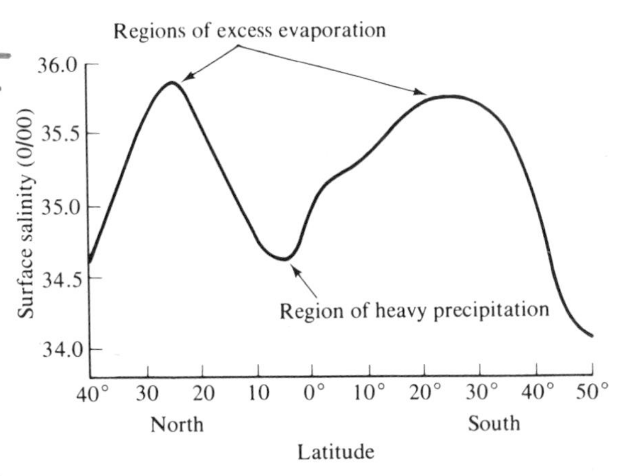

Figure 11.2 *Variations in surface salinity with latitude.*

precipitated as sediment. The net effect is that the makeup of sea water remains constant.

Resources from Seawater

We have seen that the ocean is a great storehouse of dissolved minerals. Some are presently being extracted from seawater while others remain as untapped potential resources. Products obtained commercially from seawater include common salt (sodium chloride), the lightweight metal magnesium, and bromine, which is used principally in gasoline additives and in the manufacture of fireproofing materials. For many years seawater provided a high percentage of the world's magnesium and bromine. In the late 1960s, 70 percent of the world's bromine and 61 percent of its magnesium came from the chemical processing of seawater. Since that time, more and more of the production of these two elements has shifted away from seawater for economic reasons. Today, water from Utah's Great Salt Lake, and brine wells and evaporite deposits in the United States and other areas, are the primary sources. However, in countries lacking large brine and evaporite deposits, such as Japan and the Scandinavian countries, seawater is still a primary source of these elements.

Figure 11.3 *Harvesting salt produced by the evaporation of seawater. (Courtesy of the Leslie Salt Company)*

Since Neolithic times the ocean has been an important source of salt for human consumption, and the sea remains a significant supplier. To obtain pure sodium chloride, a series of evaporating ponds is used. The idea is that different salts precipitate from seawater at different salinities. In the first pond, evaporation proceeds until the salt concentration is such that calcium carbonate and calcium sulfate precipitate out. Then the remaining brine is transferred to another pond where evaporation continues. Practically pure sodium chloride precipitates there. When most of the sodium chloride has precipitated, the remaining liquid (now enriched in potassium and magnesium) is drained off and the salt is harvested (Figure 11.3). In this manner about 30 percent of the world's salt is produced.

People have long been intrigued by the prospect of extracting gold from ocean water. For example, following World War I Germany seriously considered extracting gold from seawater to pay the war debt. In fact, a major goal of the famous *Meteor* expedition was to investigate the feasibility of gold recovery from the ocean. Unfortunately, what they found out was that the concentration of gold in ocean water is extremely low. To extract a single ounce of gold, 200,000 tons of seawater would have to be processed. Further, the gold would then have to be separated from the remaining 7000 tons of solids. Thus even though millions of tons of gold may exist in the waters of the global ocean, no one is now enjoying this wealth because the cost of recovering the gold far exceeds its value.

One of the resources derived from seawater is fresh water. The removal of salts and other chemicals from seawater is termed **desalination.** The purpose of this process is to extract low-salinity ("fresh") water suitable for drinking, industry, and agriculture from water that is so saline that it is not suitable for these purposes. Although hundreds of desalination plants are

Figure 11.4 *Nansen bottle, used for sampling seawater and temperature at depth. Once several bottles are attached to a wire, the wire is lowered and a messenger (commonly a lead weight) is released. The weight slides down the wire and releases the first bottle, resulting in the trapping of the water and the freezing-in of the thermometer readings. A second messenger is then released, tripping the second bottle, and so on. (Photo courtesy of Woods Hole Oceanographic Institution)*

now operating and others are being planned, the cost of desalinized water is still high, and the total production remains small. In 1970 desalination yielded only about 380 million liters (100 million gallons) of water per day. This quantity amounts to about a tenth of the daily consumption for the city of New York.

Although fresh water produced by various desalination technologies may become more important as a source of water for human and even industrial use, it is unlikely to be an important supply for agricultural purposes. Since the water requirements of humans are quite low, the cost of supplying areas without adequate natural supplies can often be justified. In addition, the cost of water may represent only a small fraction of the ultimate cost of many industrial goods and hence can be absorbed even in competitive markets. However, the cost of water for irrigation represents a large percentage of the total cost of the crop being produced. Thus the cost of desalinized water is still much too high for crop production. Consequently, making the deserts "bloom" by irrigating them with desalinized seawater is only a dream and, for economic reasons, is likely to remain so in the foreseeable future.

The Ocean's Layered Structure

By sampling ocean waters, oceanographers have found that temperature and salinity vary with depth (Figure 11.4). Generally, they recognize a three-layered structure in the open ocean: a shallow surface mixed zone, a transition zone, and a deep zone (Figure 11.5).

Since solar energy is received at the ocean surface, it is here that water temperatures reach their highest values. Because of the mixing of these waters by waves and the turbulence from currents, there is a rapid vertical heat transfer. Hence, as the temperature-depth curve in Figure 11.6A illustrates, this mixed surface zone is characterized by nearly uniform temperatures. The thickness and temperature of this layer

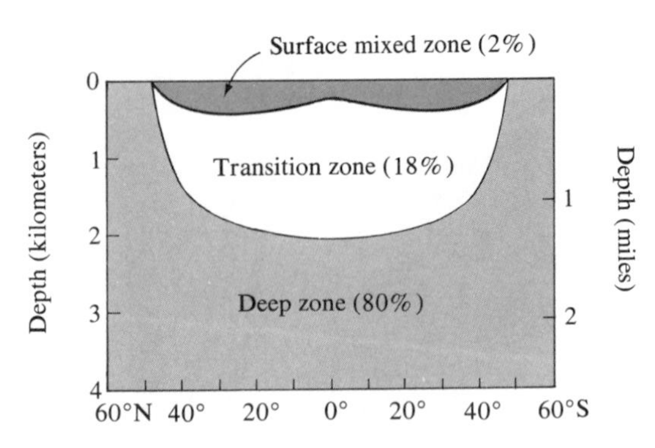

Figure 11.5 *Layered structure of the ocean.*

vary, depending upon the latitude and the season. The zone may attain a thickness of 450 meters (1500 feet) or more, and temperatures ranging between 21° C and 26° C (70° F and 80° F) are not uncommon. In equatorial latitudes the temperatures may be even higher. Below the sun-warmed zone of mixing, the temperature falls abruptly with depth (Figure 11.6A). This layer of rapid temperature change, known as the **thermocline,** marks the transition between the warm surface layer and the deep zone of cold water below. Below the thermocline temperatures fall only a few more degrees. At depths greater than about 1500 meters (5000 feet) ocean-water temperatures are consistently less than 4° C (39° F). *Since the world ocean averages more than 3800 meters (12,400 feet) deep, it is evident that the temperature of most seawater is not much above freezing.* In the high polar latitudes surface waters are cold, and temperature changes with depth are slight. Hence the three-layered structure is not present.

Salinity variations with depth correspond to the general three-layered system described for temperatures. Generally, in the low and middle latitudes a surface zone of high salinity is created when fresh water is removed by evaporation. Below the surface zone salinity decreases rapidly. This layer of rapid change, called the **halocline** (Figure 11.6B), corresponds closely to the thermocline. Below the halocline salinity variations are small.

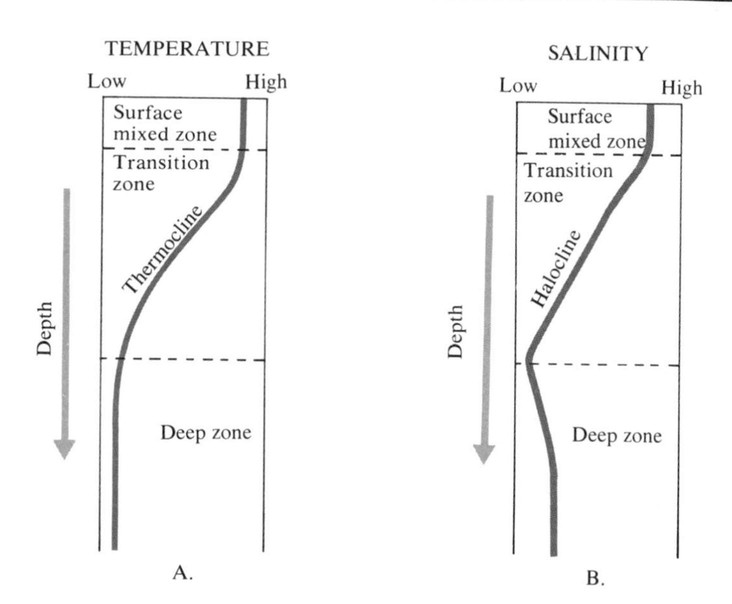

Figure 11.6 *Changes with depth of* **A.** *temperature and* **B.** *salinity.*

Surface Currents

> There is a river in the ocean. In the severest droughts it never fails, and in the mightiest floods . . . it never overflows. Its banks and its bottom are of cold water, while its current is of warm. The Gulf of Mexico is its fountain, and its mouth is in the Arctic Sea. It is the Gulf Stream. There is in the world no other such majestic flow of waters.

This quote from Maury's 1855 book, *The Physical Geography of the Sea,* describes perhaps the best-known and most-studied of the surface ocean currents. However the Gulf Stream is just a portion of a huge, slow-moving circular whirl, or **gyre,** that begins near the equator. Similar gyres may be seen in the South Atlantic, and in the North and South Pacific (Figure 11.7). What creates these large surface current systems? Although our knowledge of the processes that produce and maintain ocean currents is far from complete, we do have a general understanding of the principal factors involved.

The drag exerted by winds blowing steadily across the ocean causes the surface layer of water to move. Thus *because winds are the primary driving force of surface currents, there is a relationship between the oceanic circulation and the general circulation of the atmosphere.* A comparison of Figures 11.7 and 14.14 illustrates this. A further clue to the influence of winds on ocean circulation is provided by the currents in the northern Indian Ocean. Here there are seasonal wind shifts known as the summer and winter monsoons. When the winds change direction, the surface currents also reverse direction.

Although winds are important in generating surface currents, other factors also influence the movement of ocean waters. The most significant of these is the **Coriolis force:** *Because of the earth's rotation, currents are deflected to the right of their path of motion in the Northern Hemisphere and to the left in the Southern Hemisphere.* The Coriolis force is greater in high latitudes and diminishes toward the equator (for a more complete discussion of the Coriolis force, see Chapter 14). As a consequence of the deflective force caused by the earth's rotation, the direction of surface currents does not coincide with the wind direction. In general, the difference between wind direction and surface-

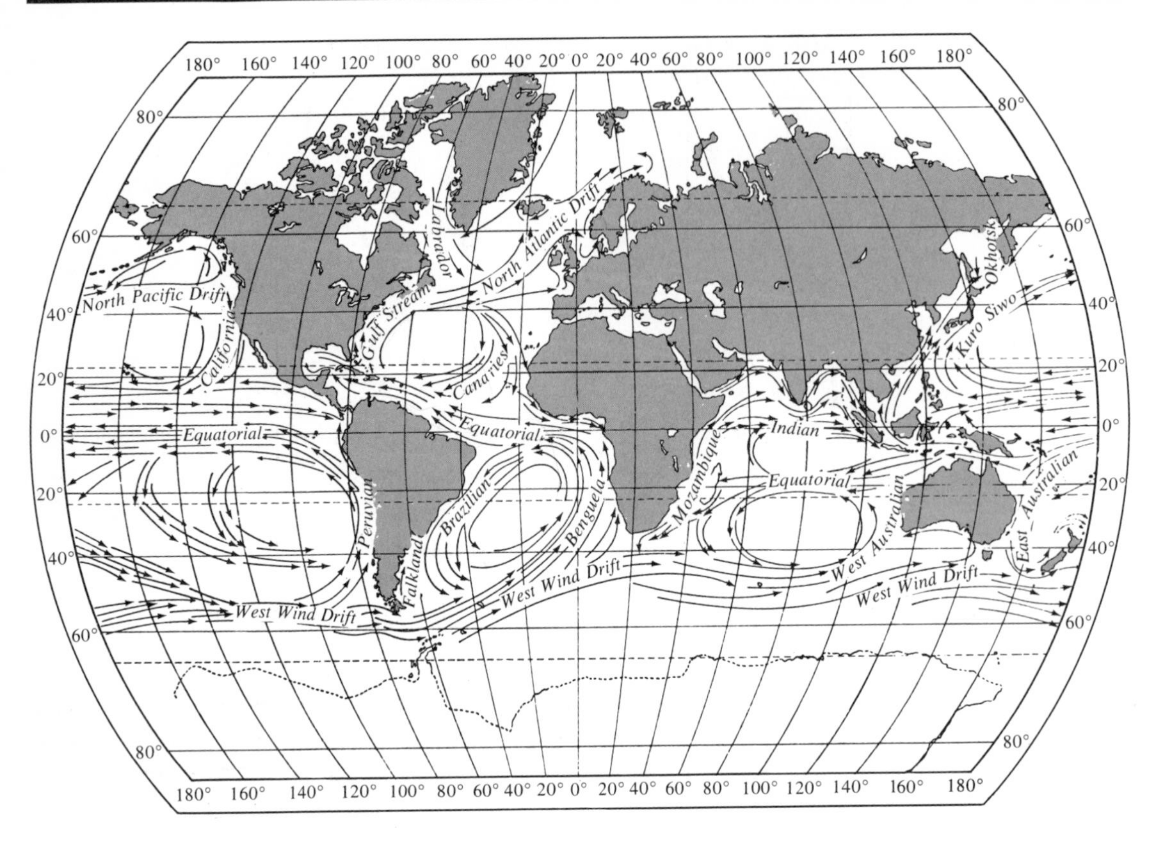

Figure 11.7 *Average position and extent of the principal ocean currents. (After* Weather and Climate *by C.E. Koeppe and G.C. DeLong, p. 167. Copyright © 1958 by McGraw-Hill Book Company. Used with permission of McGraw-Hill Book Company)*

current direction varies from about 15 degrees along shallow coastal areas to a maximum of 45 degrees in the deep oceans. Furthermore, the angle increases with depth, so at a depth of 100 meters (300 feet) the current may flow in the opposite direction to the surface waters. This phenomenon was first proposed by V.W. Ekman and has come to be called the **Ekman spiral** (Figure 11.8).

Briefly stated, the principle of the Ekman spiral is as follows. The moving surface layer of water sets the layer below in motion. This lower layer in turn exerts a frictional drag on the water immediately beneath it, causing it to flow, and so on. Each successively deeper layer moves more slowly because momentum is lost in each transfer between the layers. As each layer begins to move, it is deflected by the Coriolis force, causing it to move to the right (in the Northern Hemisphere) of the overlying layer. Because each layer is deflected to the right of the layers above, the direction of water movement shifts with increasing depth, until at a depth of about 100 meters (300 feet) the water is slowly moving in a direction opposite to the surface layer.

A closer look at the circulation of one of the major gyres—that of the North Atlantic—will serve to illustrate the general pattern taken by surface currents (Figure 11.7).

North and south of the equator are two westward-moving currents, the North and South Equatorial Currents, which derive their energy principally from the trade winds that blow from the northeast and southeast respectively toward

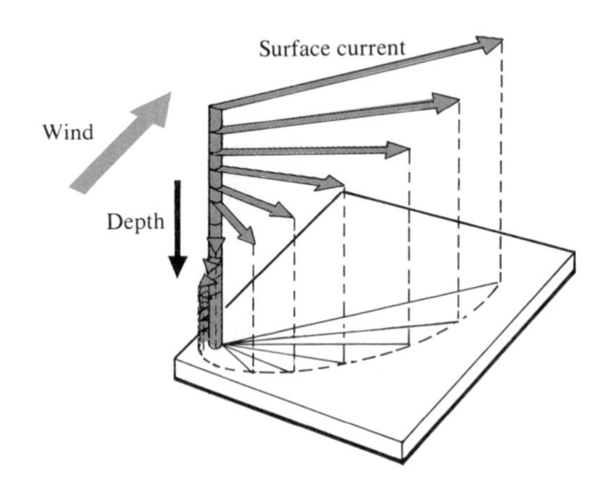

Figure 11.8 *Ekman spiral. The Coriolis force deflects the surface current to the right of the driving wind in the Northern Hemisphere. Deeper water continues to be deflected to the right and moves at a slower speed with increased depth.*

the equator. Because of the deflection created by the Coriolis force, the currents move almost due west. Between these westward-flowing currents there is a weaker, oppositely directed eastward flow, the Equatorial Countercurrent. The North and South Equatorial Currents pile water up against the eastern coast of South America, and because the trade winds are weak along the equator, some of the piled-up water flows back "downhill" (eastward), creating the countercurrent.

As the North Equatorial Current continues west, the South American continent and the Coriolis force turn the water to the north. The water then moves northwest through the Caribbean to the mouth of the Gulf of Mexico at the Straits of Yucatan. From here some curves toward the right, flowing some distance off the shore of the Gulf of Mexico, and part of it curves more sharply toward the east and flows directly toward the north coast of Cuba. These two parts reunite in the Straits of Florida to form the Gulf Stream. A tremendous volume of water flows northward in the Gulf Stream. It can often be distinguished by its deep indigo blue color, which contrasts sharply with the dull green of the surrounding water. When the Gulf Stream encounters the cold waters of the Labrador Current, in the vicinity of the Grand Banks, there is little mixing of the waters. Rather, the junction is marked by a sharp temperature change. When the warm Gulf Stream water encounters cold air with a low water-holding capacity in this region, the result is a fog that has the appearance of smoke rising from the water.

As the Gulf Stream moves along the east coast of the United States, it is strengthened by the prevailing westerly winds and is deflected to the east (to the right) between 35 degrees north and 45 degrees north latitude. As it continues northeastward beyond the Grand Banks, it gradually widens and decreases in speed until it becomes a vast, slow-moving current known as the North Atlantic Drift. As the North Atlantic Drift approaches Western Europe, it splits. Part of it moves northward past Great Britain and Norway. The other part is deflected southward as the cool Canaries Current. As the Canaries Current moves to the south, it eventually merges into the North Equatorial Current.

The clockwise circulation of the North Atlantic leaves a large central area which has no well-defined currents. This zone of calmer waters is known as the Sargasso Sea, named for the large quantities of *Sargassum,* a type of seaweed encountered there.

In the South Atlantic, surface ocean circulation is very much the same as that in the North Atlantic. The major exception is the circulation of the cold surface waters north of Antarctica. Uncomplicated by large landmasses, the currents in this region move easterly in a globe-circling pattern around the ice-covered continent. They are driven by the prevailing winds from the northwest and deflected to the left by the earth's rotation. The circulation of the Pacific generally parallels that of the Atlantic, and although the monsoons complicate the circulation of the Indian Ocean, the currents there generally coincide with the currents of the South Atlantic.

In addition to producing surface currents, winds may also cause vertical water movements.

Upwelling, the rising of cold water from deeper layers to replace warmer surface water, is a common wind-induced vertical movement. It is most characteristic along the eastern shores of the oceans, most notably along the coasts of California, Peru, and West Africa. Upwelling occurs in these areas when winds blow toward the equator parallel to the coast. Because of the Coriolis force, the surface water movement is directed offshore (Figure 11.9). As the surface layer moves away from the coast, it is replaced by water that "upwells" from below the surface. This slow upward flow from depths of 50 to 300 meters (150 to 1000 feet) brings water which is cooler than the original surface water and creates a characteristic zone of lower temperatures near the shore. Coastal upwelling also brings to the ocean surface greater concentrations of dissolved nutrients such as nitrates and phosphates than were present in the original surface water, which had been depleted by biological demands. These nutrient-enriched waters from below promote the growth of plankton, which in turn supports extensive populations of fish.

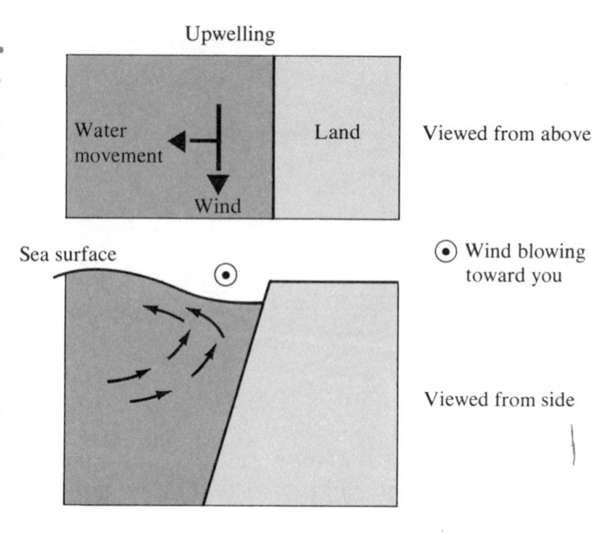

Figure 11.9 *Wind-induced coastal upwelling in the Northern Hemisphere.*

The Importance of Ocean Currents

Since ocean waters are constantly on the move, anyone who sails in the ocean soon learns about the horizontal movements we call currents. Long before Matthew Maury compiled the first global chart showing surface currents their existence was known by sailors whose ships deviated from their expected courses. Moreover, by "reading" the currents correctly, sailors realized that the time of their voyages could be considerably reduced.

In addition to being significant considerations in ocean navigation, currents have an important effect on climates. It is known that for the earth as a whole the gains in solar energy equal the losses to space of heat radiated from the earth. However this is not the case for most latitudes. There is a net gain of energy in the low latitudes and a net loss at higher latitudes. Since the tropics are not becoming progressively warmer, nor the polar regions colder, there must be a large-scale transfer of heat from areas of excess to areas of deficit. This is indeed the case. *The transfer of heat by winds and ocean currents equalizes these latitudinal energy imbalances.* Ocean-water movements account for about a quarter of this total heat transport, and winds the remaining three-quarters.

The moderating effect of poleward-moving warm ocean currents is well known. The North Atlantic Drift, an extension of the warm Gulf Stream, keeps Great Britain and much of northwestern Europe warmer than one would expect for their latitudes. Because of the prevailing westerly winds, the moderating effects are carried far inland. For example, Berlin (52 degrees north latitude) has an average January temperature similar to that experienced at New York City, which lies 12 degrees farther south, while the January mean at London (51 degrees north latitude) is 4.5° C (8° F) higher than that at New York City.

In contrast to warm ocean currents whose effects are felt most in the middle latitudes in the winter, the influence of cold currents is most pronounced in the tropics or during the summer months in the middle latitudes. Cool currents such as the Benguela Current off the west coast of southern Africa moderate the tropical heat. For example, Wavis Bay (23 degrees north latitude), a town adjacent to the Benguela Current, is 5° C (9° F) cooler in summer than Durban, which is 6 degrees latitude farther poleward but on the eastern side of South Africa, away from the influence of the current.

Deep Ocean Circulation

Deep ocean circulation cannot readily be compared with the movements of surface currents. Unlike the wind-induced movements of surface and near-surface waters, *deep ocean circulation is governed by gravity and driven by density differences.* Two factors—temperature and salinity—are most significant in creating a dense mass of water. Seawater becomes denser with decreased temperature, increased salinity, or both. Consequently, deep ocean circulation is called **thermohaline** (*thermo*—"heat", *haline*—"salt") **circulation.** After leaving the surface of the ocean, waters will not reappear at the surface again for an average of 500–2000 years.

The circulation of the Atlantic Ocean has been the most intensively studied. A simplified cross-sectional view of its deep circulation pattern is seen in Figure 11.10. Arctic and Antarctic waters represent the two major regions where dense water masses are created. Antarctic waters are chilled during the winter. The temperatures here are low enough to form sea ice, and since sea salts are excluded from the ice, the remaining water becomes saltier. The result is the densest water in all of the oceans. This cold saline brine slowly sinks to the sea floor, where it becomes Antarctic Bottom Water (ABW in Figure 11.10). It moves northward along the ocean floor, crossing the equator and reaching as far as 20 degrees north latitude.

North Atlantic Deep Water (NADW in Figure 11.10) is thought to form when the warm and highly saline waters of the Gulf Stream reach the Arctic region near Greenland. The high salinity of the warm surface current results from high evaporation in the low latitudes. In the Arctic these waters are chilled and sink to the bottom of the North Atlantic Basin. From here, the cold, dense water moves south, overriding the denser Antarctic Bottom Waters. North Atlantic Deep Water has been traced almost as far as the Ant-

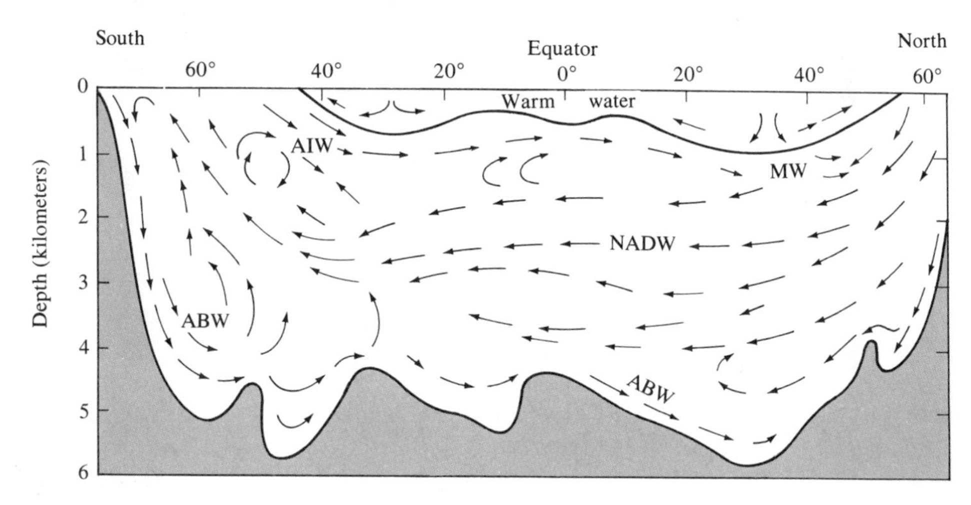

Figure 11.10 *Cross section of the deep circulation of the Atlantic Ocean. (After Gerhard Neumann and Willard J. Pierson, Jr.,* Principles of Physical Oceanography, *© 1966. Reprinted by permission of Prentice-Hall, Inc., Englewood Cliffs, N.J.)*

A.

B.

Figure 11.11 A. *High tide and* **B.** *low tide in the Bay of Fundy at Parrsboro, Nova Scotia. (Photos by G. Blouin, Information Canada Photothèque)*

arctic region, where it becomes obscured by mixing.

Other important subsurface water masses have also been identified. Antarctic Intermediate

Water (AIW in Figure 11.10) is formed when the very salty waters of the Brazil Current are chilled as they move south toward the Antarctic. Still another water mass (MW in Figure 11.10) has its source in the Mediterranean Sea. In the warm and dry climate of this region salinity is high because evaporation is great. As these highly saline waters are chilled during the winter they become even denser and sink. The water moves along the bottom to the west, passing into the Atlantic at the shallow Strait of Gibraltar. In order to replace the water lost in this outflow, less dense surface water enters the Mediterranean at Gibraltar. A similar water mass forms in the Red Sea. Since these waters are less dense than the bottom waters, they reach equilibrium after sinking to intermediate depths.

The circulation of deep waters in the Pacific and Indian oceans is different when compared to that of the Atlantic Ocean. Deep water from the Antarctic predominates in these oceans because the shallow barrier at the Bering Strait does not allow Arctic water to flow southward. Consequently, some deep water from the South Polar region may reach as far north as Japan and California.

Tides

Tides are periodic changes in the elevation of the ocean surface at a particular place. Their rhythmic rise and fall along coastlines have been known since antiquity, and next to waves, they are the easiest ocean movements to observe (Figure 11.11). Although known for centuries, tides were first explained satisfactorily by Sir Isaac Newton using the law of gravitation. Newton showed that there is a mutual attractive force between two bodies, and that since oceans are free to move, they are deformed by this force. Hence tides result from the gravitational attraction exerted upon the earth by the moon, and to a lesser extent by the sun.

Initially we will assume that the earth is a rotating sphere covered to a uniform depth with water. Since the pull of gravity is inversely proportional to the square of distance between the two objects, the moon's gravitational pull causes the water to bulge on the side of the earth nearest the moon (Figure 11.12). Less obvious is that a similar bulge, or high tide, is created on the opposite side of the earth. Here the moon's gravitational attraction on the ocean is not as great as it is on the solid earth, which is closer to the moon than the water. Consequently, the solid earth is pulled more toward the moon than the water on the far side, and the water seems to be

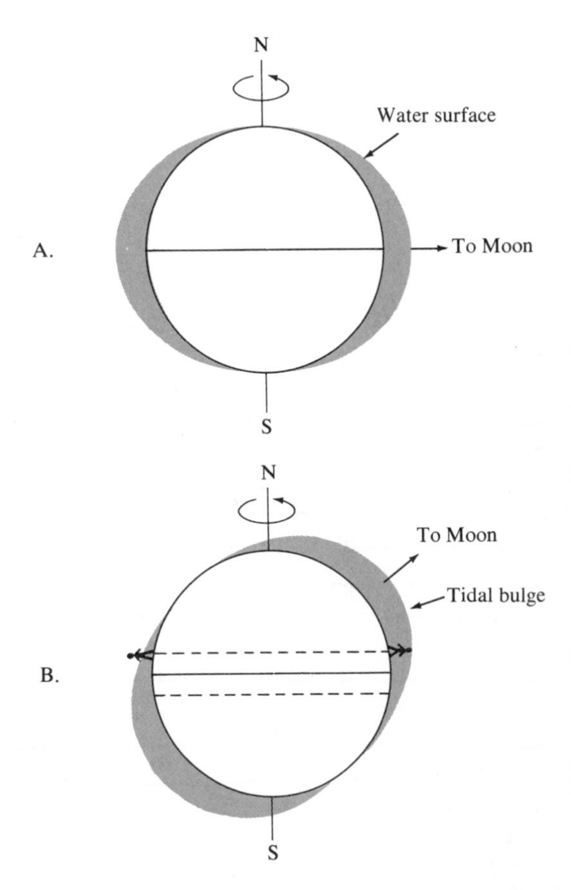

Figure 11.12 *Tidal bulges on an ocean-covered earth with the moon* **A.** *in the plane of the earth's equator and* **B.** *above the plane of the equator. In the latter situation the observer experiences unequal high tides.*

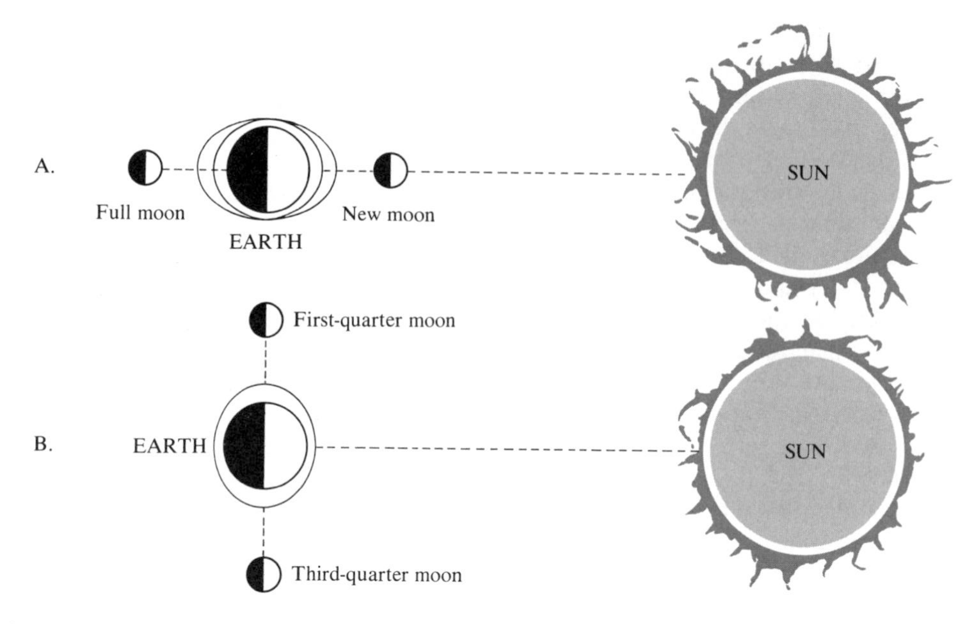

Figure 11.13 *Relationship of the moon and sun to the earth during* **A.** *spring tides and* **B.** *neap tides.*

pulled away from the earth as another tidal bulge. In addition, since the earth and moon revolve around a common center of mass, a force is created in a direction opposite to the moon's pull of gravity. The force, called centrifugal force, which is generated by the revolving earth-moon system, adds to the tidal bulge on the side of the earth farthest from the moon.

To summarize, the tide-producing force on the earth's hemisphere nearest the moon is in the direction of the moon's attraction, in other words, toward the moon. Conversely, on the opposite side of the earth, where the moon's pull of gravity is least, the tide-producing force is in the direction of the centrifugal force, or away from the moon.

The tidal bulges remain in place while the earth rotates beneath them. Therefore the earth will carry an observer at any given place alternately into areas of deeper and shallower water. When he is carried into the regions of deeper water, the tide rises, and as he is carried away, the tide falls. Therefore during one day the observer would experience two high tides and two low tides. In addition to the earth rotating, the tidal bulges also move as the moon revolves about the earth every 28 days. As a result, the tides, like the time of moonrise, occur about 50 minutes later each day. After 28 days one cycle is complete and a new one begins.

There may be an inequality between the high tides during a given day. Depending upon the position of the moon, the tidal bulges may be inclined to the equator as in Figure 11.12B. This figure illustrates that an observer in the Northern Hemisphere experiences a high tide on the side of the earth under the moon that is considerably higher than the high tide half a day later. On the other hand, a Southern Hemisphere observer would experience the opposite effect.

The sun also influences the tides, but because it is so far away, the effect is considerably less than that of the moon. In fact, the gravitational attraction of the sun causes less than half as much tidal deformation as does the moon. Near the times of new and full moons the sun and moon are aligned, and their forces are added together (Figure 11.13A). Accordingly, the two tide-producing bodies cause higher tidal bulges (high tides) and lower tidal troughs (low tides). These are called the **spring tides.** Spring tides create the largest daily tidal range, that is, the largest variation between high and low tides. Conversely, at about the time of the first and third quarters of the moon the gravitational forces of the moon and sun on the earth are at right angles, and each partially offsets the influence of the other (Figure 11.13B). As a result, the daily tidal range is least. These are the **neap tides.**

Although the discussion thus far explains the basic causes and patterns of tides, these theoretical considerations cannot be used to predict either the height or the time of actual tides at a particular place. *The shape of coastlines and the configuration of the ocean basins greatly influence the tide.* Consequently, tides at various places respond differently to the tide-producing forces. This being the case, the nature of the tide at any place can be determined most accurately by actual observation. The predictions in tidal tables and the tidal data on nautical charts are based upon such observations.

Tides are classified as one of three types according to the tidal pattern occurring at a particular locale. The **semidiurnal** type fits the twice-daily pattern described earlier. There are two high and two low tides each tidal day, with a relatively small difference in the high and low water heights. Tides along the Atlantic coast of the United States are representative of the semidiurnal type, which is illustrated by the tidal curve for Boston in Figure 11.14A. **Diurnal** tides are characterized by a single high and low water height each tidal day (Figure 11.14B). Tides of this type occur along the northern shore of the Gulf of Mexico, among other locations. More common than the diurnal tide, the **mixed** type of tide is characterized by a large inequality in high water heights, low water heights, or both (Figure 11.14C). In this case, there are usually two high and two low waters each day. Such tides are prevalent along the Pacific coast of the United States and in many other parts of the world.

Tidal current is the term used to denote the horizontal flow of water accompanying the rise and fall of the tide. As the tide rises water flows in toward the shore as a **flood tide,** submerging the low-lying coastal zone. When the tide falls, a reverse flow, the **ebb tide,** again exposes the drowned portion of the shore. The areas affected by these alternating tidal currents are **tidal flats.** Depending upon the nature of the coastal zone, tidal flats vary in size from narrow strips lying seaward of the beach to extensive areas covering many square kilometers.

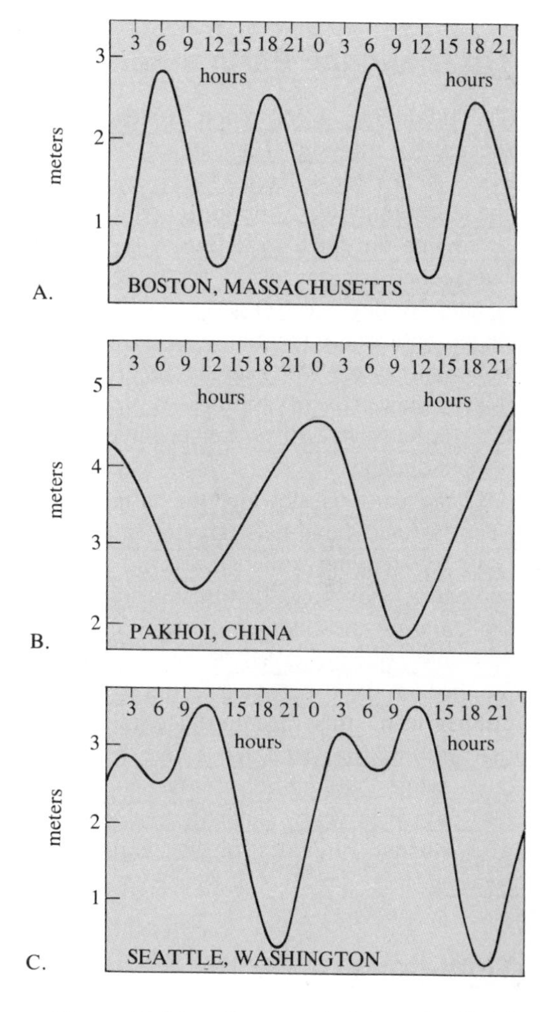

Figure 11.14 *Types of tides.* **A.** *Semidiurnal.* **B.** *Diurnal.* **C.** *Mixed.*

Although tidal currents are not important in the open sea, tides may create rapid currents in bays, river estuaries, straits, and other narrow places. Off the coast of Brittany, for example, tidal currents which accompany a high tide of 12 meters (40 feet) may attain a speed of 20 kilometers (12 miles) per hour. While it is generally believed that tidal currents are not major agents of erosion and sediment transport, notable exceptions occur where tides move through narrow inlets. Here they constantly scour the small entrances to many good harbors that would otherwise be blocked.

Tides and the Earth's Rotation

The tidal drag of the moon is steadily slowing the earth's rotation. The rate of slowing, however, is not great. It has been estimated that the weak, but steady braking action of the tide is slowing the earth's rotation at a rate between one second per day per 120,000 years and one second per day per 100,000 years. Although this may seem inconsequential, over millions of years this small effect will become very large. Eventually, thousands of millions of years into the future, the earth will no longer have alternating days and nights.

If the earth is slowing, the number of days per year must have been greater in the geologic past. By studying fossil corals and clam shells, geologists have shown that this is indeed the case. By counting the number of daily growth rings of some well-preserved fossil specimens, the number of days per year can be ascertained. Studies using this ingenious technique indicate that 350 million years ago a year had 400–410 days, while 280 million years ago there were 390 days in a year. These figures closely agree with current estimates of the earth's slowing rotation.

Tidal Power

The tides have been used as a source of power for centuries. Beginning in the twelfth century water wheels driven by the tides were used for grinding grain. During the seventeenth and eighteenth centuries much of Boston's flour was produced at a tidal mill. Today, however, tides are seen as a source of water to run electrical generators. Tidal power is harnessed by constructing a dam across the mouth of a bay or estuary (Figure 11.15). As the tide rises, water is allowed to flow through open gates into the bay behind the dam. Then when the tide wanes, the impounded water is gradually allowed to flow out, turning electrical generators as it leaves. This system, however, has the disadvantage of producing electricity only intermittently. To overcome spasmodic power production, turbines designed to operate in either direction may be used. Thus power would be generated as the bay fills during high tide, as well as when it empties at low tide.

The French developed the first commercial tidal power plant at La Rance in 1966, and plans are being made for a much larger plant not far away at Mont-Saint-Michel Bay. Near Murmansk in the Soviet Union tides are also being used to generate electricity. Although there have been plans for many years to build a plant at Passamaquoddy Bay, Maine, the United States has not yet tapped its tidal power potential.

Along most of the world's coasts it is not possible to harness tidal energy. If the tidal range is less than 8 meters (25 feet), or narrow, enclosed bays are absent, tidal power development is uneconomical. Sites having the necessary conditions are found along only about 5 percent of the world's coastlines. For this reason, tidal power can never fill a very high proportion of our ever-increasing electrical energy requirements. In fact, if all available sites were utilized, tides would provide less than 2 percent of the world's water-generated power.

Even though tidal power can only be important locally, where it is feasible it offers several environmental advantages. Electricity produced by the tides consumes no exhaustible fuels and produces no noxious wastes. Further, such facilities disturb the landscape much less than the

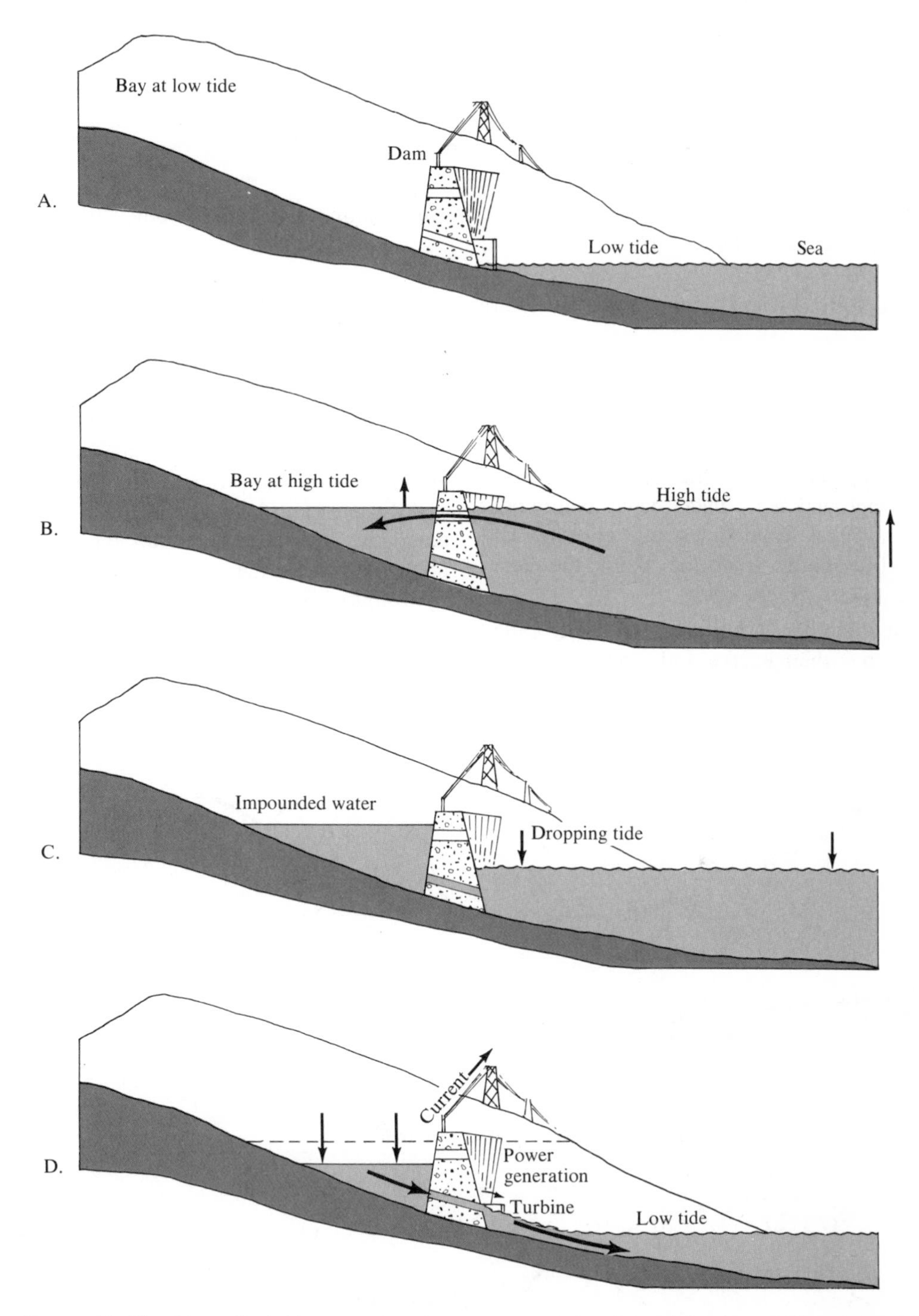

Figure 11.15 *Simplified diagram showing the principle of the tidal dam. (After John J. Fagan,* The Earth Environment, *© 1974. Reprinted by permission of Prentice-Hall, Inc., Englewood Cliffs, N.J.)*

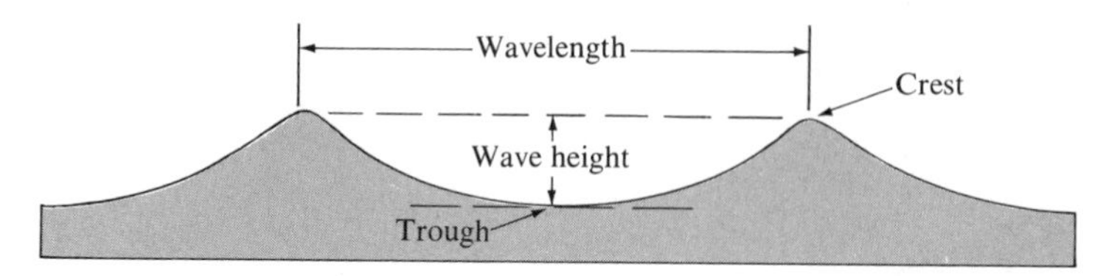

Figure 11.16 *Characteristics of a wave. (From Foster,* Physical Geology, *3rd ed., Columbus, Ohio: Charles E. Merrill, 1979)*

large reservoirs that are created when rivers are dammed.

Waves

The restless nature of ocean waters is most apparent along the shore—the dynamic interface between land and sea—where waves continually roll in and break. Sometimes waves are low and gentle, while at other times they pound the shores with great destructive force.

The undulations of the water surface, called **waves,** derive their energy and motion from the wind. If a breeze of less than 3 kilometers (2 miles) per hour starts to blow across still water, small wavelets appear almost instantly. When the breeze dies, the ripples disappear as suddenly as they formed. However if the wind exceeds 3 kilometers (2 miles) per hour, more stable waves gradually form and progress with the wind.

All waves are described in terms of the characteristics illustrated in Figure 11.16. The tops of the waves are the *crests,* which are separated by *troughs.* The vertical distance between trough and crest is called the **wave height,** and the horizontal distance separating successive crests is the

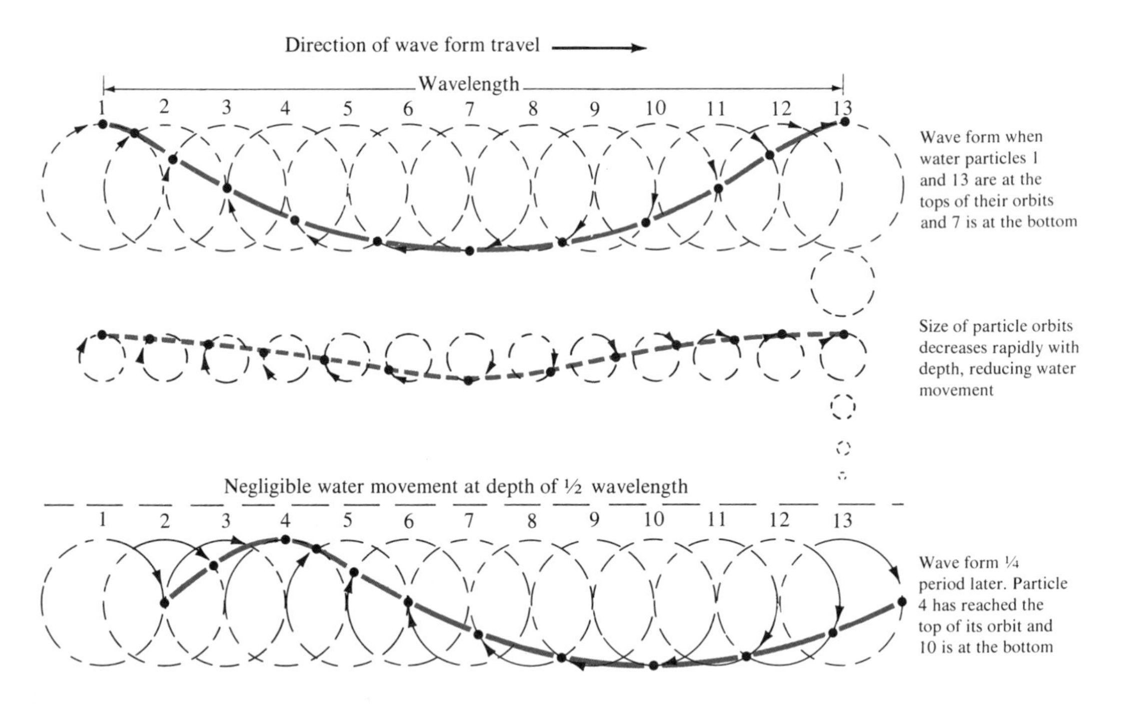

Figure 11.17 *Movement of water particles with the passage of a wave. (From Foster,* Physical Geology, *3rd ed., Columbus, Ohio: Charles E. Merrill, 1979)*

wave length. The **wave period** is the time interval between the passage of successive crests at a stationary point. The height, length, and period that are eventually achieved by a wave depend upon three factors: (1) the wind speed; (2) the length of time the wind has blown; and (3) the **fetch,** or distance that the wind has traveled across the open water. As the quantity of energy transferred from the wind to the water increases, the heights of the waves transmitting it increase. In the open ocean, wave heights of 1.5–4.5 meters (5–15 feet) are common, although storms may produce much higher waves.

Because winds are often gusty and turbulent, the waves covering the surface of the ocean are quite irregular in height and length. When the wind stops or changes direction or the waves leave the stormy area where they were created, they continue on without relation to local winds. The waves also undergo a gradual change to *swells* that are lower in height and longer in length and may carry the storm's energy to distant shores. Because many independent wave systems exist at the same time, the sea surface acquires a complex and irregular pattern. Hence the sea waves we watch from the shore are usually a mixture of swells from faraway storms and waves created by local winds.

It is important to realize that in the open sea the motion of the wave is different from the motion of the water particles within it. *It is the wave form that moves forward, not the water itself.* Each water particle moves in a circular path during the passage of a wave (Figure 11.17). As a wave passes, a water particle returns almost to its original position. This is demonstrated by observing the behavior of a floating cork as a wave passes. The cork merely seems to bob up and down and sway slightly to and fro without advancing appreciably from its original position. Because of this, waves in the open sea are called **waves of oscillation.**

However, as Figure 11.18 illustrates, the wind does drag the water slightly forward, causing the surface circulation of the oceans. The energy contributed by the wind to the water is transmitted not only along the surface of the sea but downward as well. Because of friction, there is a progressive loss of energy with an increase in depth until at a depth equal to about one-half the wave length the movement of water particles becomes negligible. This is shown by the rapidly diminishing diameters of water-particle orbits in Figure 11.17.

As long as a wave is in deep water it is unaffected by water depth. However when a wave approaches the shore the water becomes shallower and influences wave behavior. The wave begins to "feel bottom" at a water depth equal to about one-half its wave length. Since some energy is used in moving small particles of sediment back and forth, the wave slows. As the wave continues to advance toward the shore the slightly faster seaward waves catch up, decreasing the wave length. As the speed and length of the wave diminish, the wave steadily grows higher. Finally a critical point is reached when the steep wave front is unable to support the wave, and it collapses, or *breaks* (Figure 11.19). What had been a wave of oscillation now becomes a **wave of translation** in which the water advances up the shore. The turbulent water created by breaking waves is called **surf.** Following the uprush of water onto the beach, a seaward backwash occurs. The water flows back in a complex pattern.

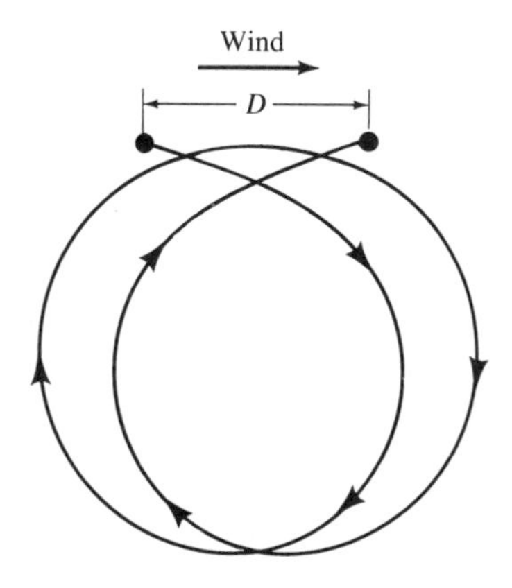

Figure 11.18 *Circular motion and displacement (D) of a particle on the surface of deep water during two wave periods.*

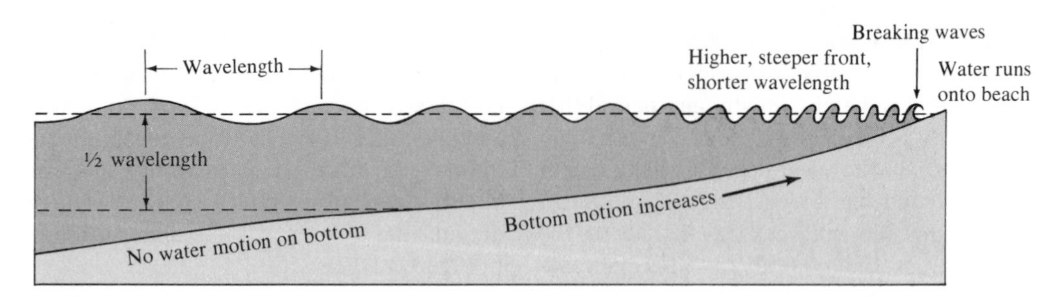

Figure 11.19 *Changes that occur when a wave moves onto the shore. (From Foster,* Physical Geology, *3rd ed., Columbus, Ohio: Charles E. Merrill, 1979)*

As surface water moves onshore the water from expended waves moves outward in a broad sheet, or sometimes in narrow, localized channels, which when strong may pull swimmers into deep water.

Wave Erosion

During periods of calm weather wave action is at a minimum. However just as streams do most of their work during floods, so waves do most of their work during storms. The impact of high, storm-induced waves against the shore can at times be awesome in its violence. Each breaking wave may hurl thousands of tons of water against the land, sometimes causing the earth to literally tremble. The pressures exerted by Atlantic waves, for example, average nearly 10,000 kilograms per square meter (more than 2000 pounds per square foot) in winter. During storms the force is even greater. During one such storm a 1350-ton portion of a steel and concreate breakwater was ripped from the rest of the structure and moved to a useless position toward the shore at Wick Bay, Scotland. Five years later the 2600-ton unit that replaced the first met a similar fate. There are many such stories that demonstrate the great force of breaking waves. It is no wonder then that cracks and crevices are quickly opened in cliffs, seawalls, breakwaters, and anything else that is subjected to these enormous shocks. Water is forced into every opening, causing air in the cracks to become highly compressed by the thrust of crashing waves. This is enough to dislodge blocks of rock, as well as to enlarge and extend preexisting fractures.

In addition to the erosion caused by wave impact and pressure, **abrasion,** the sawing and grinding action of the water armed with rock fragments, is also important. In fact, *abrasion is probably more intense in the surf zone than in any other environment.* Smooth and rounded stones and pebbles along the shore are obvious reminders of the grinding action of rock against rock in the surf zone. Further, such fragments are used as "tools" by the waves as they cut horizontally into the land (Figure 11.20).

Along shorelines composed of unconsolidated material rather than hard rock the rate of erosion by breaking waves can be extraordinary. In parts of Britain, where waves have the easy task of eroding glacial deposits of sand, gravel, and clay, the coast has been worn back 3–5 kilometers (2–3 miles) since Roman times, sweeping away many villages and ancient landmarks. A similar retreat may be seen along the cliffs of Cape Cod, which are retreating at a rate of 1 meter (3 feet) per year.

Wave Refraction

Most waves approach a shoreline at an angle. However when they reach the shallow water of a smoothly sloping bottom they are bent and tend to become parallel to the shore. Such bending of the waves is called **refraction.** The part of the wave nearest the shore touches bottom and

Wave refraction around a headland near Westport, Massachusetts. (Photo by John S. Shelton)

A growing spit. Nauset Beach east of Chatham, Cape Cod, Massachusetts. (Photo by John S. Shelton)

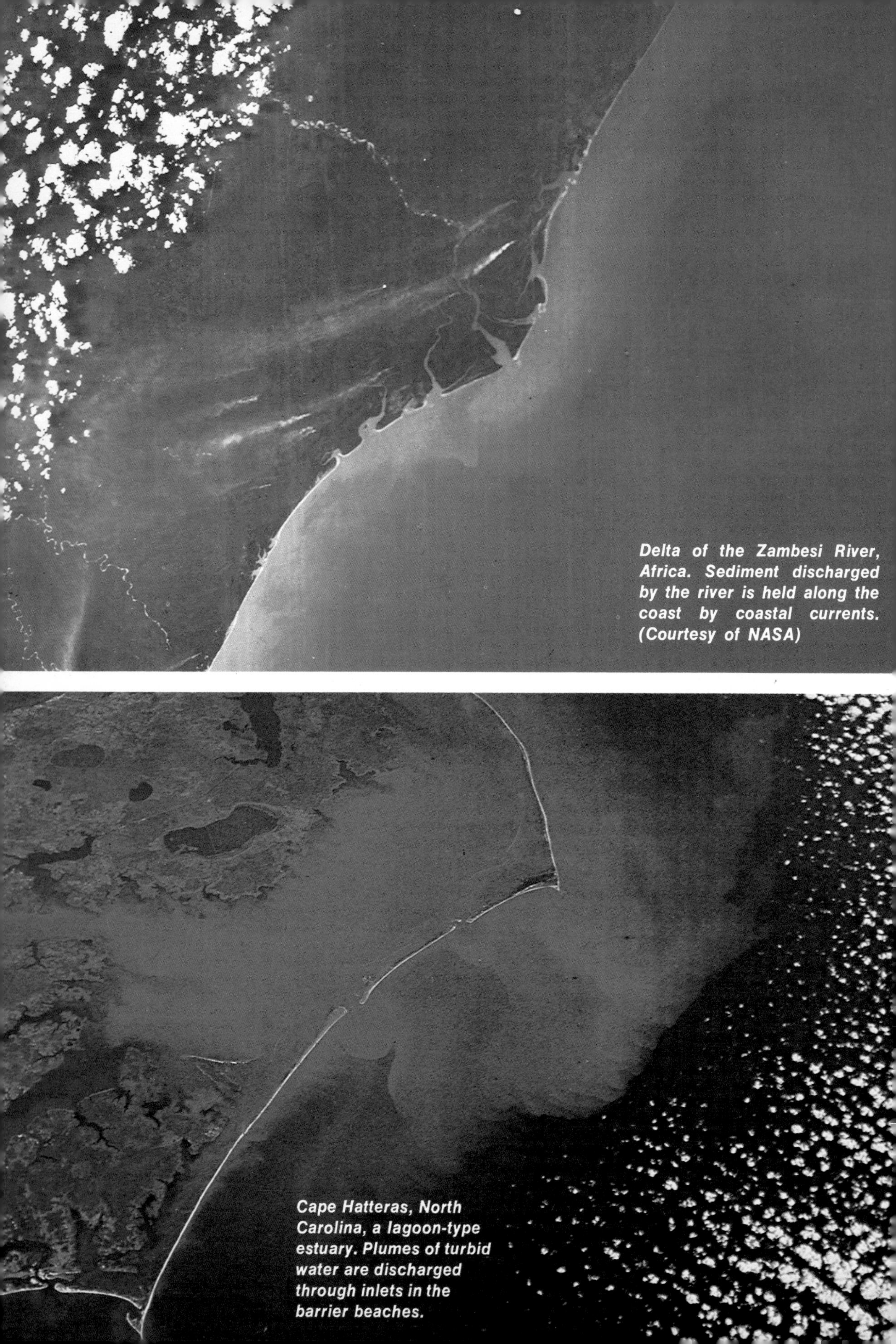

Delta of the Zambesi River, Africa. Sediment discharged by the river is held along the coast by coastal currents. (Courtesy of NASA)

Cape Hatteras, North Carolina, a lagoon-type estuary. Plumes of turbid water are discharged through inlets in the barrier beaches.

Figure 11.20 *Cliff undercut by wave erosion at Kendall Head, Moose Island, Maine. (Photo by E.S. Bastin, U.S. Geological Survey)*

slows down first, while the end that is still in deep water continues forward at its regular speed. *The net result is a wave front that may approach nearly parallel to the shore regardless of the original direction of the wave.*

Because of refraction, wave impact is concentrated against the sides and ends of headlands projecting into the water, while wave attack is weakened in bays. This differential wave attack along irregular coastlines is illustrated in Figure 11.21. Since the waves reach the shallow water in front of the headland sooner than they do in adjacent bays, they are bent more nearly parallel to the protruding land and strike it from all three sides. Over a period of time the effect of this process is to straighten irregular coastlines.

Beach Drift and Longshore Currents

Although waves are refracted, most still reach the shore at an angle, however slight. Consequently, the uprush of water from each breaking wave is oblique. However the backflow is straight down the slope of the beach. The effect of this pattern of water movement is to transport particles of sediment in a zigzag pattern along the beach (Figure 11.22). This movement, called **beach drift,** can transport sand and pebbles hundreds or even thousands of meters each day.

Oblique waves also produce currents within the surf zone that flow parallel to the shore. Since the water here is turbulent, these **longshore currents** easily move the fine suspended sand

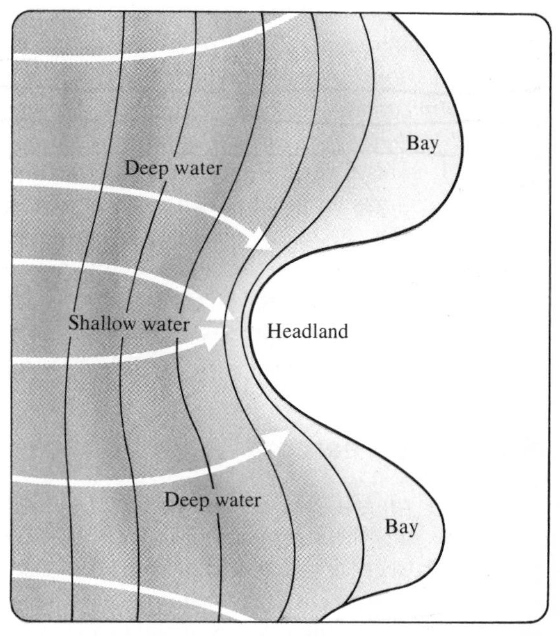

Figure 11.21 *Wave refraction along an irregular coastline.*

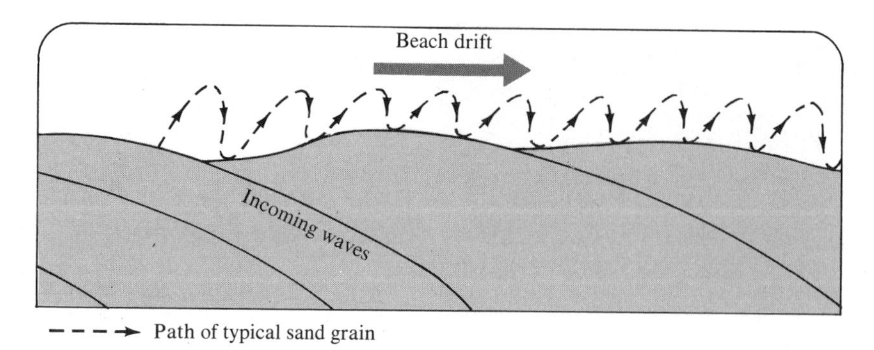

Figure 11.22 *Beach drift, caused by the uprush of water from oblique waves.*

as well as roll larger sand and gravel along the bottom. When the sediment transported by longshore currents is added to the quantity moved by beach drift the total amount can be very large. At Sandy Hook, New Jersey, for example, the quantity of sand transported along the shore over a 48-year period averaged almost 750,000 tons per year. For a 10-year period at Oxnard, California, more than 1.5 million tons of sediment moved along the shore each year.

There should be little wonder that beaches have been characterized as "rivers of sand." *At any point along a beach there is likely to be more sediment that was derived elsewhere than material eroded from the cliff immediately behind it.* It is also worth noting that much of the sediment composing beaches is not wave-eroded debris. Rather, in many areas sediment-laden rivers that discharge into the ocean are the major source of material. Hence if it were not for beach drift and longshore currents, many beaches would be nearly sandless.

Shoreline Features

Along the rugged and irregular New England coast or along the steep shorelines of the West Coast the effects of wave erosion are often easy to spot. **Wave-cut cliffs,** as their name implies, originate by the cutting action of the surf against the base of coastal land. As erosion progresses rocks overhanging the notch at the base of the cliff crumble into the surf and the cliff retreats (Figure 11.20). A relatively flat, benchlike surface, the **wave-cut platform,** is left behind by the receding cliff (Figure 11.23). The platform broadens as wave attack continues. Some of the debris produced by the breaking waves remains along the water's edge as part of the beach, while the remainder is transported farther seaward.

Headlands that extend into the sea are vigorously attacked by the waves because of refraction. The surf erodes the rock selectively, wearing away the softer or more highly fractured

Figure 11.23 *Elevated wave-cut platform along the California coast. A new platform is being created at the base of the cliff. (Photo by E. Brabb, U.S. Geological Survey)*

Figure 11.24 *Sea stacks and a sea arch along the coast of Iceland. (Courtesy of Scandinavian National Tourist Offices)*

A.

B.

C.

Figure 11.25 *Features created because of the movement of sediment by beach drift and longshore currents.* **A.** *Spit. (Photo by I.C. Russel, U.S. Geological Survey)* **B.** *Baymouth bar. (Photo by E.C. Stebinger, U.S. Geological Survey)* **C.** *Tombolo. (Photo by E.S. Bastin, U.S. Geological Survey)*

rock at the fastest rate. At first, sea caves may form. When two caves on opposite sides of a headland unite, a **sea arch** results (Figure 11.24). Finally the arch falls in, leaving an isolated remnant, or **sea stack,** on the wave-cut platform (Figure 11.24). Eventually it too will be consumed by the action of the waves.

Where beach drift and longshore currents are active, several features related to the movement of sediment along the shore may develop. **Spits** are elongated ridges of sand that project from the land into the mouth of an adjacent bay (Figure 11.25A). Often the end in the water hooks landward in response to wave-generated currents. The term **baymouth bar** is applied to a sand bar that completely crosses a bay, sealing it off from the open ocean (Figure 11.25B). Such a feature tends to form across bays where currents are weak, allowing a spit to extend to the other side. A **tombolo,** a ridge of sand that connects an island to the mainland or to another island (Figure 11.25C), forms in much the same manner as does a spit.

The gently sloping coastline found along the Gulf Coast and much of the eastern shore of the United States south of New York City is frequently characterized by **barrier islands,** which are low offshore ridges of sand that parallel the coast. The lagoons that separate these narrow islands from the shore represent zones of relatively quiet water that allow small craft traveling between New York and northern Florida to avoid the rough waters of the North Atlantic.

How barrier islands originate is still not certain. It is quite possible that they form in three or more different ways. Some are thought to have originated as spits that were subsequently severed from the mainland by wave erosion or by the general rise in sea level following the last episode of glaciation. It is also possible that some barrier islands are created when turbulent waters in the line of breakers heap up sand that has been scoured from the bottom. Since these sand barriers rise above normal sea level, the piling up of sand likely resulted from the work of storm waves at high tide. Finally, some studies suggest that barrier islands may be former sand-dune ridges that originated along the shore during the last glacial period, when sea level was lower. When the ice sheets melted, sea level rose and flooded the area behind the beach-dune complex. Using

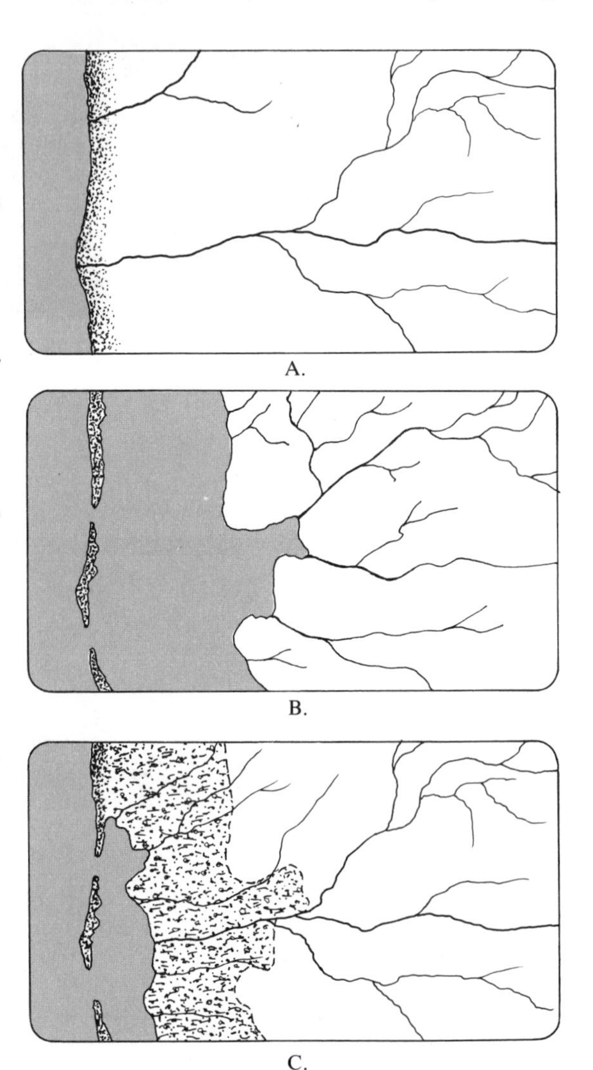

Figure 11.26 *Development of a gently sloping, smooth coastline.* **B.** *Beach-dune complex forms during a glacial period, when sea level was lower.* **B.** *Melting of the glaciers floods the area behind the dunes.* **C.** *Rivers begin to fill the lagoon with sediment. (From Foster,* Physical Geology, *3rd ed., Columbus, Ohio: Charles E. Merrill, 1979)*

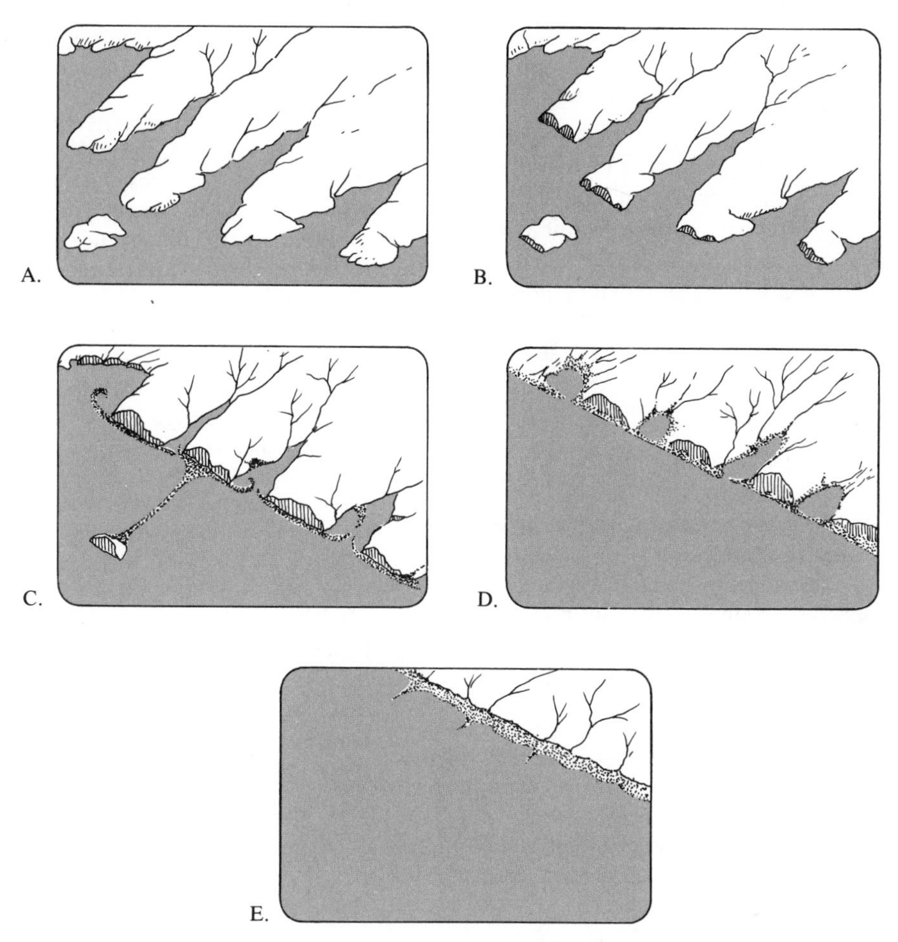

Figure 11.27 *Development of an initially irregular coastline. (From Foster,* Physical Geology, *3rd ed., Columbus, Ohio: Charles E. Merrill, 1979)*

this last explanation of barrier islands, Figure 11.26 illustrates the development of a gently sloping smooth coastline. The general sequence of events, however, would be essentially the same no matter how the barrier islands were created.

Evolution of an Initially Irregular Shoreline

There is little question that *regardless of its initial configuration a shoreline soon undergoes modification*. At first most coastlines are irregular, although the degree of and reason for the irregularity may vary considerably from place to place. Along a coastline that is characterized by varied geology the pounding surf may at first increase its irregularity because the waves will erode the weaker rocks more easily than the stronger rocks. However it is commonly agreed that if a shoreline remains stable marine erosion and deposition will eventually produce a more regular coast. Figure 11.27 illustrates the evolution of an initially irregular coast. As waves erode the headlands, creating cliffs and a wave-cut platform, sediment is carried along the shore. Some of the material is deposited in the bays, while other debris is formed into spits and baymouth bars. At the same time rivers fill the bays with sediment. Ultimately a smooth coast results.

The great variety of present-day shorelines suggests that they are complex areas. Indeed, to understand the nature of any particular coastal area, many factors must be considered, including rock types, the size and direction of waves, the number of storms, the tidal range, and the submarine profile. Moreover, recent tectonic events and changes in sea level must also be taken into account. These many variables make shoreline classification difficult.

REVIEW

1. What is meant by *salinity?* What is the average salinity of the ocean?
2. Why do variations in salinity occur? Give some specific examples to illustrate your answer.
3. What is the origin of the minerals dissolved in seawater?
4. Since the ocean contains millions of tons of gold, why is it not extracted commercially? Name some minerals that are extracted from seawater.
5. Why is desalination not likely to be a significant source of water for agriculture in the foreseeable future?
6. Although winds are the primary cause of surface currents, currents do not coincide precisely with the wind direction. Explain this phenomenon.
7. What is the Ekman spiral?
8. Briefly describe the North Atlantic gyre. How does the surface circulation of the South Atlantic and Indian oceans differ from that of the North Atlantic?
9. Describe the process of coastal upwelling.
10. How do ocean currents influence climate?
11. What is meant by *thermohaline circulation?* Sketch the deep ocean circulation of the Atlantic Ocean and label the principal water masses.
12. Discuss the origin of ocean tides.
13. Explain why an observer can experience two unequal high tides during a day (see Figure 11.12).
14. How does the sun influence tides?
15. Distinguish between semidiurnal, diurnal, and mixed tides.
16. What is meant by *flood tide? Ebb tide?*
17. How have tides affected the earth's rotation? How did geologists substantiate this theory?
18. Why will tidal power never contribute substantially to filling the world's growing energy needs? What advantages does tidal power offer?
19. List three factors that determine the height, length, and period of a wave.
20. Describe the motion of a water particle as a wave passes (see Figure 11.17).
21. Explain what happens when a wave breaks.
22. How do waves erode?
23. What is wave refraction? What is the effect of this process along irregular coastlines (see Figure 11.21)?
24. Why are beaches often called "rivers of sand"?
25. Describe the formation of the following features: wave-cut cliff, wave-cut platform, sea stack, spit, baymouth bar, tombolo.
26. Discuss three possible ways that barrier islands originate.

KEY TERMS

salinity
outgassing
desalination
thermocline
halocline
gyre
Coriolis force
Ekman spiral
upwelling
thermohaline circulation
tides
spring tides
neap tides

semidiurnal tide
diurnal tide
mixed tide
tidal current
flood tide
ebb tide
tidal flats
waves
wave height
wave length
wave period
fetch
waves of oscillation
waves of translation
surf
abrasion
refraction
beach drift
longshore current
wave-cut cliffs
wave-cut platform
sea arch
sea stack
spit
baymouth bar
tombolo
barrier islands

part 3

the atmosphere

Clouds forming along the Florida coast. The flat cloud tops represent the top of the troposphere. This lowermost layer of the atmosphere contains most of the atmosphere's mass and it is within this layer that most weather phenomena occur. (Courtesy of NASA)

Weather and Climate
Composition of the Atmosphere
Extent and Structure of the Atmosphere
Earth-Sun Relationships
Radiation
Mechanisms of Heat Transfer
Heating the Atmosphere
Temperature Measurement and Data
Controls of Temperature
World Distribution of Temperature
Air Pollution

Composition, Structure, and Temperature

The study of the atmosphere—the earth's envelope of air—goes back thousands of years. Present knowledge shows that no other planet has the exact mixture of gasses or the heat and moisture conditions necessary to sustain life as we know it. The gasses that make up the earth's atmosphere and the controls to which they are subject are vital to our very existence. In this chapter we begin our examination of the ocean of air in which we all must live. We shall attempt to answer a number of basic questions. What is the composition of the atmosphere? At what point do we leave the atmosphere and enter outer space? What causes the seasons? How is air heated, and what causes variations in this heating? How are people altering the composition of the atmosphere?

People's interest in the atmosphere is probably as old as the history of humankind. Certainly there are few other aspects of our physical environment that affect our daily lives more than the phenomena we collectively call the weather. **Meteorology,** the science of the atmosphere and its weather, goes back many centuries. The term is derived from the title of Aristotle's four-volume treatise, *Meteorologica,* literally "discourse on things above." The study of meteorology has progressed through many phases over the centuries, evolving from the days when weather lore and superstition guided explanations to what it is today—a scientific study using the most modern instruments and equipment.

Weather and Climate

If you were to search through the library for books about the atmosphere, many of the titles you would find would contain the term *weather,* while others would have the word *climate*. What is the difference between these two terms? **Weather** is a word used to denote the state of the atmosphere at a particular place for a short period of time. Weather is constantly changing—hourly, daily, and seasonally. **Climate,** on the other hand, might best be described as an aggregate or composite of weather. Put another way, the climate of a place or region is a generalization of the weather conditions over a long period of time. Therefore, a climatic description is possible only after weather records have been kept for many years.

Although weather and climate are not identical, the nature of both is expressed in terms of the same *elements*—temperature, moisture, pressure, and wind. We shall study these elements separately at first. However, keep in mind that they are very much interrelated. A change in any one of the elements will usually bring about changes in the others.

Composition of the Atmosphere

In the days of Aristotle all things were thought to be a combination of four fundamental substances—fire, air, earth (soil), and water. Today we know that matter is much more complex. For example, the earth is composed of complex minerals, which, in turn, are made of numerous elements and compounds. Furthermore, air, which appears to be a unique substance, is really a mixture of many different gasses and tiny solid particles.

The composition of air is not constant; it varies from time to time and from place to place. This variability is readily observable when we drive from the rural countryside into the city and see increases in dust, smoke, and oftentimes cloud cover. However, if the water vapor, dust, and other variable components were removed from the atmosphere, we would find that its composition is quite uniform. Clean, dry air sampled anywhere on earth is composed almost entirely of two gasses—nitrogen and oxygen. As can be seen in Table 12.1, nitrogen makes up about 78 percent of the atmosphere, and oxygen represents about 21 percent. Although these gasses are the most plentiful components of air and are of great significance to life on earth, they are of minor importance in affecting weather phenomena. The remaining percent of dry air is composed of a number of other gasses. Carbon dioxide, although present in only minute quan-

Table 12.1 ***Composition of Clean, Dry Air***

Component	Chemical Symbol	Percent of Air by Volume
Nitrogen	N_2	78.08
Oxygen	O_2	20.95
Argon	A	0.93
Carbon dioxide	CO_2	0.03
All others		Trace

tities, is nevertheless an important constituent of air, because it has the ability to absorb heat energy radiated by the earth and thus helps keep the atmosphere warm.

If meteorologists had to choose the most significant component of the atmosphere, they would undoubtedly pick water vapor. The amount of water vapor in the air varies considerably, from practically none at all up to about 4 percent by volume. Why is such a small fraction of the atmosphere so significant? Certainly the fact that water vapor is the source of all clouds and precipitation would be enough to justify the choice. However, water vapor has other roles. Like carbon dioxide, it has the ability to absorb heat energy given off by the earth as well as some solar energy. It therefore helps in the heating of the air. When water vapor changes from one state to another (Figure 13.1), it absorbs or releases heat energy (termed *latent heat*). As we shall see in later chapters, latent heat is the source of energy which drives many storms.

Another important component of the atmosphere is ozone, the triatomic form of oxygen (O_3). Ozone is not the same as the oxygen we breathe, which has two atoms per molecule (O_2). Although ozone is quite toxic if breathed, it is not present in large amounts in the lowest portion of the atmosphere. Rather, it is produced well above the earth's surface, with the greatest concentration between 20 and 50 kilometers (12 and 30 miles).

Even though ozone is poisonous, it is essential to life as we know it. This statement appears to be contradictory. How can a poisonous gas be essential to life? The answer lies in the capability of the ozone layer to absorb ultraviolet radiation. Ultraviolet rays are the ones we want to "soak up" when we go to the beach for a suntan. If ozone did not act to filter a great deal of the ultraviolet radiation, but instead the rays were allowed to reach the surface of the earth, we could be burned terribly and very likely be blinded. Without ozone in the atmosphere, life on earth could not have evolved in its present form. Land-dwelling organisms, for the most part, would not have developed at all.

The last component we shall examine is dust. Most of us probably think of dust as small visible bits of dirt. However, from a meteorological standpoint, dust is much more than that. It includes many microscopic particles that are invisible to the naked eye, among them organic materials like pollen, spores, and seeds. Dust particles are most numerous in the lower part of the atmosphere near their primary source, the earth's surface. Nevertheless, the upper atmosphere is not free of them, because some dust is carried to great heights by rising currents of air, while other particles are given to the upper atmosphere by meteors which disintegrate as they pass through the earth's envelope of air. Some particles act as surfaces upon which water vapor may condense. This function is very basic to the formation of clouds and fog. In addition, dust in the air contributes to an optical phenomenon we all have observed—the red and orange colors of sunrise and sunset.

Extent and Structure of the Atmosphere

To say that the atmosphere begins at the earth's land-sea surface and extends upward is rather obvious. However, where does the atmosphere end and outer space begin? To help us understand the vertical extent of the atmosphere, let us examine the changes in atmospheric pressure with height. The pressure of the atmosphere is simply the weight of the air above. At sea level, the average weight of the atmosphere, that is, the average pressure, is slightly more than 1 kilogram per square centimeter (14.7 pounds per square inch). Obviously the pressure at higher altitudes is less. This fact was demonstrated in a dramatic way by some nineteenth century balloonists who were attempting to study the upper atmosphere. Several passed out upon reaching heights in excess of

Table 12.2 ***Percent of Sea Level Pressure Encountered at Selected Altitudes***

Altitude (kilometers)	Percent of Sea Level Pressure
0	100
5.6	50
16.2	10
31.2	1
48.1	0.1
65.1	0.01
79.2	0.001
100	0.00003

6 kilometers; some even perished in the rarified upper air.

By examining Table 12.2 you can see that one-half of the atmosphere lies below an altitude of 5.6 kilometers. At about 16 kilometers, 90 percent of the atmosphere has been traversed, and above 100 kilometers, only 0.00003 percent of all the gasses composing the atmosphere remain. At this latter altitude, the atmosphere is so tenuous that the density of air is less than can be found in a perfect artificial vacuum at the earth's surface. However, traces of our atmosphere extend far beyond even this altitude. Thus, to say where the atmosphere ends and outer space begins is quite arbitrary and to a large extent depends upon what phenomena we are studying. It is apparent that there is no sharp boundary. Certainly the data on vertical pressure changes reveal that *the vast bulk of the gasses composing the atmosphere are very near the earth's surface and that they gradually merge with the emptiness of space.*

The atmosphere is divided vertically into four layers on the basis of temperature (Figure 12.1). The bottom layer, where temperature decreases with an increase in altitude, is known as the **troposphere.** The term literally means the region where air "turns over," a reference to the appreciable vertical mixing of the air in this lowermost zone. Practically all clouds, and certainly all of the precipitation, as well as all of our violent storms, are born in this layer of the atmosphere. There should be little doubt why the troposphere is often called the "weather sphere." The average temperature decrease in the troposphere is 6.5° C per kilometer (3.5° F per 1000 feet), a figure known as the **normal lapse rate.** The rate is not constant, however, and may vary considerably from the average value. Thus, to determine the lapse rate for any particular time and place, as well as gather information about vertical changes in pressure, wind, and humidity, radiosondes are used. The radiosonde is attached to a balloon and transmits data by radio as it ascends through the atmosphere (Figure 12.2). Although the thickness of the troposphere is not everywhere the same, the lapse rate continues to an average height of approximately 12 kilometers.

Beyond the troposphere lies the **stratosphere,** and the boundary between them is called the **tropopause.** In the stratosphere, the temperature remains constant to a height of about 20 kilometers and then begins a gradual increase which continues until the **stratopause,** at a height of about 50 kilometers above the earth's surface. While below the tropopause atmospheric properties are readily transferred by large-scale turbulence and mixing, above it, in the stratosphere, they are not. The reason for the increased temperatures in the stratosphere is that the atmosphere's ozone is concentrated in this layer. Recall that ozone absorbs ultraviolet radiation from the sun. As a consequence, the stratosphere is heated.

In the **mesosphere** temperatures again decrease with height until, at the **mesopause,** some 80 kilometers above the surface, the temperature approaches −100° C. Extending upward from the mesopause and having no well-defined upper limit is the **thermosphere,** a layer which contains only a minute fraction of the atmosphere's mass. In the extremely rarified air of this outermost layer, temperatures again increase due to the absorption of very short wave solar energy by atoms of oxygen and nitrogen. While temperatures rise to extremely high values of more than 1000° C, such temperatures are not strictly comparable to those experienced near the earth's surface. Temperature is defined in terms of the

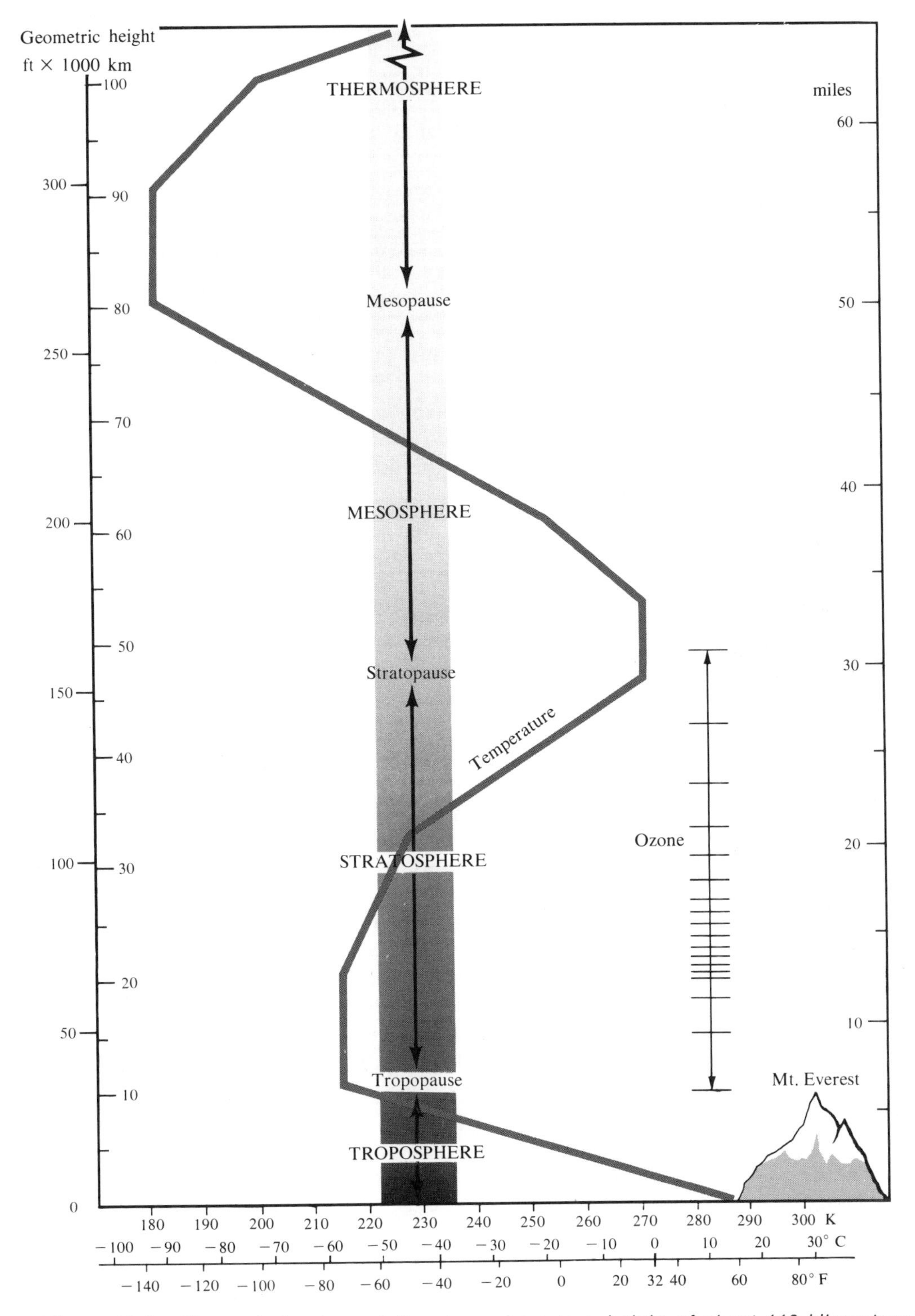

Figure 12.1 *Thermal structure of the atmosphere to a height of about 110 kilometers (70 miles).*

Figure 12.2 *Atmospheric soundings using radiosondes supply data on vertical changes in temperature, pressure, and humidity. (Courtesy of NOAA)*

average speed at which molecules move. Since the gasses of the thermosphere are moving at very high speeds, the temperature is obviously very high. However, the gasses are so sparse that only a very few of these fast-moving air molecules collide with a foreign body and hence only an insignificant quantity of energy is transferred. Thus, the temperature of a satellite orbiting the earth in the thermosphere is determined chiefly by the amount of solar radiation it absorbs and not by the temperature of the surrounding air. If an astronaut inside were to expose his or her hand, it would not feel hot.

Earth-Sun Relationships

The earth intercepts only a minute percentage of the energy given off by the sun—less than one two-billionth. This may seem to be a rather insignificant amount until we realize that it is several hundred thousand times the electrical generating capacity of the United States. *Energy from the sun is undoubtedly the most important control of our weather and climate.* Solar radiation represents more than 99.9 percent of the energy that heats the earth, drives the oceans, and creates winds. Therefore, in order to have a basic understanding of atmospheric processes, we must understand what causes the time and space variations in the amount of solar energy reaching the earth.

Motions of the Earth

The earth has two principal motions—rotation and revolution. **Rotation** is the spinning of the earth about its axis, which is an imaginary line running through the poles. Our planet rotates once in about 24 hours. During any 24-hour period, half of the earth is experiencing daylight, and the other half darkness. The line separating the dark half of the earth from the lighted half is called the **circle of illumination. Revolution** refers to the movement of the earth in its orbit around the sun. Hundreds of years ago, most people believed that the earth was stationary in space. The reasoning was that if the earth were moving, people would feel the movement of the wind rushing past them. However, today we know that the earth is traveling at nearly 113,000 kilometers per hour in an elliptical orbit about the sun. Why do we not feel the air rushing past us? The answer is that the atmosphere, bound by gravity to the earth, is carried along at the same speed as the earth.

The distance between the earth and sun averages about 150 million kilometers. However, since the earth's orbit is not perfectly circular, the distance varies during the course of a year. Each year, on about January 3, our planet is 147 million kilometers from the sun, closer than at any other time. This position is called **perihelion.** About six months later, on July 4, the earth is 152 million kilometers from the sun, farther away than at any other time. This position is

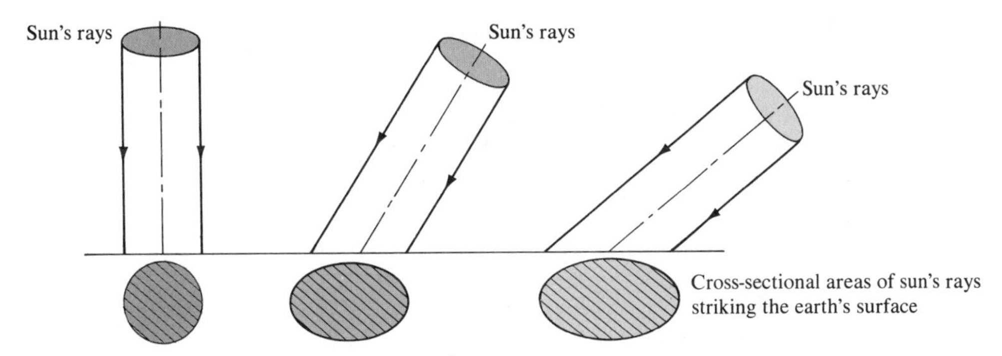

Figure 12.3 *Changes in the sun angle cause variations in the amount of solar energy reaching the earth's surface. The higher the angle, the more intense the solar radiation.*

called **aphelion.** The variations in the amount of solar radiation received by the earth as the result of its slightly elliptical orbit, however, are slight and of little consequence when explaining major seasonal temperature variations. By way of illustration, consider that the earth is closest to the sun during the Northern Hemisphere winter.

The Seasons

We know that it is colder in the winter than in the summer, but if the variations in solar distance do not cause this, what does? All of us have made adjustments for the continuous change in the length of daylight that occurs throughout the year by planning our outdoor activities accordingly. The gradual but significant change in length of daylight certainly accounts for some of the difference we notice between summer and winter. Further, a gradual change in the **altitude** (angle above the horizon) of the noon sun during the course of a year's time is evident to most people. At mid-summer the sun is seen high above the horizon as it makes its daily journey across the sky. But as summer gives way to autumn, the noon sun appears lower and lower in the sky and sunset occurs earlier each evening.

The seasonal variation in the altitude of the sun affects the amount of energy received at the earth's surface in two ways. First, when the sun is directly overhead (90-degree angle), the solar rays are most concentrated. The lower the angle, the more spread out and less intense is the solar radiation that reaches the surface. This idea is illustrated in Figure 12.3. Second, and of lesser importance, the angle of the sun determines the amount of atmosphere the rays must traverse (Figure 12.4). When the sun is directly overhead, the rays pass through a thickness of only 1 atmosphere, while rays entering at a 30-degree angle travel through twice this amount, and 5-degree rays travel through a thickness roughly equal to 11 atmospheres. The longer the path, the greater are the chances for absorption, reflection, and scattering by the atmosphere, which reduce the intensity at the surface. These same effects account for the fact that the midday sun can be literally blinding, while the setting sun can be a sight to behold.

It is important to remember that the earth is spherical. Hence, on any given day only places located at a particular latitude receive vertical (90-degree) rays from the sun. As we move either north or south of this location, the sun's rays strike at an ever-decreasing angle. Thus, the nearer a place is to the latitude receiving the vertical rays of the sun, the higher will be its noon sun.

What causes the yearly fluctuations in the sun angle and length of day? They occur because the earth's orientation to the sun continually changes. The earth's axis is not perpendicular to the plane of its orbit around the sun, but instead is tilted 23½ degrees from the perpendicular. This is termed the **inclination of the axis,** and as

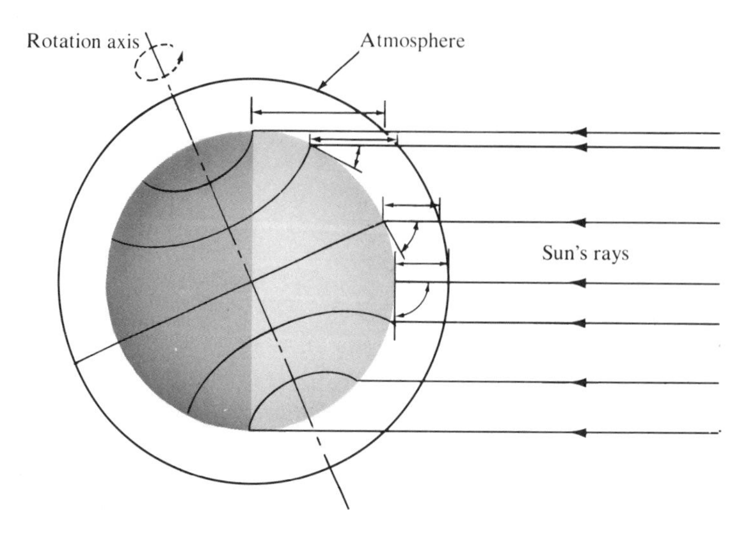

Figure 12.4 *Rays striking at a low angle must traverse more of the atmosphere than rays striking at a higher angle and thus are subject to greater depletion by reflection and absorption.*

we shall see, *if the axis were not inclined, we would have no seasonal changes*. In addition, because the axis remains pointed in the same direction (toward the North Star) as the earth journeys around the sun, the orientation of the earth's axis to the sun's rays is constantly changing (Figure 12.5). On one day each year the axis is such that the Northern Hemisphere is "leaning" 23½ degrees toward the sun. Six months later, when the earth has moved to the opposite side of its orbit, the Northern Hemisphere leans 23½ degrees away from the sun. On days between

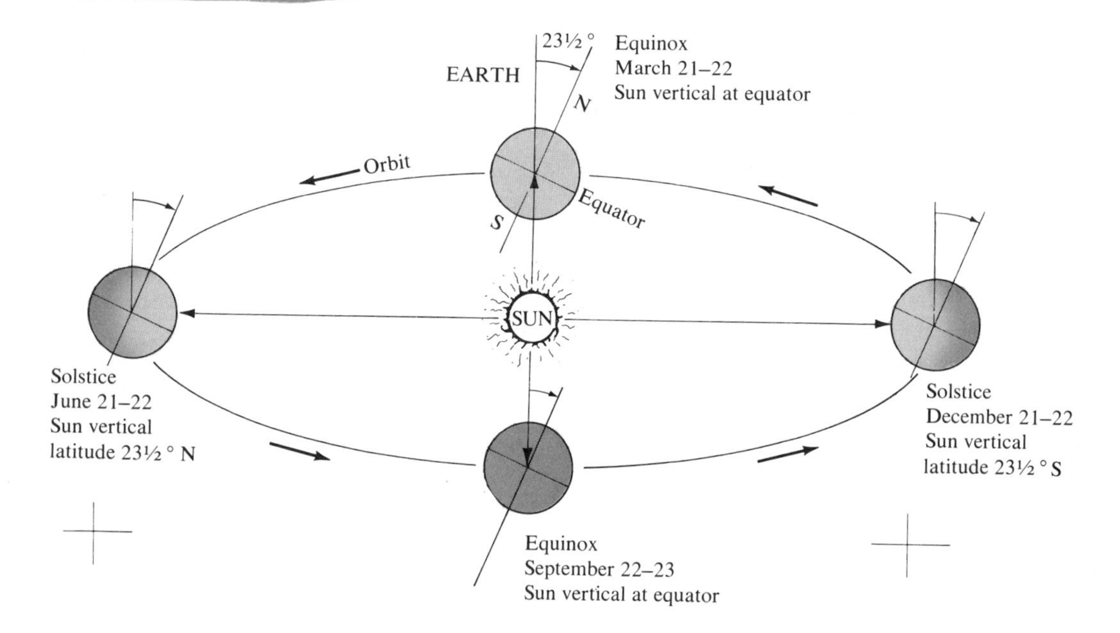

Figure 12.5 *Earth-sun relationships.*

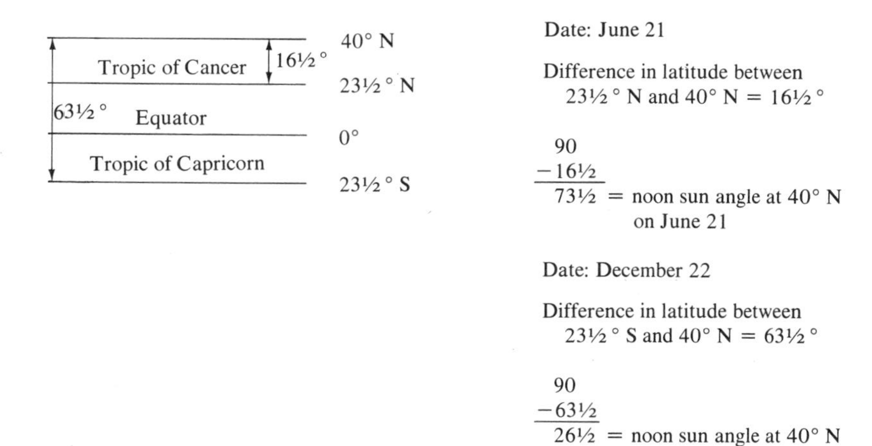

Figure 12.6 *Calculating the noon sun angle. The noon sun angle at 23½ degrees north latitude on June 21 is 90 degrees. A place located 1 degree away receives an 89-degree sun angle; a place 2 degrees away an 88-degree angle, and so forth. To calculate the noon sun angle, simply find the number of degrees of latitude separating the location receiving the vertical rays of the sun and the location in question; then subtract that figure from 90 degrees. In this figure, we see the large variation in sun angles experienced at 40 degrees north latitude for the summer and winter solstices (the extremes).*

these extremes, the earth's axis leans at amounts less than 23½ degrees to the rays of the sun. This change in orientation causes the vertical rays of the sun to make a yearly migration from 23½ degrees north of the equator to 23½ degrees south of the equator. This migration in turn causes the altitude of the noon sun to vary by as much as 47 degrees (23½ + 23½) for many locations during the course of a year. For example, a mid-latitude city like New York (about 40 degrees north latitude) has a maximum noon sun angle of 73½ degrees when the sun's vertical rays reach their furthest northward location and a minimum noon sun angle of 26½ degrees six months later (Figure 12.6).

Historically, four days a year have been given special significance based on the annual migration of the direct rays of the sun and its importance to the yearly cycle of weather. On June 21 or 22 the earth is in a position such that the axis in the Northern Hemisphere is tilted 23½ degrees toward the sun (Figure 12.7). At this time the vertical rays of the sun strike 23½ degrees north latitude (23½ degrees north of the equator), a line of latitude known as the **Tropic of Cancer.** For people living in the Northern Hemisphere, June 21 or 22 is known as the **summer solstice.** Six months later, on about December 21 or 22, the earth is in the opposite position, with the sun's vertical rays striking at 23½ degrees south latitude. This line is known as the **Tropic of Capricorn.** For those of us in the Northern Hemisphere, December 21 or 22 is the **winter solstice.** However, at the same time in the Southern Hemisphere, people are experiencing just the opposite—the summer solstice.

The equinoxes occur midway between the solstices. September 22 or 23 is the date of the **autumnal equinox** in the Northern Hemisphere, and March 21 or 22 is the date of the **vernal,** or **spring, equinox.** On these dates, the vertical rays of the sun strike at the equator (0 degrees latitude) because the earth is in such a position in its orbit that the axis is tilted neither toward nor away from the sun.

Further, the length of daylight versus darkness is also determined by the position of the earth in its orbit. The length of daylight on June 21, the

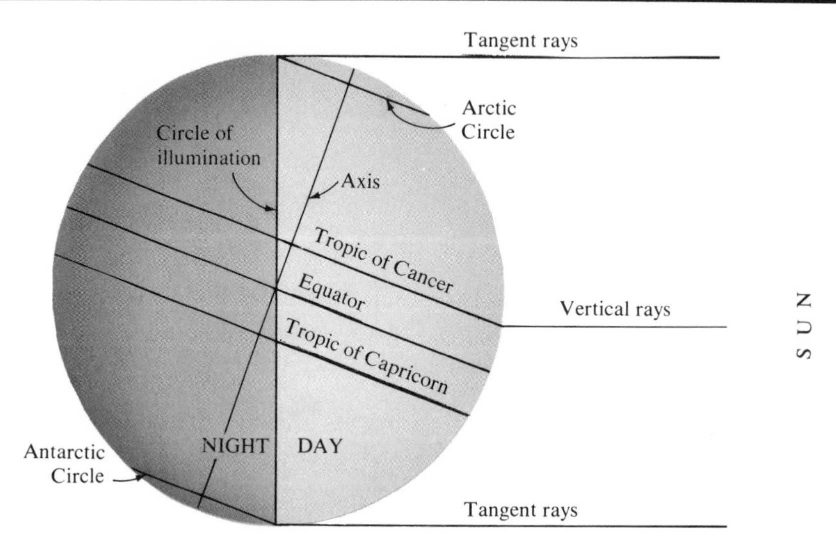

Figure 12.7 *Characteristics of the summer solstice (Northern Hemisphere).*

summer solstice in the Northern Hemisphere, is greater than the length of night. This fact can be established from Figure 12.7 by comparing the fraction of a given latitude that is on the "day" side of the circle of illumination with the fraction on the "night" side. The opposite is true for the winter solstice, when the nights are longer than the days (Table 12.3). Again for comparison let us consider New York City. It has 15 hours of daylight on June 21 and only 9 hours on December 21. Also note from Table 12.3 that on June 21 the farther you are north of the equator the longer is the period of daylight, until the Arctic Circle is reached, where the length of daylight is 24 hours. During an equinox (meaning "equal night") the length of daylight is 12 hours everywhere on earth, for the circle of illumination passes directly through the poles, dividing the latitudes in half.

Table 12.3 ***Length of Day***

Latitude (degrees)	Summer Solstice	Winter Solstice	Equinoxes
0	12 h	12 h	12 h
10	12 h 35 min	11 h 25 min	12
20	13 12	10 48	12
30	13 56	10 04	12
40	14 52	9 08	12
50	16 18	7 42	12
60	18 27	5 33	12
70	2 mo	0 00	12
80	4 mo	0 00	12
90	6 mo	0 00	12

As a review of the characteristics of the summer solstice for the Northern Hemisphere, examine Figure 12.7 and Table 12.3 and consider the following facts:

(1) The date of occurrence is June 21 or 22.
(2) The vertical rays of the sun are striking the Tropic of Cancer (23½ degrees north latitude).
(3) Locations in the Northern Hemisphere are experiencing their longest days and highest sun angles. (Opposite for the Southern Hemisphere.)
(4) The farther you are north of the equator, the longer is the period of daylight, until the Arctic Circle is reached, where the day is 24 hours long. (Opposite for the Southern Hemisphere.)

The facts about the winter solstice will be just the opposite. It should now be apparent why a midlatitude location is warmest in the summer. It is then that the days are longest and the altitude of the sun is the highest.

All places situated at the same latitude have identical sun angles and lengths of days. If the earth-sun relationships just described were the only controls of temperature, we would expect these places to have identical temperatures as well. Obviously this is not the case. Although the altitude of the sun is the main control of temperature, it is not the only control, as we shall see.

In summary, seasonal fluctuations in the amount of solar energy reaching places on the earth's surface are caused by the migrating vertical rays of the sun and the resulting variations in sun angle and length of day.

Radiation

From our everday experience we know that the sun emits light and heat as well as the rays which give us a suntan. Although these forms of energy comprise a major portion of the total energy that radiates from the sun, they are only a part of a large array of energy called **radiation,** or **electromagnetic radiation.** This array or spectrum of electromagnetic energy is shown in Figure 12–8. All radiation, whether X rays, radio, or heat waves, is capable of transmitting energy through the vacuum of space at 300,000 kilometers per second and only slightly slower through air. Nineteenth century physicists were so puzzled by the seemingly impossible task of energy traveling without a medium to transmit it that they assumed that a material, which they named ether, existed between the sun and the earth. This medium was thought to transmit radiant energy in much the same way that air transmits sound waves produced by the vibration of one person's vocal cords to another person's eardrums. Today we know that, like gravity, radiation requires no material to transmit it. Yet, physicists still do not know how this is possible.

In some respects, the transmission of radiant energy parallels the motion of the gentle swells in the open ocean. Not unlike ocean swells, these electromagnetic waves, as they are called, come

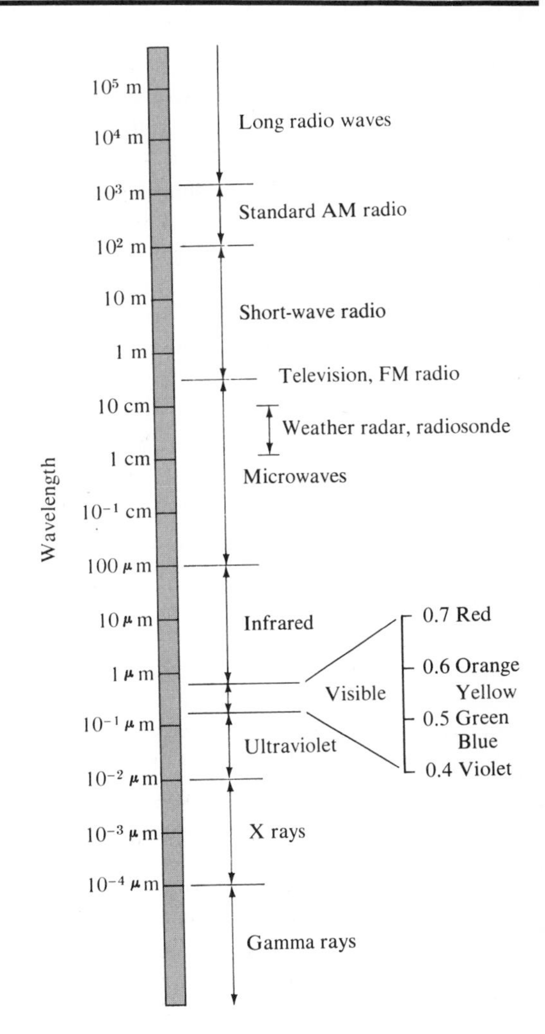

Figure 12.8 *The electromagnetic spectrum.*

in various sizes. For our purpose, the most important difference between electromagnetic waves is their wavelength, or distance from one crest to the next. Radio waves have the longest wavelengths, ranging to tens of kilometers, while gamma waves are the very shortest, being less than a one one-thousand millionth of a centimeter long. **Visible light,** as the name implies, is the only portion of the spectrum we can see. We often refer to visible light as white light since it appears "white" in color. However, it is easy to show that white light is really an array of colors, each color corresponding to a particular wavelength. Using a prism, we can divide white light into the colors of the rainbow, violet having the shortest wavelength—0.4 micrometer (1 micrometer is 0.0001 centimeter)—and red having the longest—0.7 micrometer. Located adjacent to red, and having a longer wavelength, is **infrared** radiation, which we cannot see but which we can detect as heat. The closest invisible waves to violet are called **ultraviolet** rays and are responsible for the sunburn after an intense exposure to the sun. Although we divide radiant energy into groups based on our ability to perceive them, all forms of radiation are basically the same. When any form of radiant energy is absorbed by an object, the result is an increase in molecular motion, which causes a corresponding increase in temperature.

To better appreciate how the sun's radiant energy interacts with the earth's atmosphere and surface, we must have a general understanding of the basic laws governing radiation. Although the mathematical implications of these laws are beyond the scope of this book, the concepts themselves are well within the student's grasp.

(1) All objects, at whatever temperature, emit radiant energy. Hence, not only hot objects like the sun but also the earth, including its polar ice caps, continually emit energy.

(2) Hotter objects radiate more total energy per unit area than colder objects. The sun, with a surface temperature of 6000 K, emits hundreds of thousands of times more energy than the earth, which has an average surface temperature of 288 K.

(3) The hotter the radiating body, the shorter is the wavelength of maximum radiation. This law can be easily visualized if we examine something from our everyday experience. For example, a very hot metal rod will emit visible radiation, producing a white glow. Upon cooling, it will emit more of its energy in longer wavelengths and glow a reddish color. Eventually no light will be given off, but if you place your hand near the rod, the still longer infrared radiation will be detectable as heat. The sun, with a surface temperature of 6000 K, radiates maximum energy at 0.5 micrometer, which is in the visible range. The maximum radiation for the earth occurs at a wavelength of 10 micrometers, well within the infrared (heat) range. Because the maximum earth radiation is roughly 20 times longer than the maximum solar radiation, terrestrial radiation is often called long-wave radiation, and solar radiation is called short-wave radiation.

(4) Objects that are good absorbers of radiation are good emitters as well. Technically, a perfect emitter is any object that radiates, for every wavelength, the maximum intensity of radiation possible for that temperature. The earth's surface and the sun approach being perfect radiators since they absorb and radiate with nearly 100 percent efficiency for their respective temperatures. On the other hand, gasses are selective absorbers and radiators. Thus the atmosphere, which is nearly transparent to (does not absorb) certain wavelengths of radiation, is nearly opaque (a good absorber) to others. Our experience tells us that the atmosphere is transparent to visible light; hence it readily reaches the earth's surface. This is not the case for long-wave terrestrial radiation, as we shall see later.

Mechanisms of Heat Transfer

Three mechanisms of heat transfer are recognized: radiation, conduction, and convection. Since radiation is the only one of these that can travel through the relative emptiness of space, the vast majority of energy coming to and leaving the earth must be in this form. Radiation also plays an important role in transferring heat from

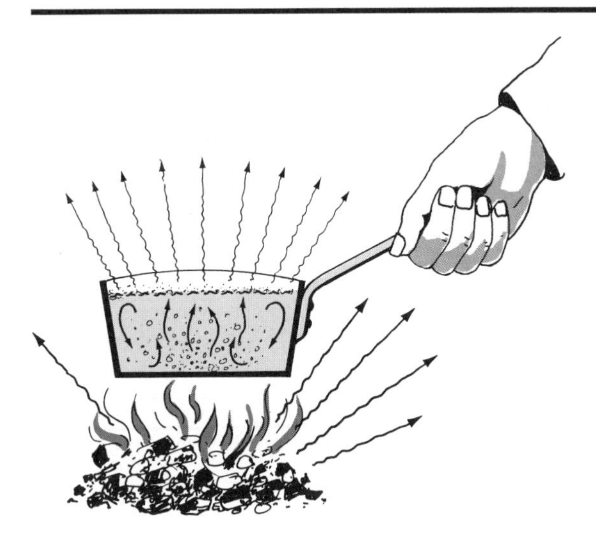

Figure 12.9 *Illustration of conduction, convection, and radiation.*

the earth's land-sea surface to the atmosphere and vice versa.

Conduction is familiar to most of us. Anyone who has attempted to pick up a metal spoon that was left in a hot pan has found that heat was conducted through the spoon. Conduction is the transfer of heat through matter by molecular activity. The energy of molecules is transferred through collisions from one molecule to another, with the heat flowing from the higher temperature to the lower temperature. The ability of substances to conduct heat varies considerably. Metals are good conductors, as those of us who have touched a hot spoon have quickly learned. Air, on the other hand, is a very poor conductor of heat. Consequently, conduction is only important between the earth's surface and the air directly in contact with the surface. As a means of heat transfer for the atmosphere as a whole, conduction is the least significant and can be disregarded when considering most meteorological phenomena.

Heat gained by the lowest layer of the atmosphere from radiation or conduction is most often transferred by convection. **Convection** is the transfer of heat by the movement of a mass or substance from one place to another. It can only take place in liquids and gasses. Convective motions in the atmosphere are responsible for the redistribution of heat from equatorial regions to the poles and from the surface upward. The term **advection** is usually reserved for horizontal convective motions such as winds, while *convection* is used to describe vertical motions in the atmosphere.

Figure 12.9 summarizes the various mechanisms of heat transfer. The heat produced by the campfire passes through the pan by conduction, warming the water at the bottom of the pan. Convection currents carry the warmed water throughout the container, heating the remaining water. Meanwhile, the camper is warmed by radiation from the fire and the pan. Furthermore, since metals are good conductors, the camper's hand is likely to be burned if a hot pad is not used. In most situations involving heat transfer, conduction, convection, and radiation typically occur simultaneously.

Heating the Atmosphere

Although the atmosphere is quite transparent to incoming solar radiation, less than 25 percent of this radiation penetrates directly to the earth's surface without some sort of interference by the atmosphere. The remainder is either *absorbed* by the atmosphere, *scattered* about until it reaches the earth's surface or returns to space, or is *reflected* back to space (Figure 12.10). What determines whether radiation will be absorbed, scattered, or reflected outward? As we shall see, it depends greatly on the wavelength of the energy being transmitted, as well as on the size and nature of the intervening material.

When light is scattered by very small particles, primarily gas molecules, it is distributed in all directions, forward as well as backward. Some of the light that is back-scattered is lost to space, while the remainder continues downward, where it interacts with other molecules which scatter it further by changing the direction of the light beam, but not its wavelength.

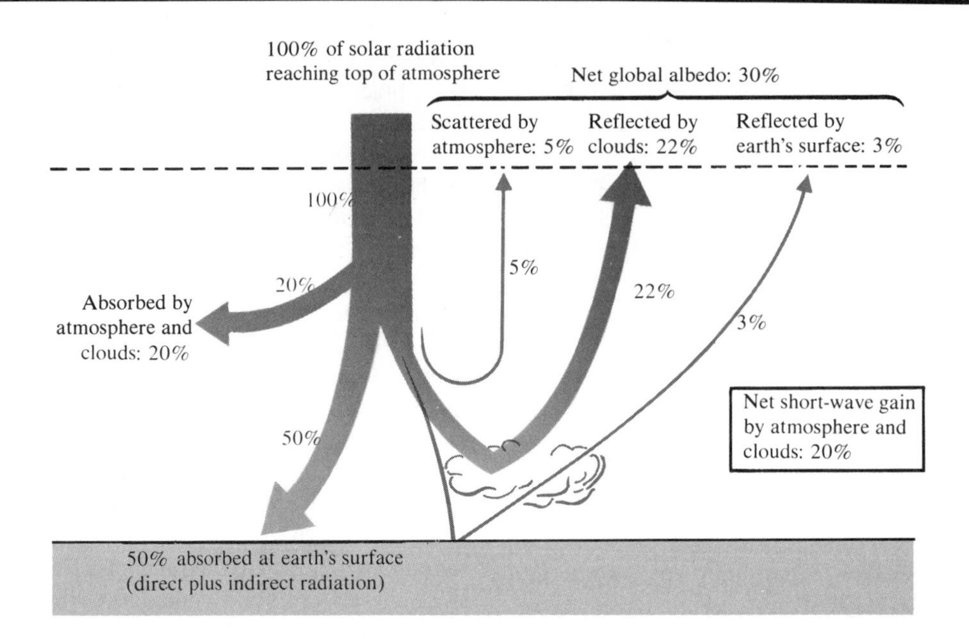

Figure 12.10 *Solar radiation budget of the earth and atmosphere. More solar energy is absorbed by the earth's surface than by the atmosphere. Consequently, the air is not heated directly by the sun, but indirectly from the earth's surface. (From Richard A. Anthes,* The Atmosphere, *2nd ed., Columbus, Ohio: Charles E. Merrill, 1978)*

The light that reaches the earth's surface after having its direction changed is called **diffused daylight.** It is the diffused light that keeps a shaded area or room lighted when direct sunlight is absent. Scattering also produces the blue color of the sky. Air molecules are more effective scatterers of the shorter wavelength (blue and violet) portion of "white" sunlight than the longer wavelength (red and orange) portion. Thus, when we look in a region of the sky away from the direct solar rays, we see predominantly blue light, which was more readily scattered. On the other hand, the sun appears to have a yellowish to reddish tint when viewed near the horizon. When the sun is in this position, the solar beam must travel through a great deal of atmosphere before it reaches your eye. Hence most of the blues and violets will be scattered out, leaving a beam of light composed mostly of reds and yellows. This phenomenon is particularly pronounced on a day when fine dust or smoke particles are present.

About 30 percent of the solar energy reaching the outer atmosphere is reflected back to space. Included in this figure is the amount sent skyward by back-scattering. This energy is lost to the earth and does not play a role in heating the atmosphere.

The fraction of the total radiation encountered that is reflected by a surface is called its **albedo.** Thus the albedo for the earth as a whole (planetary albedo) is 30 percent. However, the albedo from place to place as well as from time to time in the same locale varies considerably, depending upon the amount of cloud cover and particulate matter in the air, as well as upon the angle of the sun's rays and the nature of the surface. A lower angle means that more atmosphere must be penetrated, thus making the "obstacle course" longer, and therefore the loss of solar radiation greater (Figure 12.4). Table 12.4 gives the albedo for various surfaces and clouds. Note that the angle at which the sun's rays strike a water surface greatly affects its albedo.

As stated earlier, gasses are selective absorbers, meaning that they absorb strongly in some wavelengths, moderately in others, and only slightly

Table 12.4 *Albedo of Various Surfaces*

Surface	Percent Reflected
Clouds, stratus	
< 150 meters thick	25–63
150–300 meters thick	45–75
300–600 meters thick	59–84
Average of all types and thicknesses	50–55
Concrete	17–27
Crops, green	5–25
Forest, green	5–10
Meadows, green	5–25
Ploughed field, moist	14–17
Road, blacktop	5–10
Sand, white	30–60
Snow, fresh-fallen	80–90
Snow, old	45–70
Soil, dark	5–15
Soil, light (or desert)	25–30
Water	8 *

* Typical value for water surface, but the reflectivity increases sharply from less than 5 percent when the sun's altitude above the horizon is greater than 30 degrees to more than 60 percent when the altitude is less than 3 degrees. Rough seas have a somewhat lower albedo than calm seas.

in still others. When a gas molecule absorbs light waves, this energy is transformed into internal molecular motion, which is detectable as a rise in temperature. Nitrogen, the most abundant constituent in the atmosphere, is a rather poor absorber of all types of incoming radiation. Oxygen (O_2) and ozone (O_3) are efficient absorbers of ultraviolet radiation. Oxygen removes most of the shorter ultraviolet radiation high in the atmosphere, and ozone absorbs longer wavelength ultraviolet rays in the stratosphere between 10 and 50 kilometers. The absorption of ultraviolet radiation in the stratosphere accounts for the high temperatures experienced there. The only other significant absorber of incoming solar radiation is water vapor, which along with oxygen and ozone accounts for most of the 20 percent of the total solar radiation absorbed within the atmosphere.

For the atmosphere as a whole, none of the gasses are effective absorbers of radiation with wavelengths between 0.3 and 0.7 micrometer. This region of the spectrum corresponds to the visible range, to which a large fraction of solar radiation belongs. This explains why most visible radiation reaches the ground and why we say that the atmosphere is transparent to incoming solar radiation. Thus, direct solar energy is a rather ineffective "heater" of the earth's atmosphere. The fact that the atmosphere does not acquire the bulk of its energy directly from the sun but rather from reradiation from the earth's surface is of the utmost importance to the dynamics of the weather machine.

Approximately 50 percent of the solar energy that strikes the top of the atmosphere reaches the earth's surface directly or indirectly (diffused) and is absorbed. Most of this energy is then reradiated skyward. Since the earth has a much lower surface temperature than the sun, terrestrial radiation is emitted in longer wavelengths than is solar radiation. The bulk of terrestrial radiation has wavelengths between 1 and 30 micrometers, placing it well within the infrared range. The atmosphere as a whole is a rather efficient absorber of radiation between 1 and 30 micrometers (terrestrial radiation). Water vapor and carbon dioxide are the principal absorbing gasses in that range. Water vapor absorbs roughly five times more terrestrial radiation than do all the other gasses combined and accounts for the warm temperatures found in the lower troposphere, where it is most highly concentrated. *Because the atmosphere is quite transparent to solar (short-wave) and more absorptive of terrestrial (long-wave) radiation, the atmosphere is heated from the ground up rather than vice versa.* This explains the general drop in temperature with increased altitude experienced in the troposphere. The farther from the "radiator," the colder it gets.

When the gasses in the atmosphere absorb terrestrial radiation, they warm, but they eventually radiate this energy away. Some travels upward, where it may be reabsorbed by other gas molecules, a possibility less likely with increasing

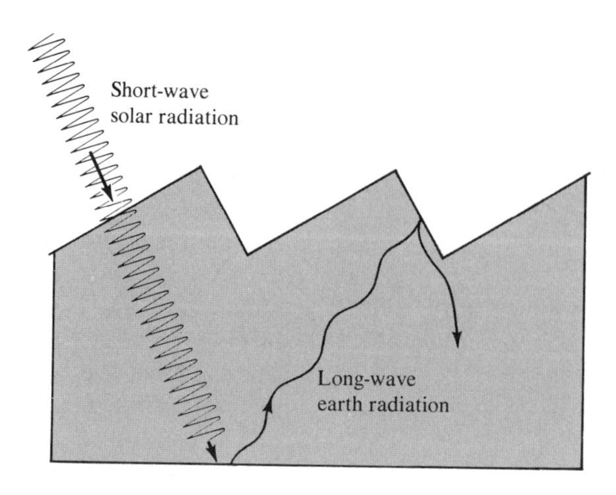

Figure 12.11 *The greenhouse effect. Short-wave solar radiation passes through the glass and is absorbed by the objects in the greenhouse. The longer wavelengths radiated by these objects cannot penetrate the glass and are trapped, thus warming the greenhouse.*

height because the concentration of water vapor decreases with altitude. The remainder travels downward and is again absorbed by the earth. Thus the earth's surface is being supplied continually with heat from the atmosphere as well as from the sun. Without these absorptive gasses in our atmosphere, the earth would not be a suitable habitat for humans and numerous other life forms.

This very important phenomenon has been termed the *greenhouse effect* because it was once thought that greenhouses were heated in a similar manner (Figure 12.11). The gasses of our atmosphere, especially water vapor and carbon dioxide, act very much like the glass in the greenhouse. They allow short-wave solar radiation to enter, where it is absorbed by the objects inside. These objects in turn reradiate the heat, but at longer wavelengths, to which glass is nearly opaque. The heat therefore is trapped in the greenhouse. However, a more important factor in keeping a greenhouse warm is the fact that the greenhouse itself prevents mixing of the air inside with cooler air outside. Nevertheless, the term *greenhouse effect* is still used.

The role that these absorbing gasses play in keeping the atmosphere warm is well known to those living in mountainous regions. More radiant energy is received on mountaintops than in the valleys below because there is less atmosphere to hinder its arrival. Yet the decrease in water vapor content with altitude allows much of the heat to escape these lofty peaks. This loss more than compensates for the extra radiation received. Hence the valleys remain warmer than the adjacent mountains even though they receive less solar radiation.

The importance of water vapor in keeping our atmosphere warm is easily demonstrated. Figure 12.12 is a map that shows the maximum and minimum temperatures for various cities in the United States on a typical July day. By comparing the temperatures for cities in the arid Southwest (Nevada and southern Arizona, for example) with temperatures in the more humid East, the following may be seen:

(1) Maximum temperatures are, for the most part, higher in the arid regions.

(2) Minimum temperatures often are lower

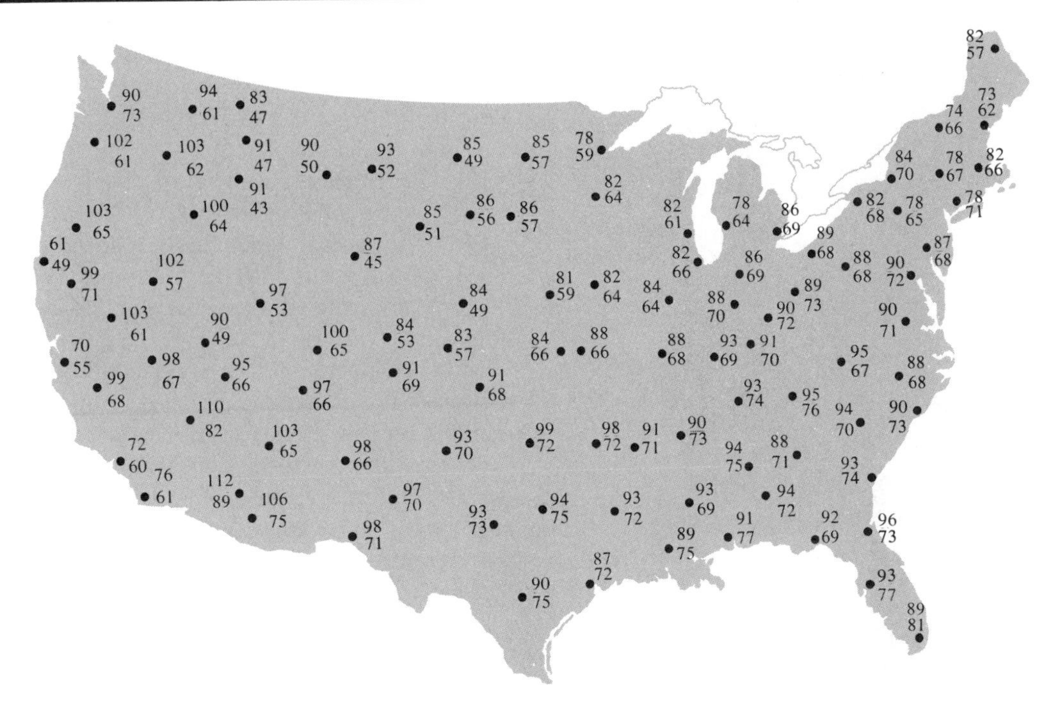

Figure 12.12 *Daily maximum and minimum temperatures (°F) for a typical July day. °C = (°F −32)/1.8.*

in the dry Southwest as compared with the humid East.

(3) The difference between the maximum and minimum temperatures (a calculation known as the **daily temperature range**) is in almost every case greater in the arid areas.

The higher maximum temperatures in arid regions are easily explained, for with little or no cloud cover, a maximum amount of sunshine is received. Further, the records of the stations in the dry regions indicate that these places cool rapidly at night, resulting in lower minimum temperatures. A hot, muggy day in a humid location often means a warm night as well, because the water vapor in the air traps some of the heat being radiated from the earth's surface. In the desert, however, there is very little water vapor in the air. Therefore, the long-wave earth radiation is lost, and the air near the surface cools rapidly. *From this example, it is quite evident that water vapor plays an important role in absorbing heat, and when it is lacking, the air cools rapidly, resulting in a high daily temperature range.*

Temperature Measurement and Data

Changes in air temperature are probably noticed by people more often than changes in any other element of weather. At a weather station, the temperature is read on a regular basis from instruments mounted in an instrument shelter (Figure 12.13). The shelter protects the instruments from direct sunlight and allows a free flow of air. In addition to a standard mercury thermometer, the shelter is likely to contain a

Figure 12.13 *Standard instrument shelter. A shelter protects instruments from direct sunlight and allows for the free flow of air. (Courtesy of Robert E. White Instruments)*

thermograph, which is an instrument that makes a continuous record of temperature, and a set of maximum-minimum thermometers. As their name implies, these thermometers record the highest and lowest temperatures during a given day.

The daily maximum and minimum temperatures are the bases for much of the temperature data compiled by meteorologists:

(1) By adding the maximum and minimum temperatures and then dividing by two, the **daily mean** is calculated.
(2) The daily temperature range, as noted earlier, is computed by finding the difference between the maximum and minimum temperatures for a given day.
(3) The **monthly mean** is calculated by adding together the daily means for each day of the month and dividing by the number of days in the month.
(4) The **annual mean** is an average of the twelve monthly means.
(5) The **annual temperature range** is computed by finding the difference between the warmest and coldest monthly means.

Mean temperatures are particularly useful for making comparisons, whether on a daily, monthly, or annual basis. It is quite common to hear a weather reporter state that, "Last month was the hottest July on record," or, "Today Chicago was ten degrees warmer than Miami." Temperature ranges are also useful statistics, because they give an indication of extremes.

Controls of Temperature

The controls of temperature are those factors that cause variations in temperature from place to place. Earlier in this chapter we examined the single greatest cause for temperature variations—differences in the receipt of solar radiation. Since variations in sun angle and length of day are a function of latitude, they are responsible for warm temperatures in the tropics and colder temperatures at more poleward locations. However, latitude is not the only control of temperature; if it were, we would expect that all places along the same parallel would have identical temperatures. This is clearly not the case. For example, Eureka, California, and New York City are both coastal cities at about the same latitude and both places have an average annual mean temperature of 11° C (52° F). However, New York City is 9° C (16° F) warmer than Eureka in July and 9° C cooler in January. Quito and Guayaquil, two cities in Ecuador, are within a relatively few kilometers of each other, yet the mean annual temperatures at these two cities differ by 12° C (21° F). To explain these situations and countless others, we must realize that factors

Table 12.5 ***Temperature Data (°C) for San Francisco, California, and Springfield, Missouri***

	Jan.	Feb.	Mar.	Apr.	May	June	July	Aug.	Sept.	Oct.	Nov.	Dec.	Annual
San Francisco, California	10	11	12	13	15	16	17	17	18	16	13	10	13
Springfield, Missouri	1	2	7	13	18	23	25	24	21	14	8	2	13

other than variations in solar radiation also exert a strong influence upon temperatures. Among the most important of these other factors are the differential heating of land and water, ocean currents, altitude, and geographic position.*

Land and Water

The heating of the earth's surface directly influences the heating of the air above. Therefore, in order to understand variations in air temperature, we must examine the nature of the surface. Different land surfaces absorb varying amounts of incoming solar energy, which in turn cause variations in the temperature of the air above. The largest contrast, however, is not between different land surfaces, but between land and water. *Land heats more rapidly and to higher temperatures than water and cools more rapidly and to lower temperatures than water. Temperature variations, therefore, are considerably greater over land than over water.*

Among the reasons for the differential heating of land and water are the following:

(1) The specific heat (amount of energy needed to raise 1 gram of a substance 1°C) is far greater for water than for land. Thus, water requires a great deal more heat to raise its temperature the same amount as an equal quantity of land.

(2) Land surfaces are opaque, so heat is absorbed only at the surface. Water, being more transparent, allows heat to penetrate to a depth of many meters.

(3) The water that is heated often mixes with water below, thus distributing the heat throughout an even larger mass.

(4) Evaporation (a cooling process) from water bodies is greater than that from land surfaces.

By examining temperature data for two locations, the moderating influence of water, and the extremes associated with land, may be demonstrated (Table 12.5). San Francisco, California, is located along a coast, while Springfield, Missouri, is in a continental position far from the influence of water. Both cities are at the same latitude and thus experience similar sun angles and lengths of day. Furthermore, both places have identical average temperatures for the entire year. Springfield, however, has an average January temperature that is 9° C colder than San Francisco's. Conversely, Springfield's July average is 8° C higher than San Francisco's. Although their average temperatures for the year are similar, Springfield, which has no water influence, experiences much greater temperature extremes than San Francisco, which does.

On a different scale, the moderating influence of water may also be demonstrated when temperature variations in the Northern and Southern hemispheres are compared. Sixty-one percent of the Northern Hemisphere is covered by water, and land represents the remaining 39 percent.

*For a discussion of the effects of ocean currents on temperature, see Chapter 11.

Table 12.6 ***Variation in Mean Annual Temperature Range with Latitude (°C)***

Latitude	Northern Hemisphere	Southern Hemisphere
0	0	0
15	3	4
30	13	7
45	23	6
60	30	11
75	32	26
90	40	31

However, the figures for the Southern Hemisphere (81 percent water, 19 percent land) reveal why it is correctly called the water hemisphere. Table 12.6 portrays the considerably smaller annual temperature variations in the water-dominated Southern Hemisphere as compared with the Northern Hemisphere.

Altitude

The two cities in Ecuador mentioned earlier—Quito and Guayaquil—demonstrate the influence of altitude upon mean temperatures. Although both cities are near the equator and relatively close to each other, the annual mean at Guayaquil is 25° C (77° F) as compared to Quito's mean of 13° C (55° F). The difference is explained in most part by the difference between the cities' elevations. Guayaquil is only 12 meters (40 feet) above sea level while Quito is high in the Andes Mountains at 2800 meters (9200 feet). Recall that temperatures drop an average of 6.5° C per kilometer in the troposphere; thus cooler temperatures are to be expected at greater heights. However, the magnitude of the difference is not explained completely by the normal lapse rate. If the normal lapse rate is used, we would expect Quito to be about 18° C cooler than Guayaquil; the difference, however, is only 12° C. Places at high altitudes, such as Quito, are warmer than the value calculated using the normal lapse rate because of the absorption and reradiation of solar energy by the ground surface.

Geographic Position

The geographic setting may greatly influence the temperatures experienced at a particular locale. For example, a windward coastal location, that is, a place that is subject to prevailing onshore winds, experiences considerably different temperatures than a coastal location where the prevailing winds are directed from the land toward the ocean. In the first situation, the place will experience the full moderating influence of the ocean—cool summers and mild winters compared with an inland station at the same latitude. However, a leeward coastal site will have a more continental temperature pattern because the winds do not carry the ocean's influence onshore. Eureka, California, and New York City, the two cities mentioned in the introduction to the section on temperature controls, illustrate this aspect of geographic position. The annual temperature range at New York City is 19° C (34° F) higher than that at Eureka.

Seattle and Spokane, both in the state of Washington, illustrate a second aspect of geographic position—mountains that act as barriers. Although Spokane is only about 360 kilometers (220 miles) east of Seattle, the towering Cascade Range separates the cities. Consequently, while Seattle's temperatures show a marked marine influence, Spokane's are more typically continental. Spokane is 7° C (13° F) cooler than Seattle in January and 4° C (7° F) warmer than Seattle in July. The annual range at Spokane is 11° C (20° F) greater than at Seattle. The Cascade Range effectively cuts Spokane off from the moderating influence of the Pacific Ocean.

World Distribution of Temperature

Temperature distribution is shown on a map by using **isotherms,** which are lines that connect places of equal temperature. The temperatures are corrected to sea level in order that altitude differences will not be a factor in the interpretation of temperature patterns. January (Figure 12.14) and July (Figure 12.15) are selected most

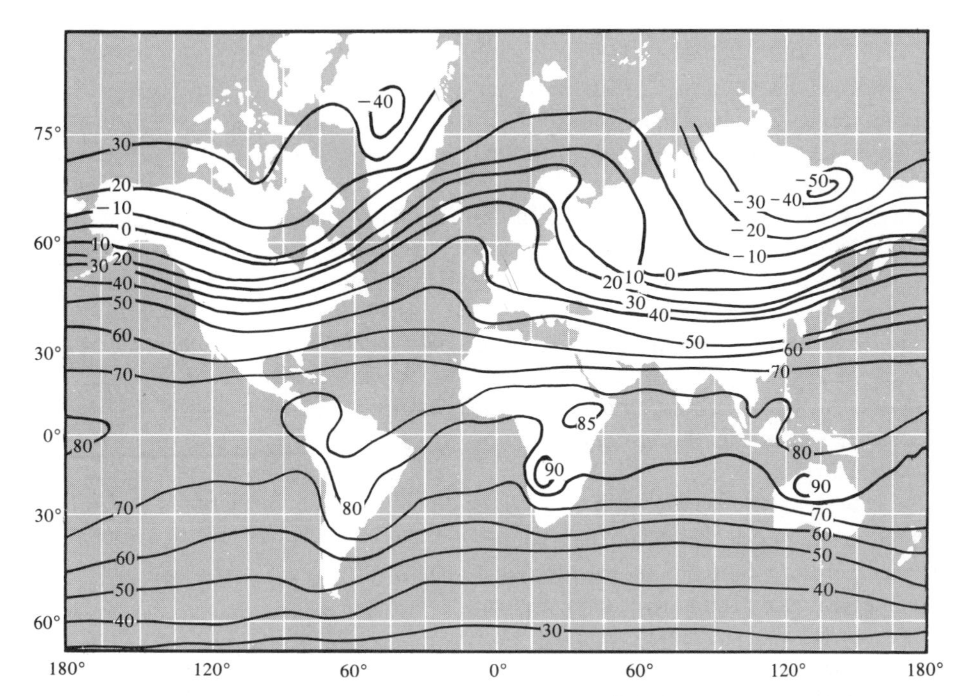

Figure 12.14 *World distribution of mean temperature (°F) for January.*

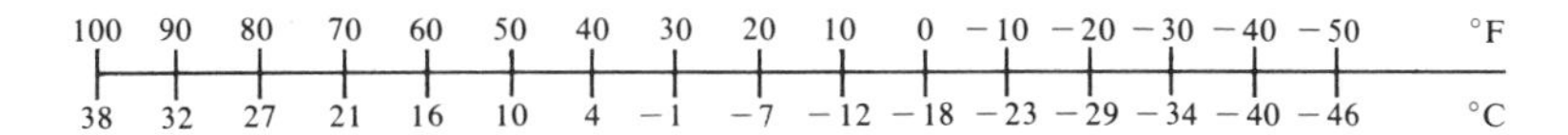

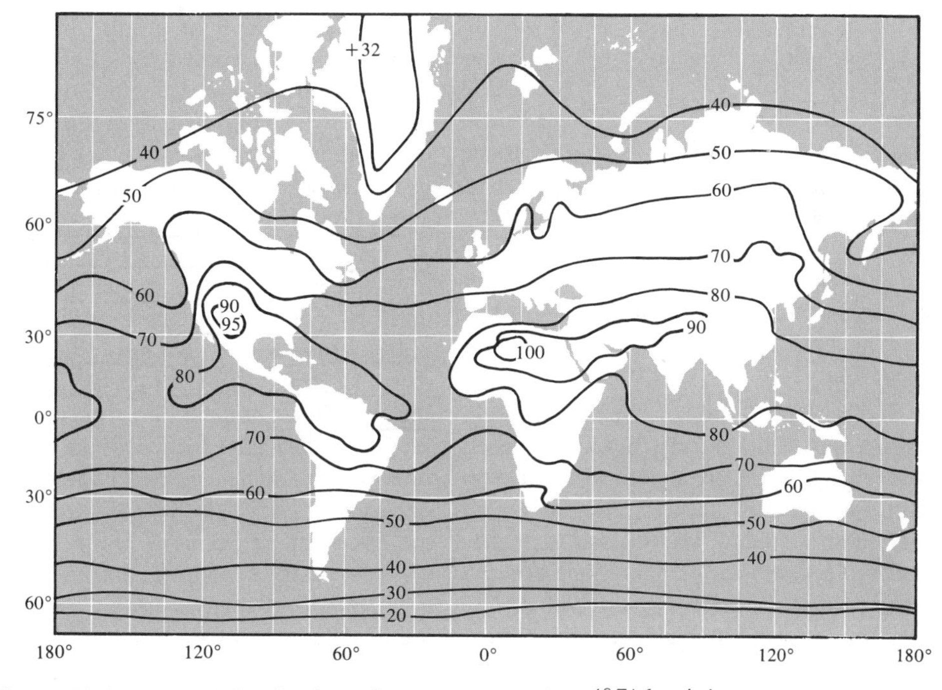

Figure 12.15 *World distribution of mean temperature (°F) for July.*

often for analysis because, for most locations, they represent the extremes of temperature.

On Figures 12.14 and 12.15 the isotherms generally trend east and west and show a decrease in temperatures poleward from the tropics, illustrating one of the most fundamental and best-known aspects of the world distribution of temperature—that the effectiveness of incoming solar radiation in heating the earth's surface and the atmosphere above is largely a function of latitude. Further, there is a latitudinal shifting of temperatures caused by the seasonal migration of the sun's vertical rays.

The added effect of the differential heating of land and water is also reflected on the January and July temperature maps. The warmest and coldest temperatures are found over land. Hence, since temperatures do not fluctuate as much over water as over land, the north-south migration of isotherms is greater over the continents than over the oceans. In addition, it is clear that the isotherms in the Southern Hemisphere, where there is little land and the oceans predominate, are much more regular than in the Northern Hemisphere, where they bend sharply northward in July and southward in January over the continents.

Isotherms also reveal the presence of ocean currents. Warm currents cause isotherms to be deflected toward the poles, while cold currents cause an equatorward bending. The horizontal transport of water poleward warms the overlying air and results in air temperatures that are higher than otherwise would be expected for the latitude. Conversely, currents moving toward the equator produce air temperatures cooler than expected.

Since Figures 12.14 and 12.15 show the seasonal extremes of temperature, they can be used to evaluate variations in the annual range of temperature from place to place. A comparison of the two maps shows that a station near the equator will record a very small annual range because it experiences little variation in the length of day and always has a relatively high sun angle. However, a station in the middle latitudes experiences much wider variations in sun angle and length of day, and thus larger variations in temperature. *Therefore, we can state that the annual temperature range increases with an increase in latitude.* Moreover, land and water also affect seasonal temperature variations, especially outside the tropics. A continental location must endure hotter summers and colder winters than a coastal location. Consequently, *the annual range will increase with an increase in continentality.*

A classic example of the effect of latitude and continentality on annual temperature range is Yakutsk, U.S.S.R. This city is located in Siberia at approximately 60 degrees north latitude and far from the influence of water. As a result, Yakutsk has an average annual temperature range of 62.2° C (112° F), one of the highest in the world.

Air Pollution

Air pollution has become a growing threat to our health and welfare because of the *ever-increasing* emissions of air contaminants into our *never-increasing* atmosphere. An average adult male requires about 13.5 kilograms (30 pounds) of air each day, as compared to about 1.2 kilograms (2.75 pounds) of food and 2 kilograms (4.5 pounds) of water. The cleanliness of the air, therefore, should certainly be as important to us as the cleanliness of our food and water.

Air is never perfectly clean. There have always been many natural sources of air pollution. Ash from volcanic eruptions, salt particles from breaking waves, pollen and spores released by plants, smoke from forest and brush fires, and wind-blown dust are all examples of "natural air pollution." Since humans have been on the earth, however, they have added to the frequency and intensity of some of these natural pollutants, particularly to the last two (Figure 12.16). With the discovery of fire came an increased number of accidental as well as intentional burnings. Even today, in many parts of the world fire is used to clear land for agricultural purposes (the so-called slash-and-burn method), filling the air with smoke and reducing visibility. When people

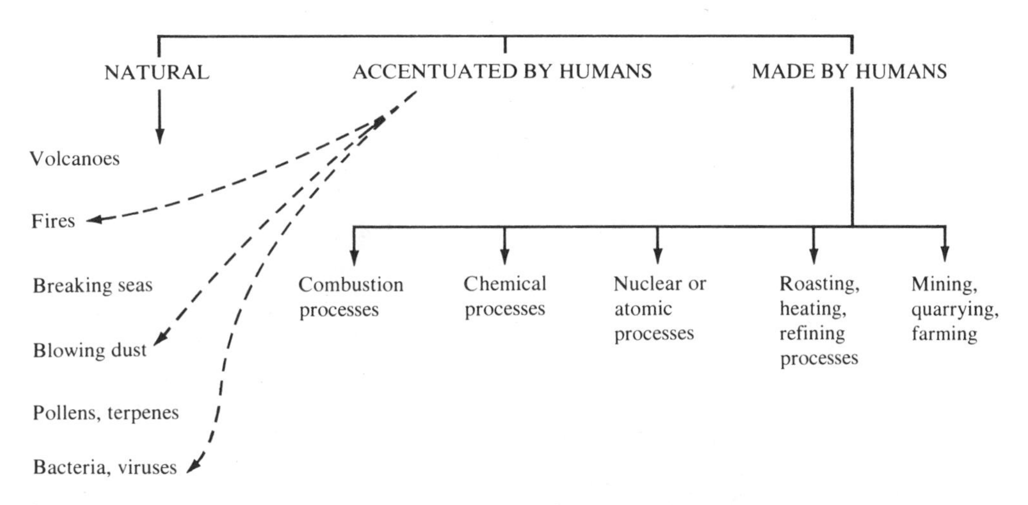

Figure 12.16 *Sources of primary pollutants. (After Reid A. Bryson and John I. Kutzbach,* Air Pollution, *Washington: Association of American Geographers, 1968, Fig. 3, p. 9)*

clear the land of its natural vegetative cover for whatever purpose, soil is exposed and blown into the air. However, when considering the air in a modern-day industrial city, these human-accentuated forms of pollution, although significant, may seem minor by comparison.

With the Industrial Revolution came "big-time" air pollution. Rather than simply accelerating natural sources, which of course continued to occur, people found many new ways to pollute the air (Figure 12.16), and many new things to pollute it with—sulfur and nitrogen oxides, carbon monoxide, and hydrocarbons, to name a few. With the rapid growth of the world's population and accelerated industrialization, the quantities of atmospheric pollutants increased drastically.

Temperature Inversions and Air Pollution

When air pollution episodes occur, they are not the result of a drastic increase in the output of pollutants but rather of changes in atmospheric conditions. Usually, when contaminants are released, they are mixed by wind and turbulence with the surrounding air and are diluted. A well-known phrase says that, "The solution to pollution is dilution." To some degree, this is true. However, if the air into which the pollutants are released is not dispersed (moved out of the area), the air will become more and more toxic. *Temperature inversions are among the most significant causes of retarding the dispersion of pollutants.* Most of the air pollution episodes cited earlier were linked to the occurrence of these phenomena. Normally, smoke and exhaust fumes are carried upward by air currents and dispersed by winds aloft. However, air will rise only as long as it is warmer and less dense than the surrounding air. When an inversion exists, warm air overlying cooler air acts as a lid and prevents further upward movement. The pollutants are thus trapped below. This is illustrated quite dramatically by the sequence of photographs of downtown Los Angeles shown in Figure 12.17.

Inversions are generally classified into one of two categories—those that form near the ground and those that form aloft. The first type, sometimes called surface inversions, develop close to the ground on clear and relatively calm nights. They form because the ground is a more effective radiator than the air above. This being the case, radiation from the ground to the clear night sky causes more rapid cooling at the surface than higher in the atmosphere. The result is that the air close to the ground is cooled more than the air above, yielding a temperature profile similar

Figure 12.17 **A.** *Los Angeles City Hall on a clear day.* **B.** *Smog engulfs City Hall as a temperature inversion at 450 meters traps pollutants below.* **C.** *A temperature inversion at 100 meters is visible as smog shrouds the lower portion of City Hall while the upper portion of the building is visible in the clear air above the base of the inversion. (A., B., and C. courtesy of the Los Angeles County Air Pollution Control District)*

to the one shown in Figure 12.18A. After sunrise the ground is heated and the inversion disappears. Although surface inversions are usually rather shallow, they may be quite deep in regions where the land surface is uneven. Because cold air is denser (heavier) than warm air, the chilled air near the surface gradually drains from the uplands and slopes into adjacent lowlands and valleys. As might be expected, these deeper surface inversions will not dissipate as quickly after

Table 12.7 ***Estimated Nationwide Emissions (millions of tons/year)***

Source	Carbon Monoxide	Particulates	Sulfur Oxides	Hydrocarbons	Nitrogen Oxides	Total
Transportation	77.4	1.3	0.8	11.7	10.7	101.9
Fuel combustion in stationary sources	1.2	6.6	26.3	1.4	12.4	47.9
Industrial processes	9.4	8.7	5.7	3.5	0.7	28.0
Solid waste disposal	3.3	0.6	<0.1	0.9	0.2	5.0
Miscellaneous	4.9	0.8	0.1	13.4	0.2	19.4
Total	96.2	18.0	32.9	30.9	24.2	202.2

SOURCE: *National Air Quality and Emissions Trends Report, 1975,* U.S. Environmental Protection Agency Publication No. EPA-450/1-76-002, November 1976.

B.

C.

sunrise. Thus, although valleys are often preferred sites for manufacturing because they afford easy access to water transportation, they are also more likely to experience relatively thick surface inversions which, in turn, will have a negative effect upon air quality.

Many extensive and long-lived air pollution episodes are linked to the second type of temperature inversion, which develops aloft in association with sinking air. As the air sinks to lower altitudes, it is compressed and because of this its temperature rises.* Since turbulence is almost always present near the ground, this lowermost portion of the atmosphere is generally prevented from participating in the general subsidence. Thus an inversion develops aloft between the lower turbulent zone and the subsiding warmed layers above (Figure 12.18B).

*Temperature changes that occur when air is compressed or allowed to expand are termed *adiabatic* temperature changes and are discussed in more detail in Chapter 13.

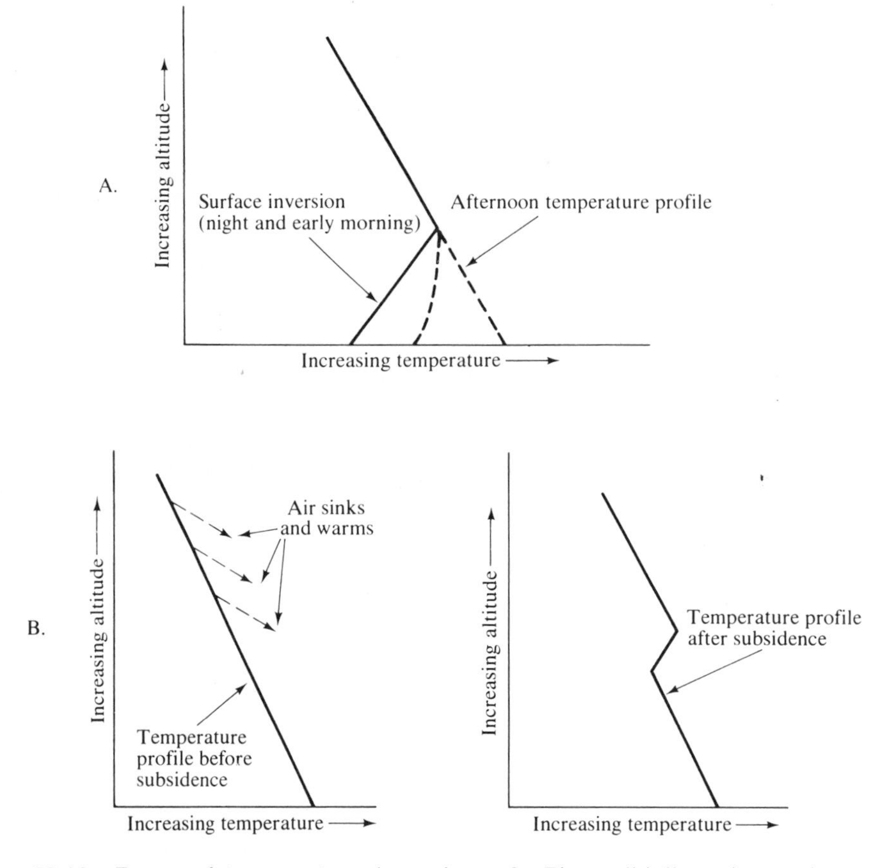

Figure 12.18 *Types of temperature inversions.* **A.** *The solid line shows the vertical temperature profile as it might appear early in the morning. The dashed lines show changes that will occur as the surface is heated during the day.* **B.** *Subsidence often creates an inversion aloft.*

Sources and Types of Air Pollution

Table 12.7 lists the major sources of air pollution as well as the types of contaminants that characterize each. The significance of the transportation category is obvious. It accounts for more than half of our air pollution (by weight). In addition to motor vehicles, this category includes trains, ships, and airplanes. However, the more than 110 million cars and trucks on U.S. roads are, without a doubt, the greatest source of air pollution.

Pollutants may be grouped into two categories—primary and secondary. The **primary pollutants** are those emitted directly from identifiable sources. These are shown in Figure 12.16. **Secondary pollutants** are produced in the atmosphere when certain chemical reactions take place among the primary pollutants. For example, when sunlight acts on nitrogen oxides and certain organic compounds, the toxic gas ozone (O_3) is produced. Reactions such as this, which are triggered by sunlight, are called **photochemical.** Sulfuric acid (H_2SO_4) is another common secondary pollutant. It is produced when sulfur dioxide combines with oxygen, yielding sulfur trioxide, which then combines with water to form this irritating and corrosive acid.

The sections that follow give brief descriptions of the major primary pollutants.

Sulfur Oxides. Sulfur dioxide, the most abundant of the sulfur oxides, is a colorless gas which is corrosive to many metals. Sulfur dioxide also creates health problems. It irritates the mucuous membranes of the respiratory tract and eyes and in high concentrations can produce serious respiratory problems. Almost three-fourths of the sulfur oxides are emitted from the burning of sulfur-bearing fuels, particularly coal and low-grade fuel oil. The primary sources are power plants and industry.

Carbon Monoxide. Carbon monoxide, an odorless and colorless gas, is the largest single air pollutant and is produced by the incomplete combustion of fuels like gasoline. It is not surprising then that highway vehicles account for 70 percent of the total emissions of this gas. Carbon monoxide reacts with the blood's hemoglobin and diminishes its capacity for transporting oxygen to the body's tissues. In large amounts, carbon monoxide is lethal; in small quantities, it may cause headache, dizziness, and nausea.

Hydrocarbons. Hydrocarbons form a complex group of pollutants that originate primarily from the inefficient combustion of gasoline, coal, and oil. Again, motor vehicles are the single greatest source. Hydrocarbons react with other contaminants to form secondary pollutants, such as ozone. When other hydrocarbons react with nitrogen oxides in the presence of sunlight, *photochemical* (Los Angeles-type) *smog* is the end product. Furthermore, some hydrocarbons have been shown to be cancer-producing agents (carcinogens).

Nitrogen Oxides. Nitrogen oxides are yellow-to-brown gasses which have a pungent odor and result from high-temperature combustion, such as that which takes place in a gasoline-powered automobile engine. The most significant impact of nitrogen oxides is that they are prime contributors to photochemical smog and therefore may lead to throat irritation, coughing, and respiratory problems.

Particulates. Particulates are a variety of solid and liquid particles which have a wide range of sizes. The primary sources are fuel (especially coal) combustion, various industrial processes, and forest fires. The influence of particulate matter in our atmosphere is quite complex. Some particles may affect various weather processes and climatic conditions, and others may damage vegetation, corrode metals, or be harmful to human health. Dustfall, especially in industrial areas, soils property and leads to higher costs for cleaning and painting.

Recently, the concern for clean air has increased, as evidenced by the passage of new laws, the establishment of new government agencies, and increased research on pollution control problems. The costs of controlling pollution may seem high; however, when compared with the costs of uncontrolled air pollution, both in dollars and in human health, the expenditures are easily justified.

REVIEW

1. How are the studies of weather and climate different? In what ways are they similar?
2. What are the major components of clean, dry air (Table 12.1)? Are they important meteorologically?
3. Why are water vapor, ozone, and dust important constituents of our atmosphere?
4. What percentage of the atmosphere is found below the following altitudes: 5.6 kilometers, 16.2 kilometers, 31.2 kilometers (see Table 12.2)?
5. Why is the troposphere called the "weather sphere?"
6. Why do temperatures rise in the stratosphere?
7. Are variations in the amount of solar energy received by the earth because of its elliptical orbit important to understanding seasonal temperature variations? Explain your answer.
8. After examining Table 12.3, write a generalization that relates the season, the latitude, and the length of daylight.
9. Describe the relationship between the temperature of a radiating body and the wavelengths it emits.
10. Distinguish among the three basic mechanisms of heat transfer.

11. Figure 12.10 illustrates what happens to incoming solar radiation. The percentages shown, however, are only global averages. In particular, the amount of solar radiation reflected (albedo) may vary considerably. What factors might cause variations in albedo?
12. How does the earth's atmosphere act as a "greenhouse"?
13. On a warm summer day, one city had a daily temperature range of 25° C (45° F), while another experienced a range of only 8.3° C (15° F). One of these cities is located in Nevada, and the other in Indiana. Which location likely had the highest daily temperature range? Explain.
14. How are the following temperature data computed? Daily mean, daily range, monthly mean, annual mean, annual range.
15. San Francisco, California, and Springfield, Missouri, have identical annual means of 13° C. After examining the data for each city (Table 12.5), explain why the average temperature does not tell the whole story. That is to say, in what ways are the annual means misleading?
16. Yakutsk, U.S.S.R., is located in Siberia at about 60 degrees north latitude. This city has the highest average annual temperature range in the world—62.2° C (112° F).
    a. Compute the noon sun angle for June 21 and December 21 (Figure 12.6).
    b. Note the length of day for the same dates (Table 12.3). Using this information and the knowledge of Yakutsk's Siberian location, explain the reasons for the very high temperature range.
17. Quito, Ecuador, is located on the equator and is not a coastal city. It has an average annual temperature of only 13° C (55° F). What is the likely cause for this low average temperature?
18. In what ways can geographic position be considered a control of temperature?
19. What is the difference between primary and secondary pollutants? What is a photochemical reaction?
20. How do temperature inversions form? Describe the role of inversions in air pollution episodes.

KEY TERMS

meteorology
weather
climate
troposphere
normal lapse rate
stratosphere
tropopause
stratopause
mesosphere
mesopause
thermosphere
rotation
circle of illumination
revolution
perihelion
aphelion
altitude (of the sun)
inclination of the axis
Tropic of Cancer
summer solstice
Tropic of Capricorn
winter solstice
autumnal equinox
vernal (spring) equinox
radiation (electromagnetic radiation)
visible light
infrared
ultraviolet
conduction
convection

advection
diffused daylight
albedo
daily temperature range
daily mean
monthly mean
annual mean
annual temperature range
isotherms
primary pollutant
secondary pollutant
photochemical

In addition to being prominent and sometimes spectacular features in the sky, clouds are of continual interest to meteorologists because they provide a visible indication of what is taking place in the atmosphere. (Courtesy of NOAA)

13 Moisture

Changes of State
Humidity
Humidity Measurement
Necessary Conditions for Condensation
Condensation Aloft: Adiabatic Temperature Changes
Stability
Forceful Lifting
Clouds
Formation of Precipitation
Sleet, Glaze, and Hail
Fog
Intentional Weather Modification

As you observe day-to-day weather changes, many questions may come to mind concerning the role of water in the air. What is humidity, and how is it measured? Why do clouds form on some occasions but not on others? What processes produce clouds and precipitation? Why are some clouds thin, white, and harmless, while others are towering, gray, and ominous? What is the difference between sleet and hail? What is fog? Are all fogs alike? This chapter investigates these and other questions as the topic of moisture in the air is examined.

An understanding of the role that water plays in the atmosphere is very important. In addition to exerting a strong influence on the heating of the air (Chapter 12), water vapor is the source of all clouds and precipitation. Areas within the Atacama Desert of South America may go years without a drop of rain; in fact, legend has it that some areas have not received rain since the arrival of the Spaniards, some 400 years ago! By contrast, Cherrapunji, India, has received 930 centimeters (366 inches) in a single month, and 2644 centimeters (1041 inches) [over 26 meters (86 feet)] in a single year. Indeed, the large variations in the amount of precipitation from place to place as well as local differences from time to time have a significant impact not only on the nature of the physical landscape but also on people's life-styles.

Changes of State

Water vapor is an odorless, colorless gas that mixes freely with the other gasses of the atmosphere. Unlike oxygen and nitrogen, the two most abundant components of the atmosphere, water vapor can change from one state of matter to another with relative ease at the temperatures and pressures experienced near the surface of the earth. It is because of this ability, which allows water to leave the oceans as a gas and return again as a liquid, that the vital hydrologic cycle exists. The processes that involve a change of state require that heat be absorbed or released (Figure 13.1). This heat energy is measured in calories. One **calorie** is the amount of heat required to raise the temperature of 1 gram of water 1° C.

The process of converting a liquid to a vapor is termed **evaporation.** It takes approximately 600 calories of energy to convert 1 gram of water to water vapor. The energy absorbed by the water molecules during evaporation is used solely to give them the motion needed to escape the surface of the liquid and become a gas. Because this energy is subsequently released as heat when the vapor changes back to a liquid, it is generally called **latent heat** (meaning hidden, or stored, heat).

Since the evaporating molecules require energy to escape, the remaining liquid must be cooled by an equivalent amount; hence the common expression "evaporation is a cooling process." You have undoubtedly experienced this cooling effect upon stepping dripping wet out of a swimming pool. On a larger scale is the cooling effect of evaporating rainwater during a summer shower.

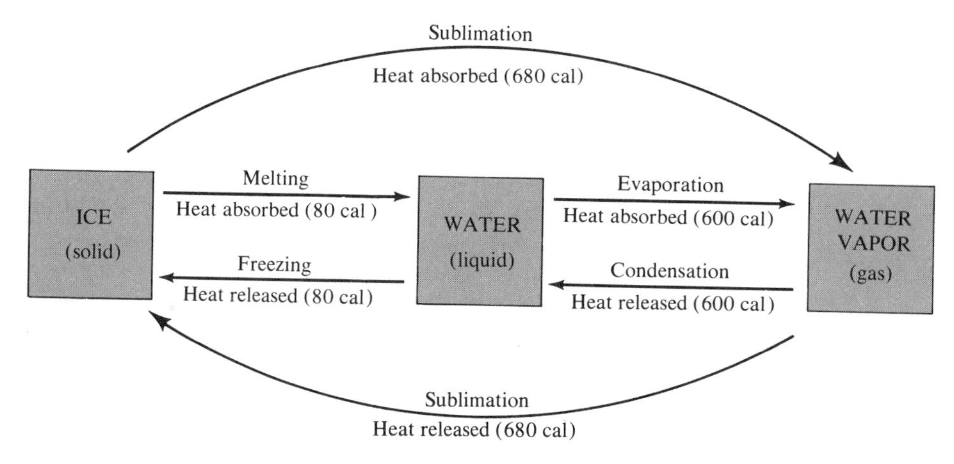

Figure 13.1 *Changes of state.*

Condensation denotes the process of water vapor changing to the liquid state. During condensation, the water molecules release energy *(latent heat of condensation)* equivalent to that which was absorbed during evaporation. This energy plays an important role in producing violent weather and can act to transfer great quantities of heat energy from tropical oceans to more poleward locations. When condensation occurs in the atmosphere, it results in such phenomena as fog and clouds.

Melting is the process by which a solid is changed to a liquid. It requires absorption of approximately 80 calories of energy per gram of water. **Freezing,** the reverse process, releases these 80 calories per gram as *latent heat of fusion.*

The last of the processes illustrated in Figure 13.1 is **sublimation.** This term is used to describe the conversion of a solid directly to a gas, without passing through the liquid state. You may have observed this change as you watched the sublimation of dry ice (solid carbon dioxide). The term *sublimation* is also used to denote the reverse process, by which a vapor is changed to a solid. This change occurs, for example, during the formation of frost. As shown in Figure 13.1, sublimation involves an amount of energy equal to the total of the other two processes.

Humidity

Humidity is the general term used to describe the amount of water vapor in the air. Several methods are used to quantitatively express humidity. Among these are absolute humidity, specific humidity, and relative humidity.

Before we consider each of these humidity measures individually, it is important to understand the concept of **saturation.** Imagine a closed jar half full of water and overlain with dry air at the same temperature. As the water begins to evaporate from the water surface, a small increase in pressure can be detected in the air above. This increase is the result of the motion of the water vapor molecules which were added to the air through evaporation. In the open atmosphere this pressure is termed **vapor pressure** and is defined as that part of the total atmospheric pressure which can be attributed to the water vapor content. In the closed container, as more and more molecules escape from the water surface, the steadily increasing vapor pressure in the air above forces more and more of these molecules to return to the liquid. Eventually the number of vapor molecules returning to the surface will balance the number leaving. At that point, the air is said to be saturated or filled to capacity. However, if we increase the temperature of the water and air in the jar, more water will evaporate before a balance is reached. Consequently, at higher temperatures more moisture is required for saturation. Stated another way, *the water vapor capacity of air is temperature dependent, with warm air having a much greater capacity than cold air.* The amount of water vapor required for saturation at various temperatures is shown in Table 13.1.

Of the methods used to express humidity, absolute and specific humidity are similar in that they both specify the amount of water vapor contained in a unit of air. **Absolute humidity** is stated

Table 13.1 ***Water Vapor Capacity (at average sea level pressure)***

Temperature (°C)	Grams/km
−40	0.1
−30	0.3
−20	0.75
−10	2
0	3.5
5	5
10	7
15	10
20	14
25	20
30	26.5
35	35
40	47

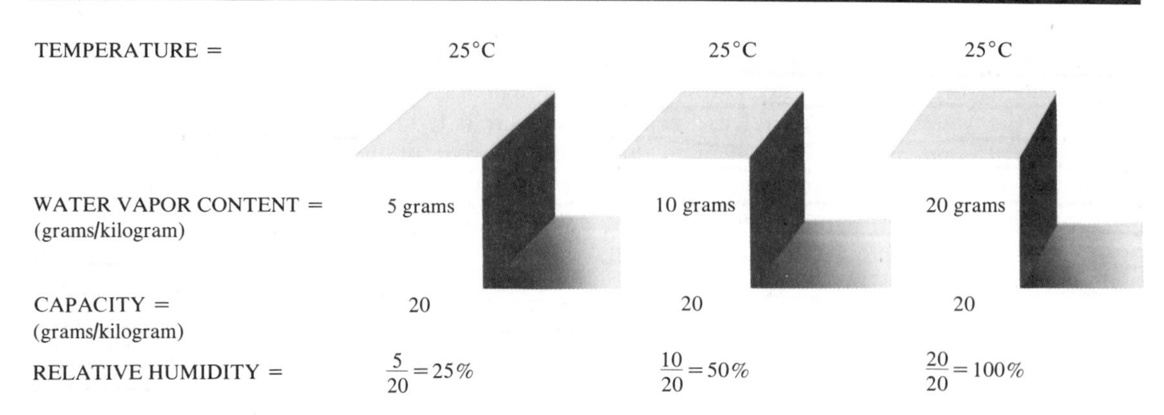

Figure 13.2 *At a constant temperature, relative humidity will increase as water vapor is added to the air. Here, the capacity remains constant at 20 grams per kilogram, while the relative humidity rises from 25 percent to 100 percent as the water vapor content increases.*

as the weight of water vapor in a given volume of air (usually as grams per cubic meter). As air moves from one place to another, even without a change in moisture content, changes in pressure and temperature cause changes in volume, and consequently in the absolute humidity, limiting the usefulness of this index. Therefore, meteorologists generally use specific humidity to express the water vapor content of air. **Specific humidity** is expressed as the weight of water vapor per weight of a chosen mass of air, including the water vapor. Since it is measured in units of weight (usually in grams per kilogram), specific humidity is not affected by changes in pressure or temperature.

The most familiar, and perhaps the most misunderstood, term used to describe the moisture content of air is relative humidity. Stated in an admittedly oversimplified manner, **relative humidity** is the ratio of the air's water vapor content to its water vapor capacity at a given temperature. From Table 13.1, we see that at 25° C, the capacity of the air is 20 grams per kilogram. If on a 25° C day the air contains 10 grams per kilogram, the relative humidity is expressed as 10/20, or 50 percent. When air is saturated, the relative humidity is 100 percent.

Since relative humidity is based on the air's water vapor content as well as on its capacity, relative humidity can be changed in either of two ways. First, if moisture is added by evaporation, the relative humidity will increase (Figure 13.2). The addition of moisture by this means occurs mainly over the oceans, but plants, soil, and smaller bodies of water do make contributions. The second method, illustrated by Figure 13.3, involves a change in temperature. We can generalize this concept as follows: *With the specific humidity at a constant level, a decrease in air temperature will result in an increase in relative humidity, and an increase in temperature will cause a decrease in the relative humidity*. In Figure 13.4 the variations in temperature and relative humidity during a typical day demonstrate rather well the relationship just described.

Another important idea related to relative humidity is the dew-point temperature. **Dew point** is the temperature to which a parcel of air would have to be cooled in order to reach saturation. Note that in Figure 13.3 unsaturated air at 20° C is cooled to 0° C before saturation occurs. Therefore, 0° C would be the dew-point temperature for this air. If this same parcel of air were cooled further, the air's capacity would be exceeded, and the excess vapor would be forced to condense.

Although relative humidity is used exclusively in our daily weather reports, it does have limitations if improperly interpreted. Recall from our earlier discussion that the capacity of air is temperature dependent. Table 13.1 illustrates that

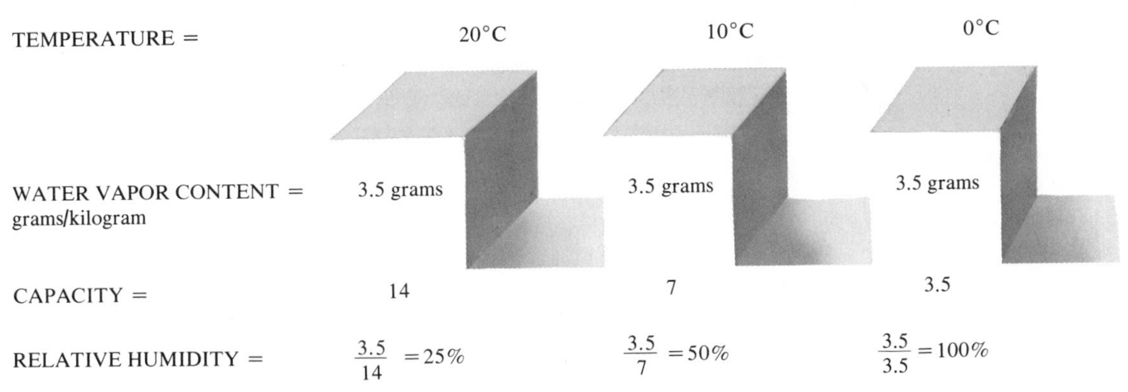

Figure 13.3 *When the water vapor content (specific humidity) remains constant, the relative humidity may be changed by increasing or decreasing the air temperature. In this example the specific humidity remains at 3.5 grams per kilogram. The reduction in temperature from 20° C to 0° C causes a decrease in capacity and thus an increase in the relative humidity from 25 percent to 100 percent.*

the capacity of air at 35° C is 35 grams per kilogram, while the capacity of air at 10° C is only 7 grams per kilogram. It should be apparent that when relative humidities are equal, air at 35° C contains five times as much water vapor as air at 10° C. This explains why cold winter air with a high relative humidity may be described as dry when compared to warm air that has an equally high relative humidity. *In summary, relative humidity indicates how near the air is to being saturated, and absolute and specific humidity denote the quantity of water vapor contained in that air.*

Humidity Measurement

Absolute and specific humidity are difficult to measure directly. However, they may be readily

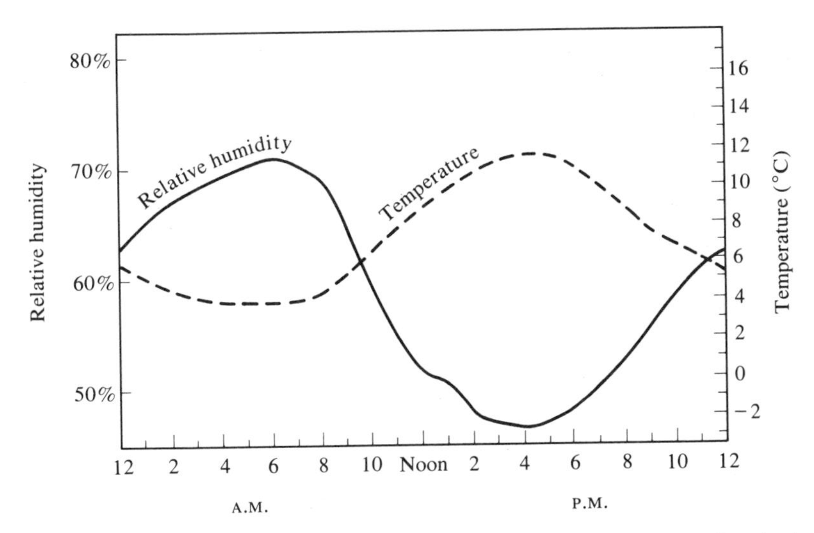

Figure 13.4 *Typical daily variations in temperature and relative humidity during a spring day at Washington, D.C.*

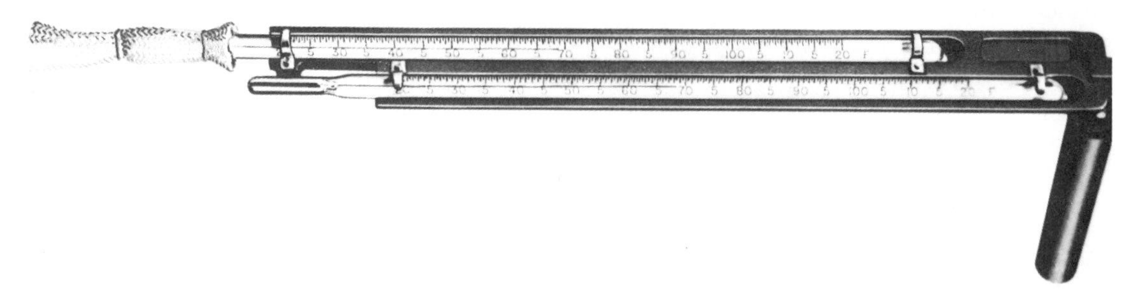

Figure 13.5 *Sling psychrometer. This instrument is used to determine relative humidity and dew point. The dry-bulb thermometer gives the current air temperature. The thermometers are spun until the temperature of the wet-bulb thermometer stops declining. Then the thermometers are read and the data used in conjunction with Tables 13.2 and 13.3. (Courtesy of WeatherMeasure Corporation)*

computed by consulting an appropriate table or graph if the air temperature and relative humidity are known.

Relative humidity is most commonly measured using a psychrometer or a hygrometer. The **psychrometer** consists of two identical thermometers mounted side by side (Figure 13.5). One of the thermometers, called the wet-bulb, has a thin muslin wick tied around the end. To use the psychrometer, the cloth sleeve is saturated with water and a continuous current of air is passed over the wick, either by swinging the instrument freely in the air or by fanning air past it. As a consequence, water evaporates from the wick and the temperature of the wet-bulb drops. The loss of heat that was required to evaporate water from the wet-bulb causes the lowering of the thermometer reading.

The amount of cooling that takes place is directly proportional to the dryness of the air. The drier the air, the more the cooling. *Therefore, the larger the difference between the thermometers, the lower is the relative humidity; the smaller the difference, the higher the relative humidity.* If the air is saturated, no evaporation will occur, and the two thermometers will have identical readings.

To determine the precise relative humidity from the thermometer readings, a standard table is used (Table 13.2). With the same information, but using a different table (Table 13.3), the dew-point temperature may also be calculated.

The second commonly used instrument for measuring relative humidity, the **hygrometer,** can be read directly, without the use of tables. The hair hygrometer operates on the principle that hair or certain synthetic fibers change their length in proportion to changes in the relative humidity, lengthening as relative humidity increases and shrinking as the relative humidity drops. The tension of a bundle of hairs is linked mechanically to an indicator that is calibrated between 0 and 100 percent. Thus we need only glance at the dial to determine the relative humidity. Unfortunately, the hair hygrometer is much less accurate than the psychrometer. Furthermore, it requires frequent calibration and is slow in responding to changes in humidity, especially at low temperatures.

A different type of hygrometer is used in remote-sensing instrument packages such as radiosondes which transmit upper-air observations back to ground stations. The electric hygrometer contains an electrical conductor coated with a moisture-absorbing chemical. It works on the principle that the passage of current varies as the relative humidity varies.

Table 13.2 ***Relative Humidity (percent)****

Dry-bulb temperature − Wet-bulb temperature = Depression of the wet bulb

Air (dry-bulb) Temperature (°C)	1	2	3	4	5	6	7	8	9	10	11	12	13	14	15	16	17	18	19	20	21	22
−20	28																					
−18	40																					
−16	48	0																				
−14	55	11																				
−12	61	23																				
−10	66	33	0																			
−8	71	41	13																			
−6	73	48	20	0																		
−4	77	54	32	11																		
−2	79	58	37	20	1																	
0	81	63	45	28	11																	
2	83	67	51	36	20	6																
4	85	70	56	42	27	14																
6	86	72	59	46	35	22	10	0														
8	87	74	62	51	39	28	17	6														
10	88	76	65	54	43	33	24	13	4													
12	88	78	67	57	48	38	28	19	10	2												
14	89	79	69	60	50	41	33	25	16	8	1											
16	90	80	71	62	54	45	37	29	21	14	7	1										
18	91	81	72	64	56	48	40	33	26	19	12	6	0									
20	91	82	74	66	58	51	44	36	30	23	17	11	5	0								
22	92	83	75	68	60	53	46	40	33	27	21	15	10	4	0							
24	92	84	76	69	62	55	49	42	36	30	25	20	14	9	4	0						
26	92	85	77	70	64	57	51	45	39	34	28	23	18	13	9	5						
28	93	86	78	71	65	59	53	47	42	36	31	26	21	17	12	8	4					
30	93	86	79	72	66	61	55	49	44	39	34	29	25	20	16	12	8	4				
32	93	86	80	73	68	62	56	55	46	41	36	32	27	22	19	14	11	8	4			
34	93	86	81	74	69	63	58	52	48	43	38	34	30	26	22	18	14	11	8	5		
36	94	87	81	75	69	64	59	54	50	44	40	36	32	28	24	21	17	13	10	7	4	
38	94	87	82	76	70	66	60	55	51	46	42	38	34	30	26	23	20	16	13	10	7	5
40	94	89	82	76	71	67	61	57	52	48	44	40	36	33	29	25	22	19	16	13	10	7

*To determine the relative humidity and dew point, find the air (dry-bulb) temperature on the vertical axis (far left) and the depression of the wet-bulb on the horizontal axis (top). Where the two meet, the relative humidity or dew point is found. For example, use a dry-bulb temperature of 20° C and a wet-bulb temperature of 14° C. From Table 13.2, the relative humidity is 51 percent, and from Table 13.3, the dew point is 10° C.

Necessary Conditions for Condensation

As we learned earlier in this chapter, condensation occurs when water vapor in the air changes to a liquid. The result of this process may be dew, fog, or clouds. Although these forms of condensation are quite different, they have two things in common. First, for any form of condensation to occur, the air must be saturated. Saturation occurs either when the air is cooled below the dew point, which most commonly happens, or when water vapor is added to the air. Second, there generally must be a surface on which the water vapor may condense. When dew occurs, objects at or near the ground serve this purpose. When condensation occurs in the air above the ground, tiny bits of particulate matter known as **condensa-**

Table 13.3 ***Dew Point Temperature (°C)****

Dry-bulb temperature − Wet-bulb temperature = Depression of the wet bulb

Air (Dry-bulb) Temperature	*1*	*2*	*3*	*4*	*5*	*6*	*7*	*8*	*9*	*10*	*11*	*12*	*13*	*14*	*15*	*16*	*17*	*18*	*19*	*20*	*21*	*22*
−20	−33																					
−18	−28																					
−16	−24																					
−14	−21	−36																				
−12	−18	−28																				
−10	−14	−22																				
−8	−12	−18	−29																			
−6	−10	−14	−22																			
−4	−7	−11	−17	−29																		
−2	−5	−8	−13	−20																		
0	−3	−6	−9	−15	−24																	
2	−1	−3	−6	−11	−17																	
4	1	−1	−4	−7	−11	−19																
6	4	1	−1	−4	−7	−13	−21															
8	6	3	1	−2	−5	−9	−14															
10	8	6	4	1	−2	−5	−9	−14	−28													
12	10	8	6	4	1	−2	−5	−9	−16													
14	12	11	9	6	4	1	−2	−5	−10	−17												
16	14	13	11	9	7	4	1	−1	−6	−10	−17											
18	16	15	13	11	9	7	4	2	−2	−5	−10	−19										
20	19	17	15	14	12	10	7	4	2	−2	−5	−10	−19									
22	21	19	17	16	14	12	10	8	5	3	−1	−5	−10	−19								
24	23	21	20	18	16	14	12	10	8	6	2	−1	−5	−10	−18							
26	25	23	22	20	18	17	15	13	11	9	6	3	0	−4	−9	−18						
28	27	25	24	22	21	19	17	16	14	11	9	7	4	1	−3	−9	−16					
30	29	27	26	24	23	21	19	18	16	14	12	10	8	5	1	−2	−8	−15				
32	31	29	28	27	25	24	22	21	19	17	15	13	11	8	5	2	−2	−7	−14			
34	33	31	30	29	27	26	24	23	21	20	18	16	14	12	9	6	3	−1	−5	−12	−29	
36	35	33	32	31	29	28	27	25	24	22	20	19	17	15	13	10	7	4	0	−4	−10	
38	37	35	34	33	32	30	29	28	26	25	23	21	19	17	15	13	11	8	5	1	−3	−9
40	39	37	36	35	34	32	31	30	28	27	25	24	22	20	18	16	14	12	9	6	2	−2

*See footnote to Table 13.2.

tion nuclei serve as surfaces for the condensation of water vapor. The importance of these nuclei should be noted, since in their absence a relative humidity of nearly 400 percent is needed to produce clouds. Once condensation occurs under such supersaturated conditions, cloud droplets can grow rapidly, producing downpours of unimaginable magnitude. Fortunately, condensation nuclei such as microscopic dust, smoke, and salt particles are profuse in the lower atmosphere. Because of this abundance of particles, relative humidity rarely exceeds 101 percent. Some particles, like salt from the ocean, are particularly good nuclei because they absorb water. These particles are termed **hygroscopic** ("water-seeking") **nuclei.**

When condensation takes place, the initial growth rate of cloud droplets is rapid but diminishes quickly because the excess water vapor is readily consumed by the numerous competing particles. This results in the formation of a cloud consisting of millions upon millions of tiny water droplets, all so fine that they remain suspended in air. The slow growth of these cloud droplets by additional condensation and the immense size difference between cloud droplets and raindrops suggest that condensation alone is not responsible for the formation of drops large enough to fall as rain.

Condensation Aloft: Adiabatic Temperature Changes

During cloud formation, and often in the formation of fog, the air is cooled to its dew point. Near the earth's surface, heat is readily exchanged between the ground and the air above. This accounts for the cooling involved in the formation of some types of fog. However, because air is a poor conductor of heat, this exchange is virtually nonexistent above a few thousand meters. Thus some other mechanism must operate during cloud formation. This process is easily understood if you have ever pumped up a bicycle tire and noticed that the pump barrel became quite warm. The heat you felt was the consequence of the work you did on the air to compress it. When energy is used to compress air, an equivalent amount of energy is released as heat. Conversely, air that is allowed to escape from a bicycle tire cools as it expands. This results because the expanding air pushes (does work on) the surrounding air and must cool by an amount equivalent to the energy expended. The temperature changes just described, in which no heat was added or subtracted, are called **adiabatic temperature changes** and result when air is compressed or allowed to expand. In summary, when air is allowed to expand, it cools, and when it is compressed, it warms.

Anytime air moves upward, it passes through regions of successively lower pressure. As a result, the ascending air expands and cools adiabatically. Unsaturated air cools at the rather constant rate of 1° C for every 100 meters of ascent (10° C per kilometer). Conversely, descending air comes under increasingly higher pressures, compresses, and is heated 1° C for every 100 meters of descent. This rate of cooling or heating applies only to unsaturated air and is known as the **dry adiabatic rate.** If air rises long enough, it will cool sufficiently to cause condensa-

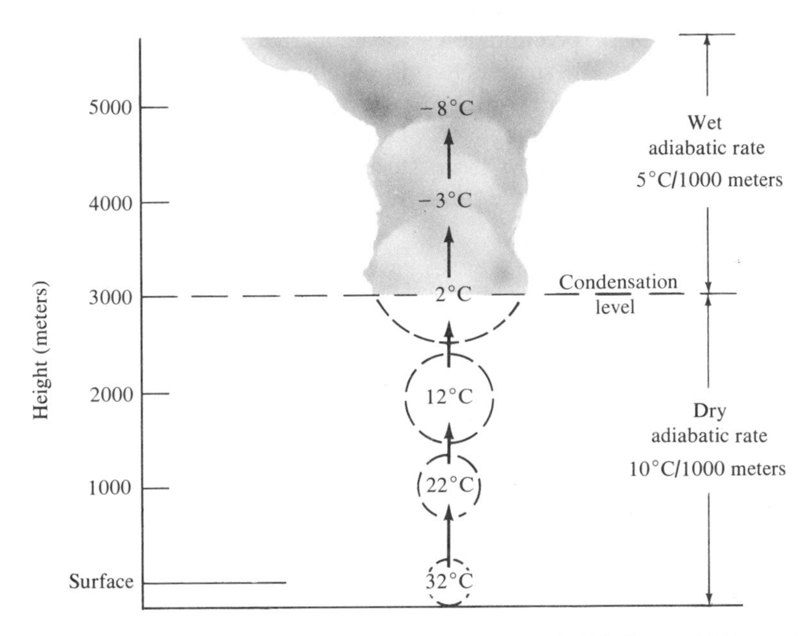

Figure 13.6 *Rising air cools at the dry adiabatic rate of 10° C per 1000 meters, until the air reaches the dew point and condensation (cloud formation) begins. As air continues to rise, the latent heat released by condensation lowers the rate of cooling. The wet adiabatic rate is therefore always less than the dry adiabatic rate.*

tion. From this point on along its ascent, latent heat stored in the water vapor will be liberated. Although the air will continue to cool after condensation begins, the released latent heat works against the adiabatic process, thereby reducing the rate at which the air cools. This slower rate of cooling caused by the addition of latent heat is called the **wet adiabatic rate** of cooling. Since the amount of latent heat released depends upon the quantity of moisture present in the air, the wet adiabatic rate varies from 0.5° C per 100 meters for air with a high moisture content to 0.9° C per 100 meters for dry air. Figure 13.6 illustrates the role of adiabatic cooling in the formation of clouds. Note that from the surface up to the condensation level the air cools at the dry adiabatic rate. The wet adiabatic rate commences at the point of condensation.

Stability

As we have learned, if air rises, it will cool and eventually produce clouds. Why does air rise on some occasions, but not on others? Why do the size of clouds and the amount of precipitation vary so much when air does rise? The answers to these questions are closely related to the stability of the air. Imagine, if you will, a large bubble of air with a thin flexible cover which allows it to expand but prevents it from mixing with the surrounding air. If the imaginary bubble were forced to rise, its temperature would decrease because of expansion. By comparing the bubble's temperature to that of the surrounding air, we can determine the stability of the bubble. If the bubble's temperature is lower than that of its environment, it will be heavier, and if allowed to move freely, it would sink to its original position. Air of this type, termed *stable air,* resists vertical displacement. On the other hand, if our imaginary bubble were warmer, and therefore lighter, than the surrounding air, it would continue to rise until it reached an altitude having the same temperature, much as a hot-air balloon rises as long as it is lighter than the surrounding air. This type of air is called *unstable air.*

Determination of Stability

In an actual situation, the stability of the air is determined by examining the temperature of the atmosphere at various heights. As you recall, this measure is termed the *lapse rate.* It is important not to confuse the lapse rate, which is the temperature of the atmosphere as determined from observations made by balloons and airplanes, with adiabatic temperature changes. The latter measure indicates the change in temperature that a parcel of air will experience as it moves vertically through the atmosphere.

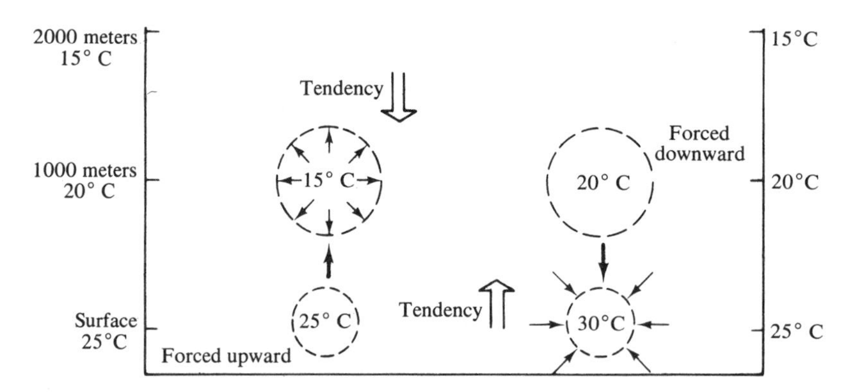

Figure 13.7 *Schematic representation of a stable atmosphere. Using the parcel of air located on the left side of this drawing, the air near the surface is potentially cooler than the air aloft and therefore resists upward motion.*

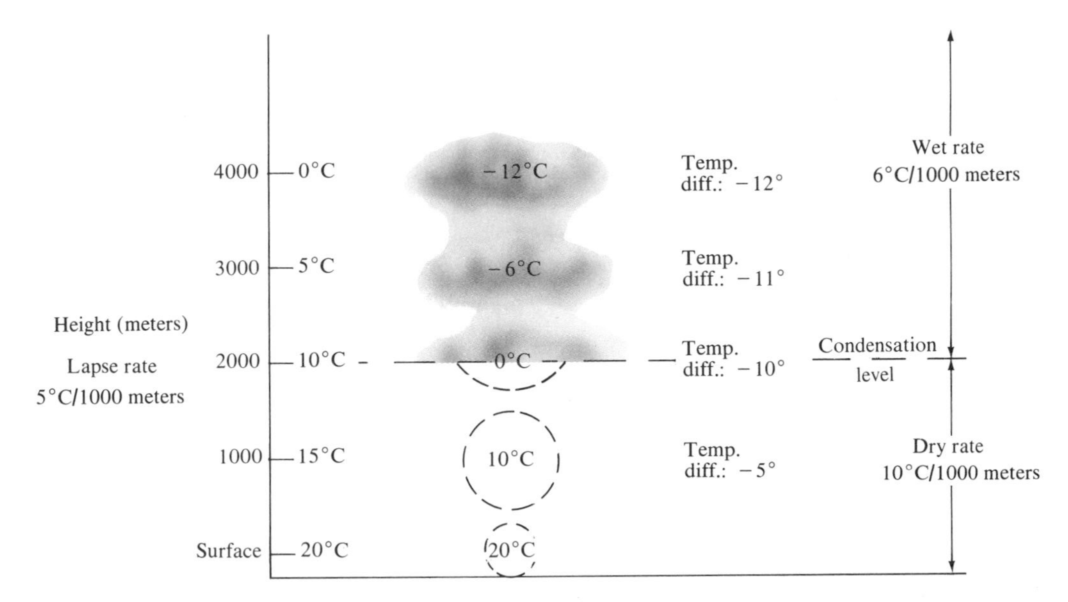

Figure 13.8 *Absolute stability prevails when the lapse rate is less than the wet adiabatic rate. The rising parcel of air is therefore always cooler and heavier than the surrounding air.*

For illustration, we will examine a situation where the prevailing lapse rate is 5° C per 1000 meters (Figure 13.7). Under this condition, when the air at the surface has a temperature of 25° C, the air at 1000 meters will be 5 degrees cooler, or 20° C, the air at 2000 meters will have a temperature of 15° C, and so forth. At first glance it appears that the air at the surface is lighter than the air at 1000 meters since it is 5 degrees warmer. However, if the air near the surface is unsaturated and were to rise to 1000 meters, it would expand and cool at the dry adiabatic rate of 1° C per 100 meters. Therefore, upon reaching 1000 meters, its temperature would have dropped a total of 10° C to 15° C. Being 5 degrees cooler than its environment, it would be heavier and tend to sink to its original position. Hence, we say that the air near the surface is potentially cooler than the air aloft and therefore will not rise. Using similar reasoning, if the air at 1000 meters subsided, adiabatic heating would increase its temperature 10 degrees by the time it reached the surface, making it warmer than the surrounding air, so its buoyancy would cause it to return. The air just described is stable and resists vertical movement.

Stated quantitatively, **absolute stability** prevails when the lapse rate is less than the wet adiabatic rate. Figure 13.8 depicts this situation using a lapse rate of 2° C per 1000 meters and a wet adiabatic rate of 3° C per 1000 meters. Note that at 1000 meters the temperature of the surrounding air is 15°C, while the rising parcel of air has cooled to 10° C and is therefore the heavier air. Even if this stable air were to be forced above the condensation level, it would remain cooler and heavier than its environment and would have a tendency to return to the surface.

At the other extreme, air is said to exhibit **absolute instability** when the lapse rate is greater than the dry adiabatic rate. As shown in Figure 13.9, the ascending parcel of air is always warmer than its environment and will continue to rise because of its own buoyancy.

Another situation in the atmosphere is called **conditional instability.** This occurs when moist air has a lapse rate between the dry and wet adia-

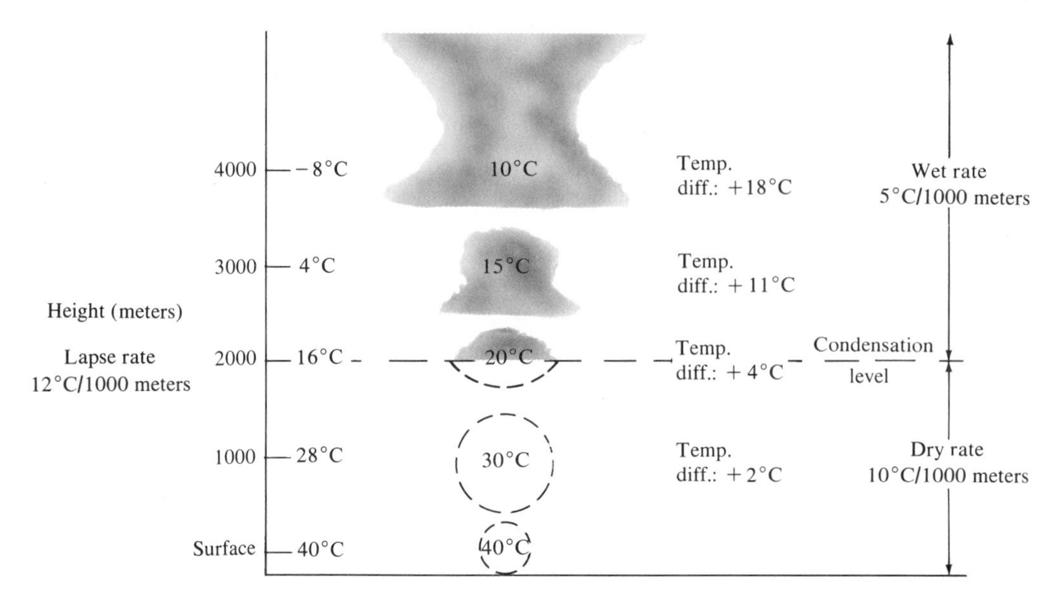

Figure 13.9 *Absolute instability illustrated using a lapse rate of 12° C per 1000 meters. The rising air is always warmer and therefore lighter than the surrounding air.*

batic rates (between 0.5° C and 1° C per 100 meters). Referring to Figure 13.10, notice that for the first 4000 meters the rising parcel of air is cooler than the surrounding air and is therefore considered stable. However, with the addition of latent heat above the condensation level, the parcel eventually becomes warmer than the surrounding air. From this point on along its ascent, the parcel will continue to rise without an outside force and is considered unstable. Conditionally unstable air can be described as air that begins its ascent as stable air but at some point above the condensation level becomes unstable. The word *conditional* is used because only if the air is forced upward initially can it become unstable. Conditional instability is perhaps the most common type of instability.

Stability and Daily Weather

From the previous discussion we can conclude that stable air resists vertical movement, whereas unstable air ascends freely because of its own buoyancy. But how do these facts manifest themselves in our daily weather? Since stable air resists upward movement, we might conclude that clouds will not form when stable conditions prevail in the atmosphere. Although this seems reasonable, processes do exist which force air aloft. These will be discussed in the following section. *On occasions when stable air is forced aloft, the clouds that form are widespread and have little vertical thickness when compared to their horizontal dimension, and precipitation, if any, is light.* By contrast, *clouds associated with unstable air are towering and are usually accompanied by heavy precipitation.* Hence we can conclude that on a dreary, overcast day with light drizzle, stable air is forced aloft. On the other hand, during a day when cauliflower-shaped clouds appear to be growing as if bubbles of hot air are surging upward, we can be relatively certain that the ascending air is unstable.

In summary, the role of stability in determining our daily weather is very important. It determines to a large degree whether or not clouds develop and produce precipitation, and whether that precipitation will come as a gentle shower or a violent downpour.

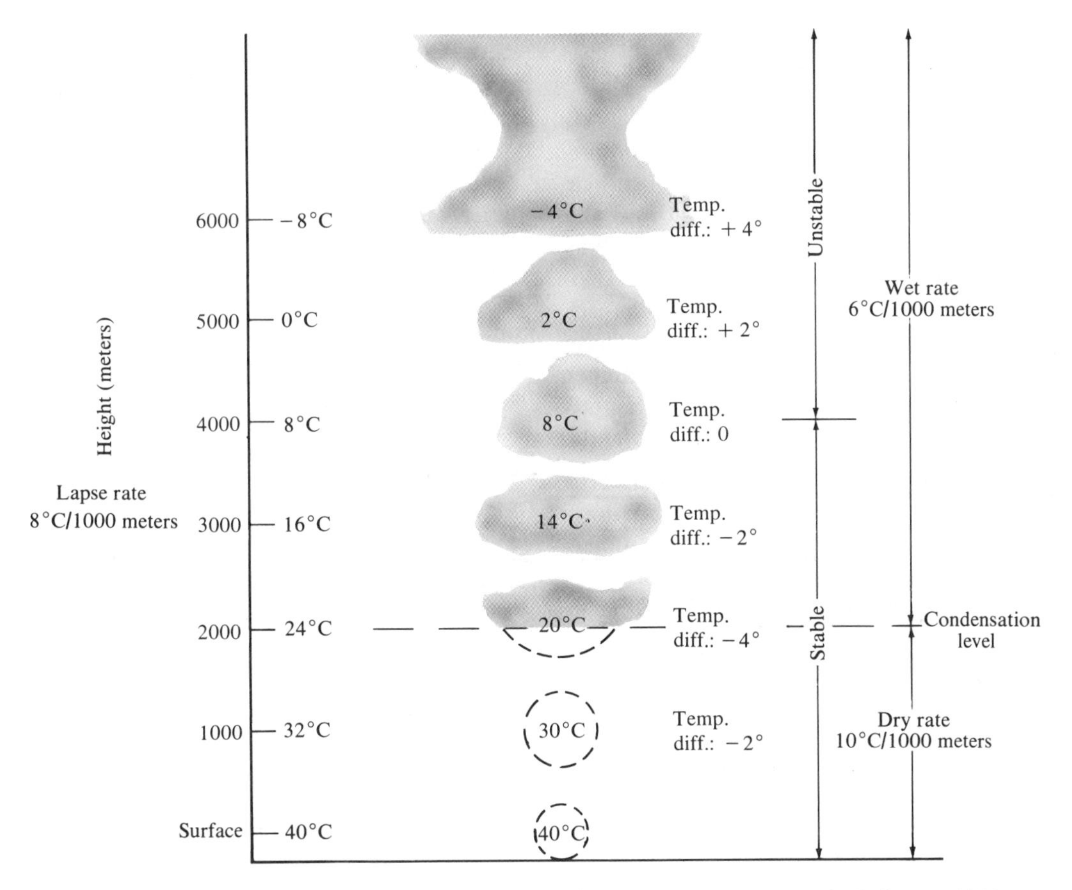

Figure 13.10 *Conditional instability illustrated using a lapse rate of 8° C per 1000 meters, which lies between the dry adiabatic rate and the wet adiabatic rate. The rising parcel of air is cooler than the surrounding air below 4000 meters and warmer above 4000 meters.*

Forceful Lifting

Earlier we demonstrated that stable and conditionally unstable air will not rise on their own; they require some mechanism to trigger the vertical movement. Three such mechanisms are convergence, orographic lifting, and frontal wedging.

Whenever air converges (flows together), it results in general upward movement. This occurs because as air converges it occupies a smaller and smaller area, which requires that the height of the air column increase. Consequently, the air within the column must move upward, enhancing instability. The Florida peninsula provides an excellent example of the role convergence plays in initiating instability. On warm days the air flow is off the ocean along both coasts of Florida, causing general convergence over the peninsula. This convergence and associated uplift, aided by the intense solar heating, causes more mid-afternoon thunderstorms in this part of the United States than in any other area.

Orographic lifting occurs when sloping terrain such as mountains act as barriers to the flow of air, forcing the air to ascend. Many of the rainiest places in the world are located on windward mountain slopes. A station at Mt. Waialeale, Hawaii, for example, records the highest average annual rainfall in the world, some 1168 centimeters. The station is located on the windward (northeast) coast of the island of Kauai, at an elevation of 1523 meters.

Besides providing the lift to render air unstable, mountains further remove more than their share of moisture in other ways. By slowing the horizontal flow of air, they cause convergence as well as retard the passage of storm systems. Also, the irregular topography of mountains enhances differential heating and surface instability. These combined effects account for the generally higher precipitation we associate with mountainous regions as compared to the surrounding lowlands.

By the time air reaches the leeward side of a mountain, much of the moisture has been lost, and if the air descends, it warms, making condensation and precipitation even less likely. The result often is a **rainshadow desert.** The Great Basin Desert of the western United States lies only a few hundred kilometers from the Pacific Ocean but is effectively cut off by the imposing Sierra Nevada. The Gobi Desert of Mongolia, the Takla Makan of China, and the Patagonia Desert of Argentina are other examples of deserts found on the leeward sides of mountains.

Frontal wedging occurs when cool air acts as a barrier over which warmer, lighter air rises. This phenomenon is quite common throughout the continental United States and is responsible for the bulk of the precipitation in many areas, as we shall see later. Figure 13.11 illustrates frontal wedging of stable and unstable air. As this figure shows, forceful lifting is important in producing clouds. However, the stability of the air determines to a great extent the type of clouds formed and the amount of precipitation that may be expected.

Clouds

Clouds are a form of condensation best described as visible aggregates of minute droplets of water or of tiny crystals of ice. In addition to being prominent and sometimes spectacular features in the sky, clouds are of continual interest to meteorologists, because they provide a visible indication of what is going on in the atmosphere. Anyone who observes clouds with the hope of recognizing different types often finds that there is a bewildering variety of these familiar white and gray masses streaming across the sky. How-

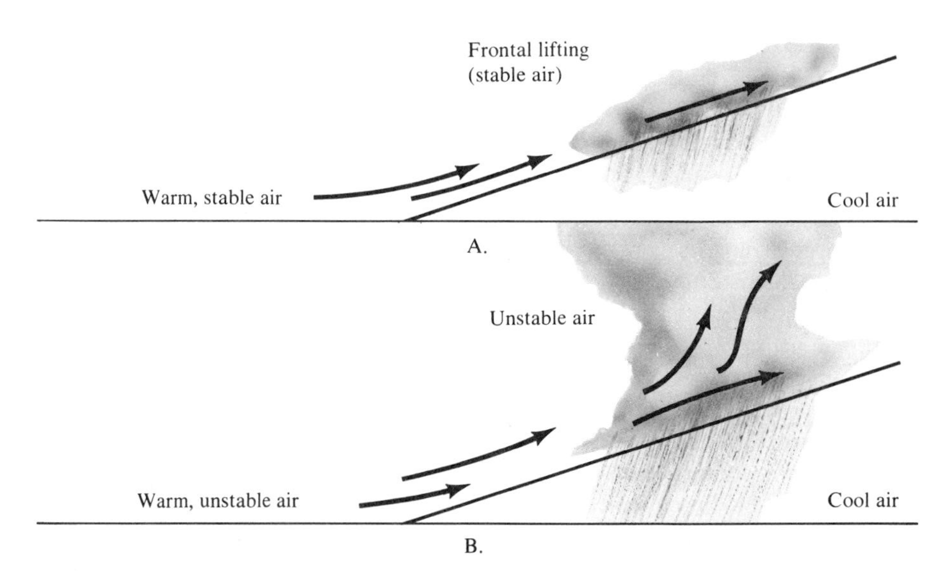

Figure 13.11 **A.** *When stable air is lifted, layered clouds usually result.* **B.** *When warm, unstable air is forced to rise over cooler air, "towering" clouds develop.*

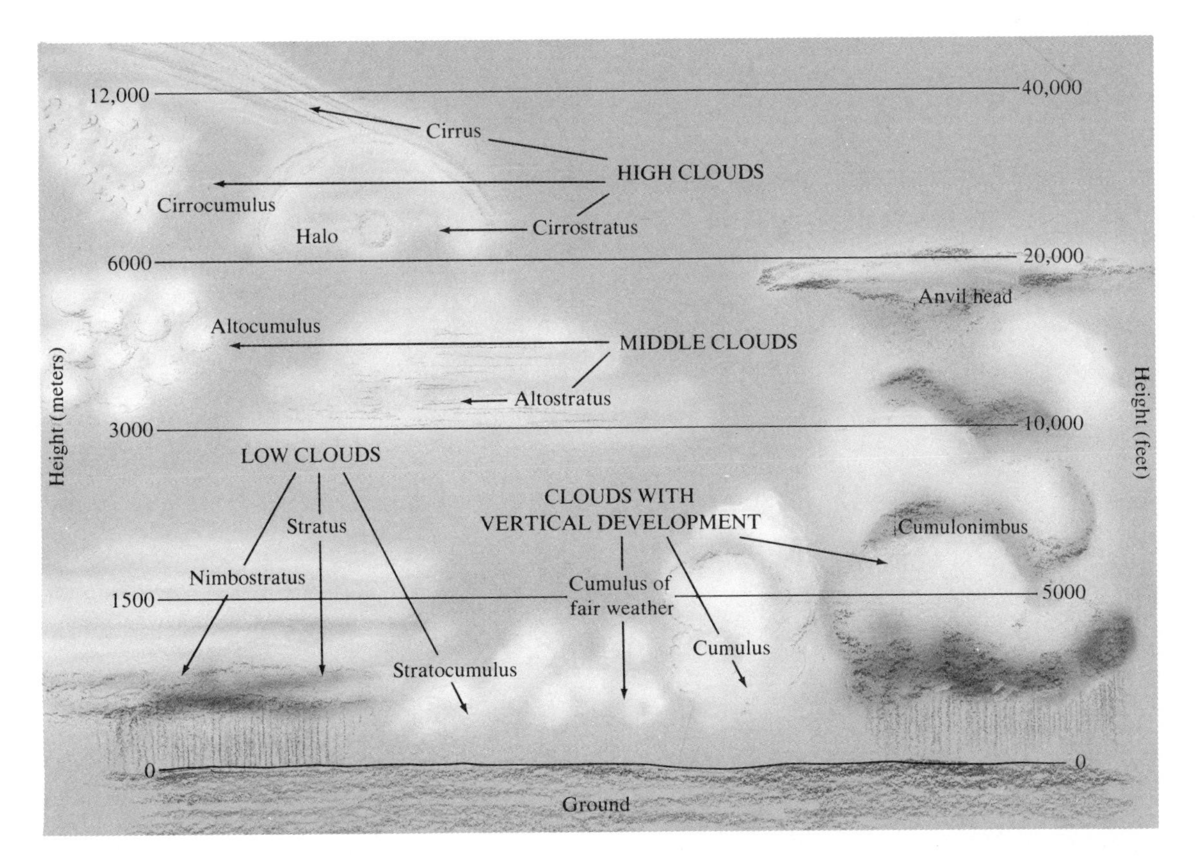

Figure 13.12 *Classification of clouds according to height and form. (After Ward's Natural Science Establishment, Inc., Rochester, N.Y.)*

ever, once one comes to know the basic classification scheme for clouds, most of the confusion vanishes.

Clouds are classified on the basis of their appearance and height (Figure 13.12). Three basic forms are recognized: cirrus, cumulus, and stratus. **Cirrus** clouds are high, white, and thin. They are separated or detached and form delicate veil-like patches or extended wispy fibers and often have a feathery appearance. The **cumulus** form consists of globular individual cloud masses. Normally they exhibit a flat base and have the appearance of rising domes or towers. Such clouds are frequently described as having a cauliflower-like structure. **Stratus** clouds are best described as sheets or layers that cover much or all of the sky. While there may be minor breaks, there are no distinct individual cloud units. All other clouds reflect one of these three basic forms or are combinations or modifications of them.

Three levels of cloud heights are recognized: high, middle, and low. **High clouds** normally have bases above 6000 meters; **middle clouds** generally occupy heights from 2000 to 6000 meters; and **low clouds** form below 2000 meters. The altitudes listed for each height category are not hard and fast. There is some seasonal as well as latitudinal variation. For example, at high latitudes or during cold winter months in the mid-latitudes, high clouds are often found at lower altitudes.

Because of the low temperatures and small quantities of water vapor found at high altitudes, all of the high clouds are thin, white, and com-

Figure 13.13 **A.** *Cirrus.* **B.** *Cirrocumulus.* **C.** *Cirrostratus with halo.* **D.** *Altocumulus.* **E.** *Altostratus.* **F.** *Stratocumulus.* **G.** *Stratus.* **H.** *Nimbostratus, one of the two major precipitation-producing clouds.* **I.** *Cumulus.* **J.** *Cumulonimbus, one of the two major precipitation-producing clouds. (A.–J. courtesy of NOAA)*

G.

H.

I.

J.

posed of ice crystals. Since much more water vapor is available at lower altitudes, middle and low clouds are thicker and denser.

Layered clouds in any of these height ranges generally indicate that the air is stable. Normally we might not expect clouds to grow or persist in stable air. However, cloud growth of this type is common when air is forced to rise, as along a front or near the center of a cyclone, where converging winds cause air to ascend. Such forced ascent of stable air leads to the formation of a stratified cloud layer that is large horizontally when compared to its depth.

Some clouds do not fit into any one of the three height categories mentioned. Such clouds have their bases in the low height range but often extend upward into the middle or high altitudes. Consequently, clouds in this category are called **clouds of vertical development.** *These clouds are all related to one another and are associated with unstable air.* While cumulus clouds are often connected with "fair" weather, they may, under the proper circumstances, grow dramatically. Once upward movement is triggered, acceleration is powerful and clouds with great vertical extent form. The end result is often a towering cloud that may produce rainshowers or a thunderstorm.

Since definite weather patterns can often be associated with particular clouds or certain combinations of cloud types, it is important to become familiar with cloud descriptions and characteristics. Table 13.4 lists the ten basic cloud types that are recognized internationally and gives some characteristics of each. The series of photos in Figure 13.13 depicts common forms of each cloud type.

Table 13.4 *Cloud Types and Characteristics*

Cloud Family and Height	Cloud Type	Characteristics
High clouds—above 6000 meters (20,000 feet)	Cirrus	Thin, delicate, fibrous ice-crystal clouds. Sometimes appear as hooked filaments called "mares' tails." (Figure 13.13A)
	Cirrocumulus	Thin, white ice-crystal clouds, in the form of ripples, waves, or globular masses all in a row. May produce a "mackerel sky." Least common of the high clouds. (Figure 13.13B)
	Cirrostratus	Thin sheet of white ice-crystal clouds that may give the sky a milky look. Sometimes produce halos around the sun or moon. (Figure 13.13C)
Middle clouds—2000–6000 meters (6500–20,000 feet)	Altocumulus	White to gray clouds often composed of separate globules; "sheep-back" clouds. (Figure 13.13D)
	Altostratus	Stratified veil of clouds that are generally thin and may produce very light precipitation. When thin, the sun or moon may be visible as a "bright spot," but no halos are produced. (Figure 13.13E)
Low clouds—below 2000 meters (6500 feet)	Stratocumulus	Soft, gray clouds in globular patches or rolls. Rolls may join together to make a continuous cloud. (Figure 13.13F)
	Stratus	Low uniform layer resembling fog but not resting on the ground. May produce drizzle. (Figure 13.13G)
	Nimbostratus	Amorphous layer of dark gray clouds. One of the chief precipitation-producing clouds. (Figure 13.13H)
Clouds of vertical development—500–1800 meters (1600–60,000 feet)	Cumulus	Dense, billowy clouds often characterized by flat bases. May occur as isolated clouds or closely packed. (Figure 13.13I)
	Cumulonimbus	Towering cloud sometimes spreading out on top to form an "anvil head." Associated with heavy rainfall, thunder, lightning, hail, and tornadoes. (Figure 13.13J)

Formation of Precipitation

Although all clouds contain water, why do some produce precipitation while others drift placidly overhead? This seemly simple question perplexed meteorologists for many years. First, cloud droplets are very small, averaging less than 10 micrometers in diameter (for comparison, a human hair is about 75 micrometers in diameter). Because of their small size, cloud droplets fall incredibly slowly. Theoretically, an average cloud droplet falling from a cloud base at 1000 meters would require about 48 hours to reach the ground. Of course, it would never complete its journey. Even falling through humid air, a cloud droplet would evaporate before it fell a few meters below the cloud base. In addition, clouds are made up of many thousands of millions of these droplets, all competing for the available water vapor; thus, their continued growth via condensation is very slow.

A raindrop large enough to reach the ground without evaporating contains roughly a million times the water of a cloud droplet. Therefore, for precipitation to form, millions of cloud droplets must somehow coalesce (join together) into drops large enough to sustain themselves during their decent. Two mechanisms have been proposed to explain this phenomenon: the Bergeron process and the collision-coalescence process.

The **Bergeron process,** named after its discoverer, relies on two interesting properties of water. First, cloud droplets do not freeze at 0° C as expected. In fact, pure water suspended in air does not freeze until it reaches a temperature of nearly −40° C. Water in the liquid state below 0° C is generally referred to as **supercooled.** Supercooled water will readily freeze if sufficiently agitated. This explains why airplanes collect ice when they pass through a liquid cloud composed of supercooled droplets. In addition, supercooled droplets will freeze upon contact with solid particles that have a crystal form closely resembling that of ice. These materials are termed **freezing nuclei.** The need for freezing nuclei to initiate the freezing process is similar to the requirement for condensation nuclei in the process of condensation. However, in contrast to condensation nuclei, freezing nuclei are very sparse in the atmosphere and do not generally become active until the temperature reaches −10° C or less. Only at temperatures well below freezing will ice crystals begin to form in clouds, and even at that, they will be few and far between. Once ice crystals form, they are in direct competition with the supercooled droplets for the available water vapor.

This brings us to the second interesting property of water. When air is saturated (100 percent relative humidity) with respect to water, it is supersaturated (relative humidity greater than 100 percent) with respect to ice. Table 13.5 shows that at −10° C, when the relative humidity is 100 percent with respect to water, the relative humidity with respect to ice is nearly 110 percent. Thus, ice crystals cannot peacefully coexist with water droplets because the air always "appears" supersaturated to the ice crystals. So the ice crystals begin to consume the "excess" water vapor, which lowers the relative humidity near the surrounding droplets. In turn, the water droplets evaporate to replenish the diminishing water vapor, thereby providing a continual source of vapor for the growth of the ice crystals (Figure 13.14).

Table 13.5 ***Relative Humidity with Respect to Ice When Relative Humidity with Respect to Water Is 100 Percent***

Temperature (°C)	Relative humidity with respect to: Water (%)	Ice (%)
0	100	100
−5	100	105
−10	100	110
−15	100	116
−20	100	121

Because the level of supersaturation with respect to ice can be quite great, the growth of ice crystals is generally rapid enough to generate crystals large enough to fall. During their descent, these ice crystals enlarge as they intercept cloud drops, which freeze upon them. Air movement will sometimes break up these delicate crystals and the fragments will serve as freezing nuclei for other liquid droplets. A chain reaction develops, producing many ice crystals, which, by accretion, form into larger crystals called snowflakes. When the surface temperature is above 4° C, snowflakes usually melt before they reach the ground and continue their descent as rain. Even a summer rain may have begun as a snowstorm in the clouds overhead.

Cloud seeding to produce precipitation utilizes the Bergeron process just described. By adding freezing nuclei (commonly silver iodide) to supercooled clouds, the growth of these clouds can be markedly changed. This process is discussed in greater detail later in this chapter.

Thirty years ago, meteorologists believed that the Bergeron process was responsible for the formation of most precipitation, with the exception of light drizzle. However, it was discovered that copious rainfall is often associated with clouds located well below the freezing level (warm clouds), particularly in the tropics. This led to the proposal of a second mechanism thought to produce precipitation, the **collision-coalescence process.**

Clouds composed entirely of liquid droplets must contain droplets larger than 20 micrometers if precipitation is to form. These large droplets

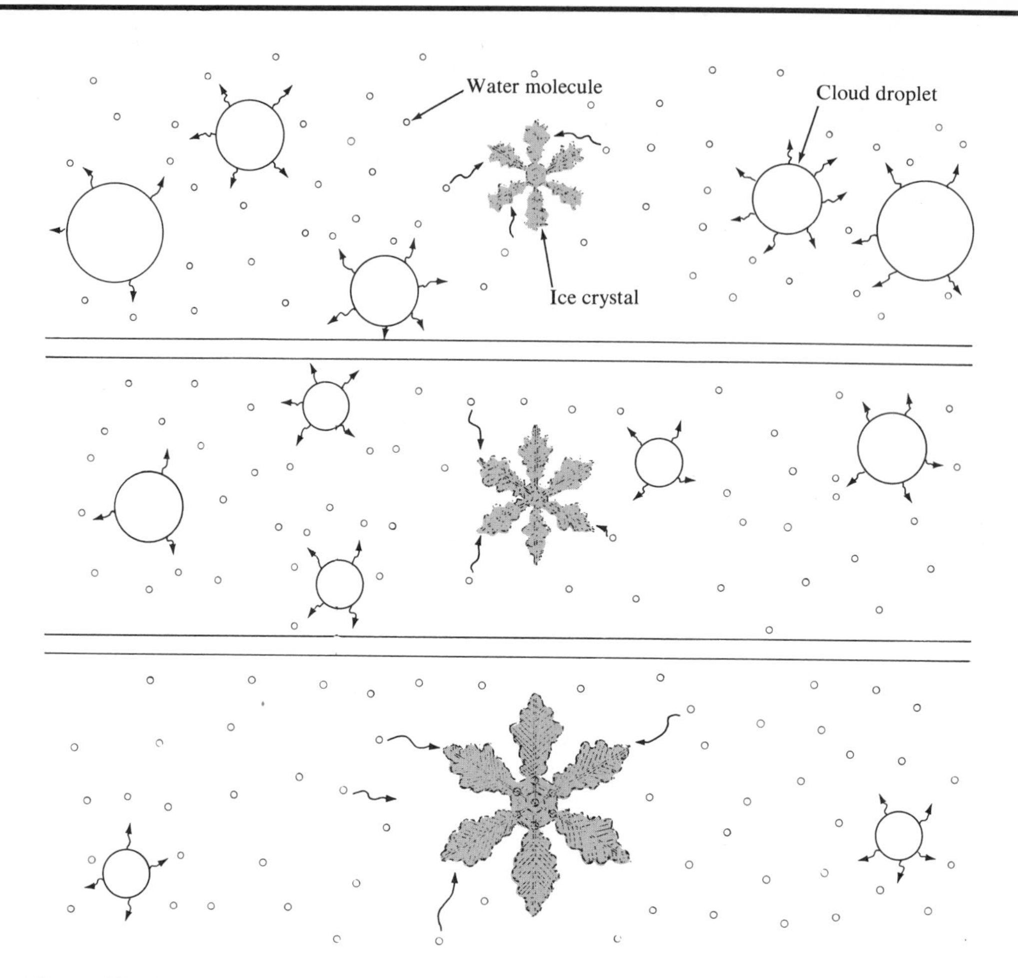

Figure 13.14 *The Bergeron process. Ice crystals grow at the expense of cloud droplets until they are large enough to fall. The size of these particles has been greatly exaggerated.*

form when "giant" condensation nuclei are present and on occasions when hygroscopic particles such as sea salt exist. Hygroscopic particles begin to remove water vapor from the air at relative humidities under 100 percent and can grow quite large. Since the rate at which drops fall is size dependent, these "giant" droplets fall most rapidly. As such, they collide with the smaller, slower droplets and coalesce. Becoming larger in the process, they fall even more rapidly (or in an updraft, rise more slowly), increasing their chances of collision and rate of growth (Figure 13.15). After a million such collisions they are large enough to fall to the surface without evaporating. Because of the number of collisions required for growth to raindrop size, droplets in clouds with great vertical thickness and abundant moisture have a better chance of reaching the required size. Updrafts also aid in this process since they allow the droplets repeatedly to traverse the cloud. Raindrops can grow to a maximum size of 5 millimeters when they fall at the rate of 30 kilometers per hour. At this size and speed the water's surface tension, which holds the drop together, is surpassed by the drag imposed by the air, which in turn succeeds in pulling the drops apart. The resulting breakup of a large raindrop produces numerous smaller drops which begin

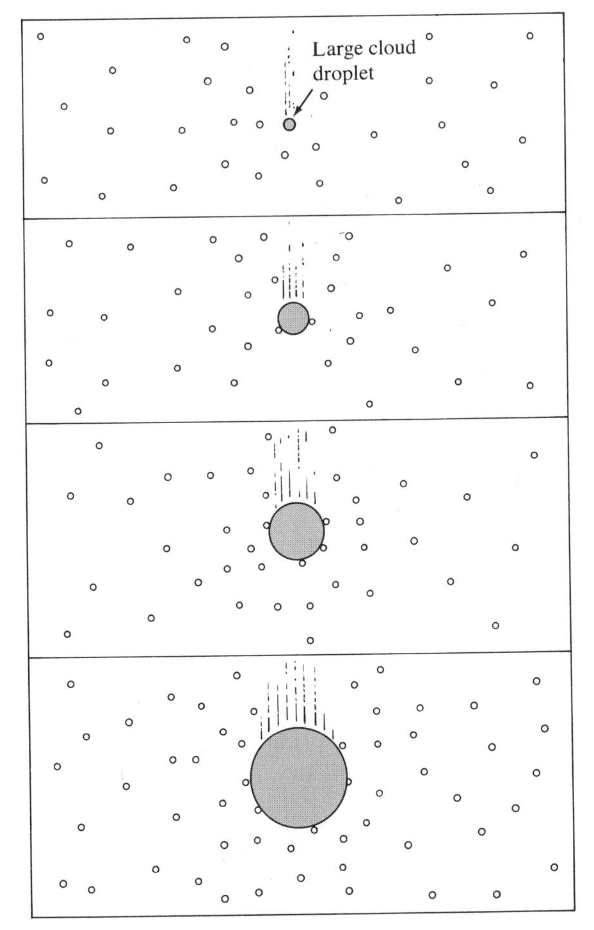

Figure 13.15 *The collision-coalescence process. Because large cloud droplets fall more rapidly than smaller droplets, they are able to sweep up the smaller ones in their path and grow. Most cloud droplets are so small that the motion of the air keeps them suspended. Even if these small cloud droplets were to fall, they would evaporate before reaching the surface.*

anew the task of sweeping up cloud droplets. Drops that are less than 0.5 millimeter upon reaching the ground are termed drizzle and require about ten minutes to fall from a cloud 1000 meters overhead.

The collision-coalescence process is not quite as simple as described, however. First, as the larger droplets descend, they produce an air stream around them similar to that produced by an automobile when it is driven rapidly down the highway. If an automobile is driven at night and we use the bugs that are often out to be analogous to the cloud droplets, it is easy to visualize how most cloud droplets are swept aside. The larger the cloud droplet (or bug), the better chance it will have of colliding with the giant droplet (or car). Second, collision does not guarantee coalescence. Experimentation has indicated that the presence of atmospheric electricity may be the key to what holds these droplets together once they collide. If a droplet with a negative charge should collide with a positively charged droplet, their electrical attraction may bind them together.

Sleet, Glaze, and Hail

While rain and snow are the most common and familiar forms of precipitation, other forms do exist and are worthy of mention. Sleet, glaze, and hail fall into this category. Although limited in occurrence and sporadic in both time and space, these forms, especially the latter two, may on occasion cause considerable damage.

Sleet is a wintertime phenomenon and refers to the fall of small, clear-to-translucent particles of ice. For sleet to be produced, a layer of air with temperatures above freezing must overlie a subfreezing layer near the ground. When the raindrops leave the warmer air and encounter the colder air below, they solidify, reaching the ground as small pellets of ice no larger than the raindrops from which they formed.

On some occasions, when the vertical distribution of temperatures is similar to that associated with the formation of sleet, freezing rain or **glaze** results instead. In such situations, the subfreezing air near the ground is not thick enough to allow the raindrops to freeze. The raindrops, however, do become supercooled as they fall through the cold air and consequently turn to ice upon colliding with solid objects. The result can be a thick coating of ice that is sufficiently heavy to break tree limbs and down power lines as well as make

Figure 13.16 *Glaze. (Courtesy of NOAA)*

walking or motoring extremely hazardous (Figure 13.16).

Hail is precipitation in the form of hard rounded pellets or irregular lumps of ice. Furthermore, large hailstones often consist of a series of nearly concentric shells of differing densities and degrees of opaqueness (Figure 13.17). Most frequently, hailstones have a diameter of about 1 centimeter, but they may vary in size from 5 millimeters to more than 10 centimeters in diameter. The largest hailstone on record fell on Coffeyville, Kansas, September 3, 1970. With a 14-centimeter diameter and a circumference of 44 centimeters, this "giant" weighed 766 grams! The destructive effects of heavy hail are well known, especially to farmers whose crops can be devastated in a few short minutes and to persons whose windows are shattered.

Hail is produced only in cumulonimbus clouds where updrafts are strong and where there is an abundant supply of supercooled water. First, rain is lifted above the freezing level by the rapidly ascending air. Once frozen, these small ice granules grow by collecting supercooled cloud droplets as they fall through the cloud. If they encounter another strong updraft, they may be carried upward again and begin the downward journey anew. Each trip above the freezing level may be represented by an additional layer of ice.

Hailstones, however, may also form from a single descent through an updraft. In this situation, the layered structure is attributed to varia-

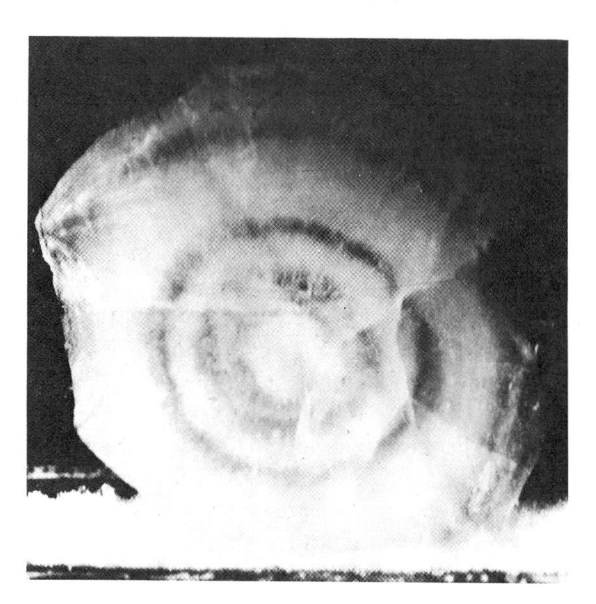

Figure 13.17 *Section through a hailstone. Note the concentric layers of ice. (Courtesy of NOAA)*

tions in the rate at which supercooled droplets accumulate and freeze, which in turn, is related to differences in the amount of supercooled water in different parts of the cumulonimbus tower.

In either case, hailstones grow by the addition of supercooled water upon growing ice pellets. The ultimate size of the hailstone depends primarily upon three factors: the strength of the updrafts, the concentration of supercooled water, and the length of the path through the cloud.

Fog

Fog is generally considered to be an atmospheric hazard. When it is light, visibility is reduced to 2 or 3 kilometers. However, when it is dense, visibility may be cut to a few tens of meters or less, making travel by any mode not only difficult but often dangerous as well. Officially, visibility must be reduced to 1 kilometer or less before fog is reported. While this figure is arbitrary, it does permit a more objective criterion for comparing fog frequencies at different locations.

Fog is defined as a cloud with its base at or very near the ground. Physically, there is basically no difference between a fog and a cloud; the appearance and structure of both are the same. The essential difference is the method and place of formation. While clouds result when air rises and cools adiabatically, fogs (with the exception of upslope fogs) are the consequence of radiation cooling or the movement of air over a cold surface. In other circumstances, fogs are formed when enough water vapor is added to the air to bring about saturation (evaporation fogs).

Fogs Caused by Cooling

When warm, moist air is blown over a cool surface, the result may be a blanket of fog called **advection fog.** Examples of such fogs are very common. The foggiest location in the United States, and perhaps in the world, is Cape Disappointment, Washington. The name is indeed appropriate, since this station averages 2552 hours of fog each year. The fog experienced at Cape Disappointment, as well as that at other West Coast locations, is produced when warm, moist air from the Pacific Ocean moves over the cold California Current and is then carried on shore by the prevailing winds. Advection fogs are also quite common in the winter season when warm air from the Gulf of Mexico is blown over cold, often snow-covered surfaces of the Midwest and East.

Radiation fog forms on cool, clear, calm nights, when the earth's surface cools rapidly by radiation. As the night progresses, a thin layer of air in contact with the ground is cooled below its dew point. As the air cools and becomes heavier, it drains into low areas, resulting in "pockets" of

Figure 13.18 *Patches of upslope fog forming on the sides of a mountain valley. (Courtesy of Ward's Natural Science Establishment, Inc., Rochester, N.Y.)*

fog. The largest "pockets" are often river valleys, where rather thick accumulations may occur.

As its name implies, **upslope fog** is created when relatively humid air moves up a gradually sloping plain or, in some cases, up the steep slopes of a mountain (Figure 13.18). Because of the upward movement, air expands and cools adiabatically. If the dew point is reached, an extensive layer of fog may form.

In the United States, the Great Plains offers an excellent example. When humid easterly or southeasterly winds move westward from the Mississippi River toward the Rocky Mountains, the air gradually rises, resulting in an adiabatic decrease of about 13° C. When the difference between the air temperature and dew point of westward-moving air is less than 13° C, an extensive fog often results in the western plains.

Evaporation Fogs

When cool air moves over warm water, enough moisture may evaporate from the water surface to produce saturation. As the rising water vapor meets the cold air, it immediately recondenses and rises with the air that is being warmed from below. Since the water has a steaming appearance, the phenomenon is called **steam fog** (Figure 13.19). Steam fog is fairly common over lakes and rivers in the fall and early winter, when the water may still be relatively warm, and the air is rather crisp. Steam fog is often quite shallow because as the steam rises it reevaporates in the unsaturated air above.

When frontal wedging occurs, warm air is lifted over colder air. If the resulting clouds yield rain, and the cold air below is near the dew point, enough rain will evaporate to produce fog. A fog formed in this manner is called **frontal,** or **precipitation, fog.** The result is a more or less continuous zone of condensed water droplets reaching from the ground up through the clouds.

In summary, both steam fog and frontal fog result from the addition of moisture to a layer of air. As we learned, the air is usually cool or cold and already near saturation. Since the capacity of air to hold water vapor at low temperatures is small, only a relatively modest amount of evaporation is necessary to produce saturated conditions and fog.

Intentional Weather Modification

The old adage "Everybody talks about the weather, but nobody does anything. . . ." seems as true today as it ever did. Attempts to alter the weather have been made, but the fruits of these labors have been meager. People have probably always had a desire to change the weather. The Hopi Indians of the American Southwest performed rain dances, and European inventors devised all sorts of gadgetry, including cannons that were fired to suppress the development of

Figure 13.19 *Early-morning steam fog on a small lake. (Courtesy of Ward's Natural Science Establishment, Inc., Rochester, N.Y.)*

The chimney (center) of a mature thunderstorm, capped by an anvil in which air currents have spread horizontally. (Courtesy of Ron Holle)

The anvil can extend hundreds of miles from the core of the cumulonimbus. (Courtesy of NASA)

Fog, formed in cool air, often drains into valleys. (Courtesy of Richard Anthes)

hailstones. The success of these early attempts is questionable, at best.

Cloud Seeding

The first scientific breakthrough in weather modification came in 1946 when V.J. Schaeffer discovered that dry ice dropped into a supercooled cloud spurred the growth of ice crystals. This discovery sparked the practice of **cloud seeding** for the purpose of rainmaking. Recall that once ice crystals form, they grow larger at the expense of the remaining liquid cloud droplets and, upon reaching a sufficient size, fall as precipitation. Shortly after Schaeffer's discovery, it was learned that silver iodide could also be used for cloud seeding. Unlike dry ice, silver iodide crystals act as freezing nuclei rather than as a cooling agent. This substance has an advantage over dry ice, in that it can be supplied to the clouds from burners located on the ground. However, for either material to be successful, certain atmospheric conditions must exist. Clouds must be present, as seeding can not generate them. In addition, at least the top portion of the cloud must be supercooled, that is, made of liquid droplets having a temperature below freezing. The object, then, is to produce just the correct number of ice crystals. Overseeding will simply produce millions upon millions of minute ice crystals, none large enough to fall.

Attempts have also been made to trigger rainfall from warm clouds. The precipitation-producing mechanism in warm clouds is the collision-coalescence process. In seeding these clouds, the object is to produce large droplets that will collide with the smaller cloud droplets and grow by coalescence. Large hygroscopic particles like salt, and even water itself, have been used for this purpose. These "giant" condensation nuclei are introduced at the base of the cloud, where updrafts carry them into the cloud. The results obtained using this method are open to question at this time.

Numerous studies have been conducted to determine the effectiveness of silver iodide in cloud seeding. In a test conducted in Missouri, it was actually determined that unseeded clouds produced more precipitation than seeded ones. But in other tests, a 10–20 percent increase in precipitation could be attributed to cloud seeding. It seems that more knowledge in the field of cloud physics is needed before cloud seeding can become as successful as first thought possible. Nevertheless, it is certain that seeding of supercooled clouds does alter their "normal" development. Because of this, cloud seeding has many other applications, including the dispersal of

Figure 13.20 *Results of dry-ice seeding at Elmendorf AFB, Alaska, on January 24, 1969. One hour and 20 minutes after seeding, visibility is greater than 1.5 kilometers. (Courtesy of NOAA)*

clouds and fogs, the suppression of hail and lightning, and hurricane modification. Cloud seeding's role in hurricane modification will be discussed in a later chapter.

Fog Dispersal. The most successful use of cloud seeding has been in the dispersal of supercooled fog at airports. An application of dry ice triggers the formation of ice crystals which fall out, leaving a "hole" in the fog and provide a clear runway for takeoffs and landings (Figure 13.20). Other cold substances such as liquid propane have been used with similar success and are somewhat easier to apply.

Unfortunately, most fog is of the warm type which is harder to combat. Warm fog can be dispersed by mechanical mixing of the fog with drier, warmer air from above, or by heating the air. When the layer of fog is very shallow, helicopters are sometimes used. By flying just above the fog, the helicopter creates a strong downdraft which pulls dry air down and mixes it with the saturated air near the ground. If the air aloft is dry enough, sufficient evaporation will take place to eliminate the foggy condition. At many airports where fog is a common hazard, it has become more common to heat and hence evaporate the fog. At Orly Airport, Paris, a sopisticated thermal fog dissipation system, called Turboclair, was installed in 1970. This system consists of eight jet engines located in underground chambers alongside the upwind edge of the runway. Although rather expensive to install, this system is capable of improving visibility for a distance of about 900 meters along the approach and touchdown zones.

Cloud Dispersal. Cloud seeding used for cloud dispersal may be of far greater importance in the future. Because of the energy crisis, the use of solar radiation for such things as home heating is becoming a more viable option. One of the major obstacles to the use of solar heat is the layer of supercooled stratus clouds often present in the winter season. This type of supercooled cloud can be dissipated by cloud seeding. The increase in the amount of winter sunshine provided by extensive seeding programs would not only increase the effectiveness of solar heating, but it might improve our psychological outlook as well.

Hail Suppression. Hail damage to crops is a concern of farmers worldwide. In France, a cloud-seeding project used silver iodide burners to combat hail damage with encouraging results. The object here is to prevent the accumulation of large amounts of supercooled water by providing freezing nuclei, which will convert the water to ice. Without ample supplies of supercooled water, hailstones cannot grow. Russian scientists have claimed great success with this method, using rockets and artillery shells to carry the freezing nuclei to the clouds.

Lightning Suppression. Because lightning is a primary cause of forest fires, the U.S. Forest Service conducted tests to determine whether cloud seeding could reduce the number of cloud-to-ground discharges. For unknown reasons, it appears that cloud seeding did reduce the number of lightning strokes, but the actual number of forest fires did not seem to change appreciably. Other attempts to discharge clouds by dropping metallic objects into them have had varying degrees of success.

Frost Prevention

Frost, the fruit-growers plight, occurs when the air temperature falls to 0° C (32° F) or below. It may be accompanied by deposits of ice crystals commonly called *white frost*. This, however, happens only if the air becomes saturated. White frost is not a requirement for crop damage.

Frost hazards exist when a cold air mass moves into a region, or when ample radiation cooling occurs on a clear night. The conditions accompanying the invasion of cold air are characterized by low daytime temperatures, long periods of effective frost, strong winds, and widespread damage. Frost induced by radiation loss is a nighttime phenomenon associated with a surface temperature inversion and is much easier to combat.

Several methods of frost prevention have been used with varying degrees of success. Generally, these attempts are directed at reducing the

Figure 13.21 *Two types of freeze controls used in Florida citrus groves. The wind machine gives protection down to −3° C, while the sprinkler system (thin pole) gives protection down to −4° C. Notice the ice on the plants. (Courtesy of U.S. Department of Agriculture)*

amount of heat lost during the night or at adding heat to the lowermost layer of air. Heat conservation methods include covering plants with material having a low thermal conductivity, such as paper and cloth, and producing particles which, when suspended in air, reduce the radiation loss. Smudge fires have been used for particle production but have generally proven unsatisfactory. In addition to the pollution problems created by the dense clouds of black smoke, the carbon particles impede daytime warming by reducing the amount of solar radiation that can reach the surface. This reduction in daytime warming offsets the benefits gained during the night.

Methods of warming include sprinkling, air mixing, and the use of orchard heaters (Figure 13.21). Sprinklers distribute water to the plants and add heat in two ways, first from the warmth of the water, but more importantly from latent heat of fusion, which is released when the water freezes. As long as an ice-water mixture remains on the plant, the latent heat released will keep the temperature from dropping below 0° C (32° C). Air mixing is successful when the temperature at 15 meters (40 feet) above the ground is 5° C (10° F) higher than the surface temperature. By using a wind machine, the warmer air aloft is mixed with the colder surface air. Orchard heaters probably produce the most successful results. As many as 30 or 40 heaters per acre are required, and the fuel cost can be significant. However, the effectiveness of the method seems to warrant the cost.

REVIEW

1. Summarize the processes by which water changes from one state to another. Indicate whether heat energy is absorbed or liberated.
2. After studying Table 13.1, write a generalization relating temperature and the capacity of air to hold water vapor.
3. How do absolute and specific humidity differ? What do they have in common? How is relative humidity different from both?

4. Referring to Figure 13.4, answer the following questions:
   a. During a typical day, when is the relative humidity highest? Lowest?
   b. At what time of day would dew most likely form?

   Write a generalization relating air temperature and relative humidity.
5. If the temperature remains unchanged and the specific humidity decreases, how will the relative humidity change?
6. On a cold winter day when the temperature is $-10°$ C and the relative humidity is 50 percent, what is the specific humidity (refer to Table 13.1)? What is the specific humidity for a day when the temperature is 20° C and the relative humidity is 50 percent?
7. Explain the principle of the sling psychrometer. The hair hygrometer.
8. Using the standard tables (Tables 13.2 and 13.3), determine the relative humidity and dew-point temperature if the dry-bulb thermometer reads 16° C and the wet-bulb thermometer reads 12° C. How would the relative humidity and dew point change if the wet-bulb thermometer read 8° C?
9. On a warm summer day when the relative humidity is high, it may seem even warmer than the thermometer indicates. Why do we feel so uncomfortable on a "muggy" day?
10. What is the function of condensation nuclei in the formation of clouds? The function of the dew point?
11. As you drink an ice-cold beverage on a warm day, the outside of the glass or bottle becomes wet. Explain.
12. Why does air cool when it rises through the atmosphere?
13. Explain the difference between the lapse rate and adiabatic cooling.
14. If unsaturated air at 23° C were to rise, what would its temperature be at 500 meters? If the dew-point temperature at the condensation level were 13° C, at what elevation would clouds begin to form?
15. Why does the adiabatic rate of cooling change when condensation begins? Why is the wet adiabatic rate not a constant figure?
16. The contents of an aerosol can are under very high pressure. When you push the nozzle on such a can, the spray feels cold. Explain.
17. How do orographic lifting and frontal wedging act to force air to rise?
18. Explain why the Great Basin area of the western United States is so dry. What term is applied to such a situation?
19. How does stable air differ from unstable air? Describe the general nature of the clouds and precipitation expected with each.
20. What is the basis for the classification of clouds?
21. Why are high clouds always thin?
22. Which cloud types are associated with the following characteristics? Thunder, halos, precipitation, hail, mackerel sky, sleet, lightning, mares' tails.
23. List five types of fog and discuss the details of their formation.
24. What is the difference between precipitation and condensation?
25. List the forms of precipitation and the circumstances of their formation.
26. Why will ice form on the wings of an airplane as it passes through a supercooled cloud?
27. What is the principle behind seeding supercooled clouds?
28. How do frost and white frost differ?

KEY TERMS

calorie
evaporation
latent heat
condensation
melting
freezing
sublimation
humidity
saturation
vapor pressure
absolute humidity
specific humidity

relative humidity
dew point
psychrometer
hygrometer
condensation nuclei
hygroscopic nuclei
adiabatic temperature changes
dry adiabatic rate
wet adiabatic rate
absolute stability
absolute instability
conditional instability
orographic lifting
rainshadow desert
frontal wedging
clouds
cirrus
cumulus
stratus
high clouds
middle clouds
low clouds
clouds of vertical development
Bergeron process
supercooled
freezing nuclei
collision-coalescence process
sleet
glaze
hail
fog
advection fog
radiation fog
upslope fog
steam fog
frontal (precipitation) fog
cloud seeding

This aneroid barograph accurately measures and records changes in atmospheric pressure. (Courtesy of WeatherMeasure Corporation)

Pressure Measurement
Factors Affecting Wind
Cyclones and Anticyclones
Wind Measurement
General Circulation of the Atmosphere
Circulation in the Mid-latitudes
Local Winds

14 Pressure and Wind

We have already dealt with the two elements of weather and climate that generally are of the greatest interest to people—temperature and moisture. In this chapter we shall investigate the remaining two elements—wind and pressure—which may not seem to be as important, but indeed are. It is the wind that often brings changes in temperature and moisture conditions, and it is pressure differences that drive the wind. Among the questions we will try to answer in this chapter are: How can the weight of the air be measured? What are the factors that control the winds? What is a prevailing wind? Why are "highs" and "lows" always shown on the weather map? How can the weather be predicted by checking the barometer?

Of the various elements of weather and climate, changes in air pressure are the least noticeable. When listening to a weather report, we are generally interested in moisture conditions (humidity and precipitation), temperature, and perhaps wind. It is the rare individual, however, who wonders about air pressure. Although the hour-to-hour and day-to-day variations in air pressure are not perceptible to human beings, they are very important in producing changes in our weather. Variations in air pressure from place to place are responsible for the movement of air (wind), as well as being the most significant factor in weather forecasting. As we shall see, air pressure is tied very closely to the other elements of weather in a cause-and-effect relationship.

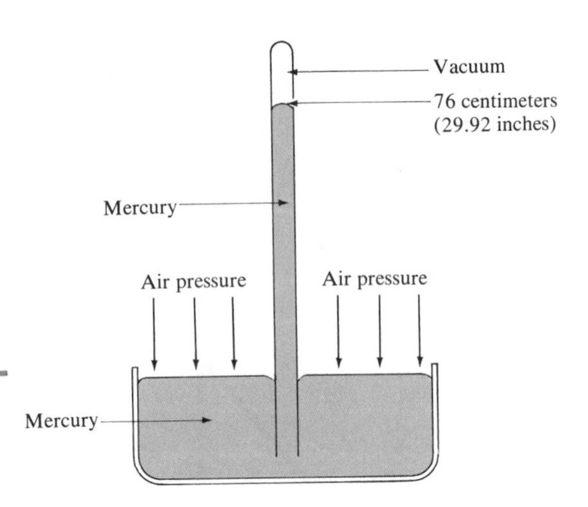

Figure 14.1 *Simple mercurial barometer. The weight of the column of mercury is balanced by the pressure exerted on the dish of mercury by the air above. If the pressure decreases, the column of mercury falls; if the pressure increases, the column rises.*

Pressure Measurement

In Chapter 12 we saw that air has weight; at sea level, it weighs 1 kilogram per square centimeter (over one ton per square foot!). *The air pressure at a particular place is simply the force exerted by the weight of the air above.* With an increase in altitude, the weight of the air above, and thus the pressure, drops, at first rapidly, then much more slowly (Table 12.2).

When meteorologists measure atmospheric pressure, they employ the unit called the millibar. Standard sea level pressure is expressed as 1013.2 millibars. Although the millibar has been the unit of measure on all United States weather maps since January 1940, you might be better acquainted with the expression "inches of mercury," which is used by the media to describe atmospheric pressure. In the United States the National Weather Service converts millibar values to inches of mercury for public and aviation use. Using inches of mercury dates from 1643, when Torricelli, a student of the famous Italian scientist Galileo, invented the **mercurial barometer.** Torricelli correctly described the atmosphere as a vast ocean of air which exerts pressure on us and all things about us. To measure this force he filled a glass tube that was closed at one end with mercury. The tube was then inverted into a dish of mercury (Figure 14.1). Torricelli found that the mercury flowed out of the tube until the weight of the column was balanced by the pressure exerted on the surface of the mercury by the air above. In other words, the weight of the mercury in the column equalled the weight of a similar diameter column of air which extended from the ground to the top of the atmosphere. Torricelli noted that when air pressure increased, the mercury in the tube rose; conversely, when air pressure decreased so did the height of the column of mercury. The length of the column of mercury, therefore, became the measure of the air pressure. With some refinements the mercurial barometer invented by Torricelli is still the standard pressure-measuring instrument used today. Standard atmospheric pressure at sea level equals 29.92 inches of mercury. A scale comparing millibars and inches is shown in Figure 14.2.

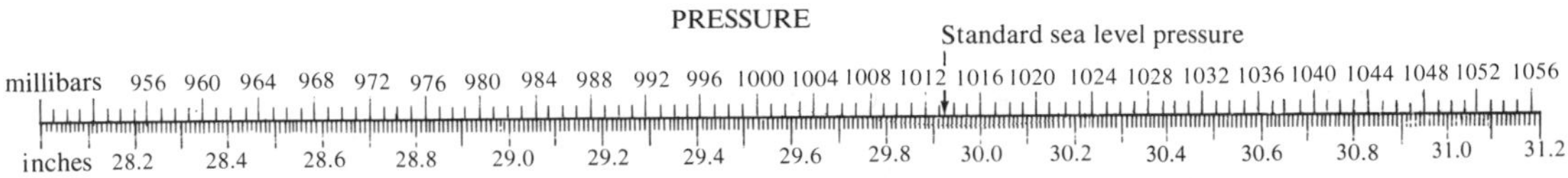

PRESSURE

Standard sea level pressure

millibars 956 960 964 968 972 976 980 984 988 992 996 1000 1004 1008 1012 1016 1020 1024 1028 1032 1036 1040 1044 1048 1052 1056

inches 28.2 28.4 28.6 28.8 29.0 29.2 29.4 29.6 29.8 30.0 30.2 30.4 30.6 30.8 31.0 31.2

Figure 14.2 *Millibars and inches. This scale compares two of the most common units of pressure measurement. Standard sea level pressure is equal to 29.92 inches or 1013.2 millibars. (After NOAA)*

The need for a smaller and more portable instrument for measuring air pressure lead to the development of the **aneroid** ("without liquid") **barometer.** Based on a different principle than the mercurial barometer, this instrument consists of partially evacuated metal chambers that have a spring inside, keeping them from collapsing. The metal chambers, being very sensitive to variations in air pressure, change shape, compressing as the pressure increases and expanding as the pressure decreases. Aneroids are often used in making **barographs,** instruments that continuously record pressure changes (see chapter-opening photo). Another important adaptation of the aneroid has been its use as an **altimeter** in aircraft. Recall that air pressure decreases with altitude and that the pressure distribution with height is well established. By marking an aneroid in meters rather than millibars, we have an altimeter.

Factors Affecting Wind

We have discussed the upward movement of air and its importance in cloud formation. As important as vertical motion is, far more air is involved in horizontal movement, the phenomenon we call **wind.** Although we know that air will move vertically if it is warmer, and consequently more buoyant, than the surrounding air, what causes air to move horizontally? Simply stated, wind is the result of horizontal differences in air pressure. *Air flows from areas of high pressure to areas of low pressure.* You may have experienced this when opening a vacuum-packed can of coffee. The noise you hear is caused by air rushing from the higher pressure outside the can to the lower pressure inside. Wind is nature's attempt to balance similar inequalities in air pressure. *Because unequal heating of the earth's surface generates these pressure differences, solar radiation is the ultimate driving force of wind.*

If the earth did not rotate, and if there were no friction, air would flow directly from areas of high pressure to areas of low pressure. But because both of these factors exist, wind is controlled by a combination of forces. These are: (1) the pressure gradient force; (2) the Coriolis force; (3) friction; and (4) the tendency of a moving object to continue moving in a straight line. The last factor is often referred to as centrifugal force. The magnitude of centrifugal force is small when compared with the other forces and thus of minor importance, except in rapidly rotating storms such as tornadoes and hurricanes. Discussions of the other three factors follow.

Pressure Gradient Force

Pressure differences cause the wind to blow, and the greater these differences, the greater the wind speed. Over the earth's surface, variations in air pressure are determined from barometric readings taken at hundreds of weather stations. These pressure data are shown on a weather map using **isobars,** which are lines connecting places of equal air pressure (Figure 14.3). The spacing of the isobars indicates the amount of pressure change occurring over a given distance and is expressed as the **pressure gradient.**

You might find it easier to visualize the concept of a pressure gradient if you think of it as being analogous to the slope of a hill. A steep

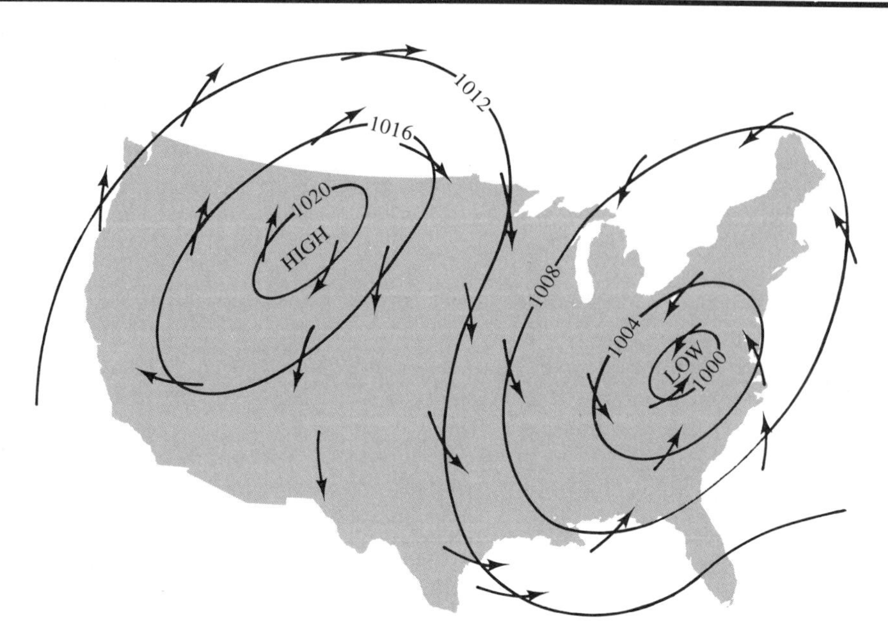

Figure 14.3 *Isobars, which are lines connecting places of equal barometric pressure, are used to show the distribution of pressure on daily weather maps. The lines usually curve and often join where cells of high and low pressure exist. The arrows indicate the expected air flow surrounding cells of high and low pressure.*

pressure gradient, like a steep hill, causes greater acceleration of a parcel than does a weak pressure gradient. Thus, the relationship between wind speed and the pressure gradient is rather simple: *Closely spaced isobars indicate a steep pressure gradient and high winds, while widely spaced*

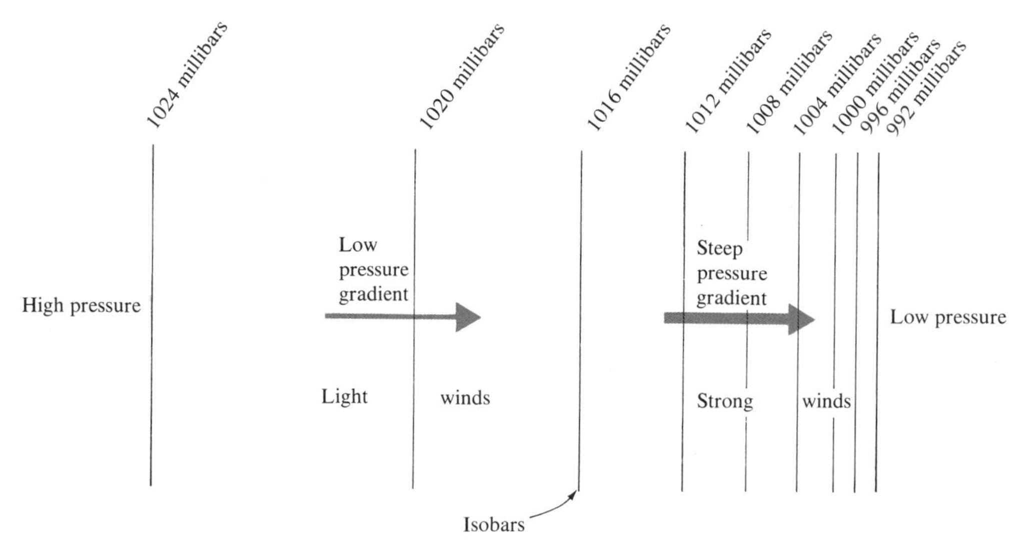

Figure 14.4 *Pressure gradient force. Closely spaced isobars indicate a steep pressure gradient and high wind speeds, and widely spaced isobars indicate low wind speeds.*

isobars indicate a weak pressure gradient and light winds. Figure 14.4 illustrates the relationship between the spacing of isobars and wind speed.

The pressure gradient is the driving force of wind. It has both magnitude and direction. While its magnitude is determined from the spacing of isobars, *the direction of force is always from areas of high pressure to areas of low pressure and at right angles to the isobars.* Once the air starts to move, the Coriolis force and friction come into play, but then only to modify the movement, not to produce it.

Coriolis Force

Figure 14.3 shows the typical air movements associated with high- and low-pressure systems. As expected, the air moves out of the regions of high pressure and into the regions of low pressure. However, the wind does not cross the isobars at right angles as the pressure gradient force directs. This deviation is the result of the earth's rotation and has been named the **Coriolis force** after its discoverer.* *All free-moving objects, including the wind, are deflected to the right of their path of motion in the Northern Hemisphere and to the left in the Southern Hemisphere.* The reason for this deflection can be illustrated by imagining the path of a rocket launched from the North Pole toward a target located on the equator (Figure 14.5). If the rocket took an hour to reach its target, during its flight the earth would have rotated 15 degrees to the east. To someone standing on the earth, it would look as if the rocket veered off its path and hit the earth 15 degrees west of its target. The true path of the rocket was straight and would appear so to someone out in space looking down at the earth. It was the earth turning under the rocket that gave it its *apparent* deflection. The same deflection is experienced by wind regardless of the direction it is moving.

For convenience, we attribute this apparent shift in wind direction to the Coriolis force. It is hardly a "real" force, but rather the effect of the earth's rotation on a moving body. *This deflecting force (1) is always directed at right angles to the direction of airflow; (2) affects only wind direction, not wind speed; and (3) is affected by wind speed. The stronger the wind, the greater the deflecting force.*

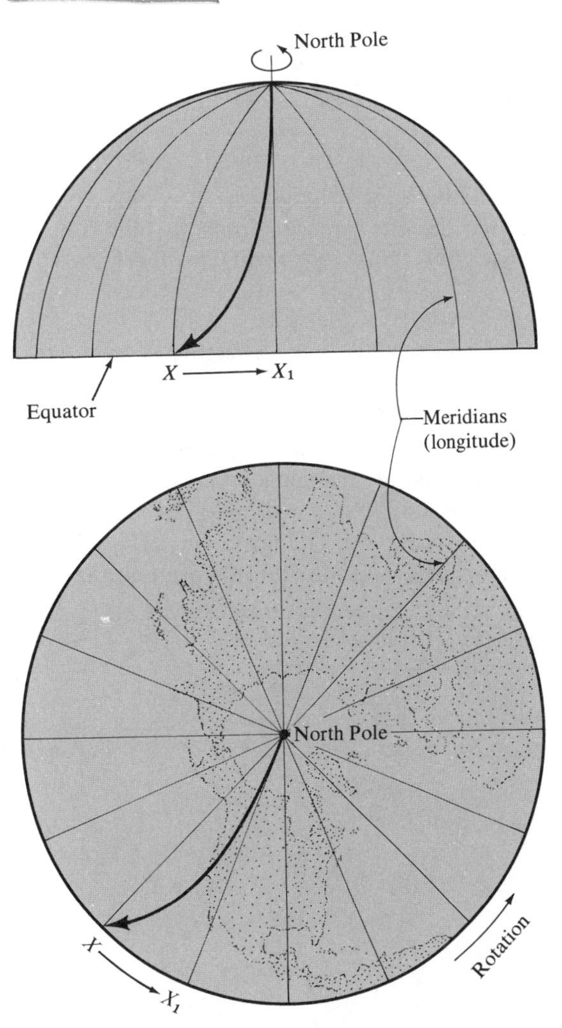

Figure 14.5 *The Coriolis force. During the rocket's flight from the North Pole to point X, the earth rotates eastward, moving point X to point X_1. The rotation gives the rocket's trajectory a curved path when plotted on the earth's surface. This deflection is termed the Coriolis force.*

*Since this is not an actual force but only an apparent one, it is sometimes referred to as the *Coriolis effect.*

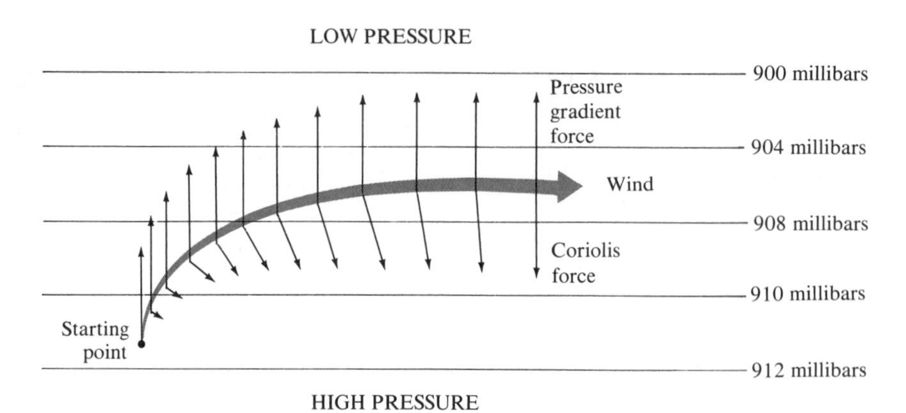

Figure 14.6 *The geostrophic wind. Upper-level winds are deflected by the Coriolis force until the Coriolis force just balances the pressure gradient force. Above 600 meters, where friction is negligible, these winds will flow nearly parallel to the isobars and are called geostrophic winds.*

Friction

Friction as a factor affecting wind is only important within the first few kilometers of the earth's surface. It acts to slow the movement of air and, as a consequence, alters wind direction. To illustrate friction's effect on wind direction, let us look first at a situation in which it has no role. Above the friction layer, the pressure gradient force and the Coriolis force are directing the flow of air. Under these conditions, the pressure gradient force will cause the air to start moving across the isobars. As soon as the air starts to move, the Coriolis force will act at right angles to this motion. The faster the wind speed, the greater the deflecting force. Eventually, the Coriolis force will balance the pressure gradient force and the wind will blow parallel to the isobars (Figure 14.6). Upper-air winds generally

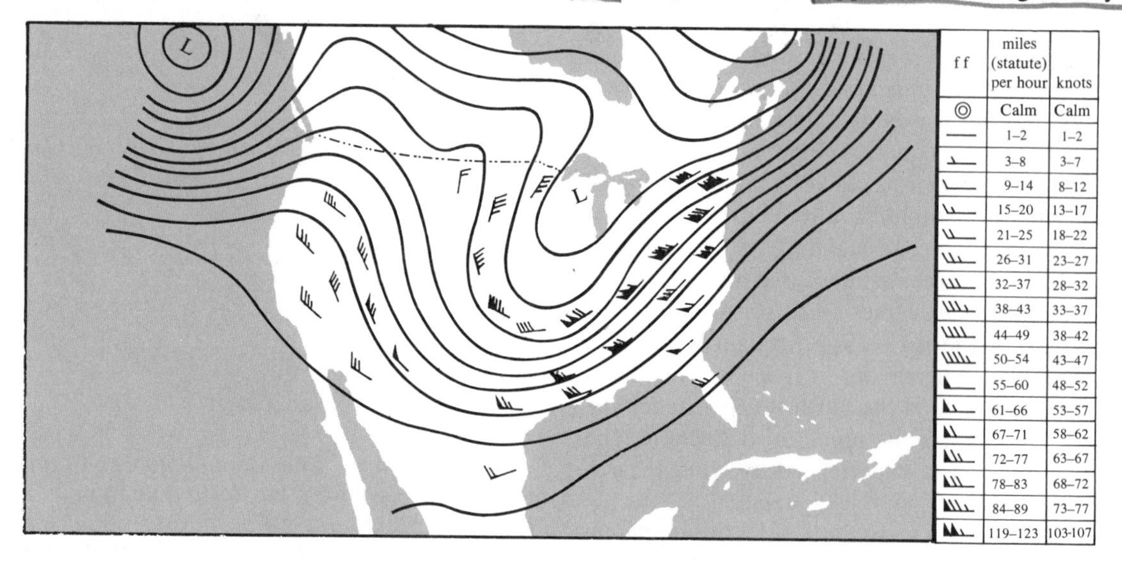

ff	miles (statute) per hour	knots
◎	Calm	Calm
	1–2	1–2
	3–8	3–7
	9–14	8–12
	15–20	13–17
	21–25	18–22
	26–31	23–27
	32–37	28–32
	38–43	33–37
	44–49	38–42
	50–54	43–47
	55–60	48–52
	61–66	53–57
	67–71	58–62
	72–77	63–67
	78–83	68–72
	84–89	73–77
	119–123	103-107

Figure 14.7 *Upper-air winds. This map shows the direction and speed of the upper-air wind of February 24, 1974. Note that the airflow is nearly parallel to the contours. These isolines are height contours for the 500-millibar level.*

take this path and are called **geostrophic winds.** Because of the lack of friction, geostrophic winds travel at higher speeds than surface winds (Figure 14.7).

The most prominent features of upper-level flow are the **jet streams.** First encountered by high-flying bombers during World War II, these fast-moving "rivers" of air travel between 120 and 240 kilometers (75 and 150 miles) per hour in a west-to-east direction. One such stream is situated over the polar front, which is the zone separating cool polar air from warm subtropical air.

Below 600 meters (2000 feet), friction complicates the airflow just described. Recall that the Coriolis force is proportional to wind speed. By lowering the wind speed, friction reduces the Coriolis force. Since the pressure gradient force is not affected by wind speed, it wins the tug-of-war shown in Figure 14.8. The result is a movement of air at an angle across the isobars toward the area of low pressure. The roughness of the terrain determines the angle of airflow across the isobars. Over the smooth ocean surface, friction is low and the angle is small. Over rugged terrain, where friction is higher, the angle could be as great as 45 degrees from the isobars. In summary, *upper airflow is nearly parallel to the isobars, while the effect of friction causes the surface winds to move more slowly and cross the isobars at an angle.*

Cyclones and Anticyclones

Among the most common features on any weather map are areas designated as pressure centers. **Cyclones,** or **lows,** are centers of low pressure, and **anticyclones,** or **highs,** are high-pressure centers. As Figure 14.3 illustrates, the pressure decreases from the outside toward the center in a cyclone, while in an anticyclone, just the opposite is the case—the pressure increases from the outside toward the center. By knowing just a few basic facts about centers of high and low pressure, you can greatly increase your understanding of current and forthcoming weather.

Cyclonic and Anticyclonic Winds

From the preceding section, we learned that the two most significant factors that affect wind are pressure differences and the Coriolis force. Winds blow from high pressure to low pressure and are deflected to the right or left by the earth's rotation. When these controls of wind are applied to pressure centers in the Northern Hemisphere, the result is that *winds blow in and counterclockwise around a low and out and clockwise around a high* (Figure 14.3). Of course, in the Southern Hemisphere the Coriolis force deflects the winds to the left, and therefore winds about a low are blowing clockwise, and winds around a high are moving counterclock-

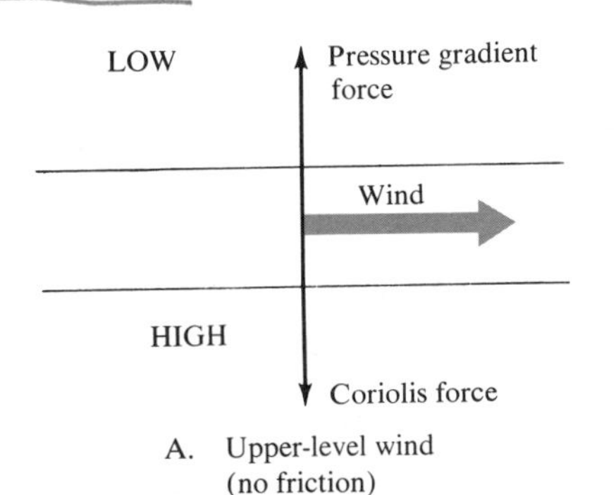

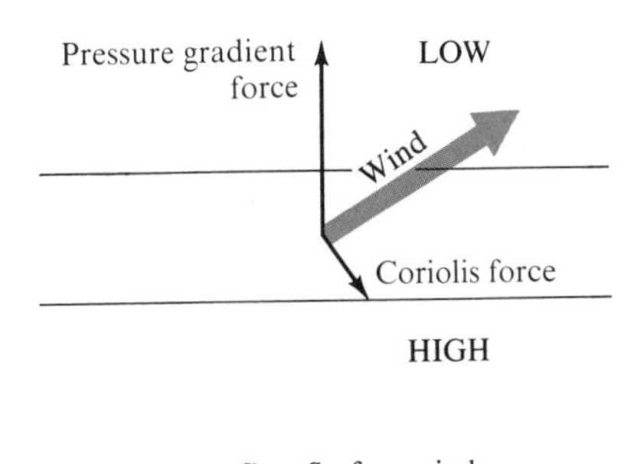

Figure 14.8 *Comparison between upper-level winds and surface winds showing the effects of friction on airflow. Friction slows surface wind speed, which weakens the Coriolis force, causing the winds to cross the isobars.*

wise. However, in whatever hemisphere, friction causes a net inflow **(convergence)** around a cyclone and a net outflow **(divergence)** around an anticyclone.

Weather Generalizations about Highs and Lows

Rising air is associated with cloudy conditions and precipitation, whereas subsidence produces adiabatic heating and clearing conditions. In this section we will learn how the movement of air can itself create pressure change and hence generate winds. Upon doing so, we will examine the interrelationship between horizontal and vertical flow, and their effects on the weather.

Let us first consider the situation around a surface low-pressure system where the air is spiraling inward. Here the net inward transport of air causes a shrinking of the area occupied by the air mass, a process which is termed horizontal convergence. Whenever air converges horizontally, it must pile up, that is, increase in height to allow for the decreased area it now occupies. This generates a "taller" and therefore heavier air column. Yet a surface low can exist only as long as the column of air above remains light. We seem to have encountered a paradox—low-pressure centers cause a net accumulation of air, which increases their pressure. Consequently, a surface cyclone should quickly eradicate itself, in a manner not unlike what happens to the vacuum in a coffee can upon opening.

In light of the preceding discussion, it should be apparent that for a surface low to exist for any reasonable time, compensation must occur at some layer aloft. For example, surface conver-

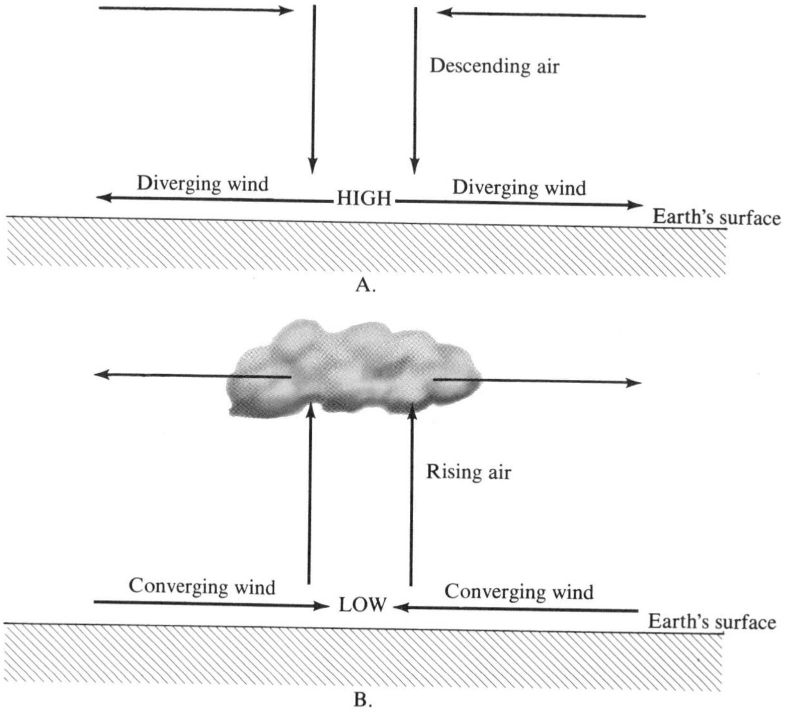

Figure 14.9 **A.** *Cross-sectional view of a high. Highs, or anticyclones, are associated with descending air and diverging winds. As a result, clear skies and "fair" weather may be expected with the approach of such a system.* **B.** *Cross-sectional view of a low. Converging winds and rising air are associated with a low, or cyclone. Consequently, clouds and rain are often associated with the passage of such a system.*

gence could be maintained if divergence (spreading out) aloft occurred at a rate equal to the inflow below. Figure 14.9 shows diagrammatically the relationship between surface convergence (inflow) and divergence (outflow) aloft that is needed to maintain a low-pressure center. Note that surface convergence about a cyclone causes a net upward movement. The rate of this vertical movement is quite slow, generally less than 1 kilometer per day. Nevertheless, since rising air often results in cloud formation and precipitation, *the passage of a low is generally related to unstable conditions and "stormy" weather*. On occasion, divergence aloft may even exceed surface convergence, resulting in intensified surface inflow and increased vertical motion. Thus divergence aloft can intensify these storm centers as well as maintain them.

Like their counterparts, anticyclones, which are associated with surface divergence, must also be maintained from above. The mass outflow near the surface is accompanied by convergence aloft and general subsidence of the air column (Figure 14.9). Since descending air is compressed and warmed, *cloud formation and precipitation are unlikely in an anticyclone, and "fair" weather can usually be expected with the approach of a high.*

For reasons which should now be obvious, it has been common practice to write on barometers intended for household use the words "stormy" at the low end and "fair" on the high end. By noting whether the pressure is rising, falling, or steady, we have a good indication of what the forthcoming weather will be. Such a determination, called the **pressure,** or **barometric, tendency,** is a very useful aid in short-range weather prediction. The generalizations relating cyclones and anticyclones to the weather conditions just considered are stated rather poetically in the proverb that follows. Note that *glass* refers to the barometer.

When the glass falls low,
Prepare for a blow;
When it rises high,
Let all your kites fly.

In conclusion, you should now be better able to understand why local television weather broadcasters emphasize the positions and projected paths of cyclones and anticyclones. The "villian" on these weather programs is always the cyclone, which produces "bad" weather in any season. Lows move in roughly a west-to-east direction across the United States and require a few days to more than a week for the journey. Their paths can be somewhat erratic; thus accurate prediction of their migration is difficult, although essential, for short-range forecasting. Meteorologists must also determine if the flow aloft will intensify an embryo storm or act to suppress its development. Because of the close tie between conditions at the surface and those aloft, a great deal of emphasis has been placed on the importance and understanding of the total atmospheric circulation, particularly in the mid-latitudes. Once we have examined the workings of the general circulation, we will again consider the structure of the cyclone in light of these findings.

Wind Measurement

Two basic wind measurements, direction and speed, are particularly significant to the weather observer. Winds are always labeled by the direction from which they blow. A north wind blows from the north toward the south, an east wind from the east toward the west. The instrument most commonly used to determine wind direction is the **wind vane** (Figure 14.10). This instrument, which is a common sight on many buildings, always points into the wind. Often the wind direction is shown on a dial that is connected to the wind vane (Figure 14.11). The dial will indicate the direction of the wind either by points of the compass, that is, N, NE, E, SE, etc., or by a 0–360-degree scale. On the latter scale, 0 degrees or 360 degrees is north, 90 degrees is east, 180 degrees is south, and 270 degrees is west. When the wind consistently blows more often from one direction than from any other, it is termed a **prevailing wind.** Wind speed is commonly measured using a **cup anemometer** (Figure 14.10). The

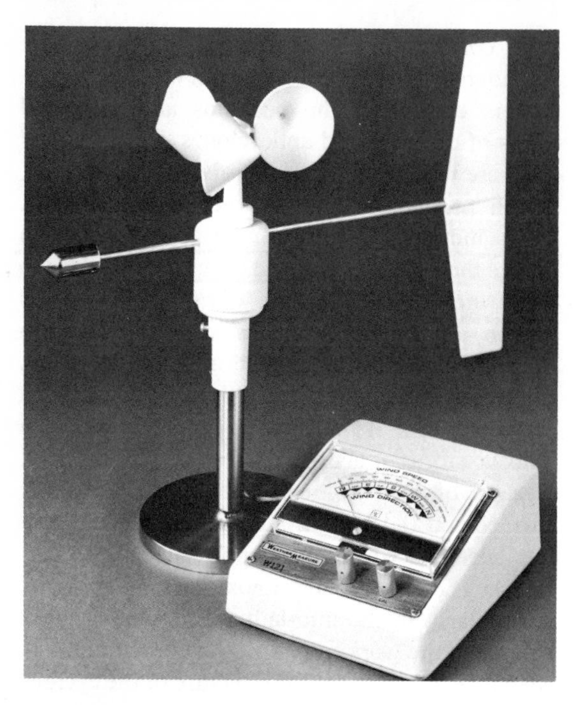

Figure 14.10 *Wind vane and cup anemometer. (Courtesy of WeatherMeasure Corporation)*

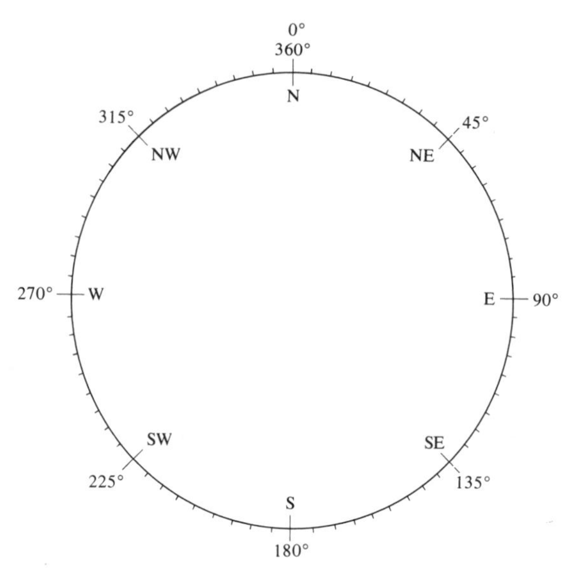

Figure 14.11 *Wind direction. Wind direction may be expressed using the points of the compass or a scale of 0–360 degrees. Winds are always labeled according to the direction from which they are blowing.*

wind speed is read from a dial much like the speedometer of an automobile.

By knowing the location of cyclones and anticyclones in relation to where you are, you can predict the changes in wind direction that will be experienced as the pressure center moves past. Since changes in wind direction often bring changes in temperature and moisture conditions, the ability to predict the winds can be very useful. In the Midwest, for example, a north wind may bring cool, dry air from Canada, while a south wind may bring warm, humid air from the Gulf of Mexico. Sir Francis Bacon summed it up nicely when he wrote, "Every wind has its weather."

General Circulation of the Atmosphere

The underlying cause of wind is the unequal heating of the earth's surface. In tropical regions more solar radiation is received than is radiated back to space. In polar regions the opposite is true—less solar energy is received than is lost. Attempting to balance these differences, the atmosphere acts as a giant heat transfer system, moving warm air poleward and cool air equatorward. On a smaller scale, but for the same reason, ocean currents also contribute. The general circulation is very complex, and there is a great deal which has yet to be explained. We can, however, approximate its major components by first considering the circulation that would occur on a nonrotating earth having a uniform surface. We will then modify this system to fit observed patterns.

On a hypothetical nonrotating planet with a smooth surface of either all land or all water, two large thermally produced cells would form (Figure 14.12). The heated equatorial air would rise until it reached the tropopause, which, acting like a lid, would deflect the air poleward. Eventually, this upper-level airflow would reach the poles, sink, and spread out in all directions at

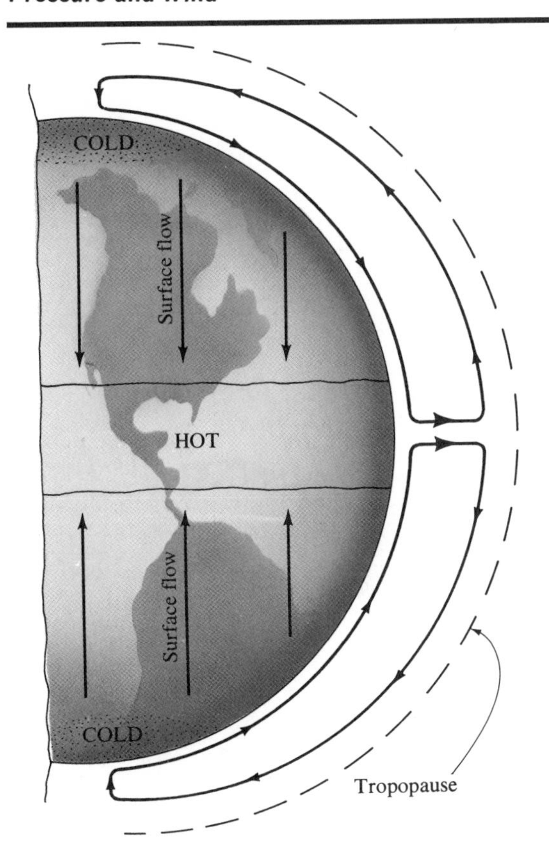

Figure 14.12 *Global circulation on a nonrotating earth. A simple convection system produced by unequal heating of the surface on a nonrotating planet.*

the surface and move back toward the equator. Once there, it would be reheated and start its journey over again. This hypothetical circulation system has upper-level air flowing poleward and surface air flowing equatorward.

If we add the effect of rotation, this simple convection system will break down into smaller cells. Figure 14.13 illustrates the three pairs of cells proposed to carry on the task of heat redistribution on a rotating planet. The polar and tropical cells retain the characteristics of the thermally generated convection described earlier. The nature of the mid-latitude circulation is complex and will be discussed in more detail in the next section.

Near the equator, the rising air is associated with the pressure zone known as the **equatorial low**—a region marked by abundant precipitation. As the upper-level air reaches 20–30 degrees latitude, it will have cooled enough to sink toward the surface. This subsidence and associated adiabatic heating produce the hot, arid regions in this latitude range. The center of this zone of subsiding dry air is the **subtropical high,** which encircles the globe near 30 degrees (Figure 14.13). Located here are extensive arid and semiarid regions. The great deserts of Australia, Arabia, and North Africa, for example, are dry primarily because of the stable conditions associated with the subtropical highs. At the surface, airflow is outward from the center of this high-pressure system. Some of the air travels equatorward and is deflected by the Coriolis force, producing the rather constant **trade winds.** The remainder travels poleward and is also deflected, generating the prevailing **westerlies** of the mid-latitudes. As the westerlies move poleward, they encounter the cool **polar easterlies** in the region of the **subpolar low.** The interaction of these warm and cool winds produces the stormy belt known as the **polar front.** The source region for the variable polar easterlies is the **polar high.** Here, polar air is subsiding and spreading equatorward.

In summary, *this simplified global circulation is dominated by four pressure zones. The subtropical and polar highs are areas of dry subsiding air which flows outward at the surface, producing the prevailing winds. The low-pressure zones of the equatorial and subpolar regions are associated with inward and upward airflow accompanied by precipitation.*

Up to this point, we have described the surface pressure and associated winds as continuous belts around the earth. The only true zonal distribution of pressure exists in the subpolar low of the Southern Hemisphere, where the ocean is continuous. At other locations, particularly in the Northern Hemisphere, where the bulk of the land exists, large seasonal temperature differences disrupt this zonal pattern. Figure 14.14

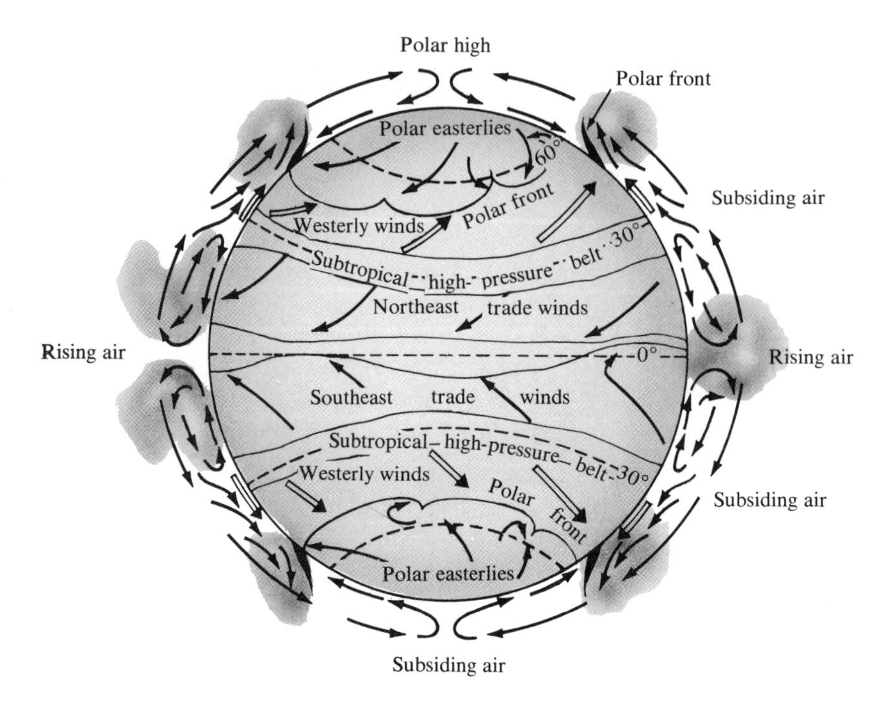

Figure 14.13 *Idealized global circulation.*

shows the resulting pressure and wind patterns for July and January. The circulation over the oceans is dominated by semipermanent cells of high pressure located in the subtropics and cells of low pressure over the subpolar regions. The subtropical highs are responsible for the trade winds and westerlies, as mentioned earlier. The large landmasses, on the other hand, particularly Asia, become cold in the winter and develop a seasonal high-pressure system from which surface flow is directed off the land (Figure 14.14). In the summer, the opposite occurs; the landmasses are heated and develop a low-pressure cell, which permits air to flow onto the land. These seasonal changes in wind direction are known as the **monsoons.** During warm months, areas such as India experience a flow of warm, water-laden air from the Indian Ocean, which produces the rainy summer monsoon. The winter monsoon is dominated by dry continental air. A similar situation exists, but to a lesser extent, over North America. In summary, *the general circulation is produced by semipermanent cells of high and low pressure over the oceans and is complicated by seasonal pressure changes over land.*

Circulation in the Mid-latitudes

The circulation in the mid-latitudes, the zone of the westerlies, is complex and does not fit the convection system proposed for the tropics. Between 30 and 60 degrees latitude, the general west-to-east flow is interrupted by the migration of cyclones, low-pressure systems often associated with precipitation, and anticyclones, high-pressure systems associated with clear skies. Do not confuse these with the much larger semipermanent pressure systems that comprise the general circulation. In the Northern Hemisphere these cells move in an eastward direction around the globe, creating an anticyclonic (clockwise) flow or a cyclonic (counterclockwise) flow in their area of influence. A close correlation was recently found between the paths taken by these

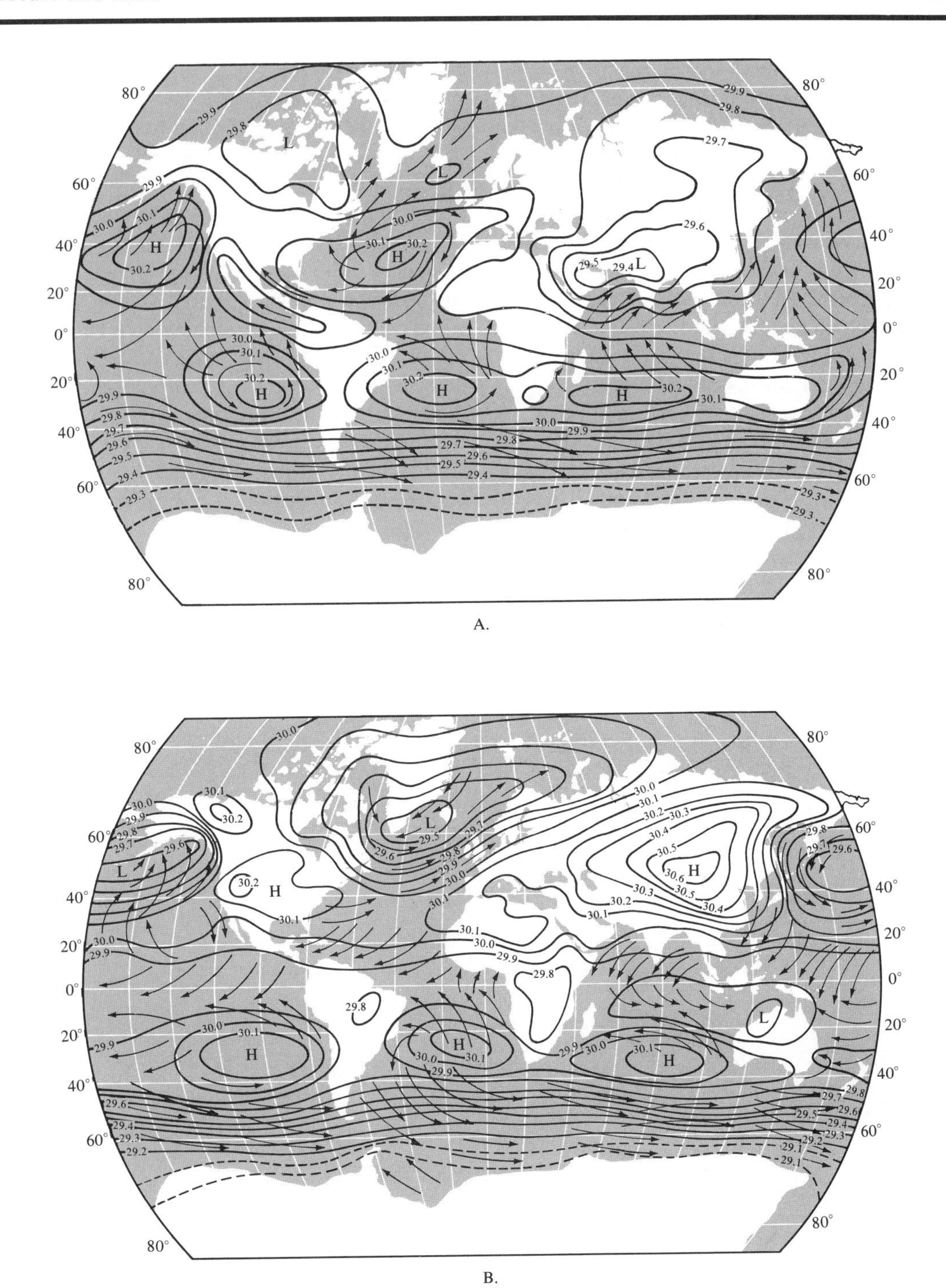

Figure 14.14 *Average surface barometric pressure in inches for* **A.** *July and* **B.** *January, with associated winds.*

surface pressure systems and the position of the upper-level airflow, indicating that the upper air is responsible for directing the movement of cyclonic and anticyclonic systems.

Although a great deal is still unknown about the wavy upper-level flow of the westerlies, some of its most basic features are understood with some degree of certainty. Among the most obvious features of the flow aloft are the seasonal changes. The change in wind speed is reflected on upper-air charts by more closely spaced contour lines in the cool season. The seasonal fluctuation of wind speeds is a consequence of the seasonal variation of the temperature gradient. The steep temperature gradient across the middle latitudes in the winter months corresponds to stronger flow aloft. In addition, the polar jet stream fluctuates seasonally such that its mean position migrates southward with the approach of winter and northward as summer nears. By midwinter, the jet core may penetrate as far south as central Florida. Since the paths of cyclonic systems are guided by the flow aloft, we can expect the southern tier of states to experience most of their severe storms in the winter season. During the hot summer months, the storm track is across the northern states, and some cyclones never leave Canada. The northerly storm track associated with summer also applies to Pacific storms, which move toward Alaska during the warm months, thus producing a rather long dry season for much of our west coast. The number of cyclones generated is seasonal as well, with the largest number occurring in the cooler months when the temperature gradients are greatest. This fact is in agreement with the role of cyclonic storms in the distribution of heat across the midlatitudes.

However, even in the cool season, the westerly flow goes through an irregular cyclic change. There may be periods of a week or more when the flow is nearly west to east as shown in Figure

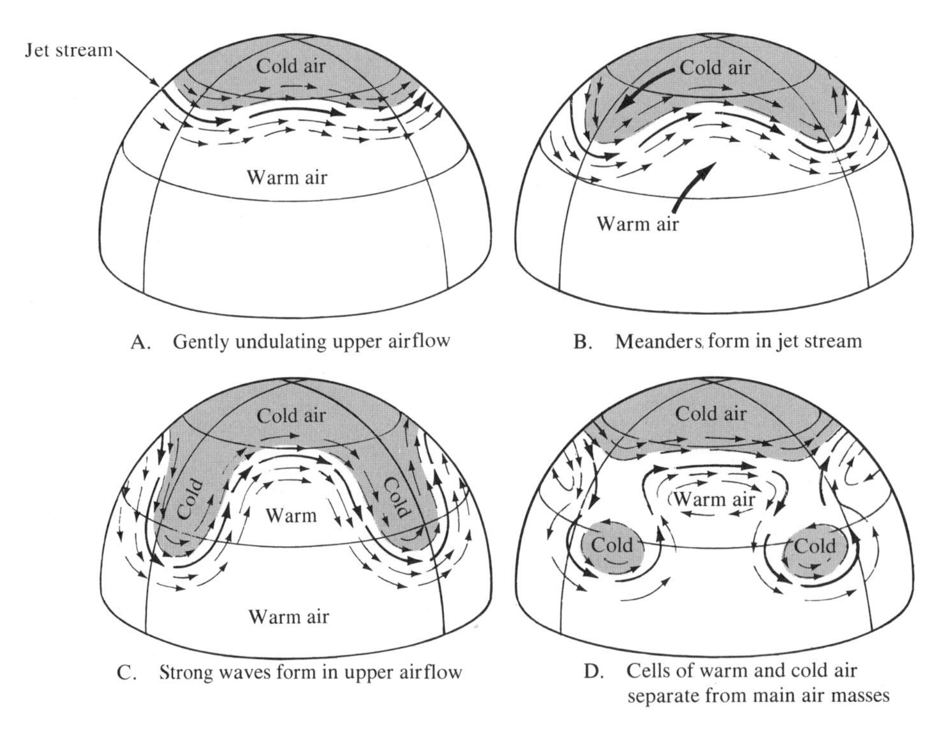

Figure 14.15 *Cyclic changes that occur in the upper-level airflow of the westerlies. The flow, which has the jet stream as its axis, starts out nearly straight, then develops meanders which are eventually cut off. (After J. Namias, NOAA)*

14.15. Under these conditions, relatively mild temperatures occur, and few disturbances are experienced in the region south of the jet stream. Then, without warning, the upper flow begins to meander wildly, producing large-amplitude waves and a general north-to-south flow (Figure 14.15). This change allows for an influx of cold air southward which intensifies the temperature gradient and the flow aloft. During these periods, cyclonic activity dominates the weather picture. For a week or more, the cyclonic storms redistribute large quantities of heat across the mid-latitudes by moving cold air southward and warm air northward. This redistribution eventually results in a weakened temperature gradient and a return to a flatter flow aloft and less intense weather at the surface. These cycles, consisting of alternating periods of calm and stormy weather, can last from one to six weeks.

Because of the rather irregular behavior of the circulation patterns of the flow aloft, long-range weather prediction still remains essentially beyond the forecaster's reach. Nonetheless, numerous attempts are being made to predict more accurately changes in the upper-level flow on a long-term basis. It is hoped that this research will answer questions such as, Will next winter be colder than normal? and, Will California experience a drought next year?

Let us consider an example of the influence of the flow aloft on the weather over an extended period. As our example, we will examine an atypical winter. In a normal January, the jet-stream core and associated westerly flow make a wavy pattern across the United States as shown in Figure 14.16. This "typical" flow pattern is believed to be caused by the mountains. During January 1977, the normal flow pattern was greatly accentuated as illustrated in Figure 14.16. The greater amplitude of the upper-level flow caused an almost continuous influx of cold air into the deep South, producing record low temperatures

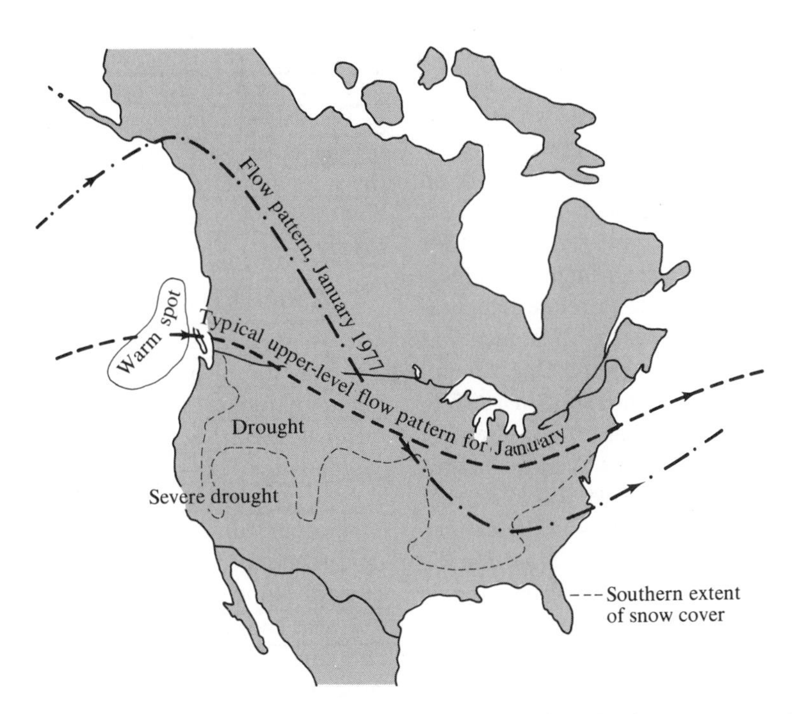

Figure 14.16 *The unusually high amplitude experienced in the flow pattern of the prevailing westerlies during the winter of 1977 brought warmth to Alaska, drought to the west, and frigid temperatures to the central and eastern United States.*

throughout much of the eastern and central United States. Because of dwindling natural gas supplies, many industries experienced layoffs. Much of Ohio was hit so hard that four-day work weeks and massive shutdowns were ordered. While most of the East was in the deep freeze, the westernmost states were under the influence of a strong ridge of high pressure. Generally mild temperatures and clear skies dominated their weather picture. But that was no blessing, since this ridge of high pressure blocked the movement of Pacific storms which usually provide much-needed winter precipitation. The shortage of moisture was especially serious in California, where January is the middle of its three-month rainy season. Throughout most of the western states, amounts of winter rain and snow, which supply water for summer irrigation, were far below normal. This dilemma was compounded by the fact that the previous year's precipitation had also been far below normal, leaving many reservoirs nearly empty. In contrast, the highly accentuated flow pattern channeled unseasonably warm air into Alaska. Even Fairbanks, which generally experiences temperatures as low as −40° C, had a rather mild January, with numerous days of above-freezing temperatures.

Although the effects of the upper-level flow on the weather are well documented, as in the example above, the causes of these fluctuations for the most part still elude meteorologists. Numerous attempts have been made to relate temperature variations to such diverse phenomena as sunspot cycles and volcanic activity. One attempt to relate the flow pattern experienced in 1977 to ocean temperatures has been given considerable attention. Jerome Namias of the Scripps Institution suggested prior to the winter of 1977 that above-average ocean temperatures in the eastern Pacific may cause a greater-than-average amplitude in the wavy pattern of the westerlies. It was also suggested that once snow is distributed farther south, the increase in albedo will further support the southward migration of the jet stream axis in this region. In other words, cold temperatures tend to perpetuate themselves by determining the position of the flow aloft.

Local Winds

Land and Sea Breezes

In coastal areas during the warm summer months, the land heats more intensely than the adjacent body of water (see Chapter 12). As a result, the air above the land surface heats and expands, creating an area of low pressure. A **sea breeze** then develops, blowing from the water (high pressure) toward the land (low pressure) (Figure 14.17). The sea breeze begins to develop shortly before noon and generally reaches its greatest intensity during the mid- to late afternoon. These cool winds are a significant moderating influence in coastal areas. At night, the land cools rapidly, eventually resulting in higher pressure. As a consequence, a **land breeze** may develop (Figure 14.17).

Mountain and Valley Breezes

Mountain and valley breezes are created for much the same reason as land and sea breezes, namely differential heating. On warm, sunny days, the floor and slopes of a valley are intensely heated. The heated air rises up the sides of the valley, creating a **valley breeze.** At night, on the other hand, the air in close proximity to the sides of the valley cools. The cooler air drains downslope into the valley, creating a **mountain breeze.**

Chinook (Foehn) Winds

Warm, dry winds are common on the eastern slopes of the Rockies, where they are called **chinooks,** and in the Alps, where the term **foehn** is used. Such winds are created when air descends the leeward side of a mountain, warming by compression. Since condensation may have occurred as the air ascended the windward side, releasing latent heat, the air descending the leeward slope will be warmer and drier than it was at a similar elevation on the windward side. The immediate effect of the chinook is often rather pronounced. A rapid temperature increase of 10° C to 15° C is not uncommon, and in some extreme cases, temperatures have been known to rise more than 30° C (60° F).

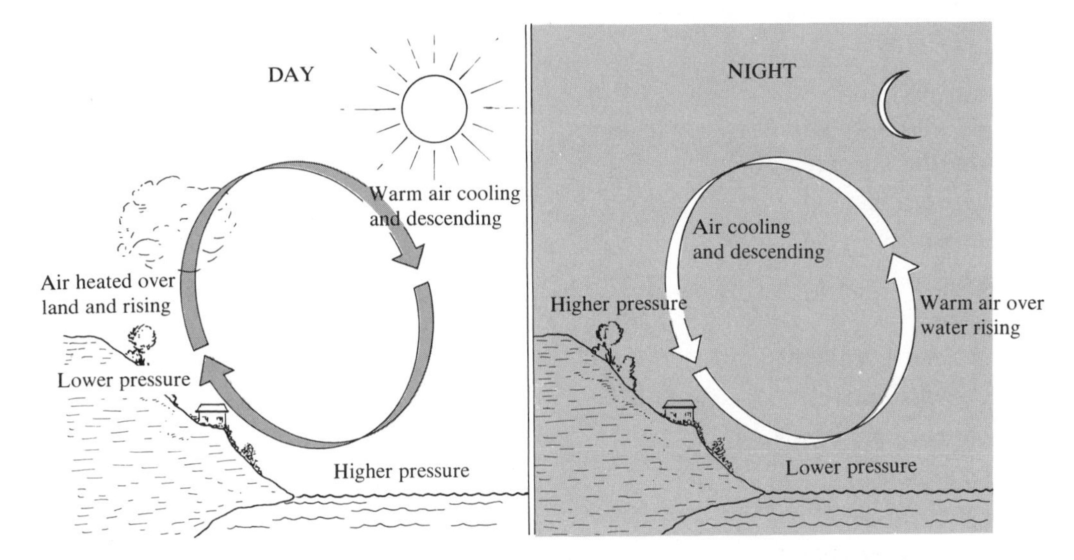

Figure 14.17 *Land and sea breezes. During the daytime, land heats more intensely than water. A low is created over the land, and the air moves from the water (higher pressure) to the land (lower pressure). At night, the land cools more rapidly, resulting in higher pressure over land and a reversal of the wind.*

REVIEW

1. What is the air pressure (weight of air) at an elevation of 5 kilometers in pounds per square inch? In inches? In millibars? (See Table 12.2.)
2. Mercury is 13 times heavier than water. If you built a barometer using water rather than mercury, how tall would it have to be in order to record standard sea level pressure (in centimeters of water)?
3. If a very flexible balloon rose to 31.2 kilometers, how many times larger would it be because of expansion than it was at sea level? (See Table 12.2.)
4. Why do your ears "pop" when you quickly descend a steep hill?
5. What force is responsible for generating wind?
6. How does the Coriolis force modify the movement of air?
7. Contrast surface and upper-air winds in terms of speed and direction.
8. Describe the weather that usually accompanies a drop in barometric pressure. A rise.
9. A northeast wind is blowing *from* the ____________ (direction) *toward* the ____________ (direction).
10. The wind direction is 225 degrees. From what compass direction is the wind blowing (Figure 14.11)?
11. Draw a diagram showing the winds associated with cyclones and anticyclones in the Southern Hemisphere.
12. If you live in the Northern Hemisphere and are directly west of a cyclone, what most probably will be the wind direction? West of an anticyclone?
13. Which global pressure system produces the tropical deserts of the world?
14. Using Figures 14.9A and 14.13, explain why some persons describe the polar regions as deserts.
15. Which global pressure systems are associated with precipitation (Figure 14.13)?
16. Describe the jet stream.
17. What influence does upper-level airflow seem to have on surface pressure systems?
18. If unequal heating produces pressure differences, which in turn drive the wind, why does the wind blow at night?
19. Describe the monsoon circulation of India (Figure 14.14).
20. Why are sea breezes in the mid-latitudes most pronounced during the summer months?

KEY TERMS

mercurial barometer
aneroid barometer
barograph
altimeter
wind
isobar
pressure gradient
Coriolis force
geostrophic winds
jet stream
cyclone (low)
anticyclone (high)
convergence
divergence
pressure (barometric) tendency
wind vane
prevailing wind
cup anemometer
equatorial low
subtropical high
trade winds
westerlies
polar easterlies
subpolar low
polar front
polar high
monsoons
sea breeze
land breeze
valley breeze
mountain breeze
chinook (foehn)

Satellite photographs like this one dramatically portray the patterned nature of our weather. Notice the swirling clouds of a middle-latitude cyclone over the central United States. A smaller, but more intense, cyclone called a hurricane can also be seen in the Pacific just a few hundred kilometers southwest of Baja California (Courtesy of NOAA)

Air Masses
Fronts
The Middle-latitude Cyclone
Thunderstorms
Tornadoes
Hurricanes

Weather Patterns and Severe Storms

Tornadoes and hurricanes rank high among nature's most destructive forces. Each spring, newspapers report the death and destruction left in the wake of a "band" of tornadoes. During the summer, we begin hearing the names of hurricanes in the news—Agnes, Betty, Camille, Doria—the list goes on and on. Thunderstorms, although less intense and more common than tornadoes and hurricanes, will also be part of our discussion in this chapter on the nature of severe weather disturbances. Before looking at violent weather, however, we shall study those atmospheric phenomena which most often affect our day-to-day weather: air masses, fronts, and traveling middle-latitude cyclones. Here, we shall see the interplay of the elements of weather discussed earlier.

Air Masses

When a portion of the lower troposphere moves slowly or stagnates over a relatively uniform surface, the air will assume the distinguishing features of that area, particularly with regard to temperature and moisture conditions. A body of air that forms in this fashion is termed an air mass. An **air mass** is a very large body of air, usually 1600 kilometers (1000 miles) or more across, and perhaps several kilometers thick, which is characterized by a homogeneity of temperature and moisture at any given altitude. When this air moves out of its region of origin, it will carry these temperatures and moisture conditions elsewhere, eventually affecting a large portion of a continent.

It may take several days for an air mass to traverse an area. As a result, the region under its influence will probably experience fairly constant weather, a situation called **air-mass weather.** Certainly, there may be some day-to-day variations, but the events will be quite unlike those in an adjacent air mass. Therefore, the boundary between two adjoining air masses having contrasting characteristics, called a *front,* marks a change in weather.

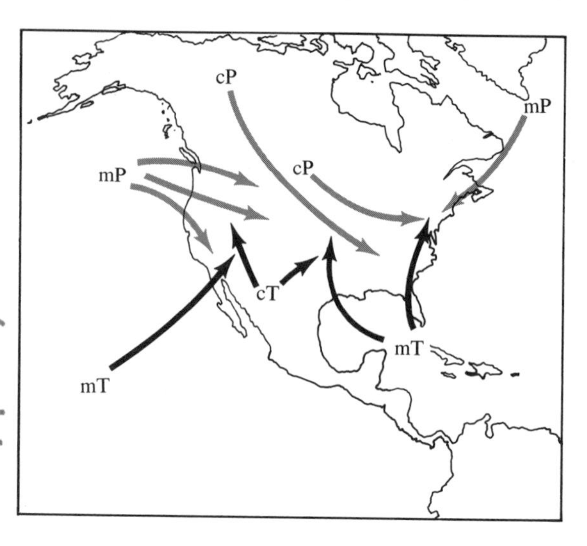

Figure 15.1 *Air masses are classified on the basis of their source region. The designation* continental *(c) or* maritime *(m) gives an indication of moisture content, while* polar *(P) and* tropical *(T) indicate temperature conditions. (After* Fire Weather, *U.S. Department of Agriculture Handbook 360, p. 129)*

Source Regions

The area where an air mass acquires its characteristic properties of temperature and moisture is called its **source region.** The source regions that produce the air masses that influence North America most are shown in Figure 15.1.

Air masses are classified according to their source region. **Polar** (P) air masses originate in high latitudes, while those which form in low latitudes are called **tropical** (T). The designation *polar* or *tropical* gives an indication of the temperature characteristics of the air mass. *Polar* indicates cold, and *tropical* indicates warm. In addition, air masses are classified according to the nature of the surface in the source region. **Continental** (c) designates land, while **maritime** (m) indicates water. The designation *continental* or *maritime* thus suggests the moisture characteristics of the air mass. Continental air is likely to be dry, and maritime air moist. *The four basic types of air masses, according to this scheme of classification, are continental polar (cP), continental tropical (cT), maritime polar (mP), and maritime tropical (mT).*

Weather Associated with Air Masses

Continental polar and maritime tropical air masses most influence the weather of North America, especially east of the Rocky Mountains. Continental polar air masses originate in northern Canada, interior Alaska, and the arctic, areas which are uniformly cold and dry in winter and cool and dry in summer. These air masses are high-pressure areas and have little cloudiness associated with them. In winter, an invasion of continental polar air brings the clear skies and cold temperatures we associate with a cold wave as it moves southward through Canada into the United States. In summer, this air mass may bring a few days of cooling relief.

Maritime tropical air masses affecting North America most often originate in the Gulf of Mexico, the Caribbean Sea, or the adjacent Atlantic Ocean. As we might expect, these air masses are warm, moisture-laden, and usually unstable. Maritime tropical air is the source of much, if not most, of the precipitation received in the eastern two-thirds of the United States. In summer, when this air mass invades the central and eastern United States, and occasionally southern Canada, it brings the heat and oppressive humidity typically associated with its source region. The tropical Pacific is also a source region for maritime tropical air. However, maritime tropical air seldom enters the continent. When it does, it may bring orographic rainfall to the southwestern United States and northern Mexico.

Of the two remaining air masses, maritime polar and continental tropical, the latter has the least influence upon the weather of North America. Hot, dry continental tropical air, originating in the Southwest and Mexico during the summer, seldom affects the weather outside its source region.

During the winter, maritime polar air masses coming from the North Pacific often originate as continental polar air masses in Siberia. As they move across the Pacific, they warm and gradually accumulate moisture. Upon entering North America, these air masses drop much of their moisture because of orographic lifting in the western mountains. By the time maritime polar air reaches the Great Plains, it is quite dry and is difficult to distinguish from continental polar air. Maritime polar air also originates off the coast of eastern Canada and occasionally influences the weather of the northeastern United States. When New England is on the northern or northwestern edge of a passing low, the cyclonic winds draw in maritime polar air. The result is a storm characterized by snow and cold temperatures, known locally as a *northeaster*.

Fronts

Fronts are defined as boundary surfaces that separate air masses of different densities, one warmer and often higher in moisture content than the other. Ideally, fronts can form between any two contrasting air masses. Considering the vast size of the air masses involved, these 15- to 200-kilometer-wide bands of discontinuity are relatively narrow. On the scale of a weather map, they are generally narrow enough to be represented satisfactorily by a broad line.

Above the ground, the frontal surface slopes at a low angle so that warmer air overlies cooler air. In the ideal case, the air masses on both sides of the front move in the same direction and at the same speed. Under this condition, the front acts as a barrier with which the air masses must move but through which they cannot penetrate. Generally, however, the pressure field across a front is such that air on one side is moving faster in the direction perpendicular to the front than the air mass on the other side of the front. Thus, one air mass actively advances into another and "clashes" with it. The boundaries were thus tagged *fronts* during World War I by Norwegian meteorologists, who visualized them as analogous to battle lines.

As one air mass moves into another, some mixing does occur along the frontal surface, but, for the most part, the air masses retain their identity as one air mass is displaced upward over the other. No matter which air mass is advancing, it is always the warmer, lighter air that is forced aloft, while the cooler, heavier air acts as the wedge upon which lifting takes place.

Warm Fronts

When the surface position of a front moves so that warm air occupies territory formally covered by cooler air, it is called a **warm front.**

On a weather map, the surface position of a warm front is denoted by a line with semicircles extending into the cooler air. East of the Rockies, warm tropical air often enters the United States from the Gulf of Mexico and overruns receding cool air. As the cold wedge retreats, friction slows the advance of the surface position of the front more so than its position aloft, so the boundary separating these air masses acquires a small slope. The average slope of a warm front is about 1:200, which means that if you are 200 kilometers ahead of the surface location of a warm front, you will find the frontal surface at a height of 1 kilometer.

As warm air ascends the retreating wedge of cold air, it cools by adiabatic expansion, producing clouds, and frequently precipitation. Typically, the sequence of clouds shown in Figure 15.2A precedes a warm front. The first sign of the approach of a warm front is the appearance of cirrus clouds overhead. These high clouds form 1000 kilometers or more ahead of the surface front where the overrunning warm air has ascended high up the wedge of cold air. As the front nears, cirrus clouds grade into cirrostratus, which blend into denser sheets of altostratus. About 300 kilometers ahead of the front, thicker stratus and nimbostratus clouds appear and rain or snow begins. Usually warm fronts produce several hours of moderate-to-gentle precipitation over a large region. This is in agreement with the rather gentle slope of a warm

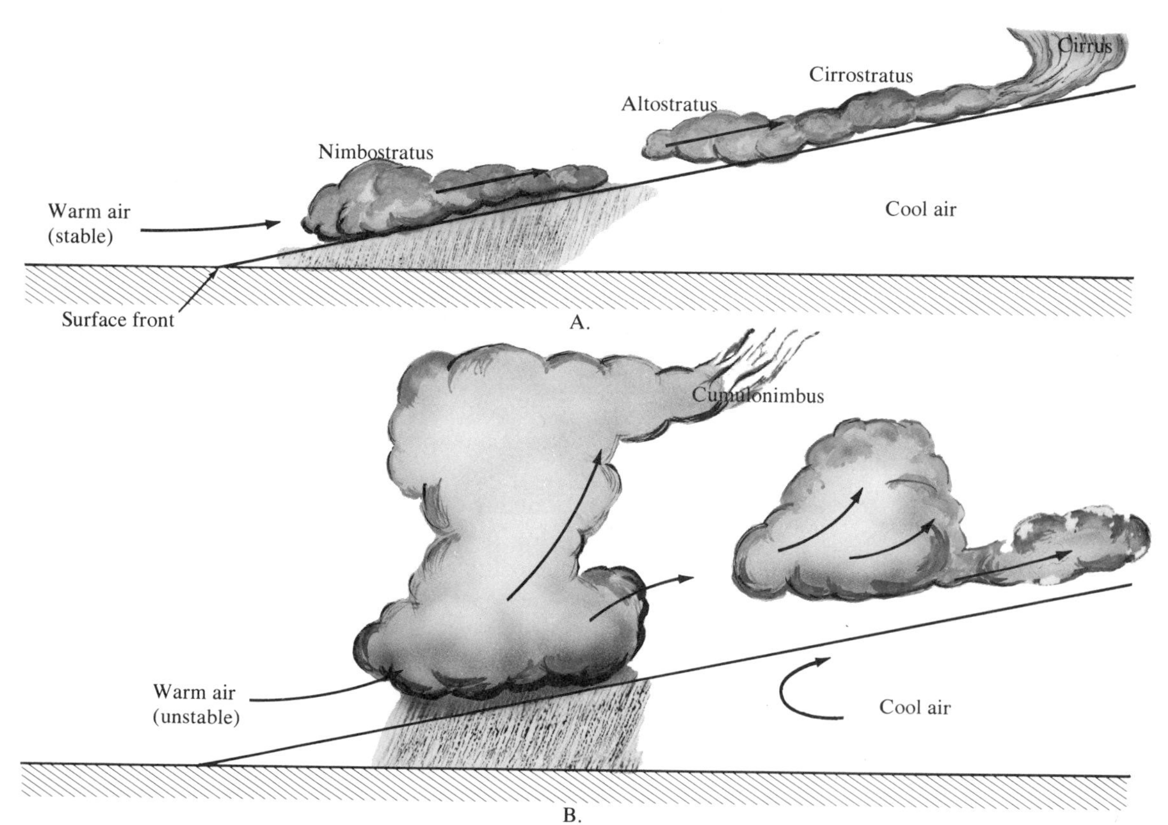

Figure 15.2 **A.** *Warm front with stable air and associated stratiform clouds. Precipitation is moderate and occurs within a few hundred kilometers of the surface front.* **B.** *Warm front with unstable air and cumuliform clouds. Precipitation is heavy near the surface front.*

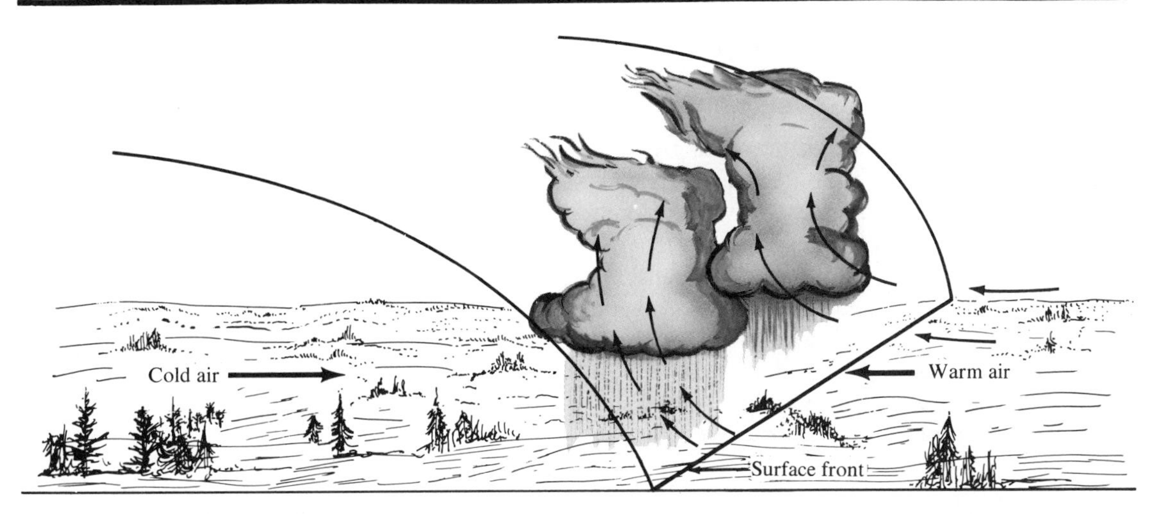

Figure 15.3 *Fast-moving cold front and cumulonimbus clouds. Often thunderstorms occur if the warm air is unstable.*

front, which does not generally encourage convectional activity. However, on some occasions, warm fronts are associated with cumulonimbus clouds and violent thunderstorms (Figure 15.2B). These result when the overrunning air is inherently unstable and the front is rather sharp. When these conditions exist, cirrus clouds are generally followed by cirrocumulus clouds, giving us the familiar "mackerel sky," which sailors saw as a warning of an impending storm, as indicated by the following proverb:

> Mackerel scales and mares' tails
> Make lofty ships carry low sails.

At the other extreme, a warm front associated with a rather dry air mass could pass unnoticed by those of us at the surface.

A rather gradual increase in temperature occurs with the passage of a warm front. As we would expect, the increase is most apparent when there exists a large temperature difference between the adjacent air masses. Furthermore, a wind shift from the east to the southwest is generally noticeable. The reason for this shift will be evident later. The moisture content and stability of the encroaching warm air mass largely determines when clear skies return. During the summer, cumulus, and occasionally cumulonimbus, are embedded in the warm unstable air mass that follows the front. Precipitation from these clouds is usually sporadic and not extensive.

Cold Fronts

When cold air is actively advancing into a region that is occupied by warmer air, the zone of discontinuity is called a **cold front.** As with warm fronts, friction tends to slow the surface position of a cold front more so than its position aloft. However, because of the relative positions of the adjacent air masses, the cold front steepens as it moves. On the average, cold fronts are about twice as steep as warm fronts, having a slope of perhaps 1:100. In addition, cold fronts advance more rapidly than warm fronts. These two differences, rate of movement and steepness of slope, largely account for the more violent nature of cold-front weather. The displacement of air along a cold front is often rapid enough that the released latent heat appreciably increases the air's buoyancy. This frequently results in the sudden downpours and vigorous gusts of wind associated with mature cumulonimbus clouds (Figure 15.3). Since a cold front produces roughly the same amount of lifting as a warm front, but over a shorter distance, the intensity of precipitation is greater, but the duration is shorter.

A cold front is sometimes preceded by altocumulus clouds. As the front approaches, generally from the west or northwest, towering clouds are often seen in the distance. Near the front, a dark band of ominous clouds foretells the ensuing weather. Usually a marked temperature drop and a wind shift from the south to west or northwest accompany the passage of the front. On a weather map, the violent weather and sharp temperature contrast are indicated by a line with triangle-shaped points that extends into the warmer air mass.

The weather behind a cold front is dominated by a subsiding and relatively cold air mass. Hence, clearing conditions prevail after the front passes. Although general subsidence causes some adiabatic heating, this has a minor effect on surface temperatures. In the winter, the clear skies associated with these cold outbreaks further reduce surface temperatures because of more rapid radiation cooling at night. If the continental polar air mass, which most frequently accompanies a cold front, moves into a relatively warm and humid area, surface heating can produce shallow convection, which in turn may generate low cumulus or stratocumulus clouds behind the front.

Occluded Fronts

Another commonly occurring front is the **occluded front.** Here, an active cold front overtakes a warm front as shown in Figure 15.4. As the advancing cold air wedges the warm front upward, a new front emerges between the advancing cold air and the air over which the warm front is gliding. The weather of an occluded front generally is quite complex. Most of the precipitation is associated with the warm air being forced aloft. However, when conditions are suitable, the newly formed front can produce precipitation of its own.

The Middle-latitude Cyclone

The cold fronts and wam fronts just described continually influence the weather in the mid-latitudes. Usually, these fronts are associated with a low-pressure system that is termed a **middle-latitude,** or **wave, cyclone.**

Life Cycle of a Wave Cyclone

According to the wave cyclone model, cyclones form along fronts where they change in a somewhat predictable way. This life cycle can last for a few hours or for several days. Figure 15.5 is a schematic representation of the stages in the development of a "typical" wave cyclone. As the figure shows, cyclones originate along a front where air masses of different densities (temperatures) are moving parallel to the front in opposite directions. In the classic model, continental polar air associated with the polar easterlies would be north of the front, and maritime tropical air of the westerlies south of the front. The result of this opposing airflow is counterclockwise (cyclonic) rotation. (To better visualize this effect, place a pencil between the palms of your hands. Now move your right hand ahead of your left hand and notice that your pencil rotates in a counterclockwise fashion.) Under the correct conditions, the frontal surface will take on a wave shape. These waves are analogous to the waves that are produced on the surface of a water body by moving air, except the scale is different. The waves generated between two contrasting air masses are usually several hundred kilometers long. Some waves tend to dampen out while others become unstable and grow in amplitude. The latter ones change in shape with time much like a gentle ocean swell does as it moves into shallow water and becomes a tall breaking wave.

Once a small wave forms, warm air invades this weak spot along the front and extends itself poleward, while the surrounding cold air moves equatorward. This change causes a readjustment

Cirrus clouds at sunset. (Photo by E.J. Tarbuck)

Altocumulus clouds at sunrise. (Courtesy of Ward's Natural Science Establishment Inc. Rochester N.Y.)

When supercooled fog freezes on objects, a delicate frosting called rime results. (Photo by W.B. Hamilton, U.S. Geological Survey)

in the pressure field which results in nearly circular isobars, with the low pressure centered at the apex of the wave. The creation of the low encourages the inflow (convergence) of air and general vertical lifting, particularly where warm air is overrunning colder air. We can see in Figure 15.6 that the air in the warm sector is flowing from the southwest toward colder air flowing from the southeast. Since the warm air is moving faster than the cold air in a direction perpendicular to this front, we can conclude that warm air is invading a region formally occupied by cold air; hence, this must be a warm front. Similar reasoning indicates that in the rear of the cyclonic disturbance, cold air is underrunning the air of the warm sector, generating a cold front there. Generally, the position of the cold front advances faster than the warm front and begins to close the warm sector as shown in Figure 15.5. This process, called **occlusion,** results in an occluded front with the displaced warm sector located aloft. The cyclone enters

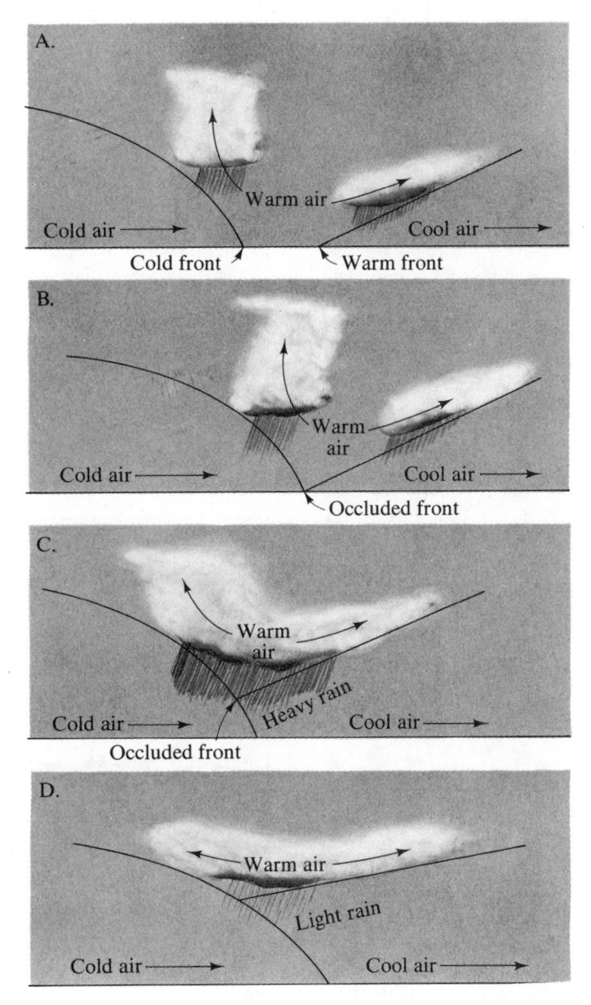

Figure 15.4 *Stages in the formation and eventual dissipation of an occluded front.*

maturity (maximum intensity) when it reaches this stage in its development. A steep pressure gradient and strong winds develop as lifting continues. Eventually all of the warm sector is forced aloft and cold air surrounds the cyclone at low levels. Once the sloping discontinuity (front) between the air masses no longer exists, the pressure gradient weakens. At this point, the cyclone has exhausted its source of energy, and the storm comes to an end.

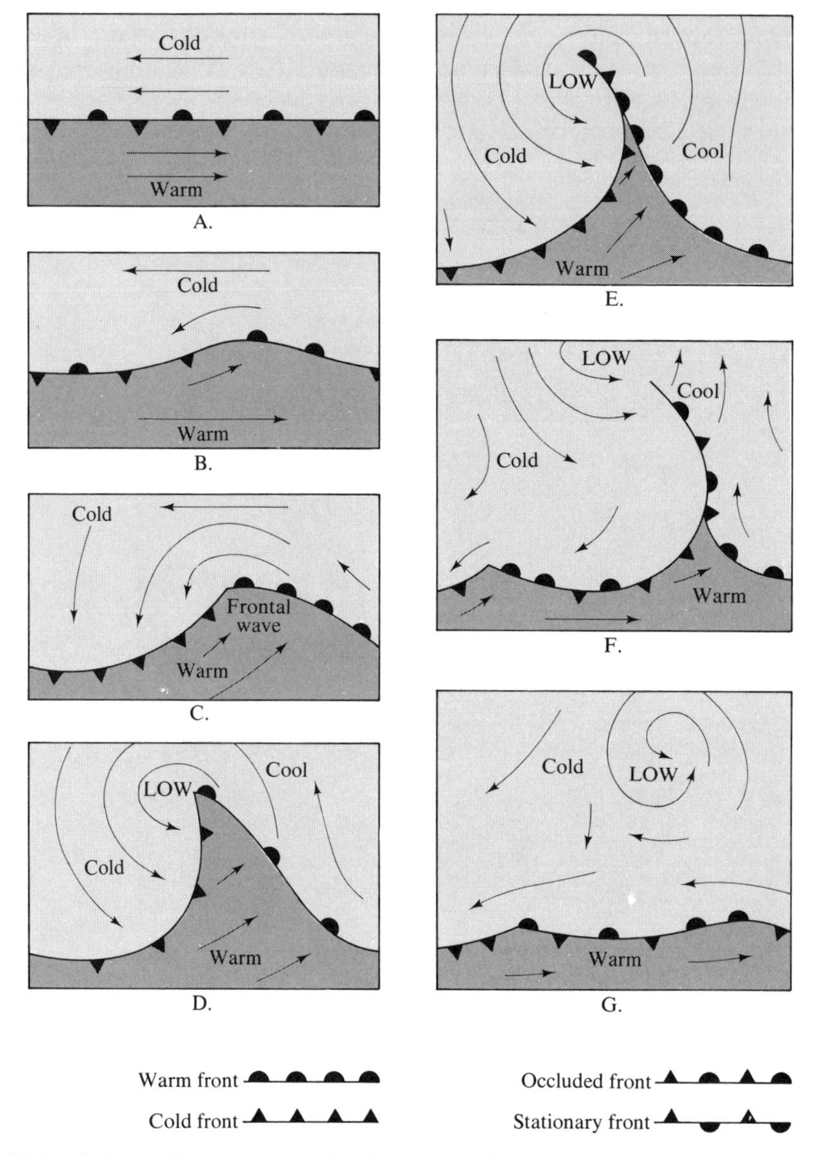

Figure 15.5 *Schematic representation (map view) of the life cycle of a hypothetical middle-latitude cyclone.* **A.** *Front develops.* **B.** *Wave appears along front.* **C.** *Cyclonic circulation is well developed.* **D.** *Cold front invades warm sector.* **E.** *Occlusion begins.* **F.** *Occluded front is fully developed.* **G.** *Cyclone dissipates (As proposed by J. Bjerknes)*

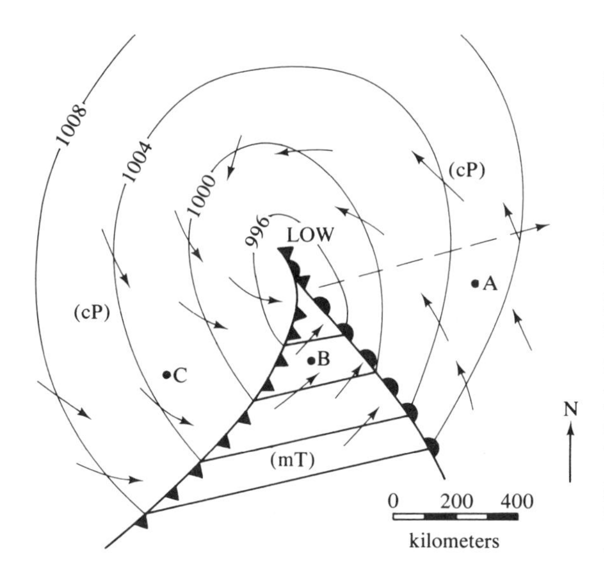

Figure 15.6 *Idealized circulation of a mature cyclone. Study this figure to determine the wind shifts that would be expected for a city as the storm moved so that the city's relative position shifted from location* A *to* B, *and then from* B *to* C.

Idealized Weather of a Wave Cyclone

As stated earlier, the cyclone model provides a useful tool for examining the weather patterns of the middle latitudes. Figure 15.7 illustrates the distribution of clouds and thus the regions of possible precipitation associated with a mature wave cyclone. Compare this drawing to the satellite photo of a cyclone shown in Figure 15.8. Guided by the westerlies aloft, cyclones generally move eastward across the United States, so we can expect the first signs of their arrival in the west. However, often in the region of the Mississippi valley, cyclones begin a more northeasterly trajectory and occasionally move directly northward. Typically, a mid-latitude cyclone requires two to four days to pass over a given region. During that relatively short time, rather abrupt changes in atmospheric conditions may be experienced. This is particularly true in the spring, when the largest temperature contrasts occur across the mid-latitudes.

Using Figure 15.7 as a guide, we will now consider these weather producers and what we should expect from them as they pass an area during the spring. To facilitate our discussion, profiles are provided along lines A–B and C–D. First, imagine the change in weather as you move along profile A–B. Here the sighting of high cirrus clouds would be the first sign of the approaching cyclone. These high clouds can precede the surface front by 1000 kilometers or more and they generally will be accompanied by falling pressure. As the warm front advances, a lowering and thickening of the cloud deck is noticed. Usually within 12 to 24 hours after the first sighting of cirrus clouds, light precipitation begins. As the front nears, the rate of precipitation increases, a rise in temperature is noticed, and winds begin to change from an easterly to a southerly flow. With the passage of the warm front, the area is under the influence of the maritime tropical air mass of the warm sector. Generally the region affected by this sector of the cyclone experiences warm temperatures, southerly winds, and generally clear skies, although fair-weather cumulus or altocumulus are not uncommon here. The rather pleasant weather of the warm sector passes quickly and is replaced by gusty winds and precipitation generated along the cold front. The approach of a rapidly advancing cold front is marked by a wall of rolling black clouds. Severe weather accompanied by heavy precipitation, hail, and an occasional tornado is a definite possibility at this time of year. The passage of the cold front is easily detected by a wind shift; the southerly flow is replaced by winds from the west to northwest and by a pronounced drop in temperature. Also, rising pressure hints of the subsiding cool, dry air behind the front. Once the front passes, the skies clear quickly as the cooler air invades the region. Usually a day or two of almost cloudless deep-blue skies can be expected unless another cyclone is edging into the region.

A very different set of weather conditions will prevail in those regions that encounter the portion of the cyclone containing the occluded front as shown along profile C–D. Here the temperatures remain cool during the passage of the storm; however, a continual drop in pressure and increasingly overcast conditions strongly hint at the approach of the low-pressure center. This sector of the cyclone most often generates snow or icing storms during the cool months. Further, the occluded front often moves more slowly than the other fronts; hence, the wishbone shaped frontal structure shown in Figure 15.7 rotates in a counterclockwise manner, such that the occluded front appears to "bend over backwards." This effect adds to the misery of the region influenced by the occluded front since it remains over the area longer than the other fronts. Also, the storm reaches its greatest intensity during occlusion; consequently, the area affected by the developing occluded front receives the brunt of the storm's fury.

To reinforce your knowledge of the weather associated with a wave cyclone, carefully examine the weather map and satellite photograph of a cyclone that moved eastward across the United States in March 1975 (Figure 15.8). Basic weather data have been included for selected weather stations. Using this information, note how the temperature, amount of cloud cover, type of precipitation, wind speed, and wind

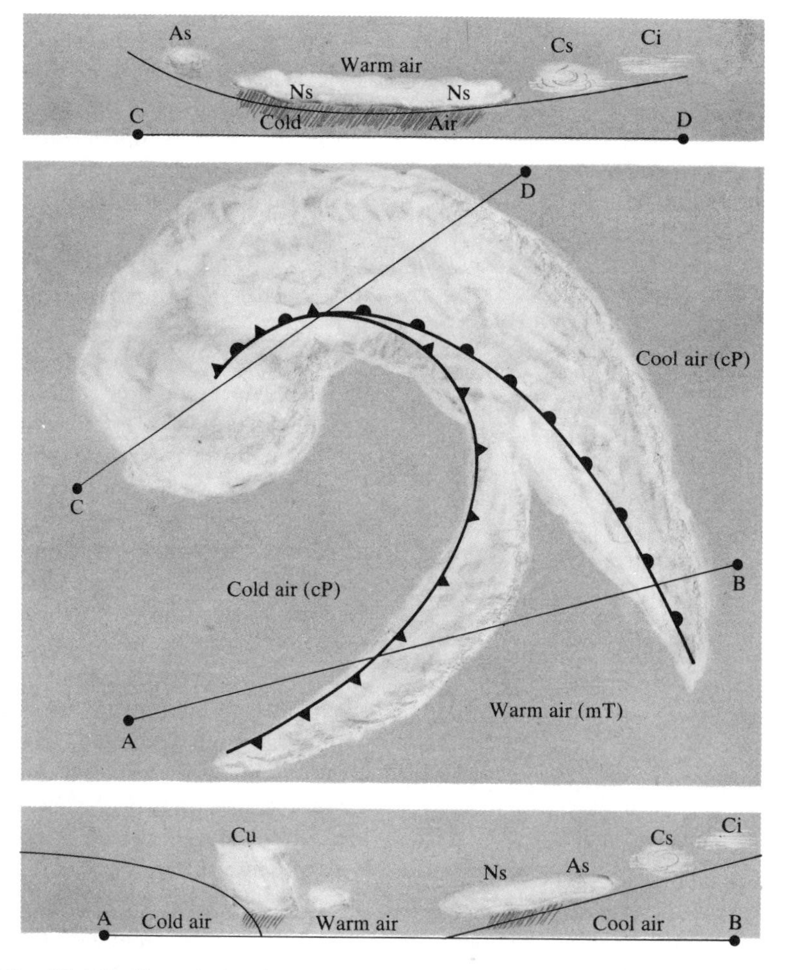

Figure 15.7 *Distribution of clouds associated with an idealized mature wave cyclone.*

direction vary with respect to the individual station location relative to the cyclone's position. In particular, locate places that are experiencing thunderstorms, snowfall, high winds, clear skies, and so forth.

Thunderstorms

Most everyone has observed small-scale phenomena that are caused by the vertical motion of warm, unstable air. You may have seen a small dust devil which formed over an open field on a hot day when it whirled its dusty load to great heights. Or perhaps you have noticed a bird glide skyward effortlessly upon an invisible thermal of hot air. These examples serve to illustrate the much more dynamic thermal instability that occurs during the development of a **thunderstorm.** At any given time over the earth's surface, nearly 2000 thunderstorms are in progress, mainly in tropical regions. Thunderstorm activity is associated with massive cumulonimbus clouds which generate torrential downpours, thunder, lightning, and occasionally hail.

Three types of thunderstorms exist: frontal, orographic, and air-mass. Frontal and orographic thunderstorms are triggered by forceful lifting, while air-mass thunderstorms seem to be initiated by intense surface heating and general atmospheric convergence. Near the equator, air-mass thunderstorms frequently occur in the late afternoon as part of the general convergence along the equatorial low. In the United States, they are a summertime phenomenon associated with the maritime tropical sector of a middle-latitude cyclone.

Although frontal, orographic, and air-mass thunderstorms have different origins, they form in a similar manner. All require warm, moist air, which, when lifted, releases latent heat, becomes unstable, and rises. Because instability and buoyancy are enhanced by high surface temperatures, thunderstorms are most common in the afternoon and early evening. However, surface heating is not sufficient for the growth of towering cumulonimbus clouds. A solitary cell of rising hot air produced by surface heating could, at best, produce a small cumulus cloud, which would evaporate within 10–15 minutes. The development of 12,000-meter (or on rare occasions 18,000-meter) cumulonimbus towers requires a continual supply of moist air. Each new surge of warm air rises higher than the last, adding to the height of the cloud (Figure 15.9). These updrafts must occasionally reach speeds over 100 kilometers per hour to accommodate the size of hailstones they are capable of carrying upward. Usually within a half hour the amount and size of precipitation that has accumulated is too much for the updrafts to support, and in one part of the cloud downdrafts develop, releasing heavy precipitation. This represents the most active stage of the thunderstorm. Gusty winds, lightning, heavy precipitation, and sometimes hail, are experienced. Eventually, downdrafts dominate throughout the cloud. The cooling effect of falling precipitation coupled with the influx of colder air aloft marks the end of the thunderstorm activity (Figure 15.9). The life span of a cumulonimbus cell within a thunderstorm complex is only about an hour, but as the storm moves, fresh supplies of warm, water-laden air generate new cells to replace those that are dissipating (Figure 15.10).

One obvious feature of a thunderstorm is thunder. Because thunder is produced by lightning, lightning must also be present (Figure 15.11). **Lightning** is very similar to the electrical shock you may have experienced on a very dry day upon touching a metal object. Only the intensities differ. During the development of a cumulonimbus cloud, a similar buildup of charge is generated. The reason for this, although not fully understood, hinges on the movement of precipitation within the cloud. The upper portion of the cloud acquires a positive charge, while the lower portion of the cloud maintains an overall negative charge while containing small positively charged pockets. These charges will build to millions, and even hundreds of millions, of volts before a lightning stroke acts to discharge the cloud by striking the earth, or possibly another cloud. The lightning we see as a single

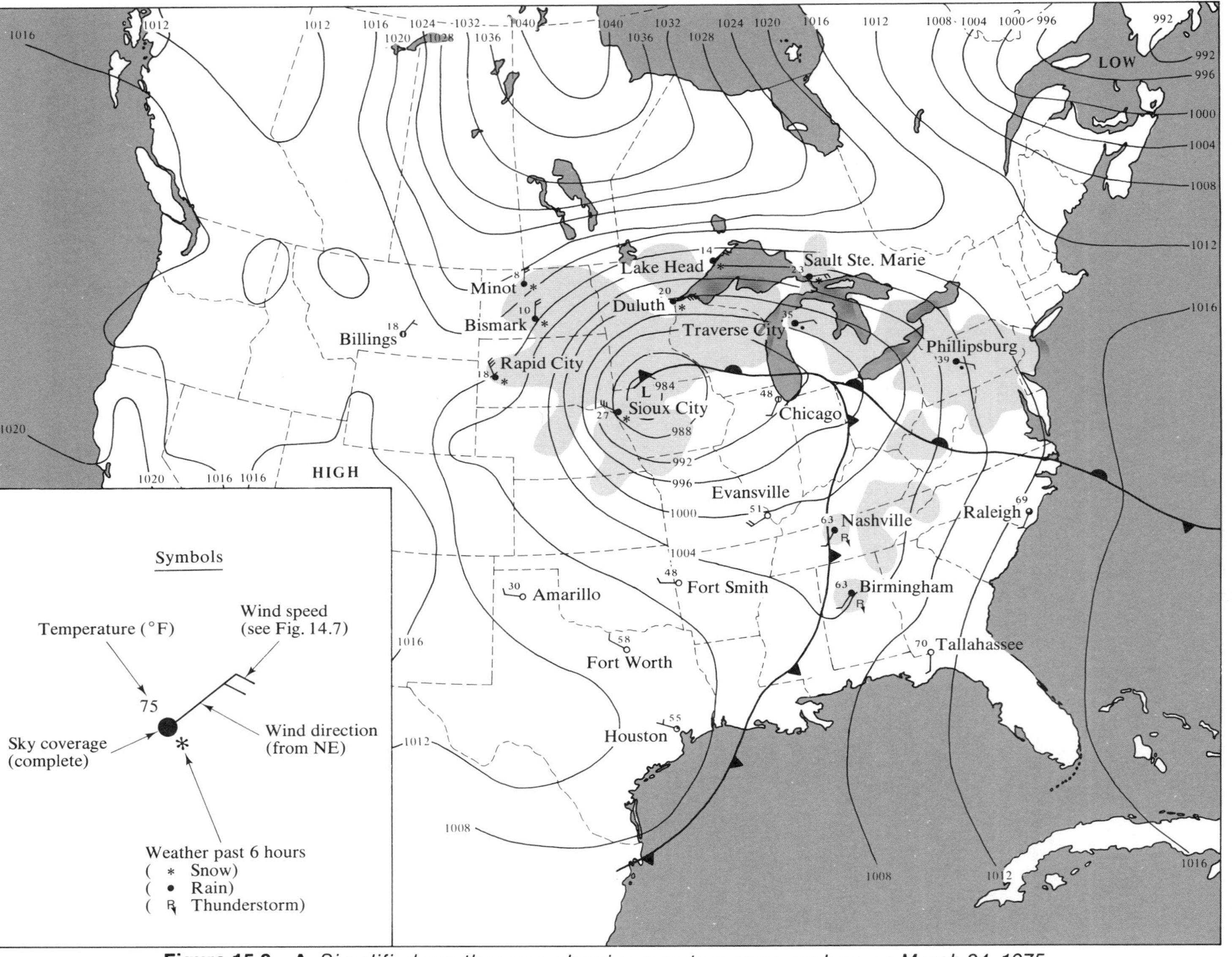

Figure 15.8 **A.** *Simplified weather map showing a mature wave cyclone on March 24, 1975.*

Figure 15.8 B. *Satellite photograph of the same cyclone shown on the weather map. (Courtesy of NOAA)*

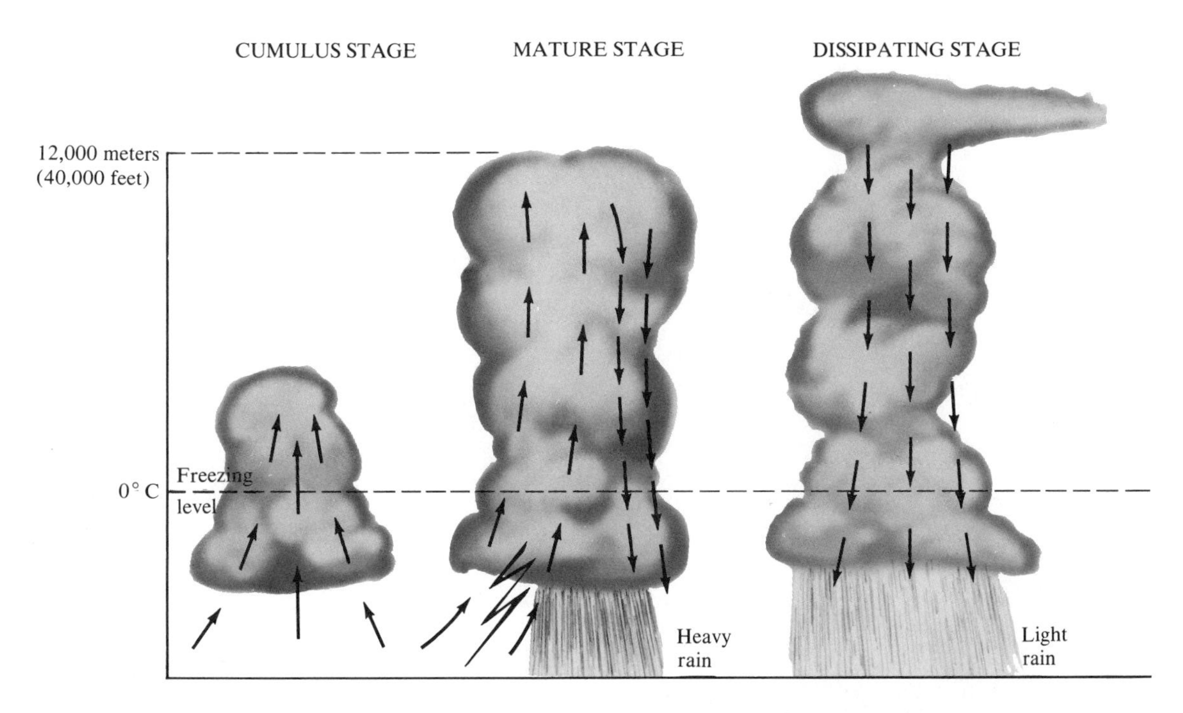

Figure 15.9 *Stages in the development of a thunderstorm. During the cumulus stage, strong updrafts act to build the storm. The mature stage is marked by heavy precipitation and cool updrafts in part of the storm. When the warm updrafts disappear completely, precipitation becomes light, and the cloud begins to evaporate.*

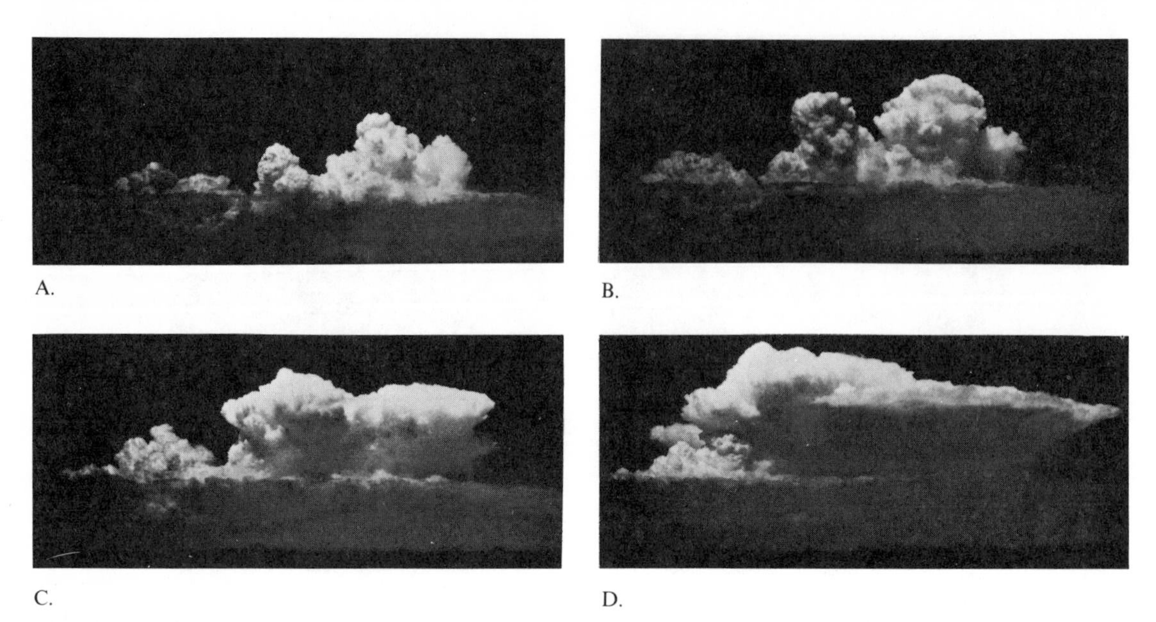

Figure 15.10 *Cumulonimbus in various stages of development. (Courtesy of R.R. Braham Cloud Physics Laboratory, The University of Chicago)*

Figure 15.11 *Lightning. (Courtesy of NOAA)*

stroke is really several very rapid strokes between the cloud and the earth, which together last only about one-tenth of a second.

Each stroke is believed to begin when the electrical field near the cloud base frees electrons in the air immediately below, thereby ionizing the air. Once ionized, the air becomes a conductive path with a length of 50 meters, called a *leader.* During this electrical breakdown, the mobile electrons in the cloud base begin to flow down this channel. This flow increases the electrical potential at the head of the leader, which causes a further extension of the conductive path through further ionization. Because this initial path extends itself earthward in short, nearly invisible bursts, it is called a *step leader.* Once this channel nears the ground, the electrical field at the surface ionizes the remaining section of the path. When the channel is complete, a wave of ionizing potential called the *return stroke* moves upward along the channel and fully ionizes it. As the wave front advances, the negative charge which was deposited on the channel is effectively lowered to the earth. It is this intense return stroke that illuminates the conductive path and discharges the lowest kilometer or so of the cloud.

The electrical discharge of lightning heats the air, causing the air to expand explosively. We hear this expansion as **thunder.** Since lightning and thunder occur simultaneously, it is possible to estimate the distance to the stroke. Lightning is seen instantaneously, while the rather slow sound waves, which travel approximately 330 meters (1000 feet) per second, reach us some time later. Therefore, if thunder is heard 5 seconds after the lightning is seen, the lightning occurred about 1500 meters (5000 feet) away (approximately 1 mile).

The thunder we hear as a rumble is produced along a rather long lightning path located at some distance from the observer. The sound that originates along the path nearest the observer arrives before the sound that originates farthest away. This lengthens the duration of the thunder. Reflection of the sound waves further delays their arrival and adds to this effect. When lightning occurs more than 20 kilometers away, thunder is rarely heard. This type of lightning, popularly called heat lightning, is no different than that which we associate with thunder.

Tornadoes

Tornadoes are local storms of short duration that may be ranked high among nature's most destructive forces. Tornadoes, sometimes called *twisters,* or *cyclones,** are intense centers of low pressure that have a whirlpool-like structure of winds rotating around a central cavity, where centrifugal force produces a partial vacuum. Pressures within the center of the tornado may be 100 millibars (3 inches) less than immediately outside the storm. With such a tremendous pressure gradient, some meteorologists have estimated that wind speeds may reach 650 kilometers (400 miles) per hour or more. Condensation occurs within the tornado, creating the pale and ominous-appearing funnel cloud, which may darken as it moves across the ground, picking up dust and debris (Figure 15.12).

Although many theories about tornado formation have been advanced, none has won general acceptance. The general atmospheric conditions that are most likely to develop into tornado activity, however, are known. *Tornadoes are most often spawned along the cold front of a middle-latitude cyclone, in conjunction with severe thunderstorms.* April through June is the period of greatest tornado frequency, although tornadoes have been known to occur during every month of the year. Throughout the spring, the air masses associated with middle-latitude cyclones are most likely to have greatly contrasting conditions.

Figure 15.12 *Funnel cloud. (Courtesy of NOAA)*

*The term *cyclone* is a poor choice here, for in meteorology it refers to any low-pressure center, not just a tornado.

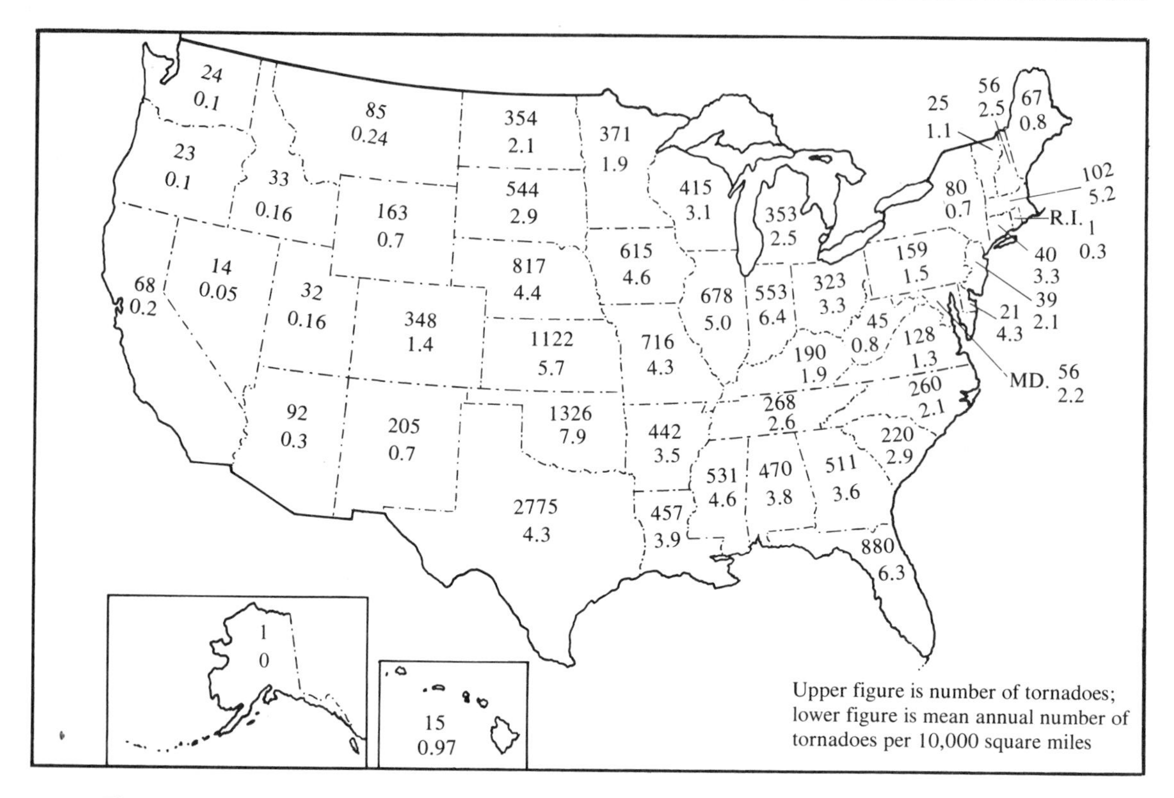

Figure 15.13 *Tornado incidence by state and area for 1953–1976. (Data from NOAA)*

Continental polar air from the Canadian arctic may still be very cold and dry, while maritime tropical air from the Gulf of Mexico is very warm and moisture laden. The greater the contrast, the more intense the storm. Since these two contrasting air masses are most likely to meet in the central United States, it is not surprising that this region generates more tornadoes than any other area of the country, and, in fact, the world. Figure 15.13, which depicts tornado incidence by state and area, readily substantiates this fact.

The average funnel cloud has a diameter of between 150 and 600 meters (500 and 2000 feet), travels across the landscape at 45 kilometers (30 miles) per hour, and cuts a path about 26 kilometers (16 miles) long. Since many tornadoes occur slightly ahead of the cold front, in the zone of southwest winds (Figure 15.6), most move toward the northeast. The Illinois example demonstrates this nicely (Figure 15.14). What Figure 15.14 also shows is that many tornadoes do not fit the description of the "average" tornado. Many have had paths a great deal longer than 26 kilometers and have traveled not at 45 kilometers per hour, but at speeds in excess of 100 kilometers (70 miles) per hour (Figure 15.15). Furthermore, there have been tornadoes that have had diameters of up to 1.5 kilometers (1 mile), more than four times the "average" size.

The near-total destruction wrought by a tornado in a populated area is linked to the combined effects of the exceedingly strong winds and the partial vacuum in the center of the storm. The winds may rip apart everything in the path of the storm, and the abrupt pressure decrease may cause some buildings to literally explode, especially if the windows are closed, because of the much lower pressure outside the structure. In addition, funnels have accomplished many seemingly impossible tasks, such as driving a piece of straw through a thick wooden plank and uprooting huge trees (Figure 15.16). In 1931,

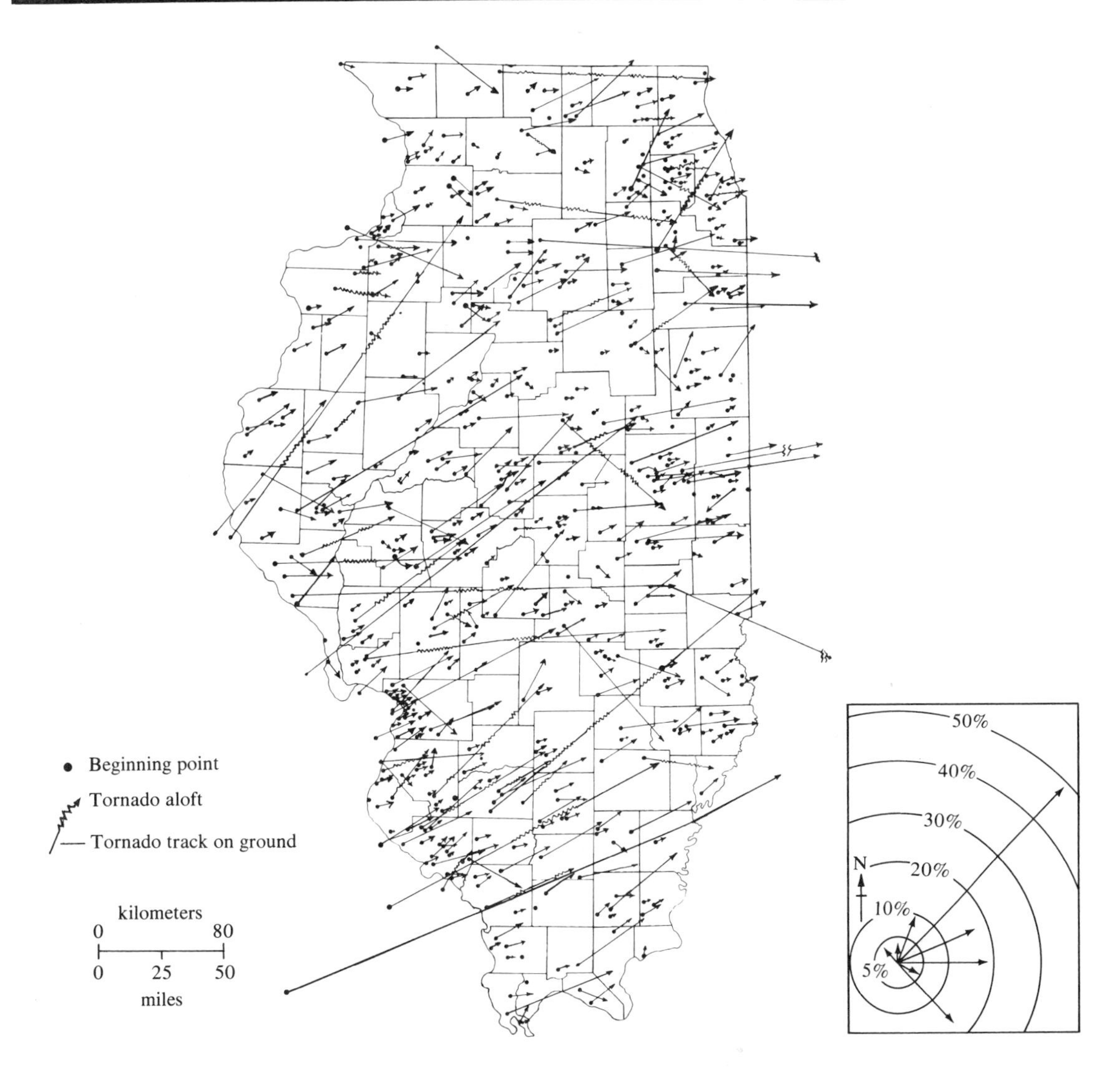

Figure 15.14 *Path of Illinois tornadoes (1916–1969). Since most tornadoes occur slightly ahead of a cold front, in the zone of southwest winds, they tend to move toward the northeast. Tornadoes in Illinois verify this. Over 80 percent exhibited directions of movement toward the northeast through east. (After J.W. Wilson and S.A. Changnon Jr.,* Illinois Tornadoes, *Illinois State Water Survey Circular 103, 1971, pp. 10, 24)*

a tornado actually carried an 83-ton railroad coach and its 117 passengers 24 meters (80 feet) through the air and dropped them in a ditch.

Tornadoes take many lives each year, sometimes hundreds in a single day. When tornadoes struck an area stretching from Canada to Georgia on April 3, 1974, the death toll exceeded 300, the worst in half a century. If there is some question as to the causes of tornadoes, there certainly is no question about the destructive effects of these violent storms. A quotation from the government publication *Tornado* puts it this way:

> Their time on earth is short, and their destructive paths are rather small. But the march of these

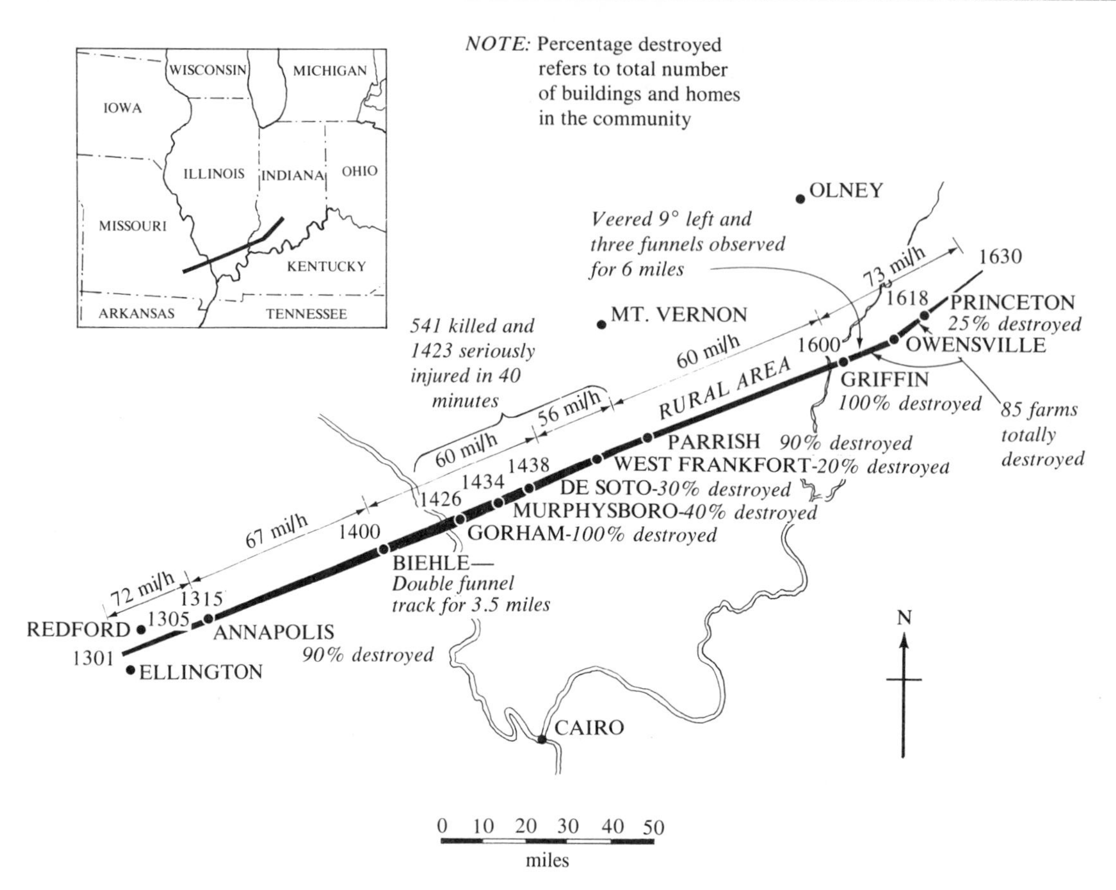

Figure 15.15 *"One tornado among the more than 13,000 which have occurred in the United States since 1915 easily ranks above all others as the single most devastating storm of this type. Shortly after its occurrence on 18 March 1925, the famed Tri-State tornado was recognized as the worst on record, and it still ranks as the nation's greatest tornado disaster. The tornado remained on the ground for 219 miles. The resulting losses included 695 dead, 2027 injured, and damages equal to $43 million in 1970 dollars. This represents the greatest death toll ever inflicted by a tornado and one of the largest damage totals." (From J.W. Wilson and S.A. Changnon Jr.,* Illinois Tornadoes, *Illinois State Water Survey Circular 103, 1971, p. 32)*

shortlived, local storms through populated areas leaves a path of terrible destruction. In seconds, a tornado can transform a thriving street into a ruin, and hope into despair.

Hurricanes

The whirling tropical cyclones that on occasion have wind speeds reaching 300 kilometers per hour are known in the United States as **hurricanes** and are the greatest storms on earth. Out at sea, they can generate 15-meter waves capable of inflicting destruction hundreds of kilometers from their source. Should a hurricane smash into land, gale-force winds coupled with extensive flooding can impose millions of dollars in damages and great loss of life. These storms form in all tropical waters (except those of the South Atlantic) between the latitudes of 5 degrees and 20 degrees (Figure 15.17) and are known in each

Figure 15.16 *The force of the winds during a tornado in Clarendon, Texas, in 1970 was enough to drive a wooden stick through a 4-centimeter pipe. (Courtesy of NOAA)*

locale by a unique name. In the western Pacific, they are called *typhoons,* in Australia, *willy-willy's,* and in the Indian Ocean, *cyclones.* The North Pacific has the greatest number of storms, averaging 20 per year. Fortunately for those living in the coastal regions of the southern and eastern United States, fewer than 5 hurricanes, on the average, develop each year in the warm sector of the North Atlantic.

Although many tropical storms develop each year, only a few reach hurricane status, which by international agreement requires wind speeds in excess of 119 kilometers (74 miles) per hour and a rotary circulation. Hurricanes average 600 kilometers (400 miles) in diameter and often extend 12,000 meters (40,000 feet) above the ocean surface. From the outer edge to the center, the barometric pressure has on occasion dropped 60 millibars, from 1010 millibars to 950 millibars (Figure 15.18). The lowest pressure ever recorded in the United States was 892.31 millibars, measured during a hurricane in September 1935. A steep pressure gradient generates the rapid, inward-spiraling winds of a hurricane. As the inward rush nears the core of the storm, it is whirled upward (Figure 15.19). Upon ascending, the air condenses and produces a thick cloud cover while simultaneously releasing large quantities of latent heat, which is the energy that drives the storm. Near the top of the hurricane the airflow is outward, carrying the rising air away from the storm center, thereby providing room for more inward flow at the surface.

At the center of the storm is the spectacular **eye** (Figure 15.20). Averaging 20 kilometers (14 miles) in diameter, this zone of calm and scattered cloud cover is unique to the hurricane.

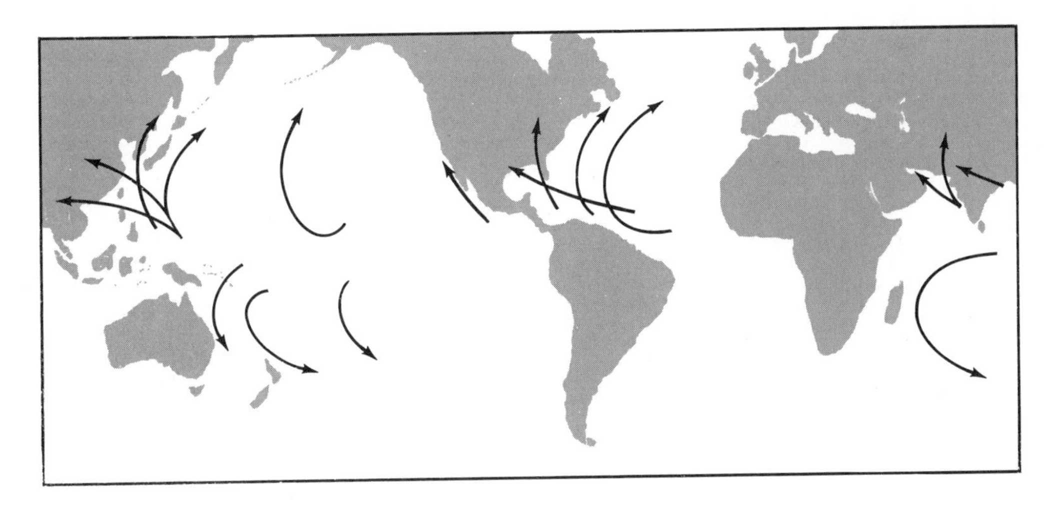

Figure 15.17 *Regions of hurricane formation and paths of movement.*

Separating periods of torrential rains, the eye, with its blue sky and intermittent sunlight against a background of circling clouds which are 12 kilometers (8 miles) high, is an awesome sight. The air within the eye slowly descends and heats by compression, making it the warmest part of the storm.

A hurricane can be described as a heat engine that is fueled by the energy liberated during the condensation of water vapor (latent heat). The energy from an average hurricane is equal to the total amount of electricity consumed in the United States over a six-month period. The release of latent heat warms the air and provides buoyancy for its upward flight. The result is to reduce the pressure near the surface, which encourages a more rapid inward flow of air. To get this engine started, a large quantity of warm, moisture-laden air is required, and a continual supply is needed to keep it going.

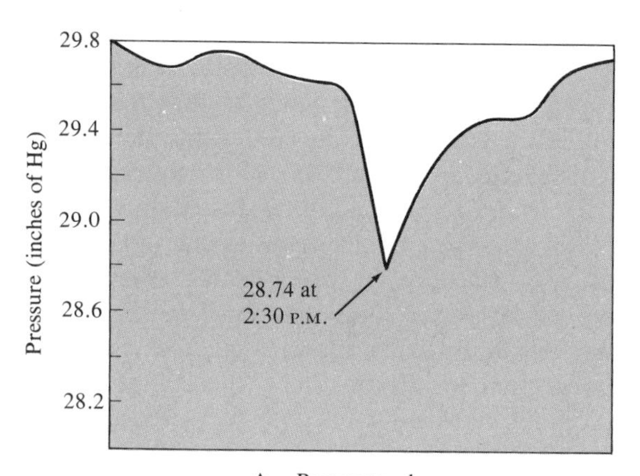

A. Barometer changes

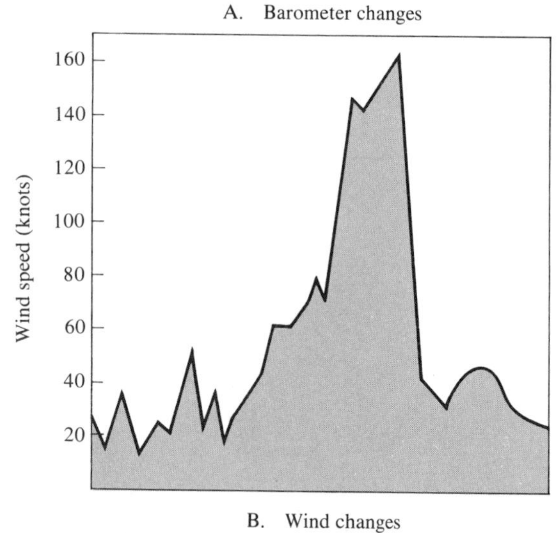

B. Wind changes

Figure 15.18 *Changes in* ***A.*** *pressure and* ***B.*** *wind speed during the passage of a hurricane, San Juan, Puerto Rico.*

Hurricanes develop most often in the late summer when water temperatures have reached 27° C (80° F). Although the exact mechanism of formation is not completely understood, we know that smaller tropical storms initiate the process. These initial disturbances are regions of low-level convergence and lifting. Many tropical disturbances of this type occur each year, but only a few develop into full-fledged hurricanes. It is believed that the upper-level airflow acts to further intensify selected storms by "pumping out" the rising air as it reaches the top of the storm, thus encouraging the influx of warm, moist air at the surface, which ascends and releases latent heat to fuel the storm. It seems that if the air is "pumped out" at the top faster than it is being replaced at the surface, the storm intensifies. However, if the rising air is not removed, the convergence at the surface will "fill" the storm center, equalizing the pressure differences, and the storm will die.

North Atlantic hurricanes develop in the trade winds, which generally move these storms from east to west at about 25 kilometers (15 miles) per hour. Then, almost without exception, hurricanes curve poleward and are deflected into the westerlies, which increases their forward motion up to a maximum of 100 kilometers per hour. Some move toward the mainland, but their irregular paths make prediction of their movement difficult.

A location only a few hundred kilometers from a hurricane—just one day's striking distance away—may experience clear skies and virtually

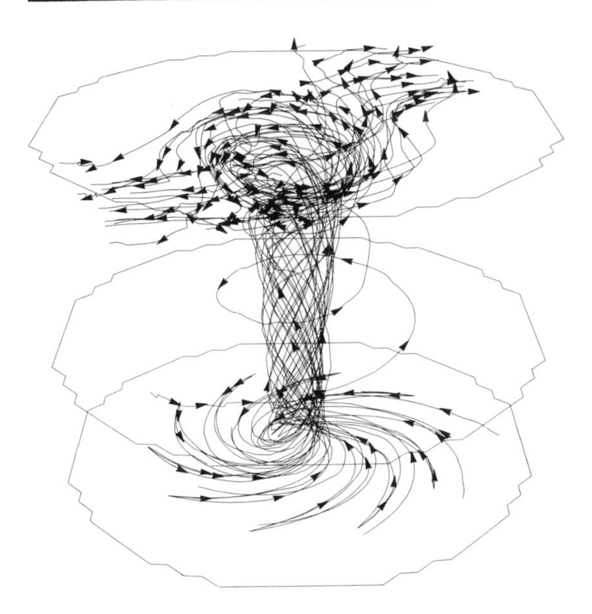

Figure 15.19 *Circulation in a well-developed hurricane. Cyclonic winds are experienced at the surface. At the core of the storm, the air is whirled in great spirals upward. Outward flow dominates near the top of the storm and seems to be aided by upper-level airflow.*

no wind. Prior to the age of weather satellites, such a situation made difficult the job of warning people of impending storms. Since the successful launching of *Tiros I* in April 1960, which inaugurated the era of weather observation by satellites, meteorologists have been able to identify and track tropical storms, even before they become hurricanes.

As a typical storm approaches land, wind speeds increase, first to gale intensity of 60 kilometers (40 miles) per hour, and eventually to over 160 kilometers (100 miles) per hour at the eye wall of the storm. With the increased wind speed comes torrential rain, which drops from 15 to 30 centimeters (6 to 12 inches) of water as it passes overhead. Flooding usually inflicts more death and destruction than wind. The greatest flooding is caused by the rising sea level, which accompanies the low pressure of the storm. It may raise the ocean a meter, and the resulting surge of water may reach 5 meters above sea level as it moves into shallow waters. We can easily visualize the damage this surge of water could inflict on low-lying coastal towns. In the delta

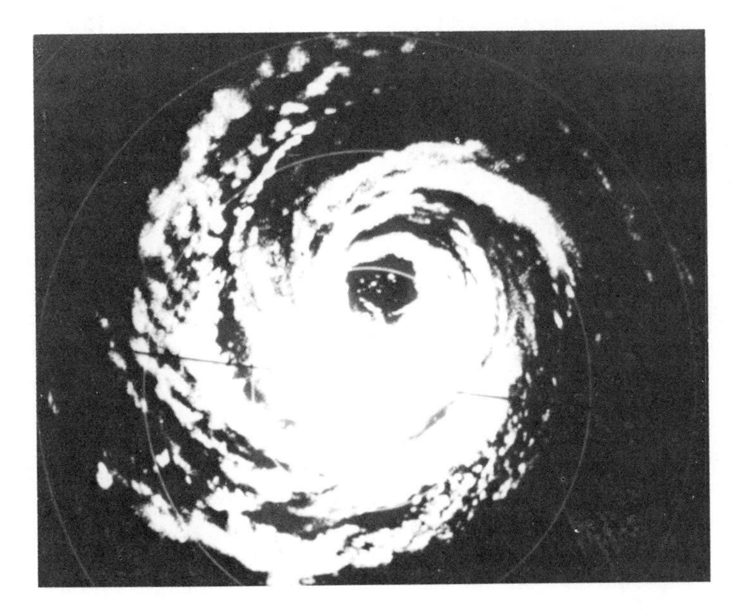

Figure 15.20 *Radar view of Hurricane Betsy showing the eye. (Courtesy of NOAA)*

areas of Bangladesh, for example, the land is generally less than 2 meters above sea level. When a storm surge, superimposed upon normal high tide, inundated that area on November 13, 1970, the official death toll was 200,000, and unofficial estimates ran to 500,000. This was one of the worst disasters of modern times. In the United States, early-warning systems have reduced the death toll substantially. However, the total property damage is on the upswing, mainly because of increased development in coastal areas. To date, the most destructive hurricane in the United States was Camille. In 1969, this storm caused an estimated $1500 million in damages.

Once on land, a hurricane loses its punch rapidly because its source of warm, water-laden air is cut off. Also, the added frictional effect of land causes the wind to move more directly into the center of the low, tending to "fill" the storm and eliminate the large pressure differences.

Methods devised to lessen hurricane destruction cannot rely on tracking alone. They must also be designed to disarm the storm. Project Stormfury is an attempt by the National Oceanic and Atmospheric Administration and the U.S. Navy to do just that. Recall that the energy for hurricanes comes from the release of latent heat in the region of the eye wall, thus creating the steep pressure gradient, which drives the winds. The object of hurricane modification is to disrupt the pressure gradient by seeding portions of the hurricane that are not in the area of maximum wind velocity. Theoretically, at least, this would cause the supercooled droplets to freeze, releasing latent heat of fusion, which would then stimulate additional cloud growth in several regions away from the eye wall. The artificial development of these secondary centers would reduce the pressure gradient between them and the eye wall of the storm and thus reduce the wind speeds. Experimental tests on a few selected hurricanes indicate that this mechanism is an effective method of reducing some of the fury of a hurricane. Tests are being conducted each hurricane season to refine and improve the technique.

REVIEW

1. What are the characteristics of a maritime tropical air mass?
2. Describe the weather associated with a continental polar air mass in the winter. In the summer. When would this air mass be most welcome in the United States?
3. Where are the source regions for the maritime tropical air masses which affect North America? The maritime polar air masses?
4. Describe the weather along a cold front where very warm, moist air is being displaced.
5. Explain the basis for the following weather proverb:

   Rain long foretold, long last;
   Short notice, soon past.
6. The formation of an occluded front marks the beginning of the end of a wave cyclone. Why is this true?
7. Refer to Figure 15.6. Describe the weather conditions that city *B* will be experiencing several hours hence. Do the same for city *A*.
8. For each of the weather elements that follow, describe the changes that an observer experiences when a wave cyclone passes with its center north of the observer: wind direction, pressure tendency, cloud type, cloud cover, precipitation, temperature.
9. Describe the weather conditions an observer would experience if the center of a wave cyclone passed to the south.
10. What is the primary requirement for the formation of thunderstorms?
11. Based on your answer to 10, where would you expect thunderstorms to be most common on the earth? In the United States?
12. How is thunder produced? How far away is a lightning stroke if thunder is heard 15 seconds after the lightning is seen?
13. Why do tornadoes have such high wind speeds?

14. The winds of a tornado have never been measured directly. How might this be explained?
15. What general atmospheric conditions are most conducive to the formation of tornadoes?
16. Why should you open the windows and doors of a house if you expect a tornado to strike?
17. A hurricane has slower wind speeds than a tornado, yet it inflicts more total damage. How might this be explained?
18. Why is upper-level airflow thought to be important to the formation of hurricanes?
19. Once on land, a hurricane's winds quickly diminish. Why?

KEY TERMS

air mass
air-mass weather
source region
polar (P) air mass
tropical (T) air mass
continental (c) air mass
maritime (m) air mass
front
warm front
cold front
occluded front
middle-latitude (wave) cyclone
occlusion
thunderstorm
lightning
thunder
tornado
hurricane

Members of the scientific staff of the Deep Sea Drilling Project study samples of sea-floor sediments recovered from a depth of 4845 meters. Sea-floor sediments provide many important clues to our planet's climatic history. By deciphering past climatic changes, scientists hope to better understand what lies ahead. (Courtesy of Deep Sea Drilling Project, Scripps Institution of Oceanography)

Is Our Climate Changing?
Theories of Climatic Change
The Climate of Cities

The Changing Climate

This chapter explores climatic change, a topic that has become the focus of increasing interest and concern throughout the world. Why the interest? The answer is that a variable climate could have a dramatic effect upon people and such critical activities as food production and energy use. If global climate is indeed changing, what factors are responsible, and on what time scales do the variations occur? Do people cause or contribute to any of these changes? These are some of the questions we shall investigate in the pages that follow.

Is Our Climate Changing?

Not too many years ago, the concept of climatic change was considered to have little but academic importance because the problems most often investigated related primarily to the remote past. "What caused the Ice Age?" seemed to be the most debated question.* Indeed, in the nineteenth and early twentieth centuries, it was assumed that climatic changes were a thing of the past, or at least that such changes occurred only over vast periods of geologic time. All of the observed departures from mean values were thought to be nothing more than what we now call statistical noise. However, since that time, and especially in the past decade or two, scientists have come to recognize that climate is inherently variable on virtually all time scales. It is no longer considered static, but rather dynamic. Furthermore, it is not just the scientific community that has shown increasing interest in climatic change. In recent years, governments as well as the general public have become aware of and have shown concern about the possible variability of our planet's climate.

Among the reasons that climatic change has become a much-discussed topic and a matter of considerable concern for the future are the following:

(1) Recent detailed reconstructions of past climates show that climate has varied on all time scales, which suggests that the climate of the future will more likely be different from that of the present than the same.
(2) Increased research into human activities and their effect on the environment has led to suspicions that people have changed or will inadvertently change the climate.
(3) There is observational evidence that global temperatures have been dropping for the past three decades and, in addition, that world climate has become more variable.

The last statement is perhaps the most important reason for the public's awareness and the government's increasing concern for the issue. For several decades, the climate had been generally favorable for crop production. For the most part, stockpiles of food grew faster than the population, and energy consumption (at least in developed countries) was not a problem—costs were relatively low and supplies seemed limitless. Then, in the 1970s, several events focused attention on global climate and its possible change. Devastating droughts in Africa, India, China, and the Soviet Union, for example, caused a serious depletion of grain reserves and led to sharply higher food prices. At the same time, other world regions were recording excessive rainfalls and floods. In January 1977, many cities in the United States experienced the coldest winter on record, and natural gas (at any price) became critically scarce; many schools and factories were forced to close to conserve fuel. Concurrently, several western states experienced severe drought conditions; water rationing became the rule in several locales. A year later, the Midwest and East again experienced record-breaking cold, compounded by some of the most severe blizzards of the century. On the other hand, the western states, in particular California, received record rainfalls which not only ended the drought but also caused massive mudslides and extensive flooding in many areas.

Are the foregoing examples proof that our climate is becoming more variable? The answer is a definite "maybe!" It should be emphasized that the fact that one winter is warmer than the last or that a particular summer is among the driest on record does not prove that climate is changing. The list of weather peculiarities over the past several years is lengthy, yet it provides no clear-cut indication of our planet's immediate or long-term climatic future. Many years of data are required to indicate a trend in the climate,

*It is still a much-discussed problem that has not yet been answered sufficiently.

but even then, atmospheric scientists often disagree as to its meaning and cause.

As was mentioned, observational data indicate that the mean global temperature has dropped approximately 0.5° C since the 1940s. Many scientists who have studied climatic history think that this cooling trend will continue into the twenty-first century. Some even believe quite strongly that the downward trend in global temperatures signals the beginning of another ice age. Other scholars, however, think the opposite. They argue that the cooling has, or soon will, "bottom out." They then foresee a gradual rise in world temperatures. Thus, rather than an impending ice age, these advocates of global warming believe that existing glaciers may shrink and that sea level may rise. Such an occurrence would be devastating because many densely populated coastal regions would be flooded. Still another group of climatologists is reserving judgment about the nature and intensity of any future climate change on earth. Although they realize that climatic change is an historic fact and indeed inevitable, they believe the current evidence of change is inadequate. They feel that predicting the future trend of global climate now is premature because the interrelationships within the earth's climatic system between the atmosphere, the oceans, and the continents are far too complex to be characterized by existing data.

How can so many scientists disagree about the facts? Who is right? Will our planet become *Hothouse Earth,* or will it experience *The Cooling?** The fact is, no one can really be sure. Richard J. Kopec has summed up the situation this way:

> The facts, per se, are irrefutable. It is their interpretation and the predictions resulting from these interpretations which have caused divergent views among climatologists. As yet, there is no deterministic, predictive model of our planet's climate, and until one is developed, predictions are as valid as the logic producing them. In addition, the periods of time involved in climatic predictions cover centuries and the validity of climate forecasting is not easily tested. As a result, there exist several opposing, and equally convincing, viewpoints concerning this very important topic.*

Tens of thousands of pages have been written about climatic change and tens of thousands more can be expected. In the space available here, we cannot discuss sufficiently the large number of past climate changes that science has documented, nor can we deal fully with the many thoughts about the future. Such an encyclopedic recitation of facts would serve little purpose here. Rather, we shall briefly review several of the major theories that have been proposed as possible answers to the question of why climates change.

Theories of Climatic Change

The theories that have been proposed to explain climatic change are many and varied, to say the least. Several have gained relatively wide support, only to subsequently lose it, and then, in some cases, regain it. It is safe to say that most, if not all, theories are controversial. This is to be expected when we consider them in light of a point made earlier, namely, that presently there is no deterministic predictive model of the earth's climate. Hence, each theory is as valid as the logic behind it. In this section, we shall examine several current theories which, to varying degrees, have gained some degree of support from a portion of the scientific community. Four of the theories deal with "natural" causes of climatic change, that is, causes unrelated to human activi-

* The author of *Hothouse Earth*, H. A. Wilcox (New York: Praeger Publishing, 1975, 182 pages), focuses on theories that predict a global warming, and L. Ponte, author of *The Cooling* (Englewood Cliffs, N.J.: Prentice-Hall, 1976, 306 pages), presents a case for a possible new ice age.

* *Atmospheric Quality and Climatic Change*, Richard J. Kopec, editor, University of North Carolina at Chapel Hill, Department of Geography, Studies in Geography No. 9, 1976, p. 2.

ties. These include changes brought about by continental drift and volcanic activity as well as variations caused by fluctuations in solar output and changes in the earth's orbit. Two other theories relate primarily to possible human-generated climatic changes. They consider the possible effects of rising carbon dioxide levels caused primarily by combustion processes and the influence of increased dust loading of the atmosphere caused by human activities.

As you read this section, you will find that there is often more than one logical way to explain the same climatic change. Furthermore, no single theory explains climatic change on all time scales. A theory that explains variations over millions of years, for example, is generally not satisfactory when dealing with fluctuations over a span of a hundred years. When (or, perhaps, if) the time comes when our atmosphere and its changes through time are fully understood, we will likely see that many of the factors discussed here, as well as others, influence climatic change.

Drifting Continents and Climatic Change

Over the past 15 to 20 years, the revolutionary theory of plate tectonics has emerged from the science of geology (see Chapter 6). This theory, which has gained very wide acceptance among scientists, states that the outer portion of the earth is made up of several individual pieces, called plates, which move in relation to one another upon a partially molten zone below. With the exception of the plate that encompasses the Pacific Ocean Basin, the large plates are composed of both continental and oceanic crust (Figure 6.3). Thus, as plates move, the continents also change positions. This theory not only provides the geologist with explanations about many previously misunderstood processes and features, but it also provides the climatologist with a probable explanation for some hitherto unexplainable climatic changes. For example, glacial features in present-day Africa, Australia, South America, and India indicate that these regions experienced an ice age near the end of the Paleozoic era, about 230 million years ago. For many years, this puzzled scientists. Was the climate in these relatively tropical latitudes once like it is today in Greenland and Antarctica? Until the plate tectonic theory was formulated and proven, there was no reasonable explanation. Today scientists realize that the areas containing these ancient glacial features were joined together as a single "supercontinent" that was located at high latitudes far to the south of their present positions. Later, this landmass broke apart, and its pieces, each moving on a different plate, drifted toward their present locations. Hence, large fragments of glaciated terrain ended up in widely scattered subtropical locations (Figure 16.1).

It is now believed that during the geologic past, continental drift accounted for many other equally dramatic climate changes as landmasses shifted in relation to one another and moved to different latitudinal positions. Changes in oceanic circulation also must have occurred, altering the transport of heat and moisture, and consequently the climate. Since the rate of plate movement is very slow, on the order of a few centimeters per year, appreciable changes in the positions of the continents occur only over great spans of geologic time. Thus, climatic changes brought about by continental drift are extremely gradual and happen on a scale of millions of years. As a result, the theory of plate tectonics is not useful for explaining climatic variations that occur on shorter time scales such as tens, hundreds, or thousands of years. Other explanations must be sought to explain these changes.

Volcanic Dust Theory

Like the plate tectonic theory, the **volcanic dust theory** is a "geological" theory of climatic change. First proposed many years ago, it is still regarded as a plausible explanation for some aspects of climatic variability.

Explosive volcanic eruptions emit great quantities of fine-grained debris into the atmosphere. Some of the biggest are sufficiently powerful to

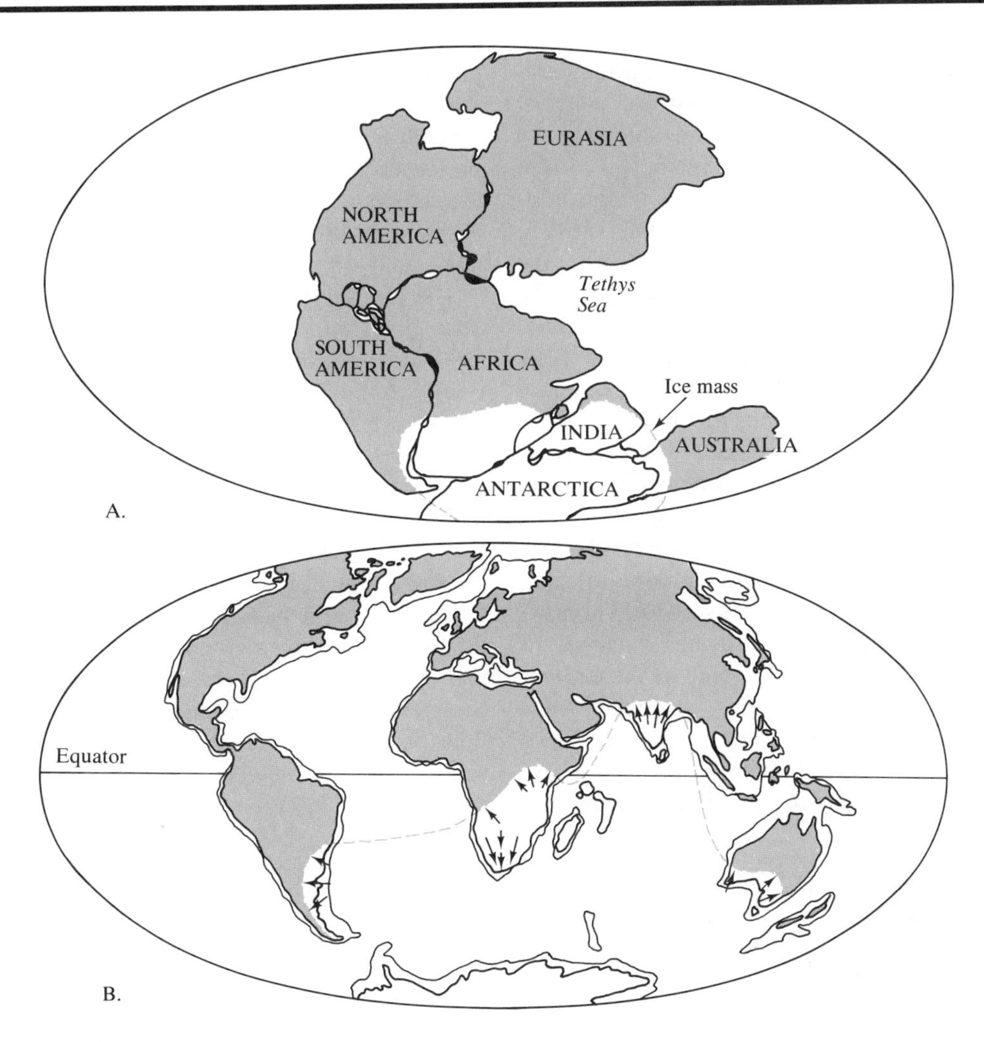

Figure 16.1 **A.** *The supercontinent Pangaea showing the area covered by glacial ice 300 million years ago.* **B.** *The continents as they are today. The shading outlines areas where evidence of the old ice sheets exists. The dashed line joining the glaciated regions indicates how large the ice sheet would have had to be if the continents had been in their present positions at the time of glaciation. (A. and B. after Dietz and Holden, 1971, and Du Toit, 1937, in R. F. Flint and B. J. Skinner,* Physical Geology, *2nd ed., p. 418, New York: Wiley, 1977)*

inject dust and ash high into the stratosphere, where it is spread around the globe and may remain for several years. The basic premise of the volcanic dust theory is that this suspended volcanic material filters out a portion of the incoming solar radiation which, in turn, leads to lower air temperatures. Thus, many scientists have speculated that periods of high volcanic activity should be accompanied by cool climatic periods, perhaps even ice ages, whereas times of reduced volcanic activity should coincide with warmer eras. This idea is, in part, supported by temperature trends in the twentieth century. From 1915 until the explosive eruption of Mt. Agung on the island of Bali in 1963, there had been little volcanic activity. It is well known that global temperatures increased during much of that period. As a result, a number of scientists have postulated that this era of rising temperatures was connected to the lack of fine volcanic debris in the atmosphere. Some studies have tended to support this conclusion but have also shown that the cooling that began in the mid-1940s cannot be explained by volcanic activity. The drop in global temperatures began before the increased explosive volcanism.

The volcanic dust theory is most often mentioned as a possible cause for the Ice Age. Although a veil of volcanic dust may indeed cause world temperatures to drop, for many years no one could substantiate the theory with facts. In 1965, Arthur Holmes, a distinguished British geologist, stated that no correlation had yet been demonstrated between glacial periods and times of prolonged or exceptionally intense volcanic activity. The results of recent investigations, however, have been more promising. Analysis of deep-sea sediment cores recovered by the *Glomar Challenger* as a part of the National Science Foundation's Deep Sea Drilling Project have shown that there was a much higher rate of explosive volcanism during the past two million years than during the previous eighteen million years (Figure 16.2). Consequently, the study reveals a reasonably good correlation between a period of greatly increased volcanic activity and an interval characterized by rapidly changing climatic conditions, namely, the most recent ice age.

The Ice Age was not a single period of continuous glaciation. Rather, it was characterized by alternating periods of glacial advance and retreat (see Figure 4.3). The warm periods between advances are termed interglacial periods. Thus, if volcanic activity was the primary mechanism causing the Ice Age, there must have been alternating periods of explosive volcanism and relatively quiet conditions. At present, the data are not sufficient to prove this. Hence, the full importance of volcanic activity as a possible cause for the Ice Age remains something of a mystery and a matter for continued speculation. Stephen H. Schneider, an active researcher and writer in the field of climatic change, summarizes the uncertainty of many scientists regarding the volcanic dust theory as follows:

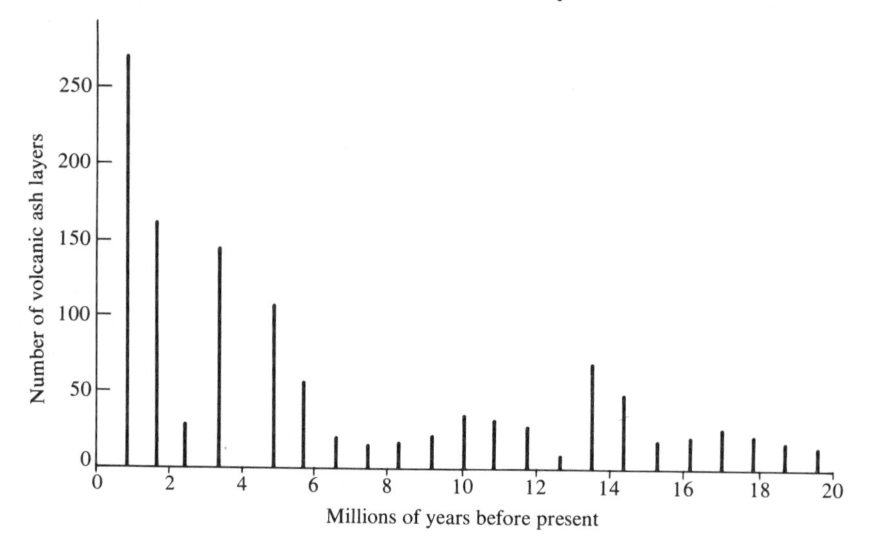

Figure 16.2 *Variations in volcanic activity over the past 20 million years are revealed by the number of volcanic ash layers. Notice the dramatic increase over the past 2 million years. [Data from James P. Kennett and Robert C. Thunell, "Global Increase in Quaternary Explosive Volcanism," Science 187 (1975): 499. Copyright 1975 by the American Association for the Advancement of Science]*

. . . although the role of volcanoes in forcing climatic change may prove critical—in which case I should be trying to learn to predict volcanic eruptions and not spend most of my time building atmospheric models—as yet we cannot be sure the extent to which the potential climatic effects of volcanic blasts are detectable in climatic records.*

Astronomical Theory

The **astronomical theory** is based on the premise that variations in incoming solar radiation are a principal factor in controlling the climate of the earth. This hypothesis was first developed and strongly advocated by the Yugoslavian astronomer and geophysicist Milutin Milankovitch. He formulated a comprehensive mathematical model based on the following elements:

(1) Variations in the shape **(eccentricity)** of the earth's orbit about the sun.
(2) Changes in **obliquity,** that is, changes in the angle that the axis makes with the plane of the ecliptic (plane of the earth's orbit).
(3) The wobbling of the earth's axis, termed **precession.**

Although variations in the distance between the earth and sun are of minor significance in understanding current seasonal temperature fluctuations, they may play a very important role in producing global climatic changes on a time scale of tens of thousands of years. A difference of only 3 percent exists between aphelion, which occurs on about July 4, in the middle of the Northern Hemisphere summer, and perihelion, which takes place in the midst of the Northern Hemisphere winter on about January 3. This small difference in distance means that the earth receives about 6 percent more solar energy in January than in July. However, this is not always the case. The shape of the earth's orbit changes during a cycle that astronomers say takes between 90,000 and 100,000 years, stretching into a longer ellipse and then returning to a more circular shape. When the orbit is very eccentric, the amount of radiation received at closest approach (perihelion) would be on the order of 20 to 30 percent greater than that at aphelion. This would most certainly result in a substantially different climate from what we now have.

The inclination of the earth's axis to the plane of the ecliptic was shown in Chapter 12 to be the most significant cause for seasonal temperature changes. At present, the angle that the earth's axis makes with the plane of its orbit is 23.5 degrees. However, this angle changes. During a cycle that averages about 41,000 years, the tilt of the axis varies between 22.1 and 24.5 degrees. Because this angle varies, the severity of the seasons must also change. The smaller the tilt, the smaller is the temperature difference between winter and summer. It is believed that such a reduced seasonal contrast could promote the growth of ice sheets. Since winters could be warmer, more snow would fall because the capacity of air to hold moisture increases with temperature. Conversely, summer temperatures would be cooler, meaning that less snow would melt. The result could be the growth of ice sheets.

Like a partly run-down top, the earth is wobbling as it spins on its axis. At the present time, the axis points toward the star Polaris (often called the North Star). However, about the year A.D. 14,000, the axis will point toward the bright star Vega, which will then be the North Star (Figure 17.23). Since the period of precession is about 26,000 years, Polaris will once again be the North Star by the year 28,000. As a result of this cyclical wobble of the axis, a very significant climatic change must take place. When the axis tilts toward Vega in about 12,000 years, the orbital positions at which the winter and summer solstices occur will be reversed. Consequently, the Northern Hemisphere will experience winter near aphelion, when the earth is farthest from the sun, and summer will occur near perihelion, when our planet is closest to the sun. Thus, sea-

* *The Genesis Strategy* (New York: Plenum Press, 1976), p. 134.

sonal contrasts will be greater because winters will be colder and summers warmer than at present.

Using eccentricity, obliquity, and precession, Milankovitch calculated variations in insolation and the corresponding surface temperature of the earth back in time in an attempt to correlate these changes with the climatic fluctuations of the Ice Age. Note that these three variables cause little or no variation in the total annual amount of solar energy that reaches the ground. Rather, their impact is felt because they change the degree of contrast between the seasons.

Since Milankovitch's pioneering work, the time scales for orbital and insolational changes have been recalculated several times, correcting past errors and adding greater precision to the measurements. Over the years, the astronomical theory has been widely accepted, then largely rejected, and now, in light of recent investigations, is quite popular again.

Among recent studies that have added credibility and support to the astronomical theory is one in which deep-sea sediments were examined.* By analyzing certain climatically sensitive microorganisms, scientists established a chronology of temperature changes going back 450,000 years. This time scale of climatic change was then compared to astronomical calculations of eccentricity, obliquity, and precession to determine if a correlation did indeed exist. It should be noted that the study was not aimed at identifying or evaluating the mechanisms by which the climate is modified by the three orbital variables, but rather the goal was to see if past climatic and astronomical changes corresponded.

Although the study was quite involved and mathematically complex, the conclusions were straightforward. The authors found that major variations in climate over the past several hundred thousand years were closely associated with changes in the geometry of the earth's orbit; that is, cycles of climatic change were shown to correspond closely with the periods of obliquity, precession, and orbital eccentricity. More specifically, they stated that, "It is concluded that changes in the earth's orbital geometry are the fundamental cause of the succession of Quaternary ice ages." * In addition, the study went on to predict the future trend of climate, but with two qualifications: (1) that the prediction apply only to the natural component of climatic change and ignore any human influence, and (2) that it be a forecast of long-term trends because it must be linked to factors which have periods of 20,000 years and longer. Thus, even if the prediction is correct, it contributes little to our understanding of climate changes over periods of tens to hundreds of years because the cycles in the astronomical theory are too long for this purpose. With these qualifications in mind, the study predicts that the long-term trend, over the next 20,000 years, will be toward a cooler climate and extensive glaciation in the Northern Hemisphere.

If the astronomical theory does indeed explain alternating glacial-interglacial periods, why have glaciers been absent throughout most of earth history? Prior to the plate tectonic theory, there was no widely accepted answer. In fact, this question was a major obstacle for the supporters of Milankovitch's hypothesis. However, today there is a plausible answer. Since glaciers can form only on the continents, landmasses must exist somewhere in the higher latitudes before an ice age can start. Long-term temperature fluctuations are not great enough to create widespread glacial conditions in the tropics. Thus, many scientists now believe that ice ages have occurred only when the earth's shifting crustal plates carried the continents from tropical latitudes to more poleward positions.

Variable Sun Theory

Two of the three theories of climatic change discussed thus far have dealt with varia-

* J.D. Hays, John Imbrie, and N.J. Shackleton, "Variations in the Earth's Orbit: Pacemaker of the Ice Ages," *Science*, 194(4270): 1121–32.

* Ibid., p. 1131. The term *Quaternary* refers to the period on the geologic calendar that encompasses the last few million years.

tions in the receipt of solar radiation. In the case of the volcanic theory, changes in the transparency of the atmosphere are believed to alter the amount of energy received at the surface. The astronomical theory is based on the fact that changes in the amount of solar energy received by the earth are caused by external factors, that is, factors unrelated to the atmosphere. The following discussion covers yet another theory that is dependent upon variations in the receipt of solar energy. Like the astronomical theory, this proposal relies upon fluctuations in solar energy that are caused by factors external to the atmosphere.

One of the most persistent theories of climatic change is based on the idea that the sun is a variable star and that its output of energy varies through time. The effect of such changes would seem to be direct and easily understood: Increases in solar output would cause the atmosphere to warm, and reductions would result in cooling. This hypothesis is very appealing because it can be used to explain climatic changes of any length or intensity. However, there is at least one major drawback. No major variations in the total intensity of solar radiation have yet been measured outside the atmosphere.

Measurements of solar radiation beyond the atmosphere were impossible until quite recently, when satellite technology became available. However, records for many years will be necessary before we begin to get a feeling for how variable (or invariable) energy from the sun really is. Thus, for the time being, science must still rely on ground-based estimates of the sun's energy output. The greatest difficulty with this method is determining the influence of the atmosphere on the amount of radiation measured. To do this, observations are made at different times of the year. As one might expect, as the sun angle gets lower, the amount of radiation that is absorbed or scattered by the atmosphere increases and the quantity received at the earth's surface declines. A graph is then constructed of the intensity of solar energy as a function of the amount of air that must be traversed (which is governed by the sun angle). This curve is then extrapolated to the point of no air at all. In this manner, the **solar constant** has been estimated to be about 2 calories/centimeter ^{2}/minute.* The difficulty with this method is that scientists cannot attain the necessary degree of precision to establish that meaningful fluctuations (which may be slight) actually occur. Further, since each ground-based observation must be corrected for large atmospheric effects, the source of any variance is extremely difficult to determine.

Theories of climatic change based on a variable sun merge with those based on sunspot cycles. The most conspicuous and best-known features on the surface of the sun are the dark blemishes called sunspots (Figure 16.3). Although their origin is uncertain, sunspots have been found to be huge magnetic storms that extend from the sun's surface deep into the interior. Furthermore, these spots are associated with the ejection from the sun of huge masses of particles which, upon reaching the upper atmosphere, interact with the gasses there to produce auroral displays.

* The solar constant is a standard energy measurement used in meteorology and refers to the amount of solar radiation striking a surface perpendicular to the sun's rays at the outer edge of the atmosphere when the earth is at an average distance from the sun.

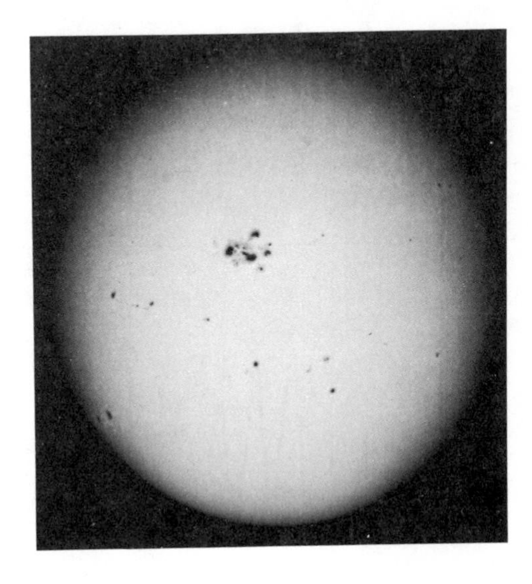

Figure 16.3 *Large sunspot group on the solar disk. (Courtesy of Yerkes Observatories)*

Along with other solar activity, the number of sunspots increases and decreases on a regular basis, creating a cycle of about eleven years. A curve of the annual number of sunspots from the early 1700s until the present appears to be quite regular (Figure 16.4). In fact, until recently, few scientists doubted that sunspots and the eleven-year cycle were not enduring features of the sun. However, articles by the solar astronomer John A. Eddy give evidence that solar activity may vary considerably, and may indeed be responsible for some climatic changes.*

Papers published by two astronomers in the late nineteenth century had indicated that the interval from about 1645 to 1715 was a period of prolonged sunspot absence. Eddy set out to see if this was indeed the case, because if it was true, it had some interesting ramifications. For instance, it would mean that the sun was not the constant star that so many believed it to be. In addition, it had implications for climatic change because the proposed sunspot minimum in the late seventeenth and early eighteenth centuries also corresponded very closely with the coldest portion of a period known in climatic history as the "Little Ice Age." This is a well-documented cold period that saw alpine glaciers in Europe advance farther than at any time since the last major glaciation. By examining historical records of observations made during this period, as well as making use of indirect evidence of solar activity, Dr. Eddy accomplished his task. His indirect evidence included, among other things, records of auroral displays and the abundance of the radioactive isotope carbon-14 (^{14}C) in tree rings. Since the number of nights when auroral displays occur has a strong positive correlation with sunspot activity, records of auroras served as an independent check of solar activity. The same thing was true for the carbon-14 record locked in trees.

> The ^{14}C history is useful in its own right as a measure of past solar activity, as has been demonstrated by a number of investigators. The isotope is continually formed in the atmosphere through the action of cosmic rays, which in turn, are modulated by solar activity. When the sun is active, some of the incoming galactic cosmic rays are prevented from reaching the earth. At these times, corresponding to maxima in the sunspot cycle, less than the normal amount of ^{14}C is produced in the atmosphere and less is found in tree rings formed then. When the sun is quiet, terrestrial bombardment by galactic cosmic rays increases and the ^{14}C proportion in the atmosphere rises.*

By comparing graphs, as shown in Figure 16.5, Eddy showed that every decline in solar activity

* Much of the remaining discussion in this section is based on three articles by John Eddy: "The Maunder Minimum," *Science*, 192(4245): 1189–1203; "The Case of the Missing Sunspots," *Scientific American*, 236(5): 80–92; and "Climate and the Changing Sun," *Climatic Change*, 1(2): 173–90.

* John A. Eddy, "The Maunder Minimum," *Science*, 192(4245): 1195.

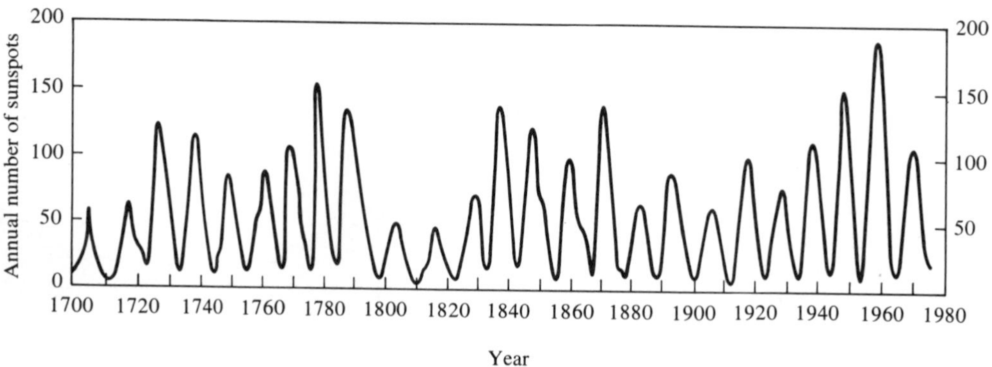

Figure 16.4 *Annual sunspot numbers, 1700–present.*

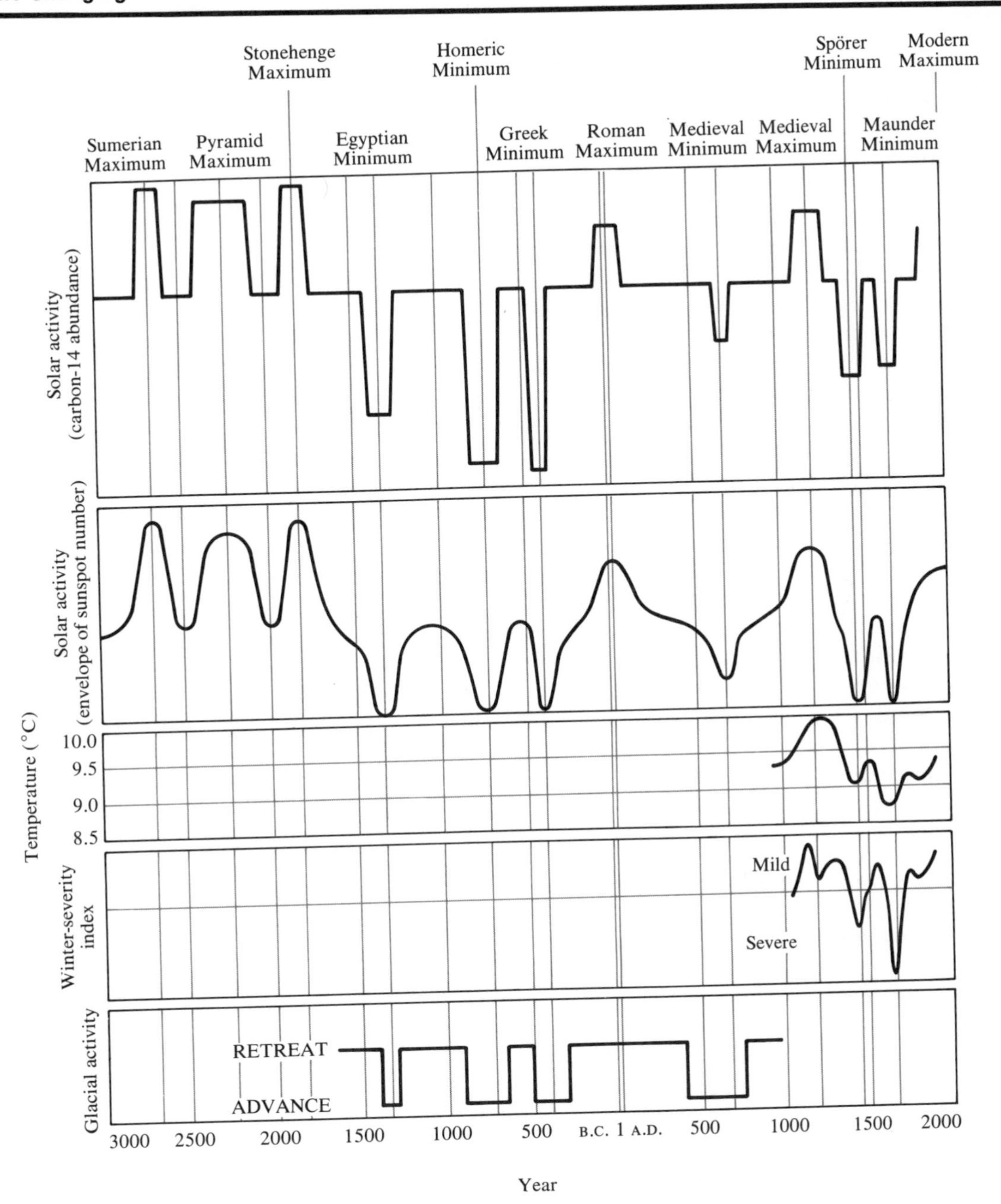

Figure 16.5 *A comparison of the carbon-14 derived history of the sun and the history of global climate based on historical records and the advance and retreat of alpine glaciers. Notice that every decrease in solar activity (minimum) matches a time of glacial advance in Europe and that each rise in solar activity (maximum) corresponds to a time of glacial retreat. The figure indicates that for almost 5000 years, all of the climatological curves seem to rise and fall in response to the long-term level of solar activity. (From "The Case of the Missing Sunspots," John A. Eddy. Copyright © 1977 by Scientific American, Inc. All rights reserved)*

corresponded to a period of glacial advance in Europe, and that each increase in solar activity was matched by a time of retreating glaciers. Furthermore, periods like the medieval maximum were found to correlate well with warm periods. Referring to these excellent matches, Eddy stated that, "These early results in comparing solar history with climate make it appear that changes on the sun are the dominant agent of climatic changes lasting between 50 and several hundred years." *

It should be remembered that Eddy's study was not concerned with short-term changes in the sunspot cycle. Thus, the investigation revealed little about how the 11-year cycle might influence short-term weather. However, others have examined sunspot cycles of various lengths in an attempt to find positive correlations and relationships. Some have been successful. For instance, one study showed a possible connection between periods of drought on the Great Plains and the minimum of the double (22-year) sunspot cycle. Nevertheless, no completely acceptable theory has yet been formulated to explain how solar variations are translated into climatic changes.

Carbon Dioxide and Climatic Change

Up to this point we have examined four potential causes of climatic change. Although each differs considerably from the others, all four have at least one thing in common—they are unaffected by human activities. They rely upon "natural" variables to produce climatic fluctuations. The two remaining theories, however, differ from these others in that they examine the possible impact of people on global climate.

In Chapter 12 we learned that although carbon dioxide (CO_2) represents only 0.03 percent of the gasses composing clean, dry air, it is nevertheless a meteorologically significant component. The importance of CO_2 lies in the fact that it is transparent to incoming short-wavelength solar radiation but is not transparent to longer-wavelength outgoing terrestrial radiation. A portion of the energy leaving the ground is absorbed by CO_2 and subsequently reemitted, part of it toward the surface, keeping the air near the ground warmer than it would be without CO_2. Thus, along with water vapor, CO_2 is largely responsible for the so-called greenhouse effect of the atmosphere. Since CO_2 is an important heat absorber, it logically follows that any change in the air's CO_2 content should alter temperatures in the lower atmosphere. This is the basis of the carbon dioxide theory of climatic change.

Paralleling the rapid growth of industrialization, which began in the nineteenth century, has been the consumption of fossil fuels (coal, natural gas, and petroleum). The combustion of these fuels has added great quantities of CO_2 to the atmosphere. Although some of the CO_2 is dissolved in the ocean, about 50 percent remains in the atmosphere. For many years it was believed that some of this excess CO_2 was also taken up by plants; that is to say, in addition to the oceans, plant life was a "sink" for CO_2. However, new studies reveal that this is not the case. In fact, there seems to be little question that the destruction of forests by people has added to the carbon dioxide content of the atmosphere. This addition of CO_2 occurs when forests are harvested and their stored carbon is released. Consequently, from 1860 to 1970, there was an increase of more than 10 percent in the CO_2 content of the air (Figure 16.6). Naturally, we would expect that as a result, global temperatures should have climbed steadily during this period. However, temperatures rose only until the mid-1940s; since then, they have slowly dropped. Is this a refutation of the CO_2 theory? Most climatologists believe that it is not. Rather, many suggest that the cooling trend since the 1940s would have been even greater had it not been for the increased greenhouse effect provided by the rising CO_2 content. On the other hand, others simply say that the effects of the CO_2 increase over more than a hundred years were not large enough to show up in any clear way in the climatic record.

* John A. Eddy, "The Case of the Missing Sunspots," *Scientific American*, 236(5): 88–92.

By examining the curve in Figure 16.6, you can see that the amount of CO_2 in the air has increased at an increasing rate. Furthermore, the graph shows the curve continuing in this manner into the twenty-first century. Current estimates now indicate that the atmosphere's present CO_2 content of almost 330 parts per million (ppm) will approach 400 parts per million by the year 2000 and will approximately double by the year 2040. With such an increase in CO_2, the enhancement of the greenhouse effect would be much greater than in the past. Thus, if the estimated increases come to pass, calculations show that we can probably expect global temperatures to increase by nearly 1° C by the year 2000 and by about 2° C by 2040.

What if the cooling trend that began in the 1940s continues? Will it mask the effect of the proposed CO_2 increases? The answer, of course, depends upon the magnitude of the temperature decline. According to some studies, this question will not have to be raised because the cooling trend will terminate soon. Studying the oxygen isotope record in the Greenland ice core, Wallace Broecker of the Lamont-Doherty Geological Observatory suggests that the present cooling is only one of a long series of similar climatic fluctuations over the past millenium (Figure 16.7). He states that,

> by analogy with similar events in the past, the present cooling will . . . bottom out during the next decade or so. Once this happens, the CO_2

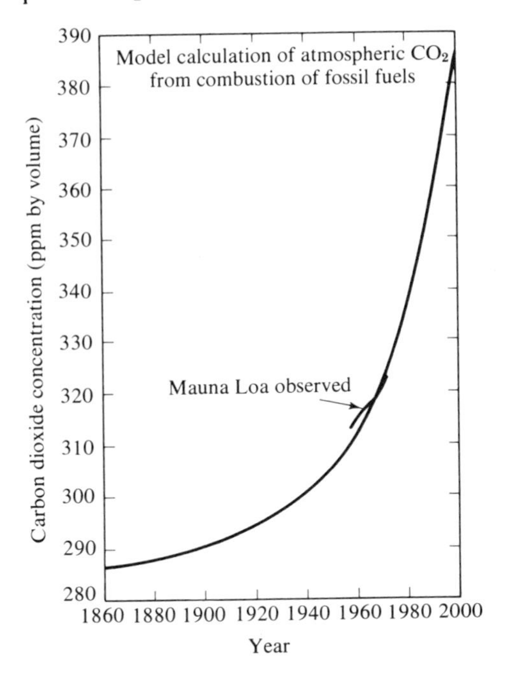

Figure 16.6 *Changes in atmospheric carbon dioxide resulting from the combustion of fossil fuels, 1860–2000. (From L. Machta and K. Telegadas, "Inadvertent Large-scale Weather Modification," in* Weather and Climate Modification, *edited by Wilmot N. Hess, p. 697, New York: Wiley, 1974)*

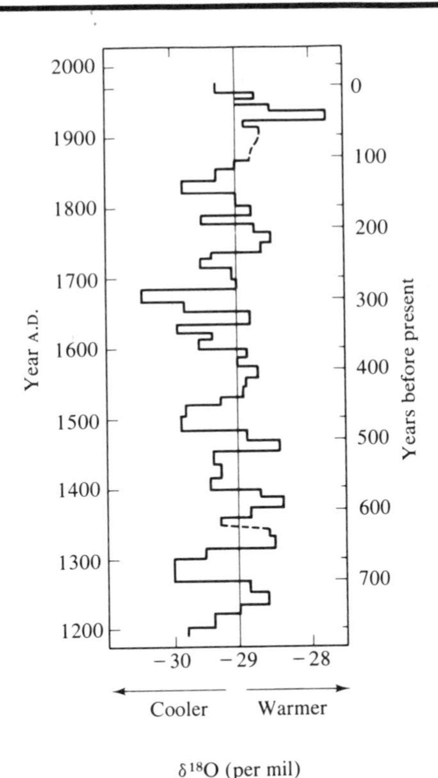

Figure 16.7 *Temperature variations as revealed by differences in the $^{18}O/^{16}O$ ratio in Greenland ice cores. A decrease of 1 per mil in the ^{18}O content corresponds to a 1.5° C drop in air temperature. (After Wallace S. Broecker, "Climatic Change: Are We on the Brink of a Pronounced Global Warming?"* Science *189(1975): 460. Copyright 1975 by the American Association for the Advancement of Science)*

> effect will tend to become a significant factor and by the first decade of the next century we may experience global temperatures warmer than any in the last 1000 years.*

If Broecker's prediction comes to pass, humankind may face an historic dilemma. On the one hand, an increase in the world's average temperature could result in an enlargement of arid lands. This in turn would have a negative impact upon agricultural production in a world where many are already malnourished and where the population continues to climb. On the other hand, action that might correct the global temperature rise, such as a major reduction in the consumption of fossil fuels, could be equally destabilizing. George M. Woodwell summarizes the dilemma this way:

> A major effort to change the balance of land use between agriculture and forest, in addition to an effort to restrict the burning of fossil fuels, would so upset established patterns of social and economic development as to be equivalent to the drastic changes in the human condition that a warming of the climate might lead to.**

In this discussion, we have assumed that global temperatures would rise if the CO_2 content increased. However, the possibility exists that increasing the amount of CO_2 in the atmosphere would ultimately result in a temperature decline. This could occur if, because of higher temperatures, the cloud cover were to increase because of increased evaporation. The increased cloud cover would, in turn, reduce the amount of solar radiation reaching the ground, which would result in a temperature decrease. Calculations show that an increase in the world's cloud cover of just 0.6 percent could decrease the low-level temperature by 0.5° C.

Because the atmosphere is a very complex interactive physical system, scientists must consider many possible outcomes, such as the one just mentioned, when one of the system's elements is altered. These various possibilities, termed **climatic feedback mechanisms,** make climatic modeling more complex and climatic predictions less certain. In the case of a global warming, as might result from increased CO_2, other possible feedback mechanisms also exist. For example, warming the lower atmosphere may also warm the surface of the ocean. Since warm water cannot hold as much CO_2 in solution as cold water, more CO_2 could be released to the air. This might result in a positive feedback because more atmospheric CO_2 would speed up the warming. Yet another possibility exists if a warming should occur. Although higher temperatures mean increased evaporation, more clouds may not result. If this is so, the greater amount of water vapor in the air could increase the absorption of long-wavelength terrestrial radiation, thus compounding the temperature increase because of an even more effective atmospheric greenhouse. Hence, unlike the first example, these latter two possibilities would reinforce the warming trend rather than offset it.

Human-Generated Dust and Climatic Change

Human activities such as transportation, solid waste disposal, industrial processes, and slash-and-burn agriculture are capable of generating great quantities of particulate matter (referred to by some as aerosols or turbidity, and by others simply as dust). Do these particles influence climate? If so, will they cause global temperatures to rise or drop? A candid answer to the first question is "Perhaps," and to the second, "We are not sure!" Indeed, some scientists believe strongly that the addition of human-generated dust will cause world temperatures to drop, whereas others present a strong case for just the opposite effect. In the discussion that follows, we shall briefly examine both points of view.

A well-known meteorologist, Reid A. Bryson, has for many years been a strong proponent of the idea that increased dustiness of the atmo-

---

* Wallace S. Broecker, "Climatic Change: Are We on the Brink of a Pronounced Global Warming?" *Science*, 189(4201): 460–61.

** "The Carbon Dioxide Question," *Scientific American*, 238(1): 34.

sphere caused by people is the primary cause of the global cooling that began in the 1940s. In part, the logic that advocates of this hypothesis use is as follows:

(1) Since the late nineteenth century and continuing until recently, sunspot activity has been increasing. If it is assumed that increased numbers of sunspots reflect an increasing solar energy output, temperatures should have continued to rise after the 1940s. Since this is not the case, other factors must be operative.

(2) A rise in atmospheric CO_2 should cause temperatures to increase. Although the CO_2 content of the atmosphere has continued to climb since the 1940s, as it had for 60 to 70 years before, temperatures have fallen. Hence, some other factor(s) must be overshadowing the CO_2 effect.

(3) A rise in atmospheric turbidity should cause global temperatures to fall by scattering away increased amounts of solar radiation. Dust concentrations have grown enormously since the 1930s, following the same trend as industrialization, mechanization, urbanization, and population. In addition, volcanic activity has increased somewhat over previous levels since the 1950s, adding to atmospheric particulate concentrations.

Thus, to explain the cooling trend in light of increased occurrence of sunspots and amounts of CO_2, which are assumed to cause temperatures to rise, Professor Bryson concludes that, ". . . increasing global air pollution, through its effect on the reflectivity of the earth, is currently dominant and is responsible for the temperature decline of the past decade or two." *

Although the proponents of the foregoing ideas presume that the net effect of increased aerosols is to scatter away a portion of the incoming solar energy, others are not as confident about this conclusion. While it seems likely that the reflectivity of the atmosphere is altered by the addition of aerosols, the nature of the change in the world's overall albedo is difficult to determine. For example, in regions where particles are bright relative to the surface below (which is believed to be the case over the oceans), the amount of light reflected is likely to increase, causing the air to cool. However, where the particles are relatively dark compared to the surface (a situation that is most probable over landmasses), the absorption of solar radiation will increase, albedo will decrease, and temperatures will rise. Thus, determining whether the world's mean temperature will rise or fall depends upon the ratio of absorption to scattering on a global scale.

Unlike volcanic dust, which may be thrown high into the stratosphere where it can remain for several years, the mean residence time for human-made dust particles, which are confined largely to the troposphere, is only a few days. Within that time, they are usually washed out of the air. Keeping this in mind, and realizing that the main source of these particles is the continents, some meteorologists reason that the primary effect of the dust must be upon the air temperatures over landmasses. If this is the case, the net influence of human-made aerosols is probably a small warming (remember that over the continents aerosols are relatively dark in comparison to the surface). If this conclusion is correct, some other explanation will have to be found for the cooling trend that began in the 1940s.

Of the many influences that pollution has on the atmospheric environment, the effects on global climate are certainly the least understood. Will increases in atmospheric CO_2 cause a global warming that, in turn, may be reinforced by positive feedback mechanisms, or will increased cloud cover negate the impact of CO_2? Is human-generated dust causing a global cooling, or is its net effect a warming? Is it possible that the climate would fluctuate without changes in any of these factors? In the preceding discussions of the possible impact of people on climate, we have seen that science thus far has been unsuccessful in providing adequate answers to these questions.

*Reid A. Bryson, "'All Others Factors Being Constant. . . .': A Reconciliation of Several Theories of Climatic Change," *Weatherwise*, 21(1968): 61.

However, although the effects of pollutants on the world's climate are not yet well understood, they may nevertheless be very significant, for even slight changes in temperature could be responsible for important shifts in weather patterns. Such changes could have a dramatic negative influence upon human activities such as agriculture, and perhaps even threaten survival in some regions.

How important is the role of people in bringing about climatic change? Although many differences of opinion and much uncertainty exist in the scientific community concerning this question, there seems to be general agreement on at least one point. W. W. Kellogg and S. H. Schneider stated it this way:

> Human influence on climate, which may already be appreciable, can only be properly assessed when the natural forces at play are understood.*

In a report resulting from a conference on human impact on climate, the point was made as follows:

> . . . the uncertainty lies in the fact that we are competing with powerful natural sources that are also causing the climate to change, as it has changed in the past. We must understand these factors well enough to be able to put man's activities into proper perspective relative to nature's.**

The Climate of Cities

In the preceding section, we saw that the nature of human impact upon global climate is, to say the least, uncertain. In this section, we shall see that there is a human-generated climatic impact on a far different scale that is much less uncertain. The most apparent human impact on climate is the modification of the atmospheric environment by the building of cities. The construction of every factory, road, office building, and house destroys existing microclimates and creates new ones of great complexity. As far back as the early nineteenth century, Luke Howard, the Englishman who is best remembered for his cloud classification scheme, recognized that the weather in London differed from that in the surrounding rural countryside, at least in terms of reduced visibility and increased temperature. Indeed, with the coming of the industrial revolution in the 1800s, the trend toward urbanization accelerated, leading to significant changes in the climate in and near most cities. At the beginning of the last century, only about 2 percent of the world's population lived in cities of more than 100,000 people. Today, not only is the total world population dramatically larger, but a far greater percentage reside in cities. On a worldwide basis, perhaps a quarter of the population is urban, and in many

* "Climate Stabilization: For Better or for Worse?" *Science*, 186(4170): 1163.

** *Inadvertent Climate Modification: Report of the Study of Man's Impact on Climate*. Boston: M.I.T. Press, 1971, p. 27.

Table 16.1 ***Average Climatic Changes Produced by Cities***

Element	Comparison with Rural Environment
Particulate matter	10 times more
Temperature	
Annual mean	0.5–1.5° C higher
Winter	1–2° C higher
Solar radiation	15–30% less
Ultraviolet, winter	30% less
Ultraviolet, summer	5% less
Precipitation	5–15% more
Thunderstorm frequency	16% more
Winter	5% more
Summer	29% more
Relative humidity	6% lower
Winter	2% lower
Summer	8% lower
Cloudiness (frequency)	5–10% more
Fog (frequency)	60% more
Winter	100% more
Summer	30% more
Wind speed	25% lower
Calms	5–20% more

SOURCE: After Landsberg, Changnon, and others.

regions (including the United States, Western Europe, and Japan), the percentage is more than two or three times that figure.

As Table 16.1 illustrates, the climatic changes produced by urbanization involve all major surface conditions. Some of these changes are quite obvious and relatively easy to measure. Others are more subtle and sometimes difficult to measure. The amount of change in any of these elements, at any time, depends on several variables, including the extent of the urban complex, the nature of industry, site factors such as topography and proximity to water bodies, time of day, season, and existing weather conditions.

The Urban Heat Island

Tables 16.2 and 16.3 illustrate the most studied and well-documented urban climatic effect—the **urban heat island.** The term simply refers to the fact that temperatures within cities are generally higher than in rural areas. This is shown in Table 16.2 by using mean temperatures and in Table 16.3 by using a less common statistic, the number of days on which the temperature either equals or exceeds 32° C, or is lower than or equal to 0° C. Although used less frequently, the data in Table 16.3 are nonetheless revealing. They show us that cities not only have higher mean temperatures than outlying locales, but also more hot days and fewer cold days. The magnitude of the temperature differences as shown by these tables is probably even greater than the figures indicate, because studies have shown that temperatures observed at suburban airports are usually higher than those in truly rural environments.

Table 16.2 ***Average Temperatures (°C) for Philadelphia Airport and Downtown Philadelphia (ten-year averages)***

	Airport	Downtown
Annual mean	12.8	13.6
Mean June maximum	27.8	28.2
Mean December maximum	6.4	6.7
Mean June minimum	16.5	17.7
Mean December minimum	−2.1	−0.4

SOURCE: After H. Neuberger and J. Cahir, *Principles of Climatology*. New York: Holt, Rinehart and Winston, 1969, p. 128.

Table 16.3 ***Average Annual Number of Hot Days (≥32° C) and Cold Days (≤0° C) at Downtown (D) and Airport (A) Stations for Five American Cities***

City	≤0° C	≥32° C
Philadelphia, Pa.		
D	73	32
A	89	25
Washington, D.C.		
D	68	39
A	72	33
Indianapolis, Ind.		
D	106	36
A	124	23
Baltimore, Md.		
D	62	35
A	96	33
Pittsburgh, Pa.		
D	96	19
A	124	9

SOURCE: After H. Neuberger and J. Cahir, *Principles of Climatology*. New York: Holt, Rinehart and Winston, 1969, p. 128.

As is typical, the data for Philadelphia show that the heat island is most pronounced when minimum temperatures are examined. While mean maximum temperatures are only 0.3–0.4° C higher in the city, minimums are from 1.2 to 1.7° C higher. Figure 16.8A, which shows the distribution of average minimum temperatures in the Washington, D.C., metropolitan area for the three-month winter period over a five-year span, also illustrates a well-developed heat island. The warmest winter temperatures occurred in the heart of the city, while the suburbs and surrounding countryside experienced average minimum temperatures that were as much as 3.3° C lower.

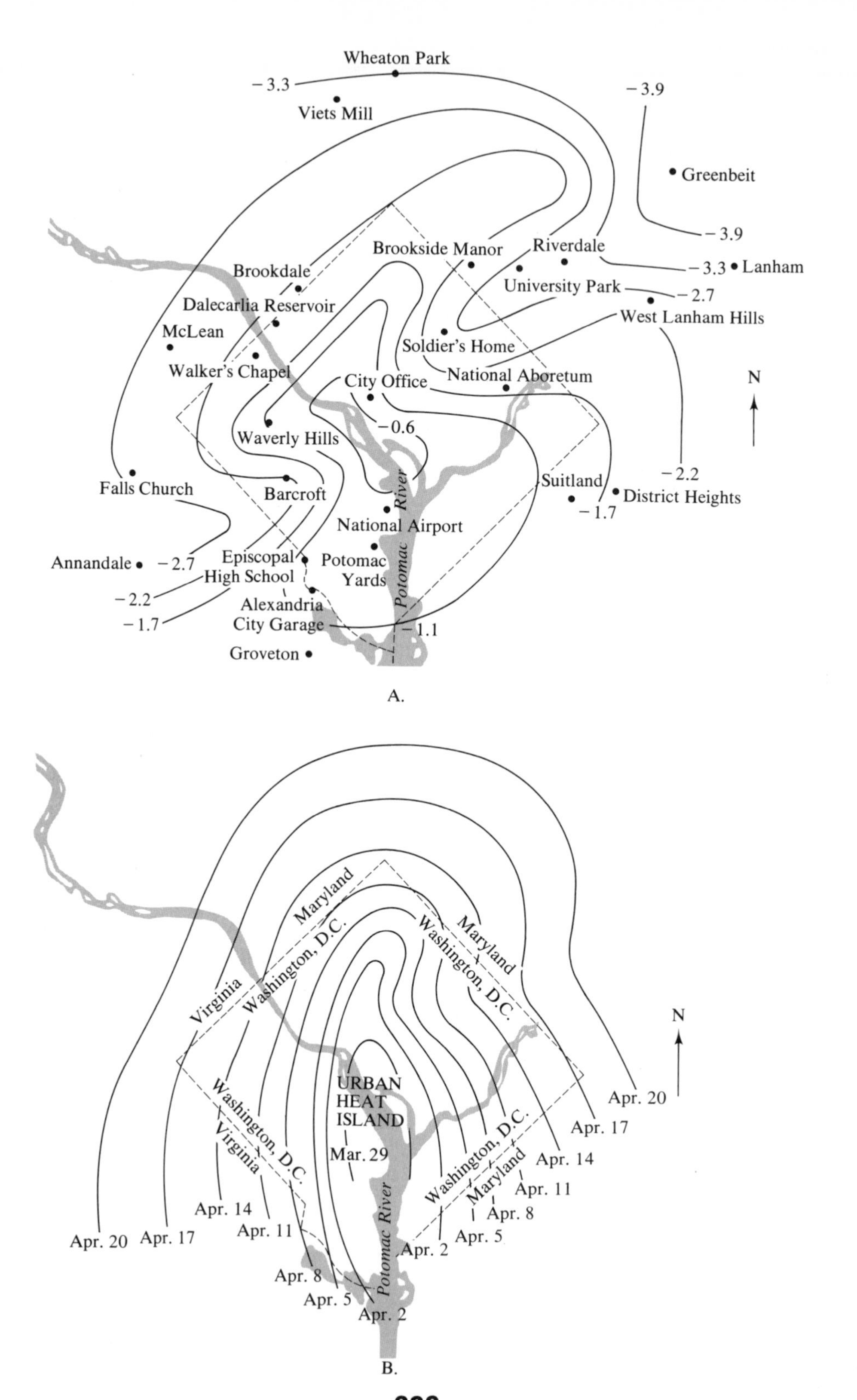
Wheaton Park
− 3.3
Viets Mill
− 3.9
Greenbeit
− 3.9
Brookside Manor
Riverdale
− 3.3
Lanham
Brookdale
University Park
− 2.7
Dalecarlia Reservoir
West Lanham Hills
McLean
Soldier's Home
Walker's Chapel
City Office
National Aboretum
N
− 0.6
Waverly Hills
River
− 2.2
Falls Church
Barcroft
Suitland
District Heights
− 1.7
National Airport
Annandale
− 2.7
Episcopal
High School
Potomac
Yards
Potomac
− 2.2
Alexandria
− 1.7
City Garage
− 1.1
Groveton
A.
Maryland
Washington, D.C.
Maryland
Washington, D.C.
Virginia
Washington, D.C.
Washington, D.C.
N
URBAN
HEAT
ISLAND
Apr. 20
Apr. 17
Virginia
Washington, D.C.
Mar. 29
Apr. 14
Maryland
Apr. 11
Apr. 14
Apr. 8
Apr. 11
Apr. 20
Apr. 17
Potomac River
Apr. 5
Apr. 2
Apr. 8
Apr. 5
Apr. 2
B.

Remember that these temperatures are averages, because on many clear, calm nights the temperature difference between the city center and the countryside was considerably greater, often 11° C or more. Conversely, on many overcast or windy nights the temperature differential approached 0 degrees. The map in Figure 16.8B resulted from the same study of metropolitan Washington, D.C., and reveals another interesting aspect of the heat island effect. It shows that over the five-year study period the last winter freeze occurred eighteen days earlier in the central part of the city than in the outlying areas.

Although the heat island is most obvious on clear, calm nights, the effect is still evident in long-term mean values. Isotherms for the Paris region, for example, show a pronounced heat island of about 1.6° C (Figure 16.9). A comparison of annual mean temperatures in and around London reveal similar results. Whereas the center of London has an annual mean of 11° C, the suburbs average 10.3° C, and the adjacent rural countryside has an annual mean of 9.6° C.

The radical change in the surface that results when rural areas are transformed into cities is a significant cause of the urban heat island. First, the tall buildings and the concrete and asphalt of the city absorb and store greater quantities of solar radiation than do the vegetation and soil typical of rural areas. In addition, because the city surface is impermeable, the runoff of water following a rain is rapid, resulting in a severe reduction in the evaporation rate. Hence, heat that once would have been used to convert liquid water to a gas now goes to further increase the surface temperature. At night, while both the city and countryside cool by radiative losses, the stonelike surface of the city gradually releases the additional heat accumulated during the day, keeping the urban air warmer than that of the outlying areas.

A portion of the urban temperature rise must also be attributed to waste heat from sources such as home heating and air conditioning, power generation, industry, and transportation. Many studies have shown that the magnitude of human-generated energy in metropolitan areas is great when compared to the amount of energy received from the sun at the surface. For example, investigations in Sheffield, England, and Berlin showed that the annual heat production in those cities was equal to approximately one-third of that received from solar radiation. Another study of densely built-up Manhattan in New York City revealed that during the winter, the quantity of heat produced from combustion alone was 2½ times greater than the amount of solar energy reaching the ground. In summer, the figure dropped to ⅙.

There are other, somewhat less influential, causes of the heat island. For example, the "blanket" of pollutants over a city, including particulate matter, water vapor, and carbon dioxide, contributes to the heat island by absorbing a portion of the upward-directed long-wave radiation emitted at the surface and reemitting some of it back to the ground. A somewhat similar effect results from the complex three-dimensional structure of the city. The vertical walls of office buildings, stores, and apartments do not allow radiation to escape as readily as in outlying rural areas where surfaces are relatively flat. As the sides of these structures emit their stored heat, a portion is reradiated between buildings rather than upward, and is therefore slowly dissipated.

In addition to retarding the loss of heat from the city, tall buildings also alter the flow of air. Because of the greater surface roughness, wind speeds within an urban area are reduced. Esti-

Figure 16.8 **A.** *The heat island of Washington, D.C., as shown by average minimum temperatures (°C) during the winter season (December–February). The city center had an average minimum that was nearly 4° C higher than in some outlying areas.* **B.** *Average dates of the last freezing temperatures in the spring for the Washington, D.C., area. (A. and B. after Clarence A. Woollum, "Notes From a Study of the Microclimatology of the Washington, D.C., Area for the Winter and Spring Seasons,"* Weatherwise *17(1964): 264, 267)*

mates from available records suggest a decrease on the order of about 25 percent from rural values. The lower wind speeds decrease the city's ventilation by inhibiting the movement of cooler outside air which, if allowed to penetrate, would reduce the higher temperatures of the city center.

Urban-induced Precipitation

Most climatologists agree that cities influence the frequency and amount of precipitation in their vicinities. Several reasons have been proposed to explain why an urban complex might be expected to increase precipitation.

(1) The urban heat island creates thermally induced upward motions which act to diminish the atmosphere's stability.
(2) Clouds may be modified by the addition of condensation nuclei and freezing nuclei from industrial discharges.
(3) Because of the obstructions to airflow caused by the increased roughness of the urban landscape, there is an increase in low-level turbulence.

Several studies comparing urban and rural precipitation have concluded that the amount of precipitation over a city is about 10 percent greater than over the nearby countryside. However, more recent investigations have shown that although cities may indeed increase their own rainfall totals, the greatest effects may occur downwind of the city center.

A striking and controversial example of such a downwind effect was examined by Stanley Changnon in the late 1960s.* Records indicated that since 1925, LaPorte, Indiana, located 48 kilometers downwind of the large complex of industries at Chicago, experienced a notable increase in total precipitation, number of rainy

* S. A. Changnon, Jr., "The LaPorte Weather Anomaly: Fact or Fiction?" *Bulletin of the American Meterological Society*, 49(1): 4–11. (*Note:* The term *anomaly* refers to something that deviates in excess of normal variation.)

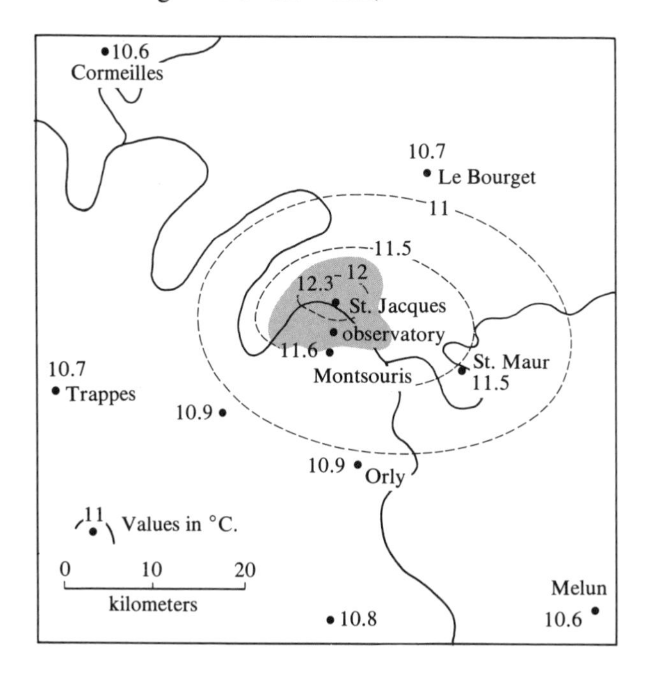

Figure 16.9 *The urban heat island of Paris as shown by mean annual isotherms (°C). (After Helmut E. Landsberg, "Man-Made Climatic Changes,"* Science *170(1970): 1270. Copyright 1970 by the American Association for the Advancement of Science)*

Table 16.4 ***Difference between Average La Porte Weather Values for 1951–1955 and the Means at Surrounding Stations Expressed as a Percentage of the Means***

Annual precipitation	31
Warm season precipitation	28
Annual number of days with precipitation $\geq$ 0.25 inch	34
Annual number of thunderstorm days	38
Annual number of hail days	246

SOURCE: After S.A. Changnon, Jr., "The LaPorte Weather Anomaly: Fact or Fiction?" *Bulletin of the American Meteorological Society*, 49(1).

days, number of thunderstorm days, and number of days with hail (Table 16.4). The magnitude of the changes and the absence of such change in the surrounding area led to widespread public attention. Many questioned whether the anomaly was real or simply the result of such factors as observer errors and changes in the exposure of instruments. However, after completing his study, Changnon concluded that the observed differences were real and were likely produced by the large industrial complex west of LaPorte. Among the reasons for this conclusion was that the number of days with smoke and haze (a measure of atmospheric pollution) in Chicago after 1930 corresponded quite well with the LaPorte precipitation curve. An examination of Figure 16.10, for example, shows a marked increase in smoke-haze days after 1940, when the LaPorte graph began its sharp rise. Further, the decrease in smoke-haze days after a peak in 1947 also generally matches a drop in the LaPorte curve. A second urban-related factor, steel production, was also found to correlate with the LaPorte precipitation curve. Records showed that seven peaks in steel production between 1923 and 1962 were associated with highs in the LaPorte curve.

The conclusions were accepted very cautiously by many and criticized by others who believed that because of the high natural rainfall variability of the area the data were neither sufficient nor accurate enough to make a sound determination. Nevertheless, this study was a pioneering effort which illustrated the possible effect of human activity on local climates.

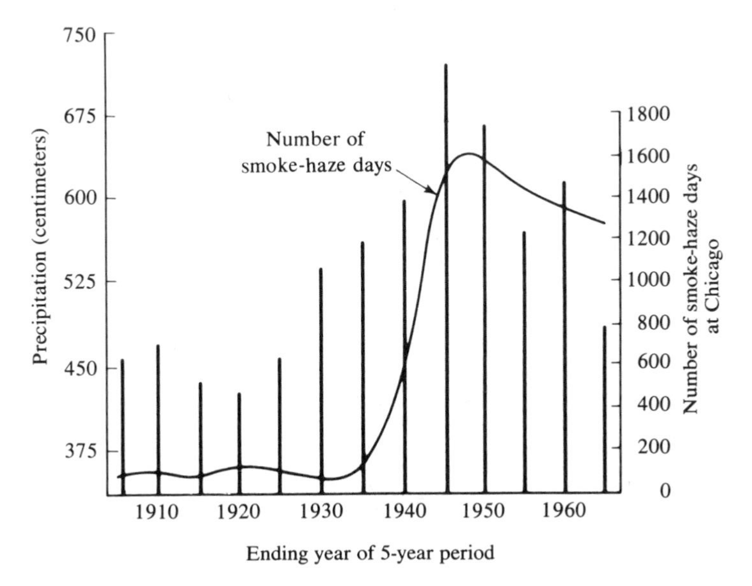

Figure 16.10 *Precipitation values at LaPorte, Indiana, and smoke-haze days at Chicago both plotted as five-year moving totals. [Data from Stanley A. Changnon, Jr., "The LaPorte Weather Anomaly: Fact or Fiction?"* Bulletin of the American Meteorological Society *49(1968)]*

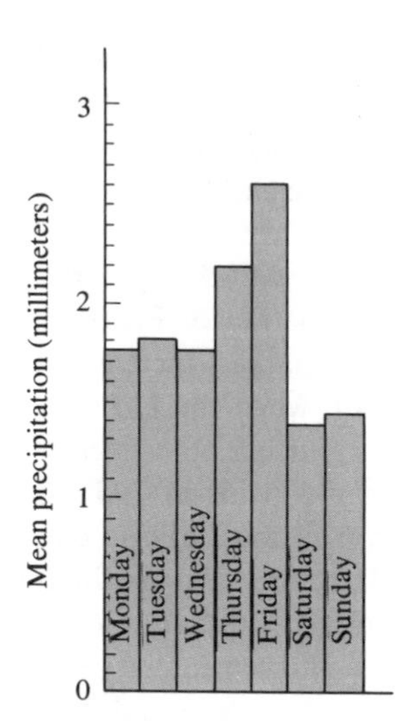

Figure 16.11 *Average precipitation by day of the week in Paris for an eight-year period (1960–1967). Notice the gradual increase from Monday to Friday and the sharp drop in precipitation totals on Saturday and Sunday. The average for weekdays was 1.93 millimeters compared to only 1.47 millimeters on weekends, a difference of 24 percent.*

Subsequent studies of the complex problem of urban effects on precipitation have generally confirmed the view that cities increase rainfall in downwind areas. One study, in the St. Louis area, showed that the magnitude of the downwind precipitation increase climbed as industrial development expanded. Another line of evidence from the St. Louis study and other studies also supports the hypothesis that cities are responsible for these increases. If cities are really modifying precipitation, the effects should be greater on weekdays, when urban activities are most intense, than on weekends, when much of the activity ceases. This is indeed the case. In the St. Louis study, analysis of precipitation on weekdays versus weekends revealed a significantly greater frequency of rain per weekday than per weekend day in the affected downwind area. The graph in Figure 16.11 shows similar findings that have been reported for Paris. Here a rise in precipitation can be noted from Monday through Friday, whereas the amounts for Saturday and Sunday are considerably smaller.

Other Urban Effects

In earlier sections, we saw that the greater air pollution in cities (1) contributes to the heat island by inhibiting the loss of long-wave radiation at night and (2) has a cloud-seeding effect that is believed to increase precipitation in and downwind of cities. These influences, however, are not the only ways in which pollutants influence urban climates. For example, the blanket of particulates over most large cities significantly reduces the amount of solar radiation

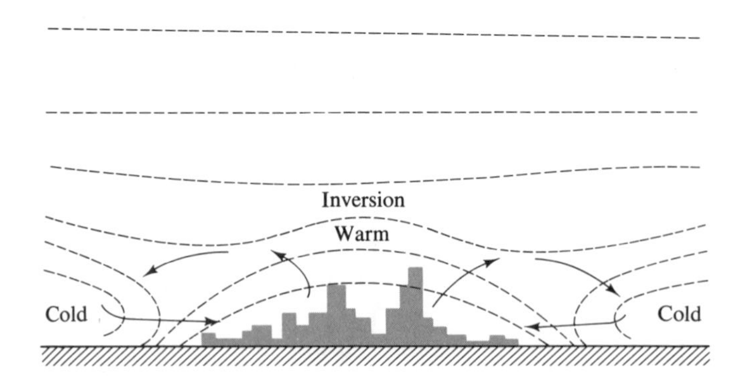

Figure 16.12 *Idealized scheme of nighttime circulation above a city on a clear, calm night. The dashed lines are isotherms and the arrows represent the wind (country breeze). (After Helmut E. Landsberg, "Man-Made Climatic Changes,"* Science *170(1970): 1271. Copyright 1970 by the American Association for the Advancement of Science)*

reaching the surface. As Table 16.1 indicates, the overall reduction in the receipt of solar energy is 15 percent, whereas short-wavelength ultraviolet is decreased by up to 30 percent. This weakening of solar radiation is, of course, variable. The decrease will be much greater during air pollution episodes than for periods when ventilation is good. Furthermore, particles are most effective in reducing solar radiation near the surface when the sun angle is low, because the length of the path through the dust increases as the sun angle drops. Hence, for a given quantity of particulate matter, solar energy will be reduced by the largest percentage at higher-latitude cities during the winter.

Even though relative humidities are generally lower in cities, because temperatures are higher and evaporation is reduced, the frequency of fogs and the amount of cloudiness are greater (see Table 16.1). It is likely that the great quantities of condensation nuclei produced by human activities in urban areas is a major contributor to these phenomena. When hygroscopic (water-seeking) nuclei are plentiful, water vapor readily condenses on them, even when the air is not yet saturated. Although cities have been shown to increase precipitation, some meteorologists are concerned that with a large oversupply of condensation nuclei too many particles will compete for available moisture. As a consequence, cloud droplets will be smaller and the coalescence process will be inhibited. However, as air pollution abatement efforts increase, this possibility should diminish, as should the city's effect on fogs and clouds.

Finally, mention should be made of another urban-induced phenomenon, the **country breeze.** As the name implies, this circulation pattern is characterized by a light wind blowing into the city from the surrounding countryside. It is best developed on relatively clear, calm nights, that is, on nights when the heat island is most pronounced. Heating in the city creates upward air motion, which in turn initiates the country-to-city flow (Figure 16.12). One investigation in Toronto showed that the heat island created a rural-city pressure difference that was sufficient to cause an inward and counterclockwise circulation centered on the downtown area. When this circulation pattern exists, pollutants emitted near the urban perimeter trend to drift in and concentrate near the city's center.

REVIEW

1. List some of the reasons that have made climatic change a topical subject.
2. How does plate tectonic theory (continental drift) help explain the previously "unexplainable" glacial features in present-day Africa, South America, and Australia?
3. Can continental drift explain short-term climatic changes? Explain.
4. What is the basic premise of the volcanic dust theory?
5. Is it possible that increased volcanic activity contributed to Ice Age cooling? What evidence supports your answer?
6. Can the volcanic dust theory presently be used to explain the alternating glacial and interglacial periods of the Ice Age?
7. List and describe each of the three variables in the astronomical theory of climatic change.
8. Do recent studies of sea-floor sediments tend to confirm or refute the astronomical theory? What do these studies predict for the future?
9. Why are scientists not confident about possible variations in the solar constant based on measurements made at the earth's surface?
10. What indirect evidence was used in John Eddy's investigations to double-check observed variations in solar activity (sunspots)?
11. What is the basic principle of the carbon dioxide theory of climatic change?
12. Is the fact that the atmosphere has been cooling for the past three decades necessarily a refutation of the CO_2 theory? Explain.
13. What are climatic feedback mechanisms? Give some examples.
14. Will human-generated dust cause global temperatures to rise or fall? Make a case for each point of view.
15. List two ways in which the difference between a "typical" rural surface and a "typical" city surface adds to the urban heat island.

16. How do each of the following factors contribute to the urban heat island: heat production, "blanket" of pollutants, three-dimensional city structure?
17. During what season does heat production most affect the heat island?
18. List the three factors that are the likely causes of greater precipitation in and downwind of cities.
19. a. What is the LaPorte anomaly?
    b. What evidence led Changnon to conclude that the LaPorte situation was related to the Chicago urban-industrial complex?
20. Describe the reasons for the following urban climatic characteristics:
    a. reduced solar radiation.
    b. reduced relative humidity.
    c. increased fog and cloud frequency.
    d. reduced wind speeds and more days when calms prevail.
21. What is a "country breeze"? Describe its cause.

KEY TERMS

climatic change
volcanic dust theory
astronomical theory
eccentricity
obliquity
precession
variable sun theory
solar constant
climatic feedback mechanisms
urban heat island
country breeze

part 4

astronomy

Stonehenge, an ancient solar observatory. (British Crown copyright, reproduced with permission of Controller of Her Britannic Majesty's Stationery Office)

Ancient Astronomy
The Birth of Modern Astronomy
Constellations and Astrology
Positions in the Sky
Motions of the Earth
Calendars

17

The Earth's Place in the Universe

The earth is one of nine planets that orbit the sun, which is a star. The sun is a part of a much larger family of perhaps 100,000 million stars that compose the Milky Way, which in turn is only one of thousands of millions of galaxies in an incomprehensibly large universe. This view of the earth's position in space is considerably different from that held only a few hundred years ago, when the earth was thought to occupy a privileged position as the center of the universe. This chapter examines some of the events that led to the unfolding of modern astronomy. In addition, it examines the earth's place in time and space.

Long before recorded history, people were aware of the close relationship between events on earth and the positions of the heavenly bodies, the sun in particular. They noted that changes in the seasons and floods of great rivers like the Nile occurred when the celestial bodies reached a particular place in the heavens. Early agrarian cultures, dependent on the weather, believed that if the heavenly objects could control the seasons, they must also have a strong influence over all earthly events. This belief undoubtedly was the reason that early civilizations, particularly the Chinese, Egyptian, and Babylonian, began keeping records of the positions of celestial objects (Figure 17.1). They recorded the location of the sun, moon, and the five planets visible to the unaided eye as they moved slowly against the background of "fixed" stars. In addition, the early Chinese also kept rather accurate records of comets and "guest stars." Today, we know that a "guest star" appears when a normal star, which before was too faint to be visible, increases its brightness as it explosively ejects gasses from its surface, a phenomenon we call a supernova. A study of their archives shows that the Chinese recorded every occurrence of the famous Halley's comet for at least ten centuries. However, because this comet reappears only once every 75–79 years, they were unable to establish that what they saw was the same object. So they, like most ancients, considered comets to be mystical in nature. Usually, comets were thought to be bad omens and were blamed for a variety of disasters, ranging from wars to plagues (Figure 17.2).

Ancient Astronomy

The "Golden Age" of early astronomy (600 B.C.–A.D. 150) was centered in Greece. The early Greeks have been criticized, and rightly so, for using philosophical arguments to explain natural phenomena. However, they did rely on observational data as well. The basics of geometry and trigonometry, which they had developed, were used to measure the sizes and distances of the largest bodies in the heavens—the sun and the moon.

The Greeks held the **geocentric** ("earth-centered") view of the universe. The earth, they believed, was a spherical, motionless body at the center of the universe. It was, in turn, surrounded by a transparent hollow sphere **(celestial sphere)** on which the stars were hung and carried on a

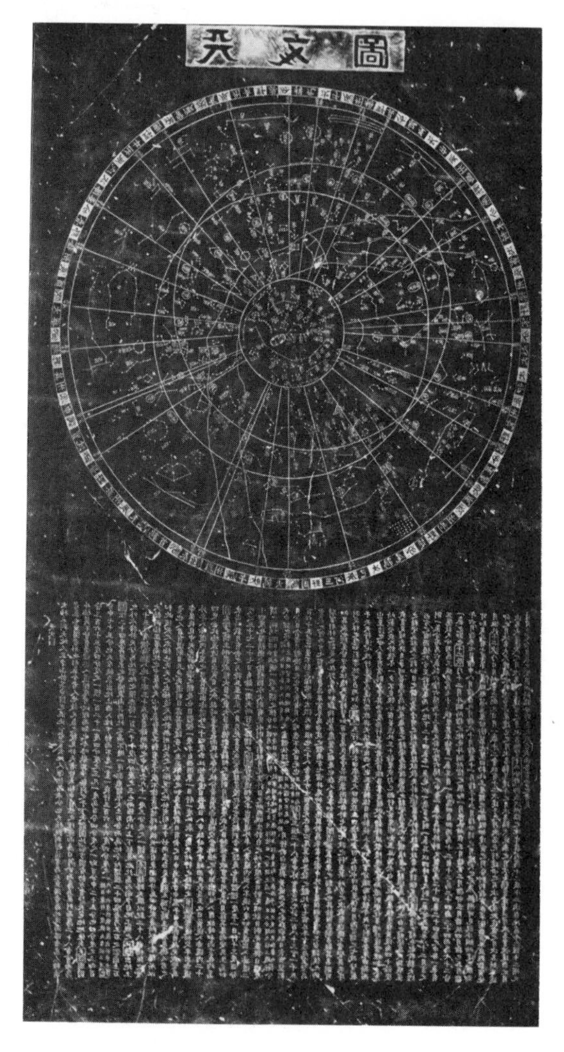

Figure 17.1 *Ancient Suchow star chart. (Rubbing and photograph property of L.W. Fredrick, Leander McCormick Observatory)*

Figure 17.2 *Bayeux tapestry illustrating the apprehension caused by Halley's comet in 1066. (Yerkes Observatory photograph)*

daily trip around the earth (an effect that is actually caused by the earth's rotation on its axis). Some early Greeks realized that the motion of the stars could be explained as easily by a rotating earth, but they rejected that idea, since the earth exhibited no feeling of motion and seemed too large to be moving about. In fact, proof of the earth's rotation was not demonstrated until 1851.

To the Greeks, all of the stars on the celestial sphere, except seven, appeared to remain in the same relative position to one another. These seven wanderers, called planets by the Greeks, included the sun, the moon, Mercury, Venus, Mars, Jupiter, and Saturn. Each was thought to have a circular orbit around the earth. Although this system was incorrect, the Greeks were able to refine it to the point that it explained the apparent movements of the celestial bodies as seen from earth.

Early Greeks

Many astronomical discoveries have been credited to the Greeks. As early as the fifth century B.C., the Greeks understood what causes the phases of the moon. Anaxagoras reasoned that the moon shines by reflected sunlight, and because it is a sphere, only half is illuminated at one time. As the moon orbits the earth, that portion of the illuminated half that is visible from the earth is always changing. Anaxagoras also realized that an eclipse of the moon occurs when it moves into the shadow of the earth.

The famous Greek philosopher Aristotle (384–322 B.C.) concluded that the earth was spherical, since it always cast a curved shadow when it eclipsed the moon. Although most of the teachings of Aristotle were passed on and, in fact, considered infallible by many, his belief in a spherical earth was somehow lost.

The first Greek to profess a sun-centered, or **heliocentric,** universe was Aristarchus (312–230 B.C.). Aristotle had suggested the idea earlier but had rejected it, reasoning that the motion of the earth around the sun would be demonstrated by a shift in the position of the stars, a phenomenon which was not observed. Aristotle's reasoning was sound, based on the data available; however, the stars are such great distances from the earth that the shift is only detectable using precision telescopes, which he of course did not have.

Aristarchus also used simple geometric relations to calculate the relative distances from the earth to the sun and the moon. He later used these data to calculate their relative sizes. Because of observational errors beyond his control, he came up with measurements that were much too small. However, he did learn that the sun was many times more distant than the moon and many times larger than the earth. The latter fact may have prompted him to suggest a sun-centered universe. Nevertheless, because of the strong influence of Aristotle, the earth-centered view dominated European thought for nearly 2000 years.

The first successful attempt to establish the size of the earth is credited to Eratosthenes (276–194 B.C.). Eratosthenes observed the angles of the noonday sun at Syene (presently Aswan) and Alexandria, two cities directly north and south of each other (Figure 17.3). Having found that the

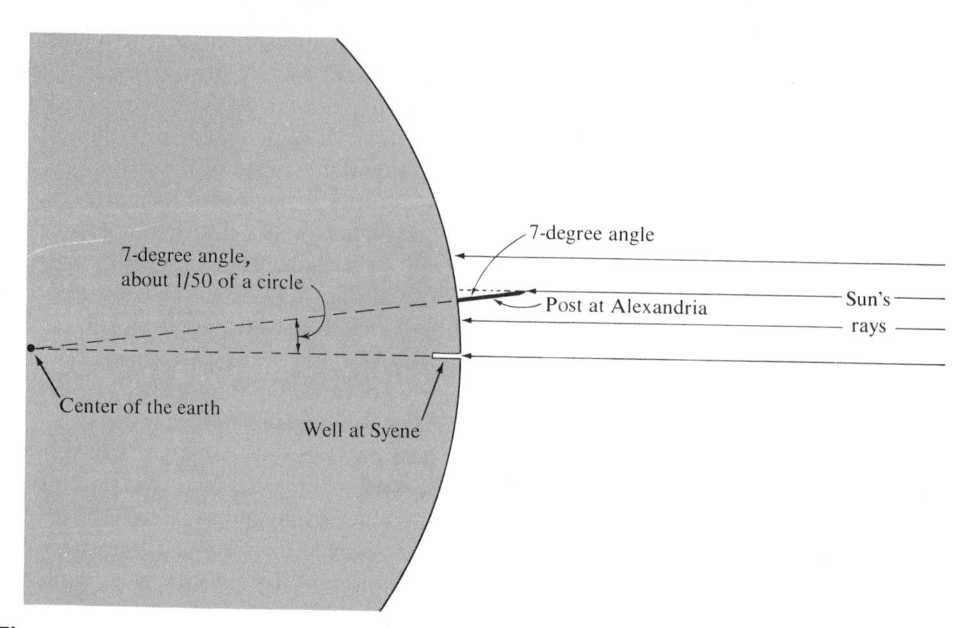

Figure 17.3 *Orientation of the sun's rays at Syene and Alexandria on June 21 when Eratosthenes calculated the earth's circumference.*

angles differed by 7 degrees, or 1/50 of a complete circle, he concluded that the circumference of the earth must be 50 times the distance between these two cities (5000 stadia), which gave him a figure of 250,000 stadia. Because authorities disagree on the size of a stadium in modern units, the accuracy of his work remains uncertain. Some historians, however, believe the stadium equals 157.5 meters (517 feet), which would make Eratosthenes' calculation of the earth's circumference 38,460 kilometers (24,662 miles), a figure very close to the modern value.

Probably the greatest of the early Greek astronomers was Hipparchus (second century B.C.), who is best known for his star catalog. Hipparchus determined the location of almost 850 stars, which he divided into six groups according to their brightness. He measured the length of the year to within minutes of the modern value and developed a method for predicting the time of an eclipse of the moon, accurate to within a few hours.

Although many of the Greek discoveries were obscured during the Middle Ages, the earth-centered view that the Greeks proposed became the established view in Europe. Presented in its finest form by Claudius Ptolemy, this geocentric outlook became known as the **Ptolemaic system.**

The Ptolemaic System

Much of our knowledge of Greek astronomy comes from a thirteen-volume treatise, *Almagest* ("the great work"), which was compiled by Ptolemy in A.D. 141. In this work, Ptolemy is credited with developing a model of the universe that accounted for the observable motions of the planets. The precision with which his model was able to predict future locations of the planets is testified to in that the model went virtually unchallenged, in principle at least, for nearly thirteen centuries.

In the Greek tradition, the Ptolemaic model had the planets moving in circular orbits around a motionless earth. (The circle was considered the pure and perfect shape by the Greeks.) Because the motion of the planets, as plotted against the background of stars, results from the combination of the motion of the earth and the planet's own motion around the sun, a rather odd apparent motion results. When viewed from the earth, the planets move slightly eastward among the stars each day. Periodically, each planet appears to stop, reverse direction for a period of time, and then resume an eastward motion. The apparent westward drift is called **retrograde motion.** Figure 17.4 illustrates the retrograde motion of Mars. Because the earth has a faster orbital speed than Mars, it overtakes Mars, and while doing so, Mars appears to be moving backward, that is, to be in retrograde motion. This is analogous to what a race-car driver sees as he passes a slower car. The slower planet, like the slower car, appears to be going backward, although its actual motion is in the same direction as the faster-moving body.

It is much more difficult to accurately represent retrograde motion using the incorrect earth-centered model, but Ptolemy was able to do just that (Figure 17.5). Rather than use one circle for an orbit, he placed the planet on a small circle **(epicycle),** which revolved around a large circle **(deferent).** By trial and error, he was able to select just the right combinations of circles to produce the amount of retrograde motion observed for each planet. (It is interesting to note that almost any closed curve can be produced by the combination of two circular motions, a fact that can be verified by persons who have used the Spirograph® design drawing toy.) In order to make Ptolemy's model fit the observed positions of the planets, circles were added and refinements made, first by Ptolemy, and later by others. These additional circles so complicated the system that King Alfonso was led to say that, had he been present at the Creation, he would have recommended a simpler plan.

It is nevertheless a tribute to Ptolemy's genius that he was able to account for the motion of the planets as well as he did, considering he used an incorrect model. It is possible, as some have suggested, that he did not mean his model to represent reality, but only to be used to calculate the positions of the heavenly bodies. (Whether he did or not, we will undoubtedly never know.) However, the Church did accept Ptolemy's theory as the correct representation of the heavens, which created problems for those who found fault with it.

The Birth of Modern Astronomy

Modern astronomy was not born overnight. Its development involved a break from deeply entrenched philosophic views and the founding of a "new and greater universe" governed by discernible laws. The most noted men involved in this transition were Nicolaus Copernicus, Tycho Brahe, Johannes Kepler, Galileo Galilei, and Sir Isaac Newton.

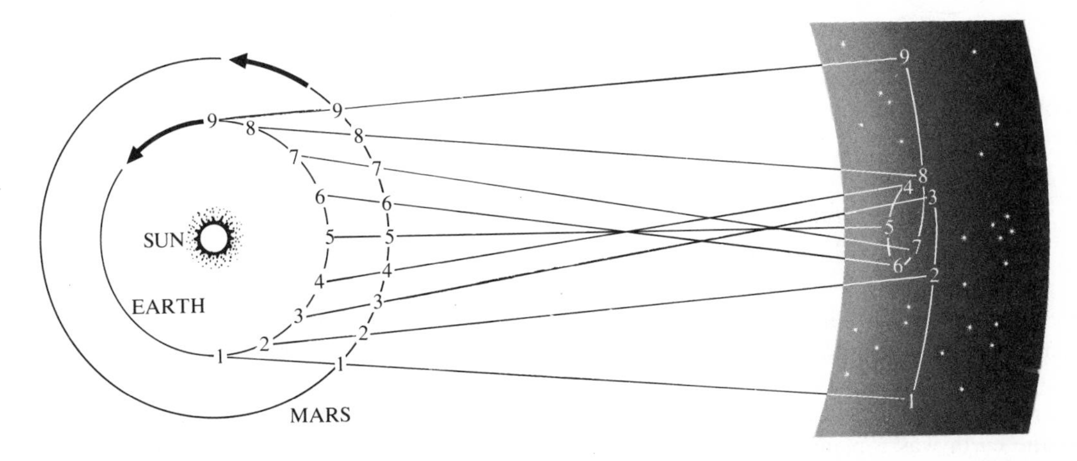

Figure 17.4 *Retrograde (backward) motion of Mars as seen against the background of distant stars.*

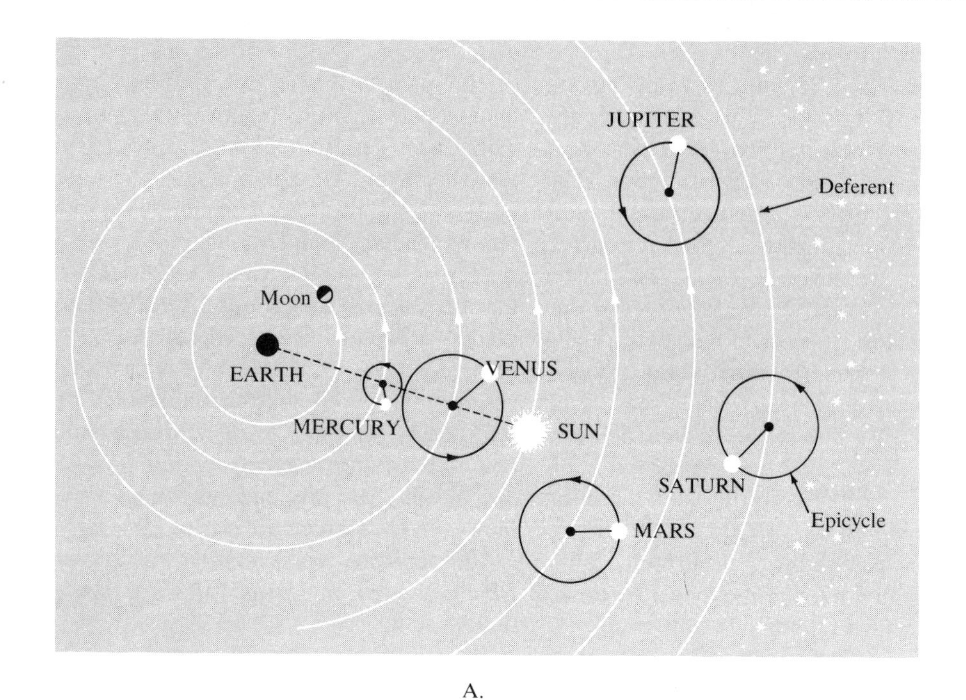

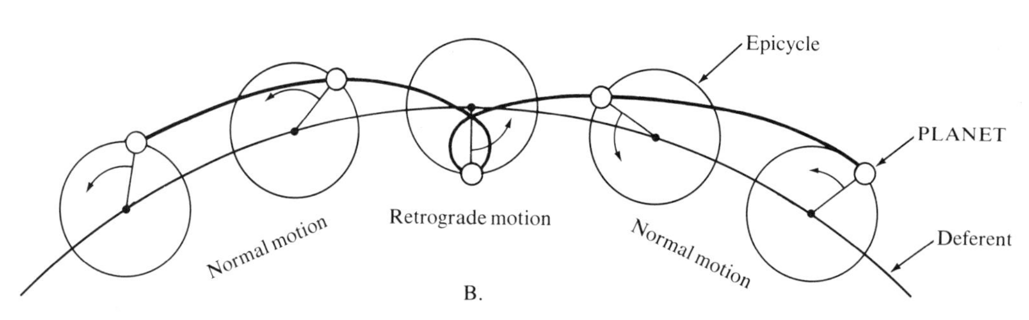

Figure 17.5 *The universe according to Ptolemy.* **A.** *The entire celestial sphere rotates around a motionless earth along with the planets, which have their own orbits.* **B.** *Retrograde motion as explained by Ptolemy.*

Nicolaus Copernicus

For almost thirteen centuries after the time of Ptolemy, very few astronomical advances were made in Europe. The first great astronomer to emerge after the Middle Ages was the Polish astronomer Nicolaus Copernicus (1473–1543) (Figure 17.6). Copernicus became convinced that the earth was a planet, just like any of the other five then known. The daily motions of the heavens, he reasoned, could be better explained by a rotating earth. To counter the Ptolemaic objection that the earth would fly apart if it rotated, he suggested that the much larger celestial sphere subjected to rotation would be even more likely to fly apart.

Having concluded that the earth was a planet, Copernicus reconstructed the solar system with the sun at the center and the planets Mercury,

Figure 17.6 *Nicolaus Copernicus. (Yerkes Observatory photograph)*

Venus, Earth, Mars, Jupiter, and Saturn orbiting it. This was a major break from the ancient idea that a motionless earth was the center of all movement. However, Copernicus retained a link to the past and used circles to represent the orbits of the planets. Unable to get satisfactory agreement between predicted locations of the planets and the observed positions, he found it necessary to add epicycles like those used by Ptolemy. (The discovery that the planets have elliptical orbits would be left to Kepler.) Also, like his predecessors, Copernicus used philosophical justifications, such as the following, to support his point of view:

> . . . in the midst of all stands the sun. For who could in this most beautiful temple place this lamp in another or better place than that from which it can at the same time illuminate the whole?

Copernicus' monumental work, *De Revolutionibus,* which set forth his controversial ideas, was published while he lay on his deathbed. Hence, he never experienced the criticisms which fell on many of his followers. The Copernican system was considered heretical, and expounding it cost at least one man his life. Giordano Bruno was seized by the Inquisition in 1600 and, refusing to denounce the Copernican theory, was burned at the stake.

Tycho Brahe

Tycho Brahe (1546–1601) was born of Danish nobility three years after the death of Copernicus. Reportedly, Tycho became interested in astronomy while viewing a solar eclipse that had been predicted by astronomers. He persuaded King Fredrich II to establish the observatory on the island of Hveen, near Copenhagen, which he headed. There he designed and built pointers (the telescope had not yet been invented), which he used for 20 years to systematically measure the location of the heavenly bodies. These observations, particularly those of Mars, which were far more precise than any made so far, were his legacy to astronomy.

Tycho did not believe in the Copernican (sun-centered) system, because, like Aristotle, he was unable to observe the shift in stars (a feat that was not accomplished until 1836). It is ironic that the data he collected to refute the Copernican view would later be used to support it.

With the death of his patron, the King of Denmark, Tycho was forced to leave his observatory. It was probably his arrogance and extravagant nature that caused the conflict with the new ruler. Tycho moved to Prague, where in the last year of his life he acquired an able assistant by the name of Johannes Kepler. Kepler retained most of the observations made by Tycho and put them to exceptional use.

Johannes Kepler

If Copernicus can be considered the person who ushered out the old astronomy, Jo-

Figure 17.7 *Johannes Kepler. (Smithsonian Institution Photo No. 56123).*

hannes Kepler (1571–1630) may then be given credit for ushering in the new (Figure 17.7). Armed with Tycho's data, a good mathematical mind, and, of greater importance, a strong faith in the accuracy of Tycho's work, Kepler was able to derive three basic laws of planetary motion. The first two laws resulted from his inability to fit Tycho's observations of Mars to a circular orbit. Unwilling to concede that the discrepancies were due to observational error, he searched for another solution. This endeavor led him to discover that the orbit of Mars is actually elliptical (Figure 17.8). About that same time, he realized that the orbital speed of Mars varies in a predictable way. As it approaches the sun, it speeds up, and as it pulls away from the sun, it slows down. In 1609, after almost a decade of work, Kepler set forth these ideas in the form that is now known as Kepler's first two laws of planetary motion: (1) *The path of each planet around the sun is an ellipse with the sun at one focus* (Figure 17.8). An elliptical orbit accounts more accurately for the motions of the planets. This fact dealt a fatal blow to the idea of circular orbits. (2) *Each planet revolves so that a line connecting it to the sun sweeps over equal areas in equal intervals of time.* This law of equal areas expresses the variations in orbital speeds of the planets geometrically. Figure 17.9 illustrates the second law. Note that in order for a planet to sweep equal areas in the same amount of time, it must travel most rapidly when it is nearest the sun and most slowly when it is farthest from the sun.

Kepler was very religious and believed that the Creator made an orderly universe. The uniformity he tried to find eluded him for nearly a decade. Then, in 1619, he published his third law in *The Harmony of the Worlds*. Simply, this law states that *the orbital periods of the planets and their distances to the sun are proportional.* In its simplest form, the rotational period is measured in years, and the planet's solar distance is expressed in terms of the earth's mean distance to the sun. The latter "yardstick" is called the **astronomical unit** (AU) and averages about 150 million kilometers (93 million miles). Using these units, Kepler's third law states that *the planet's orbital period squared is equal to its mean solar distance cubed* ($p^2 = d^3$). Consequently, the solar distances of the planets can be calculated when their periods of rotation are known. For example, Mars has a period of 1.88 years, which squared equals 3.54. The cube root of 3.54 is 1.52, and that is the distance to Mars in astronomical units (Table 17.1).

Kepler's laws assert that the planets revolve around the sun and therefore support the Copernican theory. Kepler, however, did fall short of determining the forces that act to produce the planetary motion he had so ably described. That task would remain for Sir Isaac Newton.

Table 17.1 *Periods and Solar Distances of Planets*

Planet	Solar Distance (AU)	Period (years)
Mercury	0.39	0.24
Venus	0.72	0.62
Earth	1.00	1.00
Mars	1.52	1.88
Jupiter	5.20	11.86
Saturn	9.54	29.46
Uranus	19.18	84.01
Neptune	30.06	164.8
Pluto	39.44	247.7

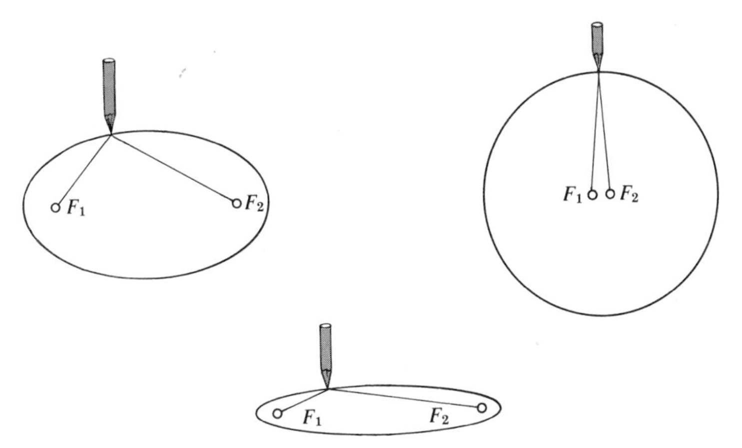

Figure 17.8 *Drawing ellipses with various eccentricities. The farther points F_1 and F_2 (foci) are moved apart, the more flattened (greater eccentricity) is the resulting ellipse.*

Galileo Galilei

Galileo Galilei (1564–1642) was the greatest Italian scientist of the Renaissance (Figure 17.10). He was a contemporary of Kepler and, like Kepler, strongly supported the Copernican theory of a sun-centered solar system. Galileo's greatest contributions to science were his descriptions of the behavior of moving objects. These he derived from experimentation. The method of using experiments to determine natural laws had essentially been lost since the time of the early Greeks. Galileo's most famous experiment occurred at the Leaning Tower of Pisa, where he dropped objects of varying weights to demonstrate that the weight of an object does not affect its rate of fall. He correctly concluded that air resistance is the cause of the

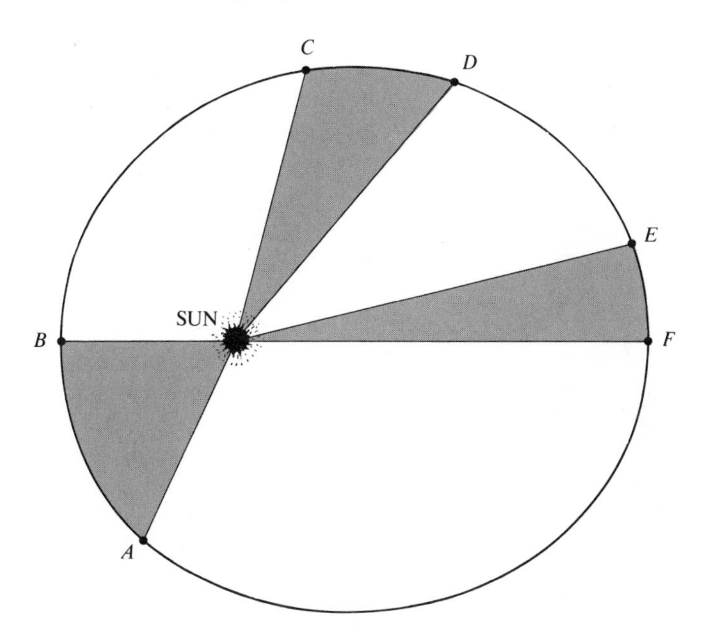

Figure 17.9 *Law of equal areas. A line connecting a planet to the sun sweeps out an area in such a manner that equal areas (shaded) are swept out in equal times.*

variations observed for very light objects. This same experiment was performed most dramatically on the airless moon when David Scott, an *Apollo 15* astronaut, showed that a feather and a hammer fall at the same rate.

All astronomical discoveries before Galileo's time were made without the aid of a telescope. In 1609, Galileo heard that a Dutch lens maker had devised a system of lenses that magnified objects. Apparently without ever seeing one, Galileo constructed his own telescope, which had a magnification of 3. He immediately made others, his best having a magnification of about 30.

With the telescope, Galileo was able to view the universe in a completely new way. He made many important discoveries that supported the Copernican view of the universe, some of which follow:

(1) He discovered four satellites or moons orbiting Jupiter and accurately determined their periods of revolution, which range from 2 to 17 days (Figure 17.11). This find dispelled the old idea that the earth was the only center of motion. It also countered the argument frequently used by those opposed to the sun-centered system that the moon would be left behind if the earth really revolved around the sun.

(2) He was able to see the planets as circular disks rather than just as points of light. This indicated that they might be earth-like.

(3) Venus, he learned, has phases just like the moon, proving that Venus orbits the sun. In the Ptolemaic system, Venus always lies between the earth and the sun, which means that only the crescent phase could be seen from earth. He also saw that the size of Venus appears smallest when it is in full phase, and thus furthest from the earth (Figure 17.12).

(4) His telescope allowed an intimate view of the moon's surface. He found that it is not a smooth glass sphere as the ancients had suspected and the Church had decreed. Rather, he saw mountains, craters, and plains. The latter he thought might be bodies of water. The idea that the moon contains seas was strongly promoted by others, as we can tell from some of the names given (Sea of Tranquility, Sea of Storms, and so forth).

(5) Viewing the sun (which may have caused the eye damage that later resulted in his total blindness), Galileo saw sunspots (dark regions caused by slightly lower temperatures). He was able to track these spots and estimate the rotational period of the sun as just under a month. Hence, another heavenly body was found to have "blemishes" and a rotational motion.

Figure 17.10 *Galileo Galilei. (Yerkes Observatory photograph)*

In 1616 the Church condemned the Copernican theory as contrary to Scripture, and Galileo was told to abandon it. Unwilling to accept the verdict of the Church, Galileo began writing his most famous work, *Dialogue of the Great World Systems*. Despite poor health and a narrow escape with death, he was able to complete the project. In 1630 he went to Rome seeking permission from Pope Urban VIII to publish. Because the book was a dialogue that expounded both the Ptolemaic and Copernican systems,

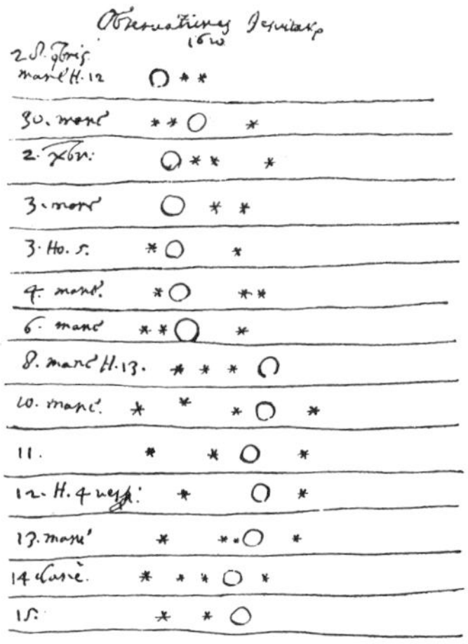

Figure 17.11 *Drawing by Galileo of Jupiter and its four largest satellites. (Yerkes Observatory photograph)*

publication was allowed. However, Galileo's enemies were quick to realize that he was promoting the Copernican view at the expense of the Ptolemaic system. Sale of the book was quickly halted, and Galileo was called before the Inquisition. Tried and convicted of believing doctrines contrary to Divine Scripture, he was sentenced to permanent house arrest, where he remained for the last ten years of his life. Despite the restrictions placed on him, and despite his age and the grief he carried after the death of his eldest daughter, Galileo continued to work. In 1637 he became totally blind, yet during the next few years he completed his most scientific work, a book on the study of motion. When Galileo died in 1642, the Grand Duke of Tuscany wanted to erect a monument in his honor, but fear that it would offend the Holy Office still prevailed, and it was never built.

Sir Isaac Newton

According to the "old" calendar, Sir Isaac Newton (1643–1727) was born in the year of Galileo's death (Figure 17.13). His many accomplishments in the fields of mathematics and physics led his successor, Lagrange, to say that, "Newton was the greatest genius that ever existed. . . ."

Although Kepler and those who followed attempted to explain the forces involved in planetary motion, they fell short of the mark. Kepler believed that some force pushed the planets along their orbits. Galileo, however, correctly reasoned that no force was required to keep an object in motion. Galileo had proposed

Figure 17.12 *Phases of Venus, illustrating that Venus appears smallest at full phase. In the Ptolemaic model, Venus is never on the opposite side of the sun and, therefore, could never appear in full phase. (Courtesy of Lowell Observatory)*

Figure 17.13 *Sir Isaac Newton. (Yerkes Observatory photograph)*

that the natural tendency for a moving object not affected by an outside force was to continue moving at a uniform speed and in a straight line. This concept was later formalized by Newton and is now known as Newton's first law of motion. It remained for Newton to explain why the planets traveled around the sun according to Kepler's laws.

The problem, then, was not to explain the force that keeps the planets moving but rather to determine the force that keeps them from going in a straight line out into space. It was to this end that Newton conceptualized the force of gravity. At the early age of 23, he envisioned a force that extended from the earth into space and held the moon in orbit around the earth. Although others had theorized the existence of such a force, he was the first to formulate and test the law of universal gravitation, which states that *every body in the universe attracts every other body with a force that is proportional to their masses and inversely proportional to the square of the distance between them.* Thus, the gravitational force decreases with distance, so that two objects 3 kilometers apart have 3^2, or 9, times less gravitational attraction than when the same objects are 1 kilometer apart. The law of gravitation also states that the greater the mass of the object, the greater its gravitational force. *The mass of an object can be considered the measurement of the total amount of matter it contains,* and by convention, mass is numerically equal to the weight of an object at sea level on the earth. However, unlike weight, the mass of an object does not change. For example, a man weighing 180 pounds on the earth would weigh 1/6 as much, or 30 pounds, on the moon, but his mass would remain unchanged.

With his newly developed calculus, Newton proved that the force of gravity, combined with the tendency of a planet to remain in straight-line motion, results in the elliptical orbits discovered by Kepler. The earth, for example, moves forward in its orbit about 30 kilometers (18½ miles) each second, and during the same second, the force of gravity pulls it toward the sun about ½ centimeter (⅛ inch). It is, therefore, as Newton concluded, the combination of the earth's forward motion and its "falling" motion that defines its orbit (Figure 17.14). If gravity should be miraculously cut off, the earth would move in a straight line out into space. On the other hand, if the earth's forward motion were suddenly stopped, gravity would pull it toward the sun.

Up to this point, we have discussed the earth as if it were the only planet. However, all bodies in the solar system have gravitational effects on the earth. Therefore, the orbit of the earth is not the perfect ellipse determined by Kepler. Any variance in the orbit of a body from its predicted path is referred to as **perturbation.** For example, Jupiter's gravitational pull on Saturn is enough to reduce Saturn's orbital period by nearly one week from the predicted period. As we shall see, the application of this concept led to the discovery of the planet Neptune because of Neptune's gravitational effect on the orbit of Uranus.

Newton used the law of gravitation to redefine Kepler's third law, which states the relationship between the orbital periods of the planets and their solar distances. When restated, Kepler's third law takes into account the mass of the bodies involved and thereby provides a method that can be used to determine the mass of a body when the orbit of one of its satellites is known. The mass of the sun is known from the orbit of the earth, and the mass of the earth has been determined from the orbit of the moon. In fact,

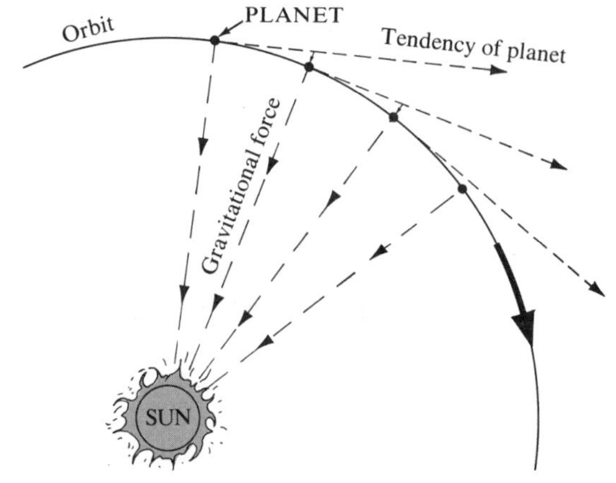

Figure 17.14 *Orbital motion of the earth.*

the mass of any body with a satellite can be determined. The masses of bodies that do not have satellites can be determined only if the bodies noticeably affect the orbit of a neighboring body or of a nearby artificial satellite.

Constellations and Astrology

At an early date, people became fascinated with the star-studded skies and began to identify the patterns they saw. These configurations, called **constellations,** were named in honor of mythological characters. It takes a good bit of imagination to make out the intended subjects, as most were probably not intended to be likenesses in the first place. Although we inherited many of the constellations from Greek mythology, it is believed that the Greeks acquired most of theirs from the Babylonians. Today, 88 constellations are recognized, and they are used to divide the sky into units, just as state boundaries divide the United States. Every star in the sky is in but is not necessarily part of one of these constellations. Constellations therefore enable astronomers to roughly identify that position of the heavens they are observing. For the student, the constellations provide a good way to become familiar with the night sky (see Appendix E).

Some of the brightest stars in the heavens were given proper names, such as Sirius, Arcturus, and Betelgeuse. In addition, the brightest stars in a constellation are generally named in order of their intensity by the letters of the Greek alphabet—alpha (α), beta (ß), and so on—followed by the name of the parent constellation. For example, Sirius, the brightest star in the constellation Canis Major (Larger Dog), is called Alpha (α) Canis Majoris. Lesser stars are systematically numbered within each constellation.

The planets, moon, and sun lie along nearly the same plane. Therefore, they move along the

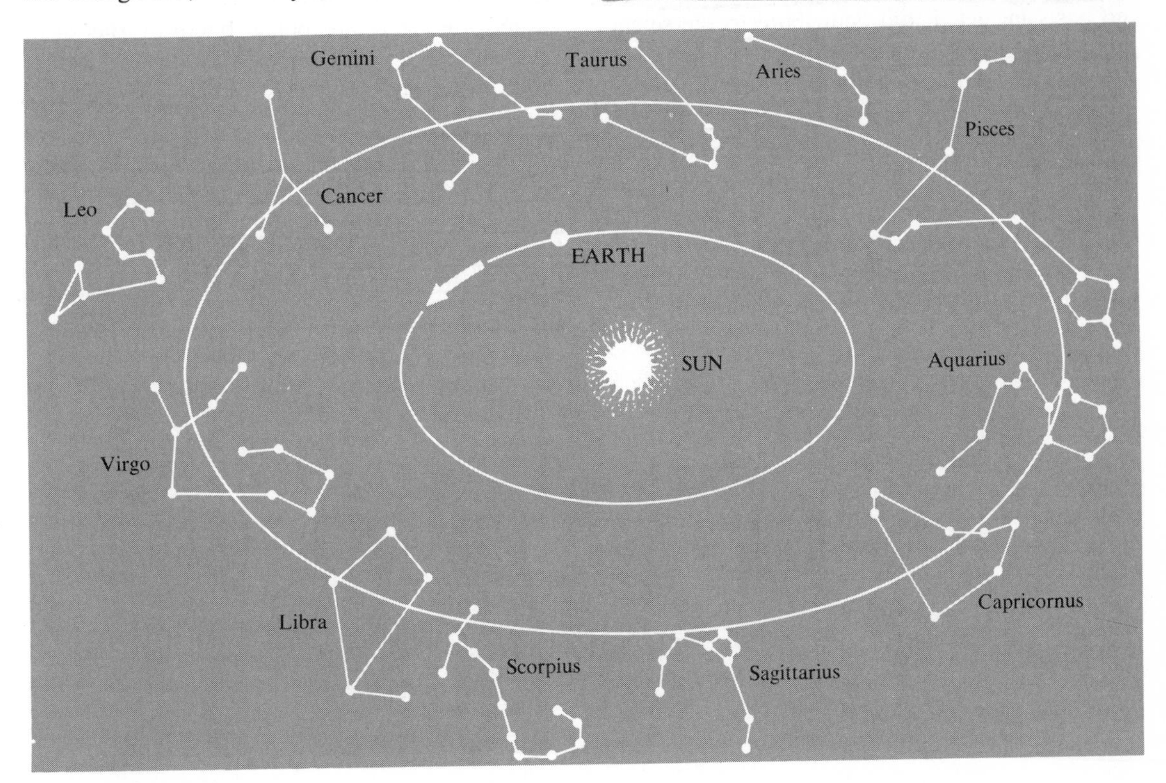

Figure 17.15 *Constellations of the Zodiac.*

same region of the sky, which has become known as the *Zodiac* ("zone of animals"). Because the moon goes through its phases about twelve times each year, the Babylonians divided the Zodiac into twelve constellations (Figure 17.15). Thus, each successive full moon occurs in the next constellation. The twelve constellations of the zodiac are Aries, Taurus, Gemini, Cancer, Leo, Virgo, Libra, Scorpio, Sagittarius, Capricorn, Aquarius, and Pisces. These names should be familiar to you as the astrological signs of the Zodiac. When first established, the vernal equinox (first day of spring) occurred when the sun was in the constellation Aries, and the signs of the Zodiac occurred in their respective constellations. Now, because of the earth's precession, the vernal equinox occurs when the sun is in Pisces. In several years, it will occur when the sun is in Aquarius. Hence, the popular "Age of Aquarius" is coming.

Although astrology is not a science and has no basis in fact, it did contribute to the science of astronomy. The position of the moon, sun, and planets at the time of a person's birth (sign of the Zodiac) were considered to have great influence on his life. Even the great astronomer Kepler was required to make horoscopes as part of his duties. In order to make horoscopes for the future, astrologers attempted to predict the future positions of the celestial bodies. Consequently, some of the improvements in astronomical instruments were probably due to the desire for more accurate predictions of events such as eclipses, which were considered highly significant in a person's life.

Even prehistoric people built observatories. That found at Stonehenge, in England, was undoubtedly an attempt at better solar prediction (Figure 17.16). Here, at the time of midsummer (June 21–22—the summer solstice), the rising sun emerges directly above the heel stone. Besides keeping this calendar, Stonehenge may also have provided a method of determining eclipses.

Positions in the Sky

If you gaze away from the city lights on a clear night, you will get the distinct impression that the stars produce a spherical shell surrounding the earth. This impression seems so real that it is easy to understand why many early Greeks regarded the stars as being fixed to the clear crystalline celestial sphere. Although we realize this sphere does not exist, it is convenient to use it for locating stars.

One method for doing this, called the **equatorial system,** has the celestial sphere divided into a coordinate system very similar to that used for locations on the earth's surface (Figure 17.17). Because the celestial sphere appears to rotate around a line extending from the earth's axis,

Figure 17.16 *Stonehenge, an ancient solar observatory. On June 21–22 (summer solstice), the sun can be observed rising above the heel stone.*

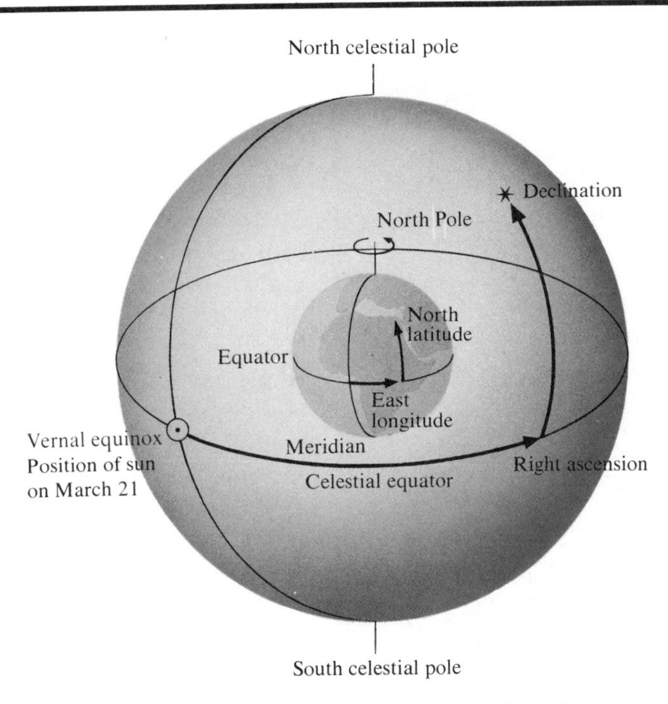

Figure 17.17 *Astronomical coordinate system on the celestial sphere.*

the north and south celestial poles are in line with the terrestrial North and South poles. The north celestial pole is very near the bright star Polaris ("pole star"), which is also called the North Star. To an observer in the Northern Hemisphere, the stars appear to circle Polaris, because it, like the North Pole, is in the center of motion (Figure 17.18). Figure 17.19 shows how you can locate the North Star from the constellation of the Big Dipper. The intersection of a plane through the earth's equator with the celestial sphere defines the celestial equator (which is 90 degrees from the celestial poles).

The two most commonly used coordinates in the equatorial system, **declination** and **right ascension,** are analogous to latitude and longitude (Figure 17.17). Like latitude, declination is the angular distance north or south of the equator (celestial). Right ascension is the angular distance measured eastward along the celestial equator from the position of the vernal equinox. (The **vernal equinox** is at the point in the sky where the sun crosses the celestial equator, at the onset of spring.) While declination is expressed in degrees, right ascension is usually expressed in hours. This is not as complex as it first appears, since a simple relationship exists between degrees and hours. One complete rotation, or 360 degrees, takes 24 hours. Hence, 15 degrees equal 1 hour, and 1 degree equals 4 minutes. If the right ascension of a star equals 2 hours, it can be expressed as 30 degrees. To visualize distances on the celestial sphere, it may be helpful to remember that the moon and sun have an apparent width of about ½ degree. The declination and right ascension of some of the brightest stars can be determined from the star charts in Appendix E.

Motions of the Earth

There are two primary motions of the earth—rotation and revolution. **Rotation** is the turning of a body on its axis, and **revolution** is the motion of a body around some point in space. The main consequences of the earth's rotation are day and night. But, as stated earlier, day and night and the apparent motions of the stars could be accounted for equally well by a revolving sun

Figure 17.18 *Star trails in the region of Polaris (north celestial pole) on a time exposure. (Courtesy of Lick Observatory)*

and celestial sphere. Copernicus realized that a rotating earth greatly simplified the existing model of the universe and, therefore, strongly advocated it as the correct view. He was, however, unable to prove that the earth rotated. The first substantial proof was presented 300 years after his death, by the French physicist Jean Foucault.

In 1851 Foucault used a free-swinging pendulum to demonstrate that the earth does, in fact, turn on its axis. To envision Foucault's arguments, imagine a large pendulum swinging at the North Pole. Keep in mind that once a pendulum is put into motion, it continues swinging in the same plane unless acted upon by some outside force. A sharp stylus is attached to the bottom of this pendulum, and it marks the snow as it oscillates. When we observe the marks made by the stylus, we note that the pendulum is slowly but continually changing position. In 24 hours it has returned to the starting position (Figure 17.20). Since no outside force acted on the pendulum to change its position, what we observed must have been the earth rotating under it. Foucault conducted a similar experiment in Paris.

The earth's rotation has become a standard method of measuring time. Each rotation equals about 24 hours. We can, however, measure the earth's rotation in two ways. Consequently, there are two kinds of days. You are most familiar with the **mean solar day,** which is the time interval from one noon to the next and averages 24 hours. Noon is determined when the sun has reached its zenith (highest point). The **sidereal**

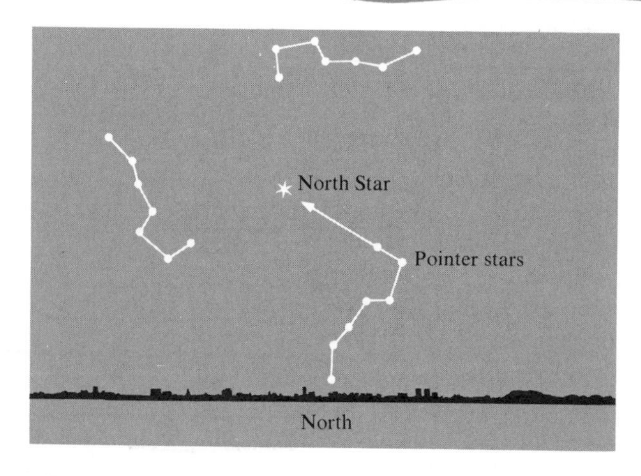

Figure 17.19 *Locating the North Star from the Big Dipper above the northern horizon.*

day, on the other hand, is the time it takes for the earth to make one complete rotation (360 degrees) with respect to a particular star. It is measured from the time required for a star to reappear at the identical position in the sky. The sidereal day has a period of 23 hours, 56 minutes, and 4 seconds (measured in solar time), which is almost 4 minutes shorter than the mean solar day. This difference results because the direction to the distant stars is only infinitesimally changed by the earth's motion around the sun, while the direction to the sun changes by almost 1 degree each day. This difference is diagrammatically shown in Figure 17.21. If it is not obvious why we use the solar day rather than the time required for one rotation as a measure of a civil day, consider the fact that in sidereal time "noon" occurs 4 minutes earlier each day. In six months it occurs at "midnight." Astronomers use sidereal time because the stars appear in the same position in the sky every 24 sidereal hours. Usually, an observatory will begin its sidereal day when the position of the vernal equinox is directly overhead, that is, over the meridian on which the observatory is located. Therefore, when the observatory's sidereal clock is the same as the star's right ascension, the star will be overhead, or at its highest point. For example, the brightest star in the heavens, Sirius, has a right ascension of 6 hours, 42 minutes, and 56 seconds and will, therefore, be overhead when the clock at the observatory indicates that time.

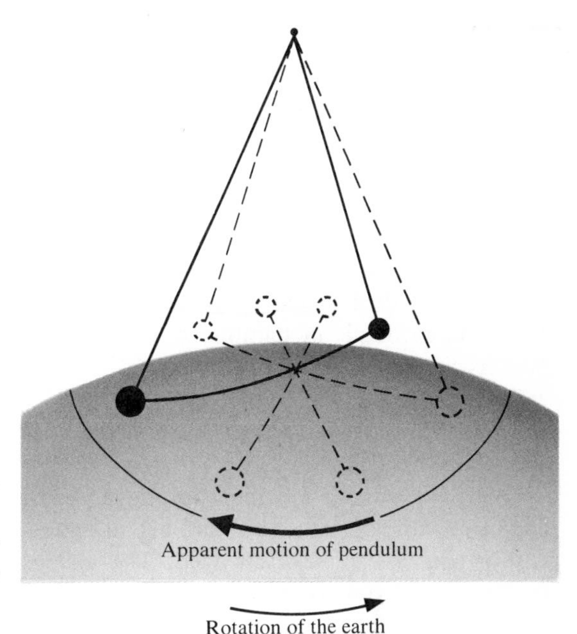

Figure 17.20 *Apparent movement of a pendulum at the North Pole caused by the earth rotating beneath it.*

The earth revolves around the sun in an elliptical orbit at a speed of nearly 100,000 kilometers (66,000 miles) per hour and at a distance that averages 150 million kilometers (93 million miles). About January 3, the earth is closest to the sun—147 million kilometers (**perihelion** position), and around July 4, it is farthest from the sun—152 million kilometers (**aphelion**

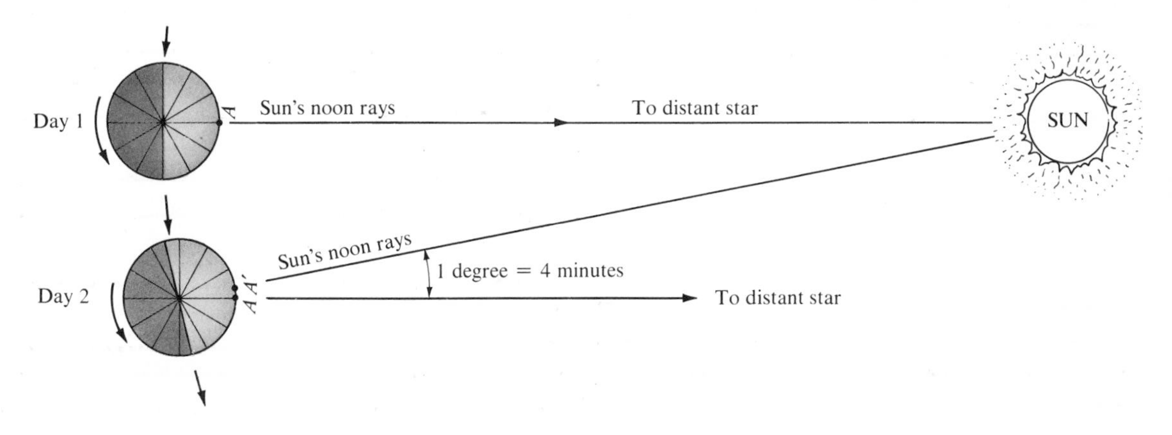

Figure 17.21 *Illustration of a solar and sidereal day.*

position). The proof of the earth's revolution was almost as difficult to find as the proof of rotation. Probably the most sought-after evidence for this motion was a shift in the position of the nearby stars. The argument went like this: If the earth does revolve, the position of a nearby star, when observed from two points in the earth's orbit six months apart, should shift with respect to the more distant stars. This apparent shift of the stars is called **stellar parallax** (see Figure 20.1). The principle of parallax is simple, even if the technique of calculating it is not.

To visualize parallax, close one eye and with your index finger in a vertical position use your other eye to line your finger up with some distant object. Now, without moving your finger, view the object with your other eye and notice that its position appears to have changed. The farther away you hold your finger, the less its position seems to shift. Analogously, because the distance to even the nearest stars is so great compared with the width of the earth's orbit, the shift that occurs is too small to have been noticed using the early telescopes, let alone the naked eye. Recall that the absence of this apparent shift was regarded as proof by Aristotle and Tycho Brahe that the earth did not orbit the sun. In 1838 the Prussian astronomer Friedrich Bessel measured stellar parallax accurately for the first time, proving that the earth has orbital motion.

Because of the earth's annual motion around the sun, each day the sun appears to be displaced eastward among the constellations a distance equal to about twice its width. The apparent annual path of the sun upon the celestial sphere is called the **ecliptic.** Most of the planets and the moon travel in nearly the same plane as the earth. Hence, their paths on the celestial sphere lie near the ecliptic. The most notable exception is Pluto, whose orbit is tilted 17 degrees to the plane of the earth's orbit.

The imaginary plane that connects the earth's orbit with the celestial sphere is called the **plane of the ecliptic.** From this reference plane, the earth's axis of rotation is tilted about 23½ degrees. The earth is not unique in this respect. Other planets have similarly tilted axes. Mars is tilted 24 degrees, and Saturn has a tilt of 27 degrees. Jupiter, however, is almost perpendicular to the plane of its orbit, and Uranus is tilted nearly 90 degrees and spins lying on its side. Because of the earth's tilt, the Northern Hemisphere leans toward the sun in June and away from the sun in December (see Figure 12.5). The main consequences of this change are the seasons, which are discussed in detail in Chapter 12. Also, because of the earth's tilt, the apparent path of the sun (ecliptic) and the celestial equator intersect each other at an angle of 23½ degrees (Figure 17.22A). Therefore, when the apparent position of the sun is plotted on the celestial sphere over a period of a year's time, its path traces a curve that intersects the celestial equator at two points (Figure 17.22B). From a Northern Hemisphere point of view, these intersections are called the vernal and autumnal equinoxes and occur about March 20–21 and September 22–23, respectively. On June 21–22, the date of the summer solstice, the sun appears 23½ degrees north of the celestial equator, and six months later, on December 21–22, the date of the winter solstice, the sun appears 23½ degrees south of the celestial equator.

A third and very slow motion of the earth is called **precession.** Although the earth's axis maintains the same angle of tilt, its direction is always changing, with the axis tracing a circle on the sky. This movement is very similar to the movement of a spinning top (Figure 17.23A). At the present time, the axis points toward Polaris. In A.D. 14,000, it will point toward the bright star Vega, which will then be the North Star (Figure 17.23B). The period of precession is 26,000 years. By the year 28,000, Polaris will once again be the North Star.

Precession has little effect on the seasons, since the angle of tilt remains unchanged. It does, however, cause the positions of the seasons (equinox and solstice) to move slightly each year among the constellations. It is this effect that causes the

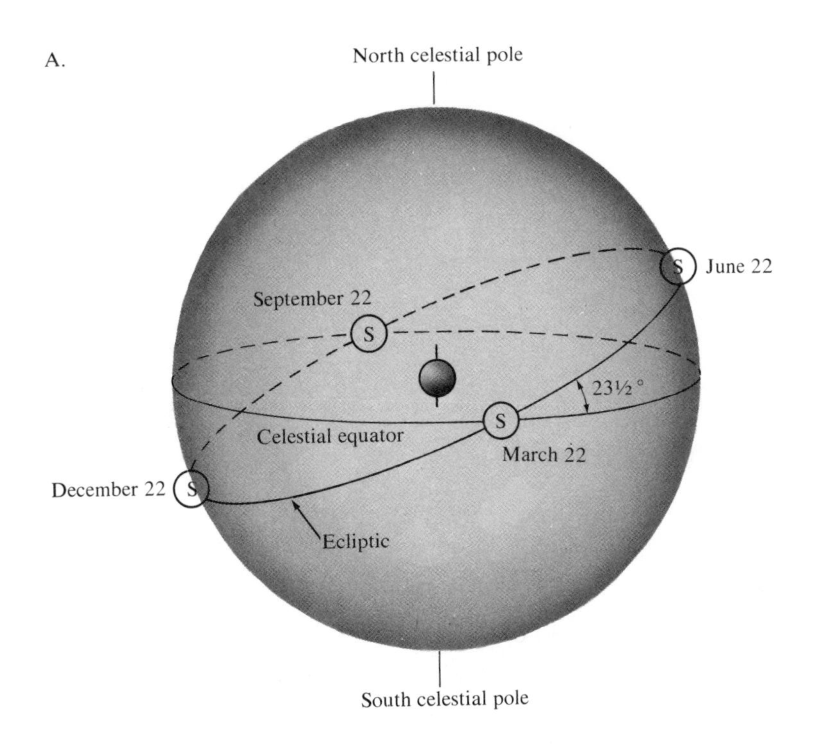

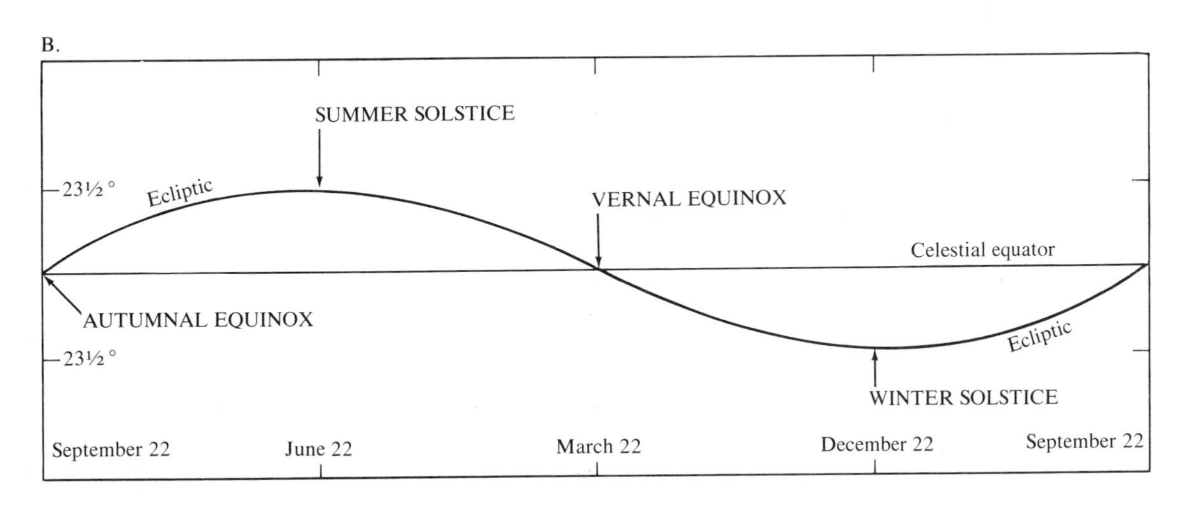

Figure 17.22 A. *The earth's orbital motion causes the position of the sun to move about 1 degree each day on the celestial sphere.* **B.** *Curved path of the ecliptic as it appears plotted on the celestial sphere.*

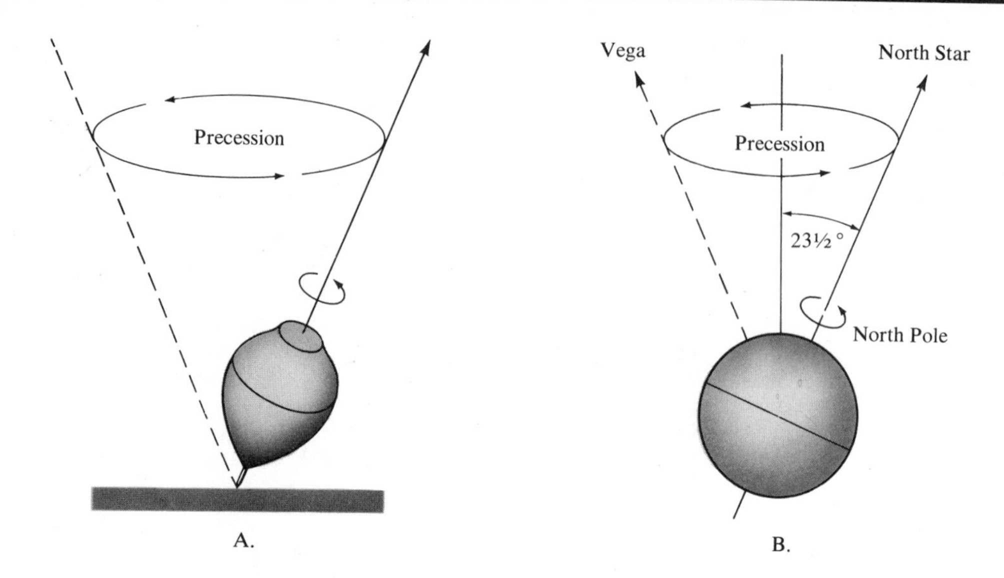

Figure 17.23 **A.** *Precession illustrated by a spinning top.* **B.** *Precession of the earth causes the location of the north celestial pole to change and, therefore, also the position of the North Star.*

signs of the Zodiac, measured in 30-degree segments from the position of the vernal equinox, to shift from the constellations of the Zodiac, which are "fixed."

Besides its own motion, the earth shares numerous motions with the sun. It accompanies the sun as it speeds in the direction of the bright star Vega at 20 kilometers (12 miles) per second. Also, the sun, like other nearby stars, revolves around the galaxy, a trip which requires 200 million years at speeds approaching 300 kilometers (200 miles) per second. In addition, the galaxie themselves are in motion. We are presently approaching one of our nearest galactic neighbors, Andromeda. In summary, the motions of the earth are many and complex, and its speed is very great. Just how rapidly the earth is moving may never be known, for according to Einstein's special theory of relativity, no reference exists from which the absolute speed of an object in space can be measured.

Calendars

One of the earliest responsibilities of astronomers was the keeping of the calendar. Although the oldest known calendars date from the eighth century B.C., others probably existed much earlier. These calendars had as their basis the movements of the heavenly bodies. Although the week has no astronomical significance, the seven days are clearly named after the seven moving celestial bodies known to the ancients; the sun, the moon, and five planets (Mercury, Venus, Mars, Jupiter, and Saturn), which are easily seen with the unaided eye. For example, Sunday is "Sun's day, "Monday is "Moon's day," and Saturday is "Saturn's day." The remaining days of the week were named in the Romance languages after Mars, Mercury, Jupiter, and Venus, in that order. For example, in Spanish, Tuesday is *Martes*.

The 29.5-day cycle of the moon through its phases is the most noticeable heavenly phenomenon except for the day itself. It became the *moonth*, now called the month. The first Western calendars based the year on the phases of the moon. Twelve *moonths* equaled a year. Because there are actually 12.4 lunar cycles in a year, the calendar had to have a full month added every three years to keep the seasons in accord. Even with this correction, the calender was still slowly falling behind. When Julius Caesar gained power, the calendar indicated spring, while the

Horsehead nebula in Orion. (Courtesy of Hale Observatories)

Veil nebula in Cygnus. (Courtesy of Hale Observatories)

The Skylab in earth orbit at an altitude of 270 miles, seen against the clouds and sea. Extended solar panels convert sunlight to electric power for use in the lab. (Courtesy of NASA)

The Martian landscape as viewed by the Viking I lander. (Courtesy of NASA)

weather indicated the middle of winter. To correct this situation, Julius Caesar ordered that 80 days be added to the year of 46 B.C. That year, for obvious reasons, was called the "year of confusion." Also at the direction of an astronomer, Caesar ordered that the calendar be based on the tropical year of 365.25 days. This was done by having 365 days each normal year and adding an extra day every fourth year. So began the tradition of leap year and the **Julian calendar.** However, the tropical year is slightly less than 365.25 days. The difference is only a little more than 10 minutes per year, but the Julian calendar worked just like a fast clock. Both would require adjustment if used long enough. By the sixteenth century, the Julian calendar was ahead by 10 days.

In 1582, the **Gregorian calendar,** which we presently use, was developed. The extra 10 days were eliminated by making Friday, October 15, the day after Thursday, October 4. To slow the calendar down, selected leap years were eliminated. A leap year is no longer added for centennial years except those divisible by 400. Hence, the years 1600, 2000, and 2400 are leap years, but all centennial years between them (e.g., 1900) are not. Our present calendar is accurate to within 1 day in 3000 years. Not all countries adopted the new calendar at the same time. When George Washington was born, the calendar indicated February 11, 1732, but when the colonies adopted the Gregorian calendar in 1752, his birth date became February 22, the day we now celebrate.

REVIEW

1. Why did the ancients think that celestial objects had some control over their lives?
2. Why was the Julian calendar revised?
3. Describe what produces the retrograde motion of Mars. What geometric arrangement did Ptolemy use to explain this motion?
4. Did Galileo invent the telescope?
5. What major change did Copernicus make in the Ptolemaic system? Why was this change philosophically significant?
6. What was Tycho's contribution to man?
7. Why was the concept of a leap year introduced?
8. Express the declination and right ascension of Arcturus (see Appendix E).
9. Explain how Galileo's discovery of a rotating sun supported the Copernican view of a sun-centered universe.
10. Does the earth move faster in its orbit near perihelion (January) or near aphelion (July)? Keeping your answer to the previous question in mind, is the solar day longest in January or in July?
11. Explain the difference between the solar and the sidereal day.
12. Of what value are constellations to modern-day astronomers?
13. Use Kepler's third law ($p^2 = d^3$) to determine the period of a planet whose solar distance is
    a. 10 AU
    b. 1 AU
    c. 0.2 AU
14. Use Kepler's third law to determine the distance from the sun of a planet whose period is
    a. 5 years
    b. 10 years
    c. 10 days
15. Newton learned that the orbits of the planets are the result of two actions. Explain these actions.
16. Using a diagram, explain why the fact that Venus appears full when it is smallest supports the Copernican view and is inconsistent with the Ptolemaic system.

KEY TERMS

geocentric
celestial sphere
heliocentric
Ptolemaic system
retrograde motion

epicycle
deferent
astronomical unit
perturbation
constellations
equatorial system
declination
right ascension
vernal equinox
rotation
revolution
mean solar day
sidereal day
perihelion
aphelion
stellar parallax
ecliptic
plane of the ecliptic
precession
Julian calendar
Gregorian calendar

The 200-inch Hale telescope dome with shutter open. (Courtesy of Hale Observatories)

The Moon, Celestial Observations, and the Sun

The celestial objects that have received the greatest attention are the moon and the sun. Several solar telescopes located throughout the world keep a constant vigil on the sun, and the moon has inspired observers since Galileo first pointed his telescope skyward and revealed its beautiful crater-marked surface. Further, the lunar space probes, culminating with the Apollo missions, have provided first-hand information on our companion in space. These data have ended years of controversy over the possible nature of the lunar surface and, as hoped, have provided clues to the early history of the solar system, for erosion has not appreciably altered the lunar surface in the last few thousand million years. This chapter will examine the basic features of the moon and the sun and provide a discussion of the tools used by astronomers to probe the universe.

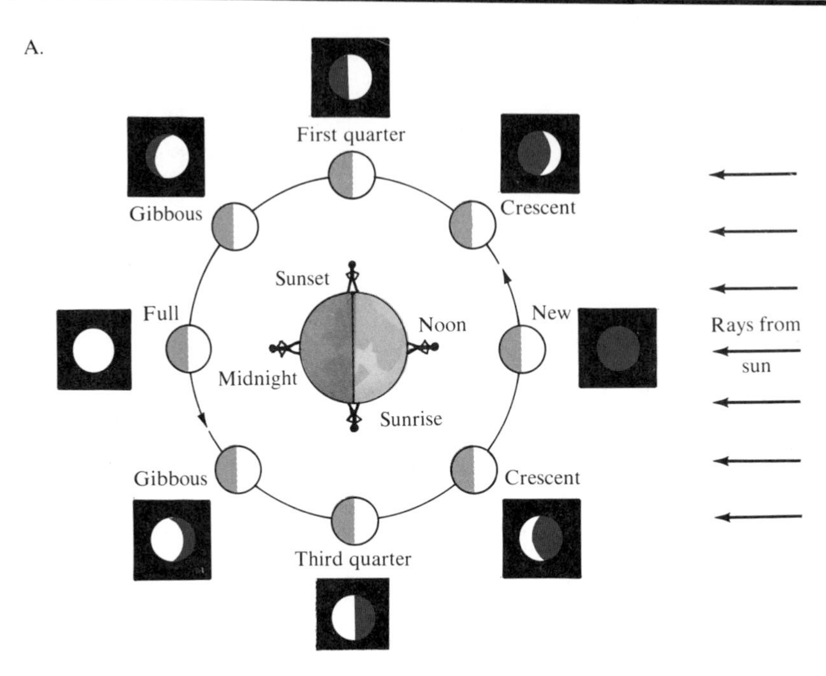

Figure 18.1 *Phases of the moon.* **A.** *The outer figures show the phases as seen from the earth.* **B.** *Compare these photographs with the diagram. (Courtesy of Lick Observatory)*

The Moon

The earth has only one natural satellite, the moon, which accompanies it on its annual flight around the sun. This planet-satellite system is unique in the solar system, because the moon is unusually large when compared with its parent planet. Consequently, the moon is often referred to as a planet.

The distance from the earth to the moon is about 384,000 kilometers, but since the orbit of the moon is elliptical, the distance has a range of 48,000 kilometers. The diameter of the moon is 3475 kilometers, and from the calculation of its mass, its density is 3.3 times that of water. This density is comparable to that of crustal rocks on earth but a fair amount less than the earth's average density. Geologists have suggested that this difference can be accounted for if the moon's iron core is rather small. The gravitational attraction at the lunar surface is one-sixth of that experienced on the earth's surface. This difference allows an astronaut to lift a "heavy" life-support system with relative ease. If he were not carrying such a load, he could jump six times higher than when on earth, easily clearing a one-story building.

Lunar Motion

Like the earth, the moon rotates on its axis but does so at a much slower rate than the earth. It also revolves around the earth with a period of about a month. The latter motion causes the relative positions of the sun, earth, and moon to change, producing the **phases of the moon** and the eclipses of the sun and moon.

Phases of the Moon. The first astronomical phenomenon to be understood was the cycle of the moon. On a monthly basis, we observe the phases as a systematic change in the amount of the moon that appears illuminated (Figure 18.1). About two days after a new cycle begins, a thin

B.

sliver *(crescent phase)* appears low in the western sky just after sunset. During the following week, it grows to a half-moon *(first-quarter phase)*, which is visible from noon to midnight. In another week, the complete disk *(full-moon phase)* can be seen rising in the east as the sun is sinking in the west. During the next two weeks, the percentage of the moon that can be seen steadily declines, until the moon disappears altogether *(new-moon phase)*. The cycle soon begins anew with the reappearance of the crescent moon.

The lunar phases are a consequence of the motion of the moon and the sunlight that is reflected from its surface (Figure 18.1B). Half of the moon is illuminated at all times, but to an earthbound observer, the percentage of the bright side facing him depends on the location of the moon with respect to the sun and the earth. When the moon lies between the sun and the earth, none of its bright side faces the earth, producing the new-moon ("no-moon") phase. Conversely, when the moon lies on the side of the earth opposite the sun, all of its lighted side faces the earth, producing the full moon. At a position between these extremes, an intermediate amount of the moon's illuminated side is visible from the earth.

Motion. The cycle of the moon through its phases requires 29½ days, a time span called the **synodic month.** Recall that this cycle was the basis for the first Roman calendar. However, this is not the true period of the moon's revolution, which takes only 27⅓ days and is known as the **sidereal month.** The reason for the difference of nearly 2 days each cycle is shown diagrammatically in Figure 18.2. Note that as the moon orbits the earth, the earth-moon system also moves in an orbit around the sun. Consequently, even after the moon has made a complete revolution around the earth, it has not yet reached its starting position, which was directly between the sun and earth (new-moon phase). This additional motion takes another 2 days.

An interesting fact concerning the motions of the moon is that the periods of rotation and revo-

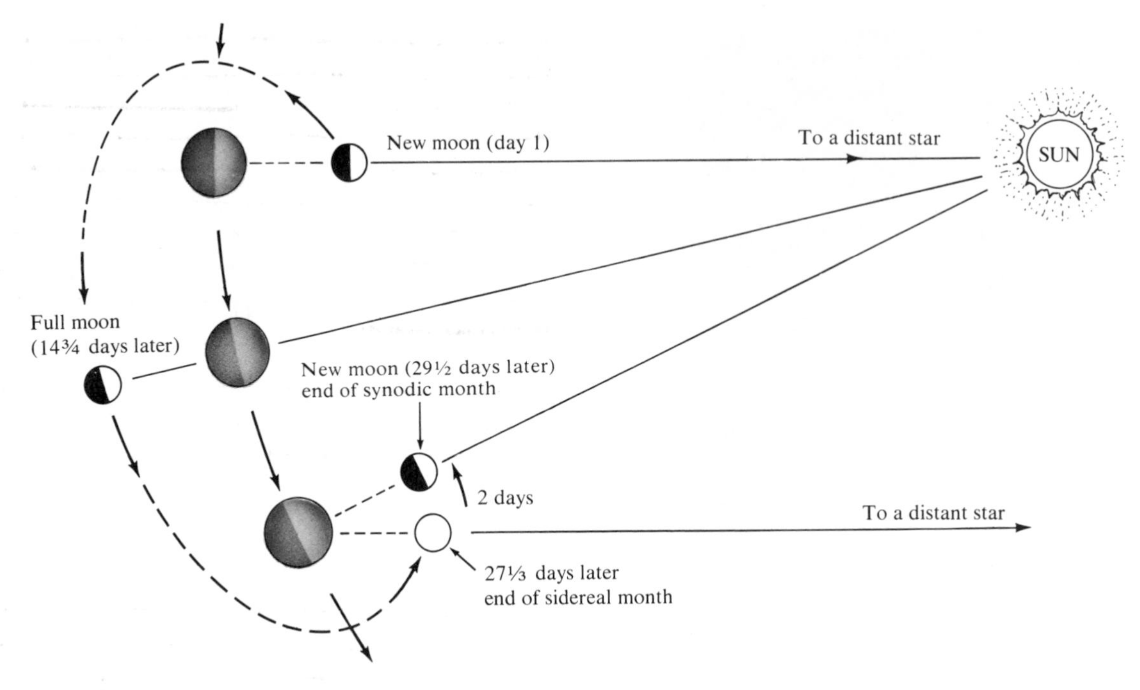

Figure 18.2 *Sidereal and synodic months.*

lution are the same—27⅓ days. Because of this, the same lunar hemisphere always faces the earth. All of the manned Apollo missions have been confined to this side, although numerous photographs have been taken of the "back" side by orbiting satellites.

Since the moon rotates very slowly, it experiences long periods of daylight and darkness, each lasting about two weeks. This, along with the absence of an atmosphere, accounts for the high surface temperature of over 100° C (212° F) on the day side of the moon and the low surface temperature of −173° C (−280° F) on its night side.

Eclipses. Along with understanding the moon's phases, the early Greeks also realized that eclipses are simply shadow effects (Figure 18.3). When the moon moves directly between the earth and the sun (new-moon phase), it casts a dark shadow on the earth, producing a **solar eclipse.** On the other hand, the moon is eclipsed **(lunar eclipse)** when it moves within the shadow of the earth, a situation that is only possible during the full-moon phase. However, if this is the case, why does a solar eclipse not occur with each new-moon phase and a lunar eclipse with each full-moon phase? They would if the orbit of the moon was lying along the plane of the earth's

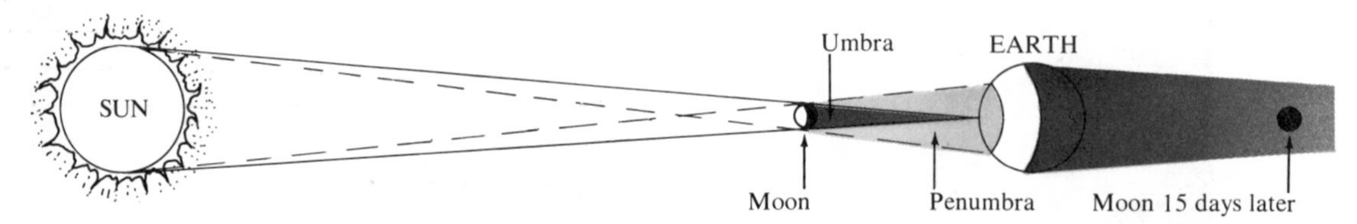

Figure 18.3 *Positions of the sun, moon, and earth during a solar and lunar eclipse (not to scale).*

orbit. However, it is inclined about 5 degrees to it. Thus, the shadow of the moon (or earth) can pass above or below the earth (or moon). Only if a new- or full-moon phase occurs when the moon lies in the plane of the ecliptic can an eclipse occur. The maximum number of eclipses in a year is seven (more of which are solar), and the minimum number is two (both solar).

During the total lunar eclipse, the circular shadow of the earth can be seen moving across the disk of the full moon. When totally eclipsed, the moon will still be visible as a coppery disk, because the earth's atmosphere bends and transmits some long-wavelength light (red) into its shadow. A total eclipse of the moon can last up to four hours and is visible to anyone on the side of the earth facing the moon.

During a total solar eclipse, the moon casts a shadow that is never wider than 275 kilometers (170 miles). Anyone observing in this region will see the moon slowly block the sun from view and the sky darken. Near totality, a sharp drop in temperature is experienced. The solar disk is blocked for at most seven minutes, and then one edge reappears. At totality, the dark moon is seen covering the complete solar disk, and only the sun's brilliant white outer atmosphere is visible (see Figure 18.26). A total solar eclipse is only visible to people in the dark part of the moon's shadow *(umbra),* while a partial eclipse can be seen by those in the light portion *(penumbra)* (Figure 18.3). Partial solar eclipses are relatively common in the polar regions, for it is there that the penumbra hits when the dark shadow of the moon just misses the earth. A total solar eclipse is a rare event at any given location. The next one that will be visible from the contiguous United States will take place on August 21, 2017.

The Lunar Surface

When Galileo pointed his telescope toward the moon, he saw two different types of terrain (Figure 18.4). The dark areas he observed are now known to be fairly smooth lowlands, while the bright regions are densely cratered highlands. Because the dark regions resembled seas on earth, they were named **maria** (singular, *mare:* Latin for "sea").

Today we know that the moon has no water or atmosphere. Thus, the processes of weathering and erosion which continually modify the earth are totally lacking. However, tiny particles (micrometeorites) continually bombard the lunar surface and smooth the landscape. Rocks, for example, can become slightly rounded on top if exposed at the surface long enough. Nevertheless, it is unlikely, except for the addition of a few large craters, that the moon has changed appreciably in the last 3000 million years.

The most obvious features on the lunar surface are craters. They are so profuse that craters *within* craters *within* craters are the rule. The larger ones seen in Figure 18.4 are about 250 kilometers (150 miles) in diameter, and they often overlap. Most craters were produced by the impact of rapidly moving debris (meteoroids), which was considerably more abundant in the early history of the solar system than it is today. By contrast, the earth has only about a dozen recognized impact craters, because our atmosphere burns up most small debris before it reaches the ground. Many of the craters which did form early in the earth's history have since been destroyed by erosion.

The process responsible for the formation of an impact crater is illustrated in Figure 18.5. Upon impact, the high-speed particle compresses the material it strikes, and the instantaneous rebound ejects material from the crater. (This process is analogous to the splash that occurs when a rock is dropped into water.) Most of the ejected material (ejecta) lands near the crater, building a rim around it. The heat generated by the impact is sufficient to melt some of this material. Astronauts have brought back samples of glass beads produced in this manner, as well as rock that was formed when angular fragments and dust were welded together by the impact. The latter material is called **lunar breccia.** A meteoroid only 3 meters (10 feet) wide can blast

Figure 18.4 *Telescopic view of the lunar surface. (Courtesy of Lick Observatory)*

out a 150-meter (500-foot) crater. A few large craters like Tycho and Copernicus have bright **rays** ("splash" marks), which extend for hundreds of kilometers from their impact structure (Figure 18.4). Glass beads are the major constituents of the rays.

The densely pock-marked highland areas make up most of the lunar surface. In fact, almost all of the "back" side of the moon is characterized by such topography. Although the highlands predominate, the less rugged maria have attracted most of the interest. The origin of maria basins as enormous impact craters was hypothesized before the turn of the century by the noted American geologist G.K. Gilbert. However, it remained for the Apollo missions to determine what filled these depressions to produce the rather flat topography. Apparently, the craters were flooded, layer upon layer, by very fluid basaltic lava, in which case they closely resemble the Columbia Plateau in the northwestern United States. The photograph in Figure 18.6 reveals the forward edge of a lava flow that is "frozen" in place. Astronauts have also viewed and photographed the layered nature of maria. The layers are over 30 meters (100 feet) thick, and the total thickness of the maria material approaches thousands of meters.

Often the lava flowed beyond the impact crater, engulfing the surrounding lowlands. (If the rim of a remnant crater can be seen above the lava, an estimate of the flow's thickness can be made.) Many geologists believe that the impacts that produced the maria basins were great enough to break the lunar crust some distance away. Examples of basins in which all the disturbed regions were filled to overflowing include Mare Tranquillitatis (Sea of Tranquility), the site where man put his first footprint on lunar soil, and Mare Imbrium (Sea of Rains). Some maria basalts fill only the central crater; these can easily be seen as dark, smooth crater floors in Figure 18.4.

An interesting fact concerning the circular maria was discovered by orbiting satellites. These regions apparently have a higher gravitational attraction than other lunar topography, for the maria tug on satellites passing overhead more than the other regions do. Because this extra pull is the result of a concentration of mass, the word **mascon (mass concentration)** has been applied to them. A few scientists believe that mascons are the buried remains of very dense meteorites whose impact created the basin. Others have suggested that they resulted from the thick basaltic masses which filled the basins and occupy the space where lighter crustal rocks formerly had been.

The lunar surface is mantled with a layer of gray unconsolidated debris derived from billions of years of meteoric bombardment (Figure 18.7). This layer, called the **regolith,** is composed of

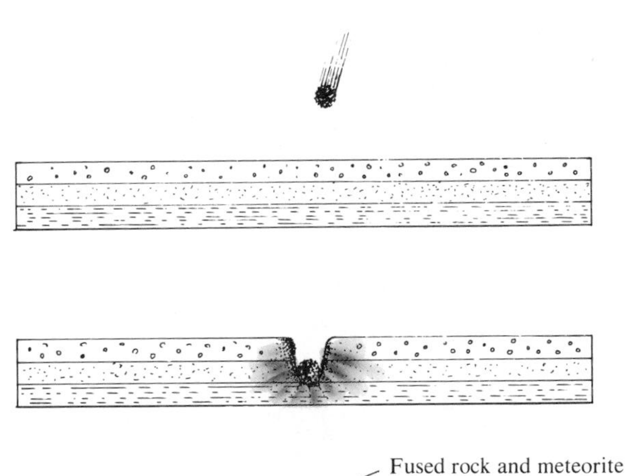

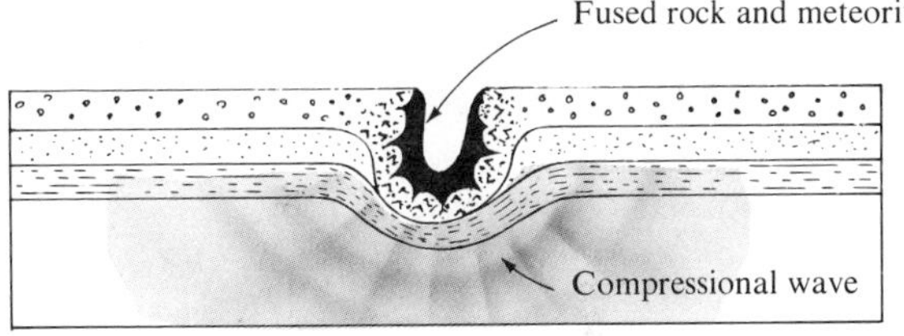

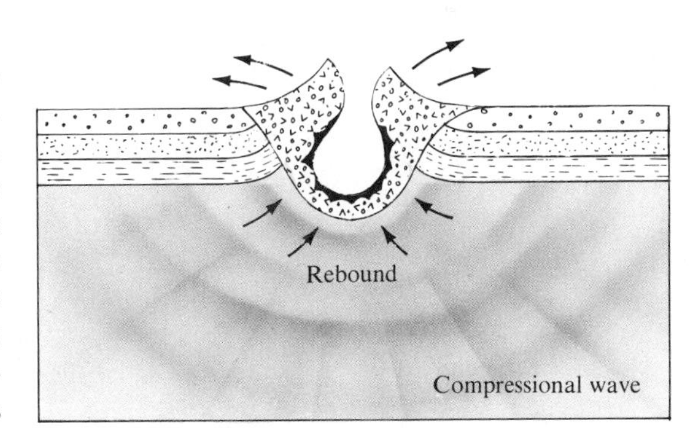

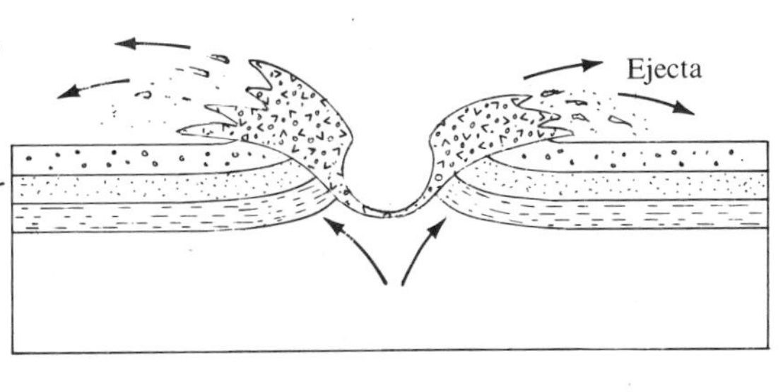

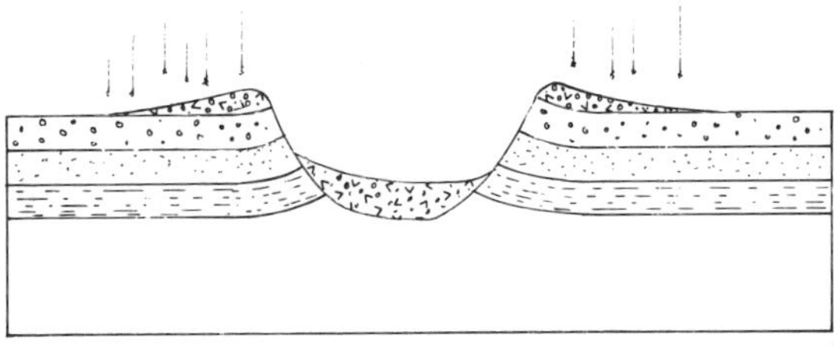

Figure 18.5 *Formation of an impact crater. The energy of the rapidly moving meteoroid is transferred into heat energy and compressional waves. The rebound of the compressed rock causes debris to be ejected from the crater, and the heat melts some material, producing glass beads. Small secondary craters are formed by the material "splashed" from the impact crater. (After E.M. Shoemaker)*

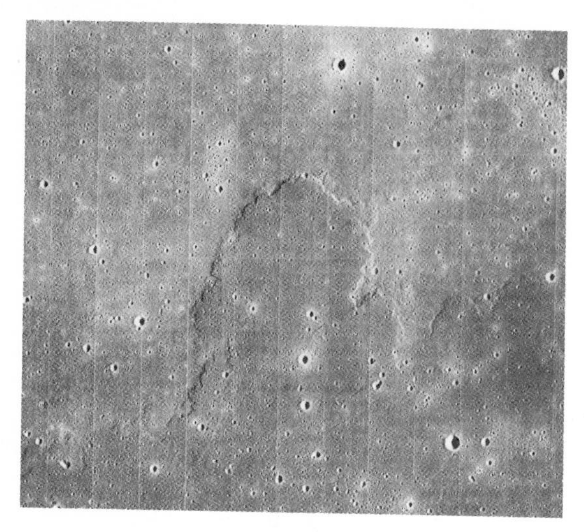

Figure 18.6 *Margin of a lava flow on the surface of Mare Imbrium. (Courtesy of National Space Data Center)*

igneous rocks, breccia, glass beads, and fine particles commonly called lunar dust. As meteoroid after meteoroid collides with the lunar surface, the thickness of the regolith layer increases, while the size of the particles diminishes. In the maria that have been explored by *Apollo* astronauts, the regolith is apparently just over 3 meters (10 feet) thick, but it is believed to be thicker in the highlands.

Lunar History

Lunar geologists have been able to work out some of the gross details of the moon's history, using among other things variations in crater density (quantity per unit area). Simply stated, the higher the crater count in an area, the longer that area has been in existence.

Although the early history of the moon can only be hypothesized, it most likely paralleled that of the earth. It is believed that all of the bodies of the solar system condensed from a large cloud of dust and gasses. Originally, the moon

Figure 18.7 *Footprint left in the lunar "soil" by an* Apollo 11 *astronaut. (Courtesy of NASA)*

was much smaller, but it swept up other debris and grew by accretion. This continual bombardment of material (or perhaps radioactive decay) generated enough heat to melt the moon's outer shell, and quite possibly the rest of the moon as well. When a large percentage of the debris had been gathered, the outer layer of the moon cooled and formed its crystalline crust. The rocks of the primitive crust are composed mostly of a calcium-rich feldspar (anorthosite). It is believed that this feldspar mineral crystallized early and, because it was lighter than the remaining melt, floated to the top. While this process was taking place, iron and other heavy material probably sank to form a small central core. Even after the crust had solidified, its surface was continually bombarded, but with less frequency. Remnants of the original crust occupy the densely cratered highlands, which have been found to be as much as 4700 million years old.

The next major event was the formation of maria basins (Figure 18.8). The meteoroids that produced these huge pits ejected mountainous quantities of lunar rock into piles rising 5 kilometers (3 miles) high. Apennine Mountain, which typifies such a pile of debris, was produced in conjunction with the formation of the Imbrium basin, the site of the exploration conducted by the *Apollo 15* astronauts. The crater density of the ejected material is greater than that of the surface of the associated mare, confirming that time elapsed between the formation and filling of these basins. Radiometric dating of the maria basalts puts their age at nearly 3500 million years, about 1000 million years younger than the initial crust. In places, the lava flows overlap the highlands, another testimonial to the lesser age of the maria.

The last large features to form on the lunar surface were the rayed craters like Copernicus (Figure 18.9). Material ejected from these large depressions is clearly seen blanketing the surface of the maria and many older rayless craters. The older craters have rounded rims, and their rays have been erased by the impact of small particles. However, even a "young" crater like Copernicus must be millions of years old. Had it formed on the earth, erosional forces would have long since removed the blemish.

Evidence carried back from lunar landings indicates that most, if not all, of the moon's activity ceased about 3000 million years ago. The youngest maria lava flows are about equivalent in age to the oldest rocks found on the earth to date. If photos had been taken several hundreds of millions of years ago, they would reveal that the moon has changed little in the intervening years. The moon is tectonically dead.

Light and Its Observation

The vast majority of our information about the universe is obtained from the study of the light emitted from celestial bodies. Although visible light is most familiar to us, it constitutes only a small part of an array of energy generally referred to as **electromagnetic radiation.** Included in this array are gamma rays, X rays, ultraviolet light, visible light, infrared light, and radio waves (see Figure 12.8). All radiant energy travels through the vacuum of space in a straight line at the rate of 300,000 kilometers (186,000 miles) per second, which equals a staggering 32,000 million kilometers per day.

Nature of Light

The properties of light do not conform to our common sense. In some instances, light behaves like waves, and in others, like discrete particles. In the wave sense, light is analogous to swells in the ocean. This motion is characterized by the wavelength, which is the distance from one crest to the next. Wavelengths vary, from several kilometers in length for radio waves to less than a billionth of a centimeter for gamma rays. Most of these waves are either too long or too short for our eyes to detect. The small band of electromagnetic radiation we can see is often called **white light.** However, even white light consists of an array of waves having various wavelengths, a fact that can easily be demonstrated with a prism (Figure 18.10). As white light is

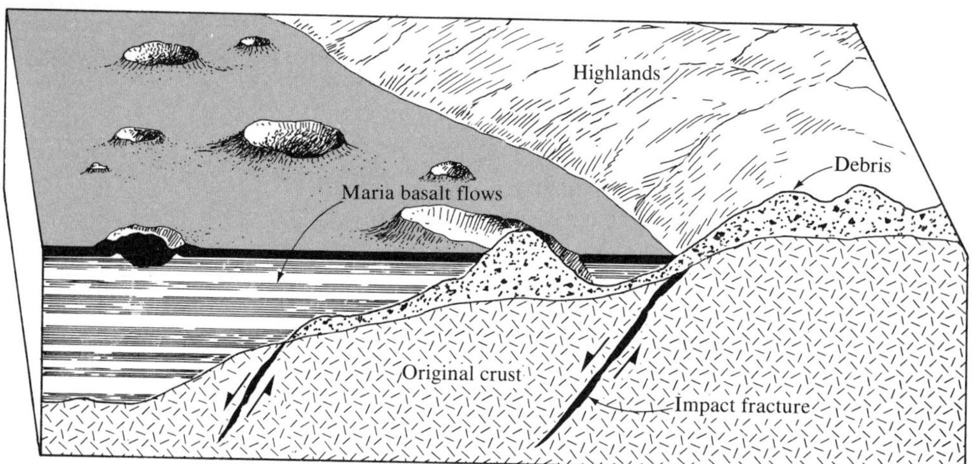

Figure 18.8 *Mare Imbrium and associated mountainous region. A few isolated peaks are standing above the lava plain, and numerous craters dot the highlands and adjacent mare. (Photo courtesy of NASA)*

passed through a prism, violet, the color with the shortest wavelength, is bent more than blue, which is bent more than green, and so forth (Table 18.1). Thus, white light can be separated

Figure 18.9 *The crater Copernicus. Notice the light-colored rays. (Courtesy of Hale Observatories)*

into its component colors in the order of their wavelengths, producing the familiar rainbow of colors.

However, some effects of light cannot be explained using the wave theory. In these cases, light acts like a stream of particles, analogous to infinitesimally small bullets fired from a machine gun. These particles, called **photons**, can exert a pressure ("push") on matter, which is called **radiation pressure**. The photons from the sun are responsible for "pushing" material away from a comet to produce its tail. Each photon has a specific amount of energy, which is related to its wavelength in a simple way: *Shorter wavelengths correspond to more energetic photons.* Thus, blue light has more energetic photons than red light.

Which theory of light—the wave theory or the particle theory—is correct? Both, since each will

Table 18.1 ***Colors and Corresponding Wavelengths***

Color	Wavelength (angstroms*)
Violet	3800–4400
Blue	4400–5000
Green	5000–5600
Yellow	5600–5900
Orange	5900–6400
Red	6400–7500

*An angstrom equals 10^{-10} centimeter.

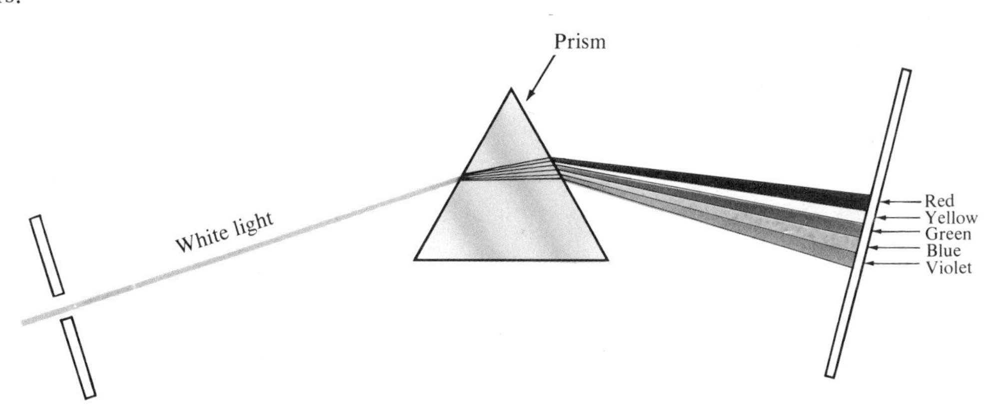

Figure 18.10 *A spectrum is produced when sunlight is passed through a prism, which bends each wavelength differently.*

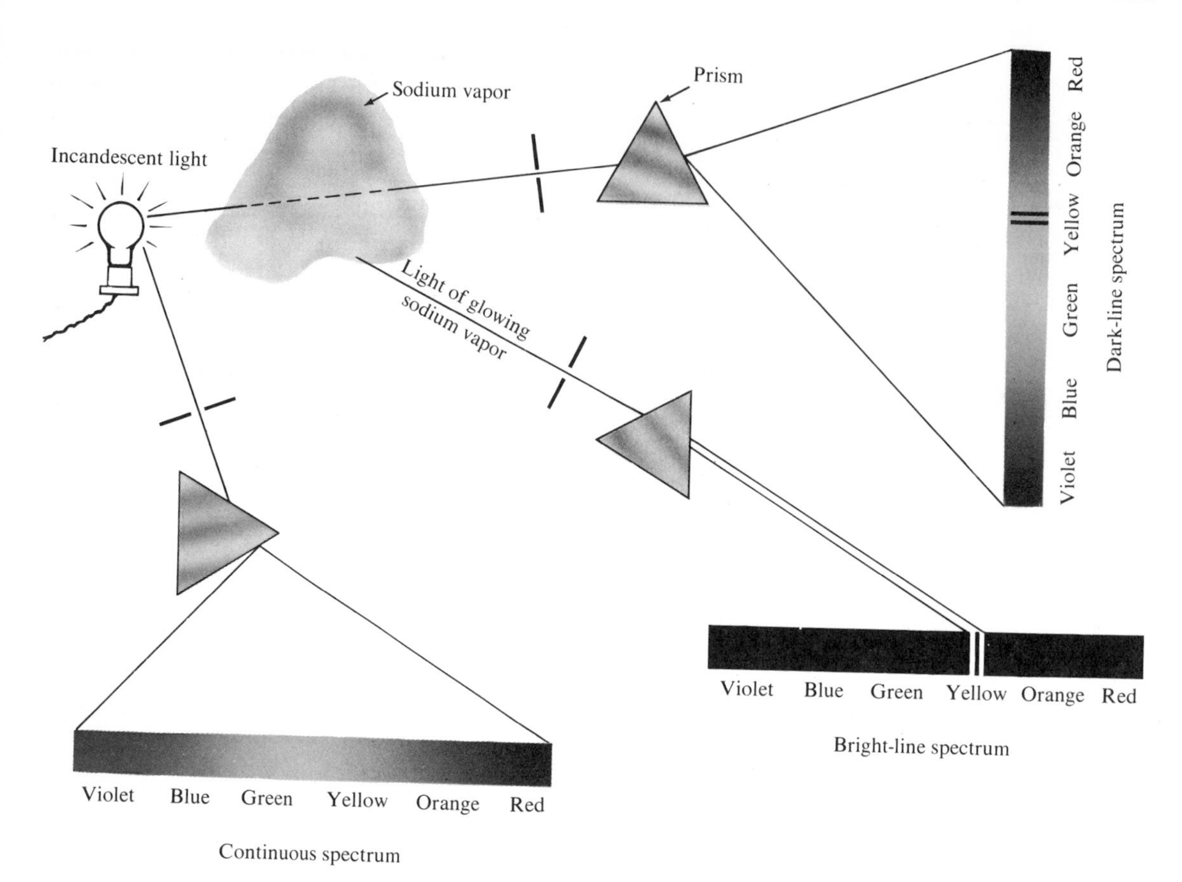

Figure 18.11 *Formation of the three types of spectra.*

predict the behavior of light for certain phenomena. As George Abell stated about all scientific laws, "The mistake is only to apply them to situations that are outside their range of validity."

Spectroscopy

When Sir Isaac Newton used a prism to disperse white light into its component colors, he unknowingly initiated the field of **spectroscopy,** which is the study of wavelength-dependent properties of light. The rainbow of colors Newton produced is called a continuous spectrum, because all wavelengths of light are included. It was later learned that two other types of spectra exist and that all three are generated under somewhat different conditions (Figure 18.11):

(1) A **continuous spectrum** is produced by an incandescent solid, liquid, or gas under high pressure. (It consists of all visible wavelengths and is generated by a common household light bulb.)

(2) A **bright-line,** or **emission, spectrum** is produced by an incandescent gas under low pressure. (It consists of a series of bright lines of particular wavelengths, which are dependent on the gas that produces them.)

(3) A **dark-line,** or **absorption, spectrum** is produced when white light is passed through a gas under low pressure, which subtracts certain wavelengths from the continuous spectrum. (It appears as a continuous spectrum with a series of dark lines at exactly the same wavelength as the bright lines which that gas emits.)

The spectrum of most stars is the dark-line type. The importance of these spectra is that each element or compound in the gaseous form (in a star, material is usually in the gaseous form) produces a unique set of spectral lines. When the spectrum of a star is studied, the spectral lines act as "fingerprints," which identify the elements present. In an admittedly oversimplified manner, we can imagine how the sun and other stars create their dark-line spectrum. A continuous spectrum is produced in the interior of the sun, where the gasses are under very high pressure. When this light passes through the less dense gasses of the solar atmosphere, they extract selected wavelengths, producing the dark lines.

When Newton studied solar light, he obtained a continuous spectrum. However, when a prism is used in conjunction with lenses, the solar spectrum can be dispersed even further. An instrument that does this is called a **spectroscope.** The spectrum of the sun, when viewed through a spectroscope, contains thousands of dark lines. Over 60 elements have been identified by matching these lines with those of elements known on earth. However, some lines are still unaccounted for.

Two other facts concerning a radiating body are important. First, if the temperature of a radiating surface is increased, the total amount of energy emitted increases at a rate known as the Stefan-Boltzmann law. Simply stated, *the total amount of energy radiated by a body is directly proportional to the fourth power of its absolute temperature*. For example, if the temperature of a star is doubled, the total radiation emitted increases 2^4 ($2 \times 2 \times 2 \times 2$), or 16 times. Second, *as the temperature of an object increases, a larger proportion of its energy is radiated at shorter wavelengths.* To illustrate this, imagine a metal rod that is heated slowly. The rod first appears dull red (long wavelengths), and later bluish white (short wavelengths). From this, it follows that a blue star is hotter than a yellow star, which is hotter than a red star.

The Doppler Effect

You may have heard the change in pitch of a train whistle as the train passes by. When the train is approaching, the sound seems to have a higher than normal pitch, and when it is moving away, the pitch appears lower than normal. This effect, which occurs for both sound and light waves, was first explained by Christian Doppler in 1842 and is called the **Doppler effect.** The reason for the difference is that it takes some time for the wave to be emitted. If the source is moving away from you, the beginning of the wave is emitted nearer to you than the end, which tends to "stretch" the wave, that is, give it a longer wavelength (Figure 18.12). The opposite would be true for an approaching source. In the case of light, when a source is moving away from you, its light will appear redder than it actually is, since its waves appear lengthened, while objects approaching will have their light waves shifted toward the blue (shorter wavelength). Thus, if a source of red light approached you at a very high speed (near the speed of light), it would actually appear blue. The same effect would be produced if you moved and the light was stationary.

Therefore, the Doppler effect reveals whether the earth is approaching or receding from a star or another celestial body. In addition, the amount of shift indicates the rate at which the relative movement is occurring. These shifts are generally measured from the dark lines in the spectrum of stars. This can be done by making a comparison

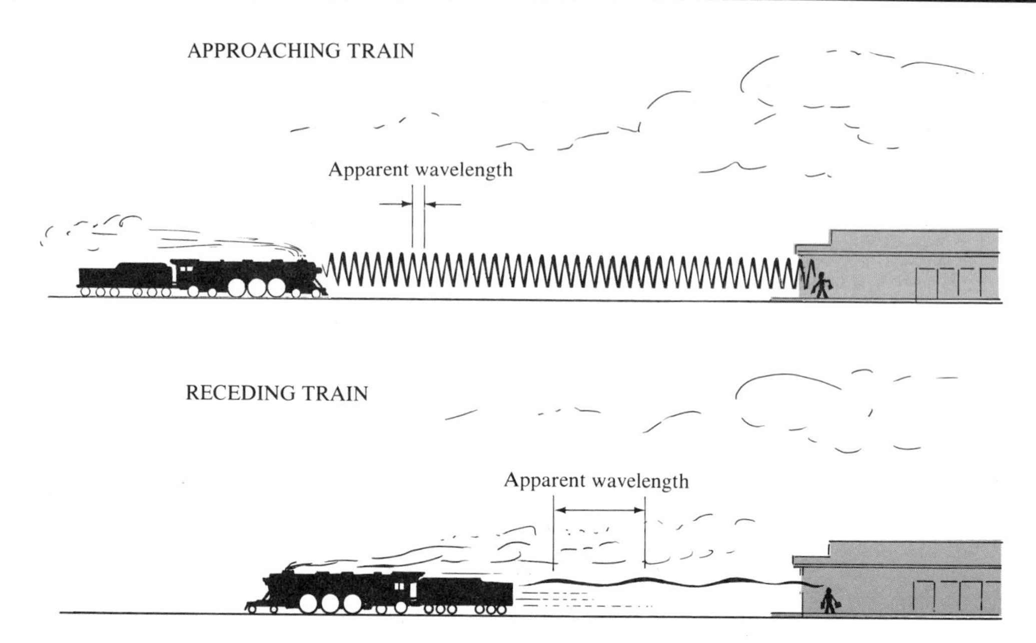

Figure 18.12 *The Doppler effect, illustrating the apparent lengthening and shortening of wavelengths caused by the relative motion of a source and an observer.*

with a standard spectrum produced in the laboratory (Figure 18.13).

Refracting Telescopes

The first telescope used simple lenses. The most important lens in a telescope, the **objective lens,** produces the image. It does so by bending the light of a distant object in such a way that the light converges at an area called the **focus** (Figure 18.14). For an object like a star, the image appears as a point of light, but for a nearby object, it appears as an inverted replica of the original. You can easily demonstrate the latter case by holding a lens in one hand and with the other hand place a white card behind the lens. Now, vary the distance between them until an image appears on the card. The distance between the focus (place where the image appears) and the lens is called the **focal length** of the lens.

Ordinarily, astronomers photograph the image, but in a telescope employed for visual use, a second lens called an **eyepiece** is required (Figure 18.14). The eyepiece magnifies the image produced by the objective lens. In this respect, it is similar to a magnifying glass used to read fine print. In summary, *the objective lens brings a very small, bright image of the subject nearer*

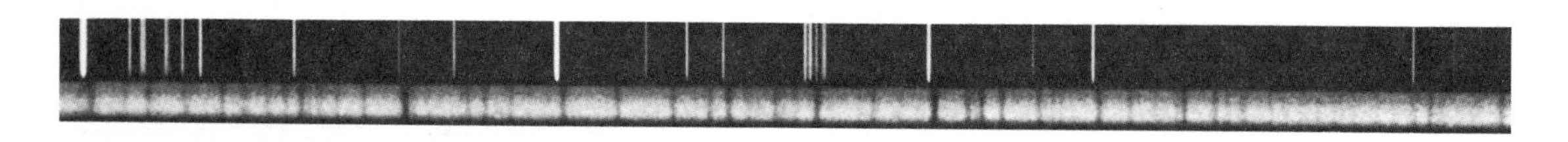

Figure 18.13 *Doppler shift in spectra of stars. The white lines are the standard spectrum. Notice that this stellar spectrum is shifted to the right (red), indicating a receding star. (Courtesy of Lick Observatory)*

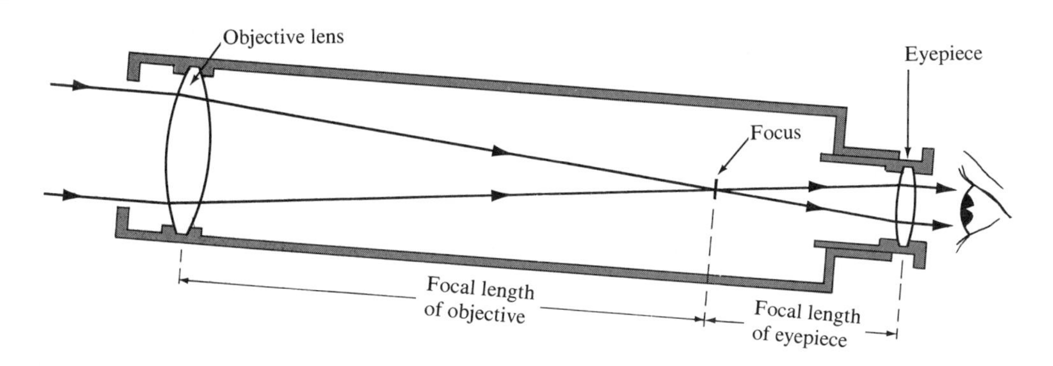

Figure 18.14 *Simple refracting telescope.*

to you, and the eyepiece "blows it up" so you can see the details.

When you think of the power of a telescope, you probably think of its magnifying power. The magnification of a simple two-lens system is found by dividing the focal length of the objective by the focal length of the eyepiece. Generally, several eyepieces are available so that astronomers can switch them and change the magnification. If a telescope has an objective lens with a focal length of 100 inches and an eyepiece with a 5-inch focal length, its magnification power is 100/5, or 20. The same telescope fitted with an eyepiece that has a focal length of ½ inch would have a magnification of 200. Hence, by choosing an eyepiece with a smaller focal length, the astronomer can increase the magnification. However, there is an upper limit to a telescope's magnification, which is about 50 times the diameter of its objective, expressed in inches. Consequently, the highest magnification of a 40-inch refractor would be 40 × 50, or 2000. Any part of the image that is not clear at this magnification will only appear as a larger blur under higher magnification. The magnifying power of a telescope is important when observing bodies in the solar system or large celestial objects.

When viewing individual stars, astronomers are more interested in the light-gathering power of their instruments than in magnification. It should be clear from Figure 18.15 that a telescope with a larger lens (or mirror) intercepts

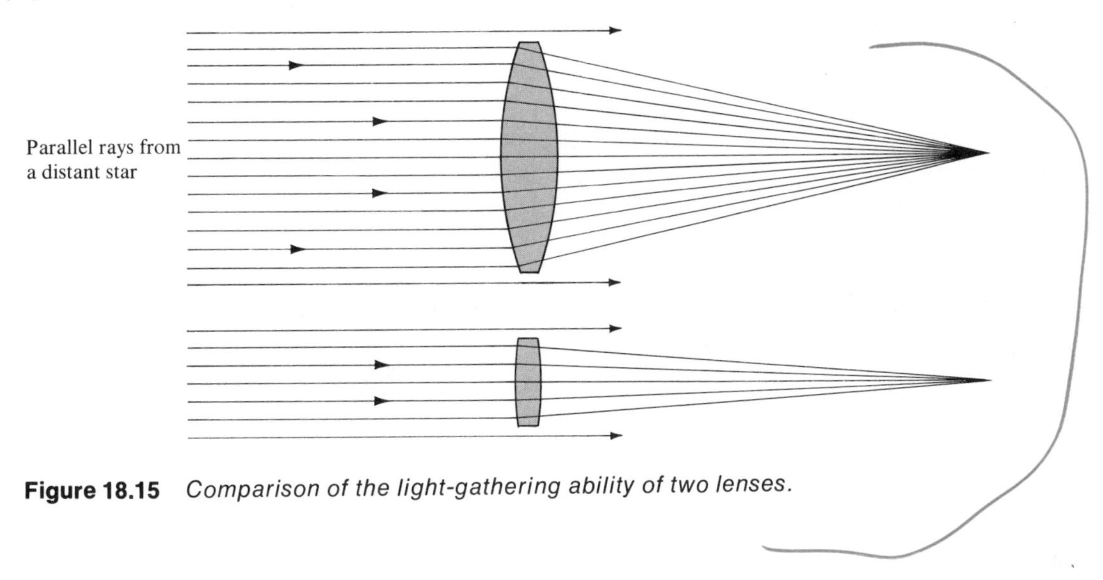

Figure 18.15 *Comparison of the light-gathering ability of two lenses.*

more light from distant objects, thereby producing brighter images. Since very distant stars appear very dim as well, a great deal of light must be collected before the image will be bright enough to be seen. Consequently, telescopes with large objectives can "see" farther into space than those with small objectives.

Most modern telescopes have supplemental devices that enhance the image. A simple, but important example is a photographic plate that can be exposed for long periods of time, thereby collecting enough light from a star to make an image that otherwise would be undetectable. Another advantage of telescopes with large objectives is their high **resolving power** (Figure 18.16), which allows for sharper images and finer detail. For example, with the unaided eye, the Milky Way appears as a band of light, but even a small telescope is capable of resolving (separating it into) individual stars. However, the condition of the earth's atmosphere *(seeing)* greatly limits the resolving power of earthbound telescopes. On a night when the stars twinkle, the seeing is poor, because the air is moving rapidly. This causes the image to move about, blurring the photograph. Conversely, when the stars shine steadily, the seeing is described as good. However, even under ideal conditions, some blurring does occur, eliminating the fine details. Thus, even the largest earthbound telescopes cannot obtain photographs of lunar features less than 0.5 kilometer (0.3 mile) in size.

The primary problem with refractors is that a lens bends light of each wavelength somewhat differently. Recall the effect of a prism on light. Consequently, when a refracting telescope is in focus for red light, blue and violet light are out of focus. This troublesome effect, known as **chromatic** ("color") **aberration,** weakens the image and produces a halo of color around it. When blue light is in focus, a reddish halo appears, and vice versa. Although this effect cannot be eliminated completely, it is reduced by using a second lens of hard glass.

Reflecting Telescopes

Newton was so bothered by chromatic aberration that he built and used telescopes that reflected light from a shiny surface (mirror). Because reflected light is not dispersed into its component colors, the problem is avoided. **Reflecting telescopes** use a concave mirror that focuses the light in front of the objective rather than behind it like a lens (Figure 18.17). The mirror is generally made of glass that is finely ground. In the case of the 200-inch Hale telescope, the grinding is accurate to about 2 millionths of an inch. The surface is then coated with a highly reflective material, usually an

A.

B.

C.

D.

Figure 18.16 *How the galaxy in Andromeda would appear using telescopes with decreasing resolution. (Courtesy of Leiden Observatory)*

aluminum-based compound. In order to focus parallel incoming light in one spot, a special curved surface called a paraboloid is used. This is the same shape as that used for the reflector in the headlights of automobiles. In the case of the auto bulb, however, the light source is at the focus, and the light goes out in parallel rays rather than coming in.

Because the focus of a reflecting telescope is in front of the mirror, provisions have to be made to view the image without blocking too much of the incoming light. Figure 18.18 illustrates the most commonly used arrangements. Most large telescopes employ more than one type. When using a very large reflecting telescope, the observer can actually get inside a viewing cage positioned at the focus to make the observations. The viewing cage blocks only about 10 percent of the total incoming light, and this is more than compensated for by the large objectives that are used (Figure 18.19). Generally, however, photographic plates are used.

Because parabolic mirrors produce sharp images for only those stars directly in the telescope's line of sight, reflectors cannot survey large sections of the sky; that is, they have a narrow field of view. A special telescope to correct for this, designed by Bernhard Schmidt, has a thin corrective lens as well as a mirror for the objective. These Schmidt cameras, as they are often called, have been used to produce a high-quality sky atlas. The Palomar Sky Survey is comprised of 1870 photographic plates covering all the northern half of the sky plus a good portion of the southern half. To the astronomer, these provide the same type of service that a good road map of a city gives to a person on a delivery route.

For a variety of reasons, all large optical telescopes built today are reflectors (or Schmidts).

A.

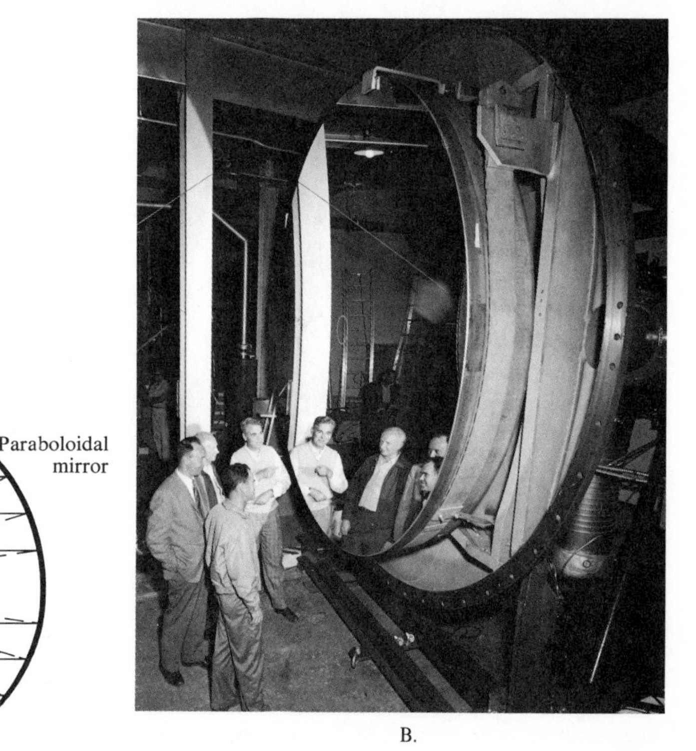

B.

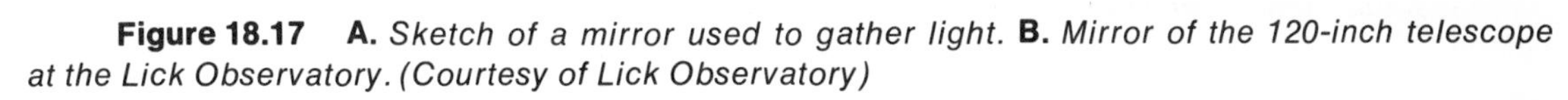

Figure 18.17 **A.** *Sketch of a mirror used to gather light.* **B.** *Mirror of the 120-inch telescope at the Lick Observatory. (Courtesy of Lick Observatory)*

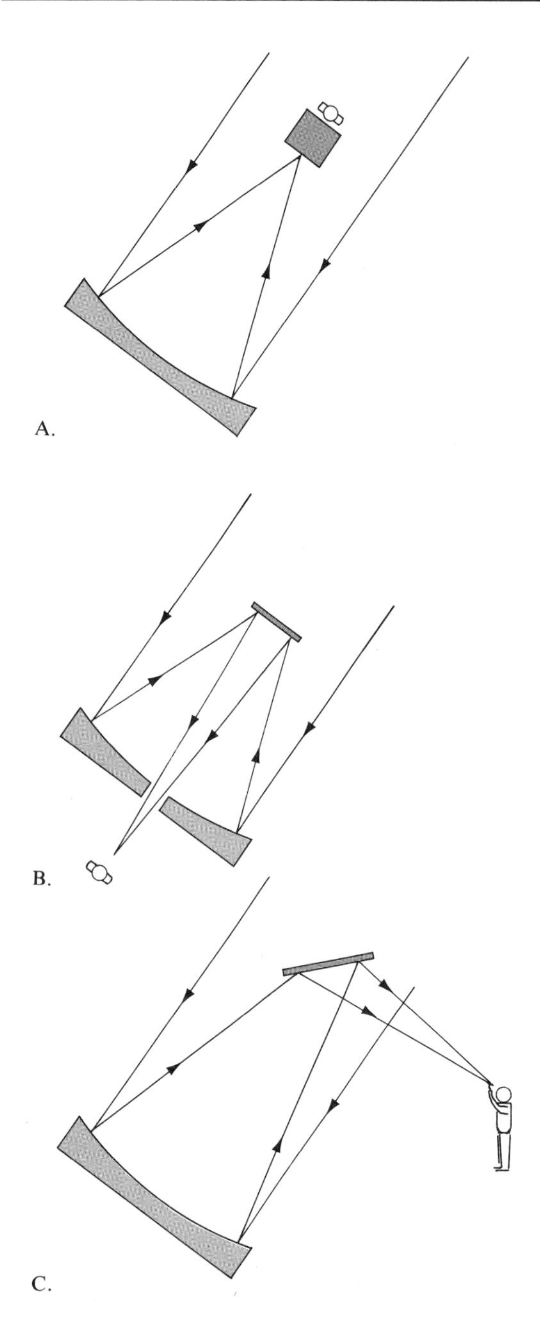

Figure 18.18 *Viewing methods used with reflecting telescopes.* **A.** *Prime focus method (only used in very large telescopes).* **B.** *Cassegrain method (most common).* **C.** *Newtonian method.*

Included among these reasons is the monumental task of producing a large piece of quality, bubble-free glass for refracting telescopes. In addition, both sides of a lens must be polished, and when a double lens is employed to reduce the chromatic aberration, a total of four sides must be polished to the highest optical standards. However, only one side of a reflector must be polished. In addition, a lens can only be supported around the edge and suffers the effects of sagging. Mirrors, on the other hand, can be fully supported from behind.

The largest telescope in the world today is the 600-centimeter (236-inch) reflector located in the Caucasus Mountains of the Soviet Union.* Right behind it in size is the 508-centimeter (200-inch) Hale telescope on Mount Palomar, in California (see chapter-opening photo). The Hale telescope is a 500-ton steel and glass marvel of modern technology which floats into position on oil bearings. To ready the instrument for the night's vigil, the astronomer uses an elevator to reach the 6-foot-wide viewing cage, located inside the telescope almost 80 feet above the floor of the dome. From his perch, he can see the 17-foot Pyrex mirror shimmering 60 feet below. In the "good old days," he would spend the night in the cold mountain air, but advances in photographic techniques and computer-enhanced pictures allow him to work in a heated room below and reduce the time required to expose the photographic film. Other large reflecting telescopes of about 381 centimeters (150 inches) are located at Kitt Peak, Arizona, and in Chile and Australia. The largest refractor in the world is the 102-centimeter (40-inch) telescope at Yerkes Observatory, in Williams Bay, Wisconsin. This instrument was built before the turn of the century.

Detecting Invisible Radiation

Light is made up of more than just the radiation that our eyes perceive. As stated

*At the time of this writing, the Soviet telescope was not yet operational.

Figure 18.19 *Observer in the viewing cage at the prime focus of the 508-centimeter (200-inch) Hale telescope. (Courtesy of Hale Observatories)*

earlier, gamma rays, X rays, ultraviolet radiation, infrared radiation, and radio waves are also produced by celestial bodies. Photographic film that is sensitive to ultraviolet and infrared radiation has been developed, thereby extending the limits of our vision. However, most of this radiation cannot penetrate our atmosphere, so balloons and rockets must be used in order to record it. Of great importance is a narrow band of radio radiation that does penetrate the atmosphere. One particular wavelength is the 21-centimeter (8-inch) line produced by neutral hydrogen. Mea-

A.

B.

Figure 18.20 **A.** *A 43-meter (140-foot) steerable radio telescope. (Courtesy of National Radio Astronomy Observatory)* **B.** *The 300-meter (1000-foot) telescope at Arecibo, Puerto Rico. (Courtesy of National Astronomy and Ionosphere Center's Arecibo Observatory, operated by Cornell University under contract with the National Science Foundation)*

surement of this radiation has permitted mapping the distribution in our own galaxy of hydrogen—the material from which stars are made.

The detection of radio waves is accomplished by "big dishes" called **radio telescopes** (Figure 18.20). In principle, the dish of one of these telescopes operates in the same manner as the mirror of an optical telescope. It is parabolic in shape and focuses the radio waves on an antenna, which "hears" the source. Because radio waves are in the order of 100,000 times longer than visible radiation, the accuracy to which the surface of the dish has to be built is greatly reduced. However, this is more than offset by the fact that radio energy from celestial sources is weak, making substantial dishes a requirement. The largest radio telescope is a bowl carved in a natural depression in Puerto Rico (Figure 18.20B). It is 300 meters (1000 feet) in diameter and has some directional flexibility in its movable antenna. The largest steerable types have about 100-meter (300-foot) dishes like that at the National Radio Astronomy Observatory, in Greenbank, West Virginia. Radio telescopes also have rather poor resolution, making it difficult to pinpoint the radio source. Pairs or groups of them are used to reduce this problem.

Radio telescopes do have a few advantages over optical telescopes. They are much less affected by turbulence in the atmosphere, clouds, and the weather in general. No protective dome is required, which reduces the cost of construction, and viewing is possible 24 hours a day. They are, however, hindered by human-made radio interference. Thus, while optical telescopes are placed on a mountaintop, radio telescopes are often hidden in a valley, or at least behind some obstruction.

Radio telescopes have revealed such spectacular events as the collision of two galaxies, but of even greater interest was the discovery of **quasars** (quasi-stellar radio sources). These perplexing objects may be the most distant things in the universe.

The Sun

The sun is one of billions of stars that make up the Milky Way galaxy, and although it is of no significance to the universe as a whole, to us on earth, it is the primary source of energy. Everything from the fossil fuels we burn in our automobiles and power plants to the breakfast we eat is ultimately derived from solar energy. The sun is also important to astronomers, since it is the only star whose surface can be observed; even with the largest telescopes, the other stars can only by resolved as points of light.

Because of the sun's brightness, it is *not safe* to observe it directly. However, a small telescope will project an image on a piece of cardboard held behind its eyepiece, and the sun may be studied from that. Several specially built telescopes around the world keep a constant vigil of the sun. One of the finest is at the Kitt Peak National Observatory (Figure 18.21). It consists of a 500-foot sloped enclosure that directs sunlight to a 60-inch, long focal length mirror situated below ground. From the mirror, a 33½-inch image of the sun is projected to an observing room, where it can be studied.

Figure 18.21 *Robert J. McMath solar telescope. (Courtesy of Kitt Peak National Observatory)*

Compared to the other stars of the universe, the sun can be considered an "average star." However, on the scale of our solar system, it is truly gigantic, having a diameter equal to 109 earth diameters (1.35 million kilometers) and a volume 1¼ million times as great as that of the earth. Yet, because of its gaseous nature, its density is only ¼ that of the earth, very closely approximating the density of water.

The Structure of the Sun

For convenience of discussion, we divide the sun into four parts: the solar interior; the visible surface, or photosphere; and the two layers of its atmosphere, the chromosphere and the corona (Figure 18.22). Since the sun is gaseous throughout, no sharp boundaries exist between these layers. The sun's interior makes up all but a tiny fraction of the solar mass, and unlike the outer three layers, it is not accessible to direct observation. We shall discuss the visible layers first.

The **photosphere** ("sphere of light") is aptly named, since it is the layer that radiates most of the sunlight we see and therefore appears as the bright disk of the sun. Although it is considered the sun's surface, it is unlike most surfaces we are accustomed to. The photosphere consists of a layer of incandescent gas 300 kilometers (200 miles) thick having a pressure less than 1/100 of our atmosphere. Furthermore, it is not smooth or uniformly bright as the ancients had imagined. It has numerous blemishes. When viewed through a telescope under ideal conditions, a grainy texture is apparent. This is the result of numerous relatively small, bright markings called **granules,** which are surrounded by narrow dark regions (Figure 18.23). Typically, granules are 1000 kilometers (600 miles) in diameter and owe their brightness to hotter gasses that are rising from below. As this gas spreads laterally, cooling causes it to darken and sink back into the interior. Although each granule lasts only a few minutes, the combined motion of all granules gives the

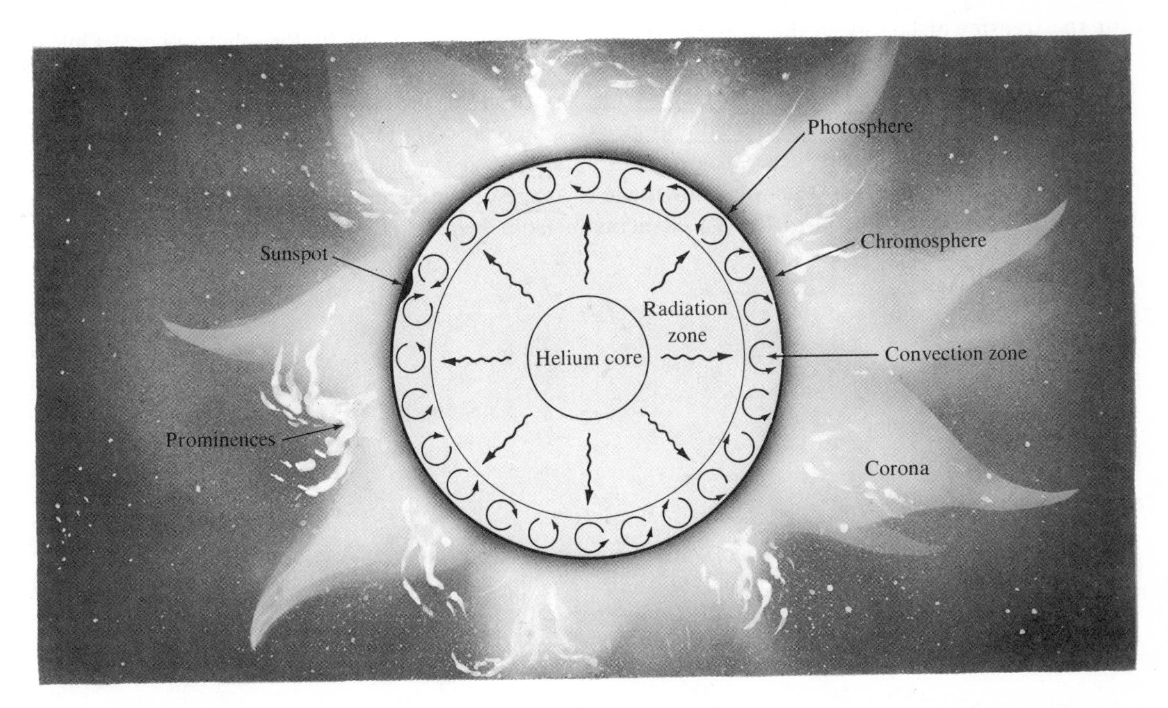

Figure 18.22 *Diagram of solar structure in cross section.*

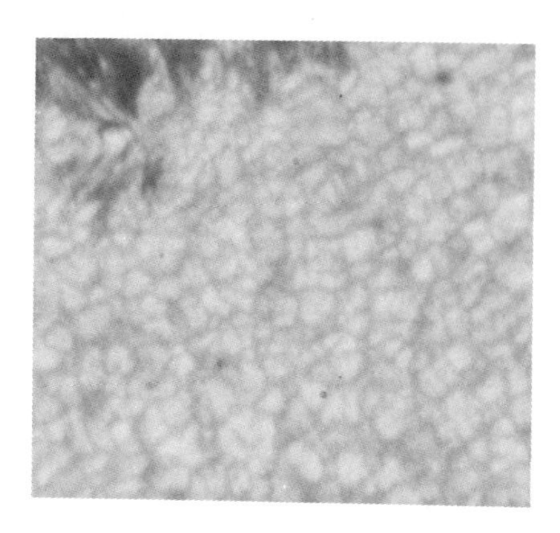

Figure 18.23 *Granules of the solar photosphere as photographed from a balloon at 24,000 meters. (Courtesy of Project Stratoscope of Princeton University, sponsored by NASA, NSF, and ONR)*

photosphere the appearance of boiling. This up-and-down movement of gas is called convection and, besides producing the grainy appearance of the photosphere, is believed to be responsible for the transfer of energy in the uppermost part of the sun's interior (Figure 18.22).

The composition of the sun's surface is revealed by the dark lines of its absorption spectrum (Figure 18.24). When these "fingerprints" are compared to the spectrum of known elements, they indicate that 60 of the 92 elements found on earth also occur on the sun. Probably all 92 elements exist there, but those that are extremely rare do not produce detectable lines. When the strengths of the absorption lines are analyzed, the relative abundance of the elements can be determined. These studies reveal that 90 percent of the sun's surface atoms are hydrogen, almost 10 percent are helium, and only minor amounts of the other detectable elements are present. Other stars also indicate similar disproportionate percentages of these two lightest elements, a fact we shall consider later.

Just above the photosphere lies the **chromosphere** ("color sphere"), a relatively thin layer of hot, incandescent gasses a few thousand kilometers thick. The chromosphere is observable for a few moments during a total solar eclipse or by using a special instrument which blocks out the light from the photosphere. On such occasions, it appears as a thin red rim around the sun. Because the chromosphere consists of hot, incandescent gasses under low pressure, it produces a bright-line spectrum which is nearly the reverse of the dark-line spectrum of the photosphere. One of the bright lines of hydrogen contributes a good portion of its total light and accounts for this sphere's red color. A study of the chromospheric spectrum conducted in 1868 revealed the existence of the heretofore unknown element helium (from *helios,* the Greek word for "sun"). Origi-

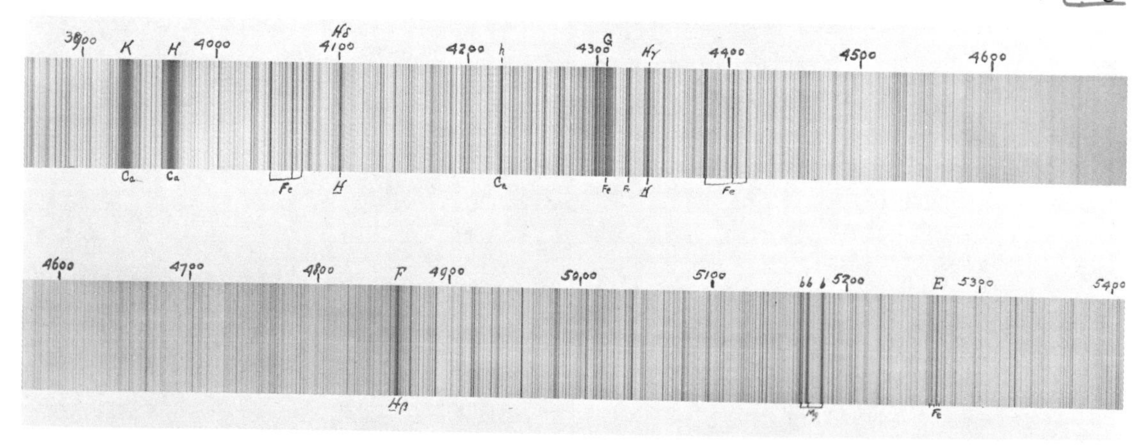

Figure 18.24 *Solar spectrum. Notice the dark lines and designated elements that produced them. (Courtesy of Hale Observatories)*

nally, helium was thought to be an element unique to the stars, but 27 years later, it was discovered in natural-gas wells on earth. The top of the chromosphere contains numerous **spicules,** which extend upward into the lower corona, almost like the trees which reach into our atmosphere (Figure 18.25. The spicules may be continuations of the turbulent motion of the granules below.

The outermost portion of the solar atmosphere, the **corona,** is very tenuous and, like the chromosphere, is visible only when the brilliant photosphere is covered (Figure 18.26). This envelope of ionized gasses normally extends a million kilometers from the sun and produces a glow about half as bright as the full moon. At the outer fringe of the corona, the ionized gasses have acquired speeds great enough to escape the gravitational pull of the sun. The streams of protons and electrons that "boil" from the corona constitute the **solar wind.** They travel outward through the solar system at very high speeds and eventually are lost to interstellar space. During their journey, the solar winds interact with the bodies of the solar system, continually bombarding lunar rocks and altering their appearance. Although the earth's magnetic field prevents the solar winds from reaching our surface, these winds do affect our atmosphere, as we shall discuss later.

Figure 18.25 *Spicules of the chromosphere as seen on the edge of the solar disk. (Courtesy of Sacramento Peak Observatory, Air Force Cambridge Research Laboratories)*

Studies of the energy emitted from the photosphere indicate that its temperature averages about 6000 K (10,000° F). Upward from the photosphere, the temperature unexpectedly increases, exceeding 1 million K at the top of the corona. It should be noted that although the coronal temperature exceeds that of the photosphere many times, it radiates much less energy because of its very low density. The high temperature of the corona is probably caused by sound waves generated by the convective motion of the photosphere. Just as boiling water makes noise, the supersonic booms similarly generated in the photosphere are believed to be absorbed by the gasses of the corona and thereby raise their temperatures.

The Active Sun

The most conspicuous features on the surface of the sun are the dark blemishes called **sunspots** (Figure 18.27A). Although sunspots were occasionally observed before the advent of the telescope, they were generally regarded as opaque objects located somewhere between the sun and the earth. In 1610 Galileo concluded that they were definitely residents of the solar surface, and from their motion, he deduced that the sun rotates on its axis about once a month. Later observations indicated that not all parts of the sun rotate at the same speed. The sun's equator rotates once in 25 days, while a place located 70 degrees from the solar equator, either north or south, requires 33 days for one rotation. If the earth rotated in a similar manner, imagine

Figure 18.26 *Solar corona photographed during a total eclipse. (Courtesy of Hale Observatories)*

the consequence! The sun's nonuniform rotation is a testimonial to its gaseous nature.

Sunspots begin as small dark pores about 1600 kilometers (1000 miles) in diameter. While most pores last for only a few hours, some grow into blemishes many times larger than the earth and last for a month or more. The largest spots often occur in pairs surrounded by several smaller spots. An individual spot contains a black center, the **umbra,** which is rimmed by a lighter region, the **penumbra** (Figure 18.27B). Sunspots appear dark only by contrast with the brilliant photosphere, a fact accounted for by their temperature, which is about 1500 K less than that of the solar surface. If these dark spots could be observed away from the sun, they would appear many times brighter than the full moon.

During the search conducted in the early 1800s for the planet Vulcan, believed to orbit between Mercury and the sun, an accurate record of sunspot occurrences was kept. Although the planet was never found, the sunspot data collected did reveal that the number of sunspots observable on the solar disk varies in an 11-year cycle. First, the number of sunspots increases to a maximum, when perhaps a hundred or more are visible at a given time, and then over a period of 5–7 years, their numbers decline to a minimum, when only a few, or possibly none, are visible (Figure 16.4). At the beginning of each cycle, the first sunspots form about 30 degrees from the solar equator, but as the cycle progresses and their numbers increase, they form nearer the equator. During the period when sunspots are most abundant, the majority form about 15 degrees from the equator. They rarely occur more than 40 degrees away from the sun's equator or within 5 degrees of it.

Another interesting characteristic of sunspots was discovered by George Hale. Hale deduced that the large spots are strongly magnetized, and when they occur in pairs, they have opposite magnetic poles. If the one on the right is a north pole, then the one on the left will be a south pole, as with the north and south poles of the earth's magnetic field. Also, every pair located in the same hemisphere will be magnetized in the same manner. However, all pairs in the other hemisphere will be magnetized in the opposite manner. At the beginning of each new sunspot cycle, the situation reverses, and the polarity of these sunspot pairs is opposite those of the previous cycle. The cause of this change in polarity, in fact the cause of sunspots themselves, is not fully explained. However, other solar activity varies in the same cyclic manner as sunspots, indicating a common origin.

Because of the brilliance of the photosphere, activity occurring above the solar surface is difficult to observe. To overcome this problem, numerous photographic methods are employed to filter solar radiation so that light of a particular spectral region can be viewed. Using such techniques, large "clouds" can be seen in the chromosphere directly above sunspot clusters (Figure 18.28). These bright centers of solar activity are called **plages** and occasionally can even be viewed before or after the sunspot occurrences. Among the more spectacular features of the active sun are the **prominences.** These huge cloudlike structures are best observed when they are on the edge, or limb, of the sun, where they often appear as great arches that extend well into the corona (Figure 18.29). Many prominences have the appearance of a fine tapestry and seem to hang motionless for days at a time, but motion pictures reveal that the material within them is continually falling like luminescent rain. Apparently, these quiet prominences are condensations of coronal material which are gracefully "sliding down" the lines of magnetic force back into the chromosphere. More rarely, the material within a prominence rises almost explosively away from the sun. These active prominences reach velocities up to 1000 kilometers (600 miles) per second and may leave the sun entirely. Prominences can also be seen

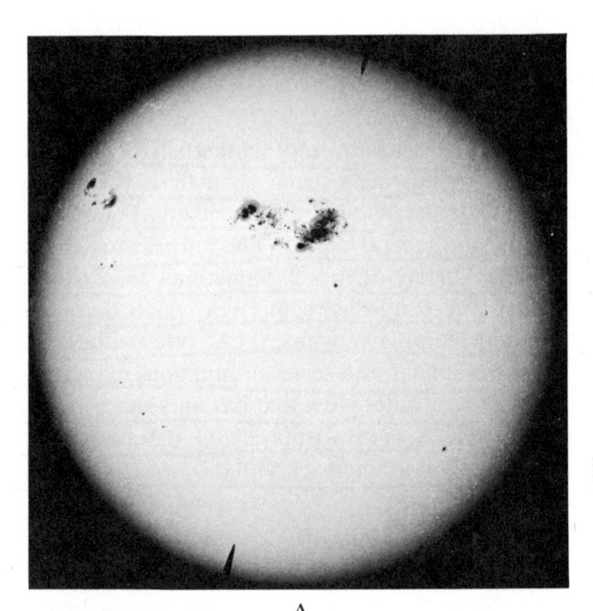

A.

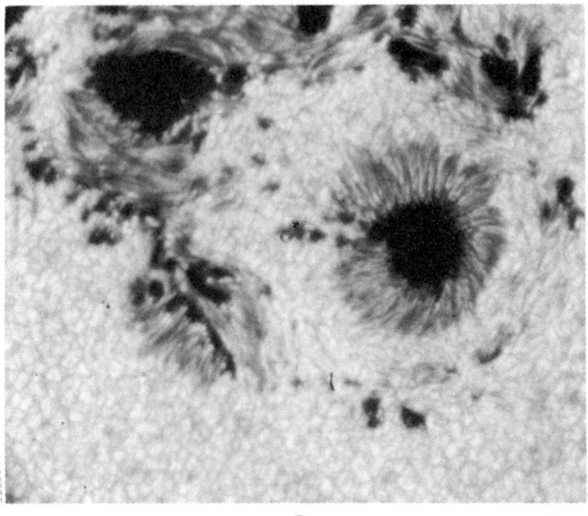

B.

Figure 18.27 **A.** *Large sunspot group on the solar disk. (Courtesy of Hale Observatories)* **B.** *View of sunspots whose umbra and penumbra are visible. (Courtesy of Project Stratoscope of Princeton University, sponsored by NASA, NSF, and ONR)*

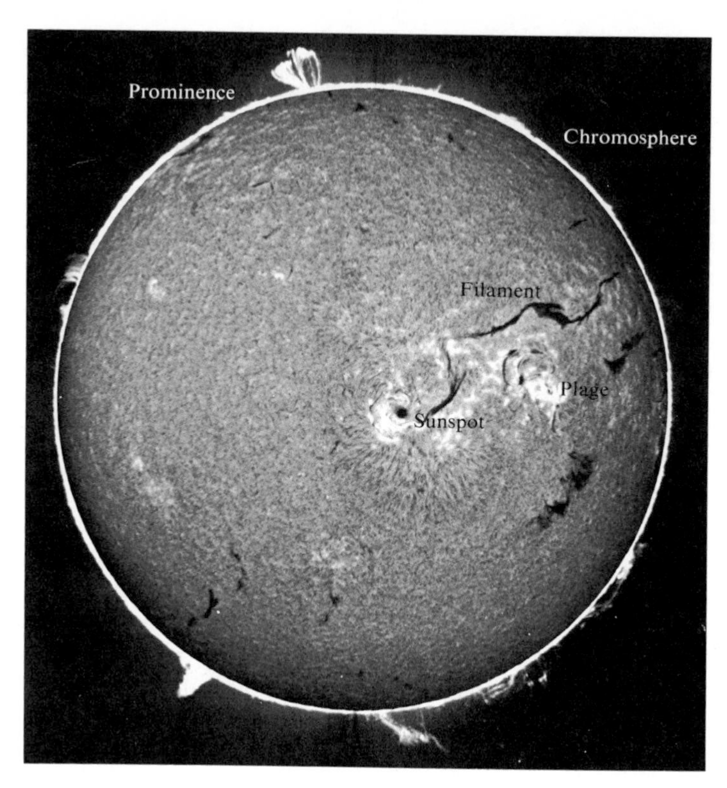

Figure 18.28 *Solar disk photographed in hydrogen alpha light, showing the manifestations of the active sun. (This composite courtesy of Hale Observatories and Sacramento Peak Observatory, Air Force Cambridge Research Laboratories)*

against the bright disk of the sun, in which case they appear as dark, thin streaks called **filaments** (Figure 18.28).

Solar flares are the most explosive events associated with sunspots. These brief outbursts normally last an hour or so and appear as a sudden brightening of the region above a sunspot cluster. During the period of their existence, enormous quantities of energy are released, much of it in the form of ultraviolet, radio, and X-ray radiation. Simultaneously, fast-moving atomic particles are ejected, causing the solar winds to intensify noticeably. Although a major flare could conceivably endanger a manned space flight, they are relatively rare. About a day after a large outburst, the ejected particles reach the earth and disturb the ionosphere, affecting long-distance radio communications. Their most spectacular effect, however, are the **auroras,** also called the northern and southern lights (Figure 18.30). Following a strong solar flare, the earth's upper atmosphere near its magnetic poles is set aglow for several nights. The auroras appear in a wide variety of forms. Sometimes the displays consist of vertical streamers in which there can be considerable movement. At other times, the auroras appear as a series of luminous expanding arcs or as a quiet glow that has an almost foglike character. Auroral displays, like other solar activities, vary in intensity with the 11-year sunspot cycle.

The Solar Interior

Since the interior of the sun cannot be observed directly, what we know about it is based on information acquired from the energy it radiates and from theoretical studies. The

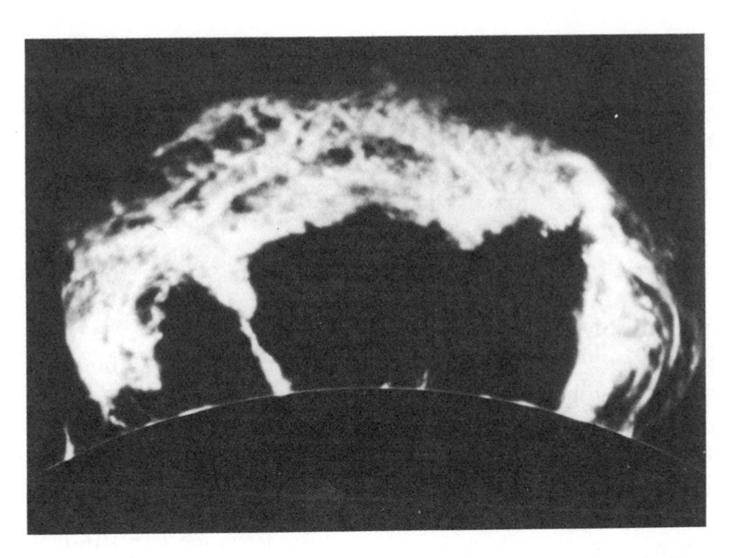

Figure 18.29 *Arched prominences. (Courtesy of Sacramento Peak Observatory, Air Force Cambridge Research Laboratories)*

source of the sun's energy, **nuclear fusion,** was not discovered until the late 1930s. Deep in its interior, a nuclear reaction called the **proton-proton chain** converts four hydrogen nuclei (protons) into a nucleus of helium. Four hydrogen atoms have a combined atomic mass of 4.032 (4 × 1.008), while the atomic mass of helium is 4.003, or 0.029 less than the combined mass of the hydrogen. The missing mass is emitted as energy according to Einstein's formula $E = mc^2$, where m equals mass and c equals the speed of light. Because the speed of light is very great, the amount of energy released from even a small amount of mass is enormous. The conversion of just a pin head's worth of hydrogen to helium can generate more energy than burning thousands of tons of coal. Most of this energy is in the form of light photons which work their way toward the solar surface, being absorbed and reemitted many times until they reach a rather opaque layer just below the photosphere. Here, convection currents transport this energy to the solar surface, where it can radiate through the transparent chromosphere and corona (Figure 18.22).

Although only a small percentage (0.7 percent) of the material involved in this reaction is actually converted to energy, the sun is still consuming itself at the rate of 4 million tons per second. As the hydrogen is consumed, the waste product of this reaction, helium, forms the solar core, which continually grows in size. Just how

Figure 18.30 *Aurora borealis (northern lights) as seen from Alaska. (Courtesy of Victor Hessler)*

long can the sun produce energy at its present rate before all of its fuel (hydrogen) is consumed? Even at the enormous rate at which the sun engulfs its fuel, it has enough to easily last another 100,000 million years. However, evidence from other stars indicates that the sun will change dramatically long before all of its hydrogen is gone. It is thought that a star the size of the sun can exist in its present form for 10,000 million years. Since the sun is already 500 million years old, it may be considered "middle aged."

In order to initiate the proton-proton reaction, the sun's temperature must have reached several million degrees. What was the source of this heat? As previously noted, the solar system is believed to have formed from an enormous cloud of dust and gasses (mostly hydrogen) which condensed gravitationally. The consequence of squeezing (compressing) a gas is to increase its temperature. Although all of the bodies in the solar system were squeezed, the sun was the only one, because of its size, that became hot enough to trigger the proton-proton reaction. Astronomers currently estimate its internal temperature at 15 million K. If Jupiter were about ten times more massive, it too might have become a star.

The idea of one star orbiting another may seem unusual, but recent evidence indicates that about 50 percent of the stars in the universe probably occur in pairs or multiples.

REVIEW

1. How much would you weigh on the moon?
2. Draw the earth-moon system to scale (approximate the diameters of the moon and the earth—3500 and 13,000 kilometers, respectively).
3. What is different about the crescent phase that precedes the new-moon phase and that which follows the new-moon phase?
4. What phase of the moon occurs about two weeks after the new moon?
5. On your drawing from question 2, show how the 5-degree inclination of the moon's orbit prevents an eclipse from occurring each month.
6. Solar eclipses are more common than lunar eclipses. Why, then, is it more likely that your region of the country will experience a lunar eclipse?
7. Why are the large-rayed craters considered relatively young features on the lunar surface?
8. Briefly outline the history of the moon.
9. How are the maria of the moon thought to be similar to the Columbia Plateau?
10. Which color has the longest wavelength? The shortest?
11. Explain how each of the three types of spectra is produced.
12. How can astronomers determine whether a star is moving toward or away from the earth?
13. What is the function of the objective lens?
14. What is the practical limit of magnification of a telescope with a 10-inch lens?
15. Determine the magnification of a telescope with a focal length of 200 inches that is fitted with a 20-inch focal length eyepiece. A 1-inch eyepiece.
16. What are the advantages of telescopes with large objective lenses?
17. Why are special viewing systems needed with reflecting telescopes?
18. Explain the statement "Photography has extended the limits of our vision."
19. Why are radio telescopes built much larger than optical telescopes?
20. What are some of the advantages of radio telescopes over optical telescopes?
21. Describe the photosphere, chromosphere, and corona.
22. List the features associated with the active sun and describe each.
23. Explain how a sunspot can be very hot and yet appear dark.
24. What is the sun's source of energy?

KEY TERMS

phases of the moon
synodic month
sidereal month
solar eclipse
lunar eclipse
maria

A Pioneer 10 view of Jupiter showing the banded cloud structure, the black dot shadow of the innermost satellite Io, and the Great Red Spot on the photo's right edge. (Courtesy of NASA)

The Great Galaxy in the constellation Andromeda, the nearest spiral galaxy 2,200,000 light-years from our own home galaxy. The bright central nucleus and the spiral arms are seen together with two dwarf companion galaxies. (Hale Observatories photo. Copyright by the California Institute of Technology and Carnegie Institution of Washington)

The Pleiades star cluster in Taurus shown embedded in the interstellar clouds which reflect the light of the brighter cluster members. (Hale Observatories photo. Copyright by the California Institute of Technology and Carnegie Institution of Washington)

lunar breccia
mascon
regolith
electromagnetic radiation
white light
photons
radiation pressure
spectroscopy
continuous spectrum
bright-line (emission) spectrum
dark-line (absorption) spectrum
spectroscope
Doppler effect
objective lens
focus
focal length
eyepiece
resolving power
chromatic aberration
reflecting telescope
radio telescope
quasars
photosphere
granules
chromosphere
spicules
corona
solar wind
sunspots
umbra
penumbra
plages
prominence
solar flare
filament
auroras
nuclear fusion
proton-proton chain

Saturn and its ring system. (Courtesy of Hale Observatories)

The Planets: An Overview
The Minor Members of the Solar System
The Planets: An Inventory

The Solar System

When people first recognized that the planets are "worlds" much like the earth, a great deal of interest was generated. The primary concern has always been the possibility of intelligent life existing elsewhere in the universe. Recent unmanned space explorations have renewed this interest. Since all the planets may have formed from the same primeval cloud of dust and gasses, they should provide valuable information concerning the earth's formation and early history.

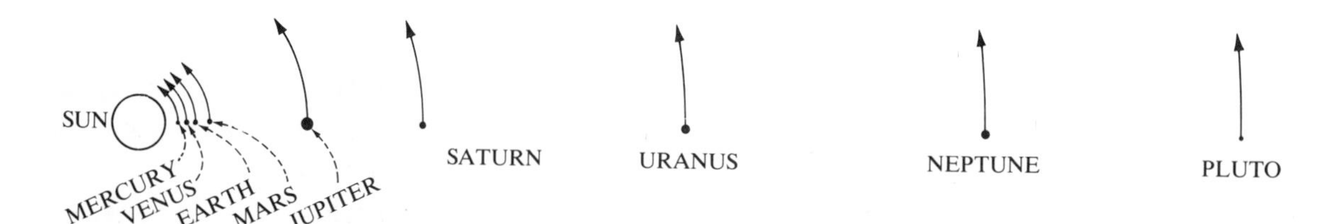

Figure 19.1 *Orbits of the planets to scale.*

The sun is the hub of a huge rotating system consisting of nine planets, their satellites, and numerous small, but nonetheless interesting bodies, including asteroids, comets, and meteoroids. About 99.85 percent of the mass of the solar system is contained within the sun, while the planets collectively make up most of the remaining 0.15 percent. The planets, in order from the sun, are Mercury, Venus, Earth, Mars, Jupiter, Saturn, Uranus, Neptune, and Pluto (Figure 19.1). Under the control of the sun's gravitational force, each planet maintains an elliptical orbit and travels in a counterclockwise direction. The planets, except for Venus and Uranus, also rotate counterclockwise about their axes.

In compliance with Kepler's laws, the nearest planet to the sun, Mercury, has the fastest orbital motion, 48 kilometers per second, and the shortest period of revolution, 88 days; and the most distant planet, Pluto, has an orbital speed of 5 kilometers per second and requires 248 years to complete one revolution. The orbits of all the planets lie within 3 degrees of the plane of the ecliptic, except for the orbits of Mercury and Pluto, which are inclined 7 degrees and 17 degrees, respectively.

The Planets: An Overview

Upon examining Table 19.1, we see that the planets fall quite nicely into two groups: the **terrestrial** (earthlike) planets of Mercury, Venus, Mars, and Earth; and the **Jovian** (Jupiterlike) planets of Jupiter, Saturn, Uranus, and Neptune. Pluto is not included in either category; because of its great distance and small size, its true nature is still a mystery. The most obvious difference between these groups is the size of their mem-

Table 19.1 ***Planetary Data***

Planet	Symbol	Mean Distance from Sun			Period of Revolution	Inclination to Ecliptic	Orbital velocity	
		AU	Millions of miles	Millions of kilometers			mi/s	km/s
Mercury	☿	0.387	36	58	88 d	7° 00′	29.5	47.9
Venus	♀	0.723	67	108	225 d	3° 24′	21.8	35.0
Earth	⊕	1.000	93	150	365.25 d	0° 00′	18.5	29.8
Mars	♂	1.524	142	228	687 d	1° 51′	14.9	24.1
Jupiter	♃	5.203	483	778	12 yr	1° 19′	8.1	13.1
Saturn	♄	9.539	886	1,427	29.5 yr	2° 30′	6.0	9.6
Uranus	⛢	19.180	1,780	2,869	84 yr	0° 46′	4.2	6.8
Neptune	♆	30.060	2,790	4,498	165 yr	1° 46′	3.3	5.4
Pluto	♇	39.440	3,670	5,900	248 yr	17° 12′	2.9	4.7

bers (Figure 19.2). The largest terrestrial planet (Earth) has a diameter only 1/4 as great as the diameter of the smallest Jovian planet (Neptune), and its mass is only 1/17 as great. Hence, the Jovian planets are often called giants. Also, because of their location, the four Jovian planets are referred to as the *outer planets,* while the terrestrial planets are called the *inner planets.*

Other dimensions along which the two groups markedly differ include density, composition, and rate of rotation. The densities of the terrestrial planets average about 5 times the density of water, and the Jovian planets have densities that average only 1.5 times that of water. In fact, one of the outer planets, Saturn, has a density which is only 0.7 that of water, which means that if a large enough ocean existed, Saturn would float in it. The composition of the planets is largely responsible for this difference.

The substances of which both groups of planets are composed have been divided into three groups based on their melting points. They are gasses, rocks, and ices. The gasses are those materials with melting points near absolute zero −273° C and consist mainly of hydrogen and helium.* The rocky materials are made of silicate minerals and metallic iron, which have high melting points, exceeding 700° C. The ices have intermediate melting points and include ammonia (NH_3), methane (CH_4), carbon dioxide (CO_2). and water (H_2O).

The terrestrial planets are composed mostly of dense rock and metallic material. The Jovian planets, on the other hand, contain a large percentage of hydrogen and helium, with varying amounts of "ice" (mostly frozen ammonia and methane), which accounts for their low densities. The outer planets are also thought to contain as much rocky material as the terrestrial planets, but whether these rocky components make up a central core or are widely scattered is uncertain.

The Jovian planets have very thick atmospheres (mostly hydrogen and helium). By comparison, the terrestrial planets have meager atmospheres at best. A planet's ability to retain an atmosphere depends on its temperature and mass. Simply stated, a gas molecule can "evaporate" from a planet if it reaches a speed known as the **escape velocity.** [For the earth, this velocity is 11 kilometers (7 miles) per second. Any material, including a rocket, must reach this speed before it can leave the earth and go into space.] The Jovian planets, because of their greater mass, have higher escape velocities than the terrestrial planets. Consequently, it is more difficult for

*Absolute zero is the lowest possible temperature. All molecular motion ceases at that point.

Planet	Period of Rotation	Diameter		Relative Mass (Earth = 1)	Average Density (g/cm^3)	Polar Flattening (%)	Eccentricity	Number of Known Satellites
		miles	kilometers					
Mercury	59^d	3,015	4,868	0.056	5.1	0.0	0.206	0
Venus	243^d	7,526	12,112	0.82	5.3	0.0	0.007	0
Earth	$23^h 56^m 04^s$	7,920	12,742	1.00	5.52	0.3	0.017	1
Mars	$24^h 37^m 23^s$	4,216	6,800	0.108	3.94	0.5	0.093	2
Jupiter	$\sim 9^h 50^m$	88,700	143,000	318.000	1.34	6.5	0.048	13
Saturn	$\sim 10^h 25^m$	75,000	121,000	95.200	0.70	10.5	0.056	10
Uranus	$10^h 45^m$	29,000	47,000	14.600	1.55	7.0	0.047	5
Neptune	16^h (?)	28,900	45,000	17.300	2.27	2.5	0.008	2
Pluto	6.4^d	~1,500	~2,400	~ 0.01(?)	~1.5(?)	?	0.250	1

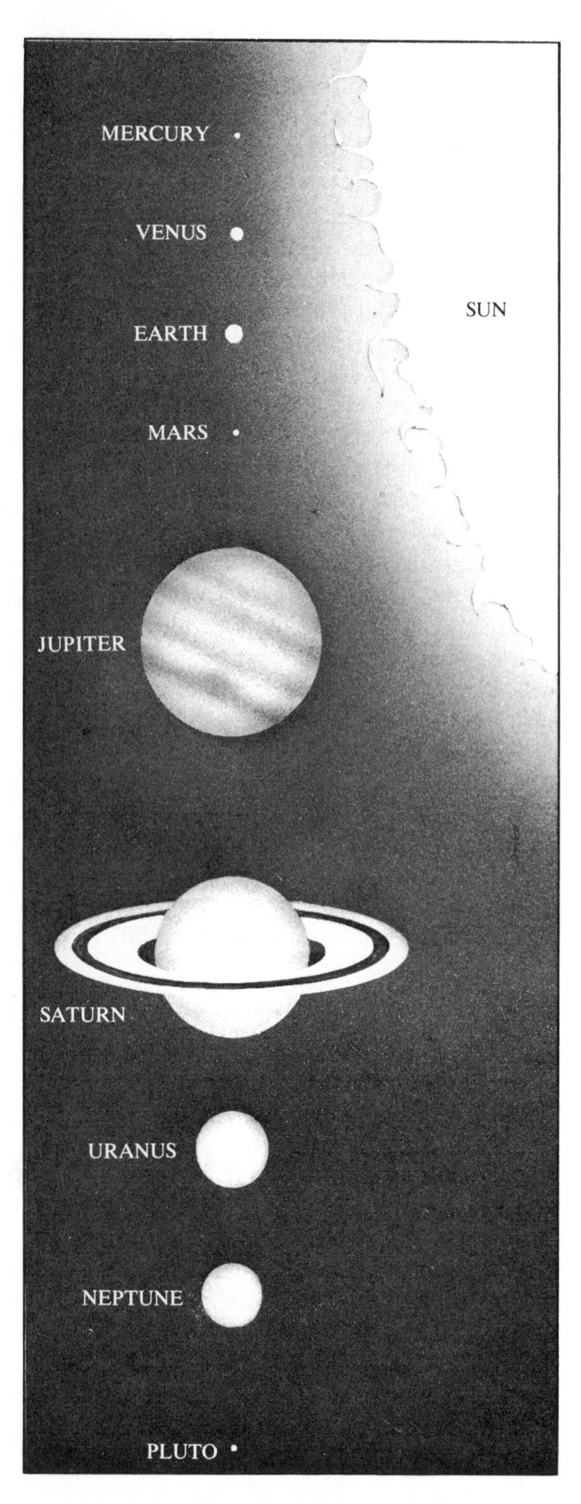

Figure 19.2 *Relative sizes of the planets.*

gasses to "evaporate" from them. Also, because the molecular motion of a gas is temperature dependent, at the low temperatures of the Jovian planets even the lightest gasses are unlikely to acquire the speed needed to escape. On the other hand, a comparatively warm body with a small mass, like our moon, is unable to hold even the heaviest gas and is thus void of an atmosphere. The slightly larger planets Earth, Venus, and Mars retain some heavy gasses (as compared to hydrogen), but even their atmospheres make up only an infinitesimally small portion of their total mass.

It is hypothesized that the mass of dust and gas from which all the planets are thought to have condensed had a uniform composition similar to that of Jupiter. However, unlike Jupiter, the terrestrial planets are nearly void of light gasses and ices. Were the terrestrial planets once much larger, and did they contain these materials but because of their close proximity to the sun lose them? Questions like these may be answered by further exploration of our solar system.

We shall consider each of the planets in more detail. First, however, a discussion of the minor members of the solar system is appropriate.

The Minor Members of the Solar System

Asteroids

When Uranus was discovered in 1781, the distance from the sun to all the known planets fit a simple arithmetic progression known as Bode's law, which is determined by adding 4 to the series of numbers 0, 3, 6, 12, 24, . . . (found by doubling the previous number) and dividing the sum by 10. The sequence of numbers so generated is given in Table 19.2, where it is compared to the actual distances from the sun to the planets in astronomical units (AU). In 1781 the only major discrepancy in Bode's law was a "gap" at 2.8 AU. So, a search was begun for a planet which occupied that orbit. In 1801 the elusive "missing planet" was discovered by an Italian monk, Giuseppe Piazzi, who named it Ceres

Table 19.2 ***Bode's Law Compared with Actual Distances***

Planet	Actual distance from Sun (AU)	Sequence
Mercury	0.39	(0 + 4) ÷ 10 = 0.4
Venus	0.72	(3 + 4) ÷ 10 = 0.7
Earth	1.00	(6 +4) ÷ 10 = 1.0
Mars	1.52	(12 + 4) ÷ 10 = 1.6
		(24 + 4) ÷ 10 = 2.8
Jupiter	5.20	(48 + 4) ÷ 10 = 5.2
Saturn	9.54	(96 + 4) ÷ 10 = 10.0
Uranus	19.18	(192 + 4) ÷ 10 = 19.6
Neptune*	30.06	(384 + 4) ÷ 10 = 38.8
Pluto*	39.44	(768 + 4) ÷ 10 = 77.2

*Unknown at the time Bode's law was "discovered."

after the protective god of Sicily. But a year later, another moving body, Pallas, was discovered at the same distance from the sun. Then, a search began for other similar small bodies, called **asteroids,** or **minor planets.** By the year 1900, over 300 asteroids had been sighted, and today the number exceeds 50,000. The orbits of 1800 of the larger asteroids have been calculated. After each orbit is established, the asteroid is given a number and the discoverer is given the opportunity to name it. At first, mythological names were used, but they were soon exhausted, and names such as Hungaria, Hilda, and Chicago became common.

The orbits of most asteroids lie between Mars and Jupiter and have periods ranging from three to six years. Some asteroids with eccentric orbits travel very near the sun, and a few large ones regularly pass close to the earth and moon. Some of the larger impact craters on the moon were probably caused by collisions with asteroids.

Asteroids are rather small bodies that have been likened to "flying mountains." The largest, Ceres, is 800 kilometers (500 miles) in diameter, but most of those that have been observed are about a kilometer or so across. Because many asteroids have irregular shapes, astronomers first speculated that they may have formed from a catastrophic breakup of a planet that once occupied an orbit at 2.8 AU. However, the total mass of the asteroids is estimated to be only 1/1000 that of the earth, itself not a large planet. What, then, happened to the remainder of the original planet? Other astronomers have hypothesized that several larger bodies once existed and that their collisions produced numerous smaller ones. The existence of several "families" of asteroids has been used to support this latter explanation. However, no conclusive evidence has been found for either hypothesis.

Comets

Comets are among the most spectacular and unpredictable bodies in the solar system. They have been compared to large, dirty snowballs, since they are made of frozen gasses (water, ammonia, methane, and carbon dioxide) which hold together small pieces of rocky and metallic materials. Many comets travel along very elongated orbits that carry them beyond Pluto (Figure 19.3). On their return, the comets are visible only after they have moved within the orbit of Saturn.

When first observed, comets appear very small, but as they approach the sun, solar energy begins to vaporize the frozen gasses, producing a glowing head called the **coma.** The size of the coma varies greatly from one comet to another. Some exceed the size of the sun, but most approximate the size of Jupiter. Within the coma, a small glowing nucleus with a diameter of only a few kilometers can sometimes be detected. As they approach the sun, some, but not all, comets develop a tail that extends for millions of kilometers. Despite the enormous size of their tail and coma, comets are thought to have an insignificant mass. They are generally less than one-thousand millionth as massive as the earth. Thus, they have been called "the nearest thing to nothing that anything can be and still be something."

The tail of a comet points away from the sun in a slightly curved manner (Figure 19.3). This fact led early astronomers to propose that the sun had a repulsive force that pushed the particles of the coma, thus forming the tail. Today, two solar forces are known to contribute to this forma-

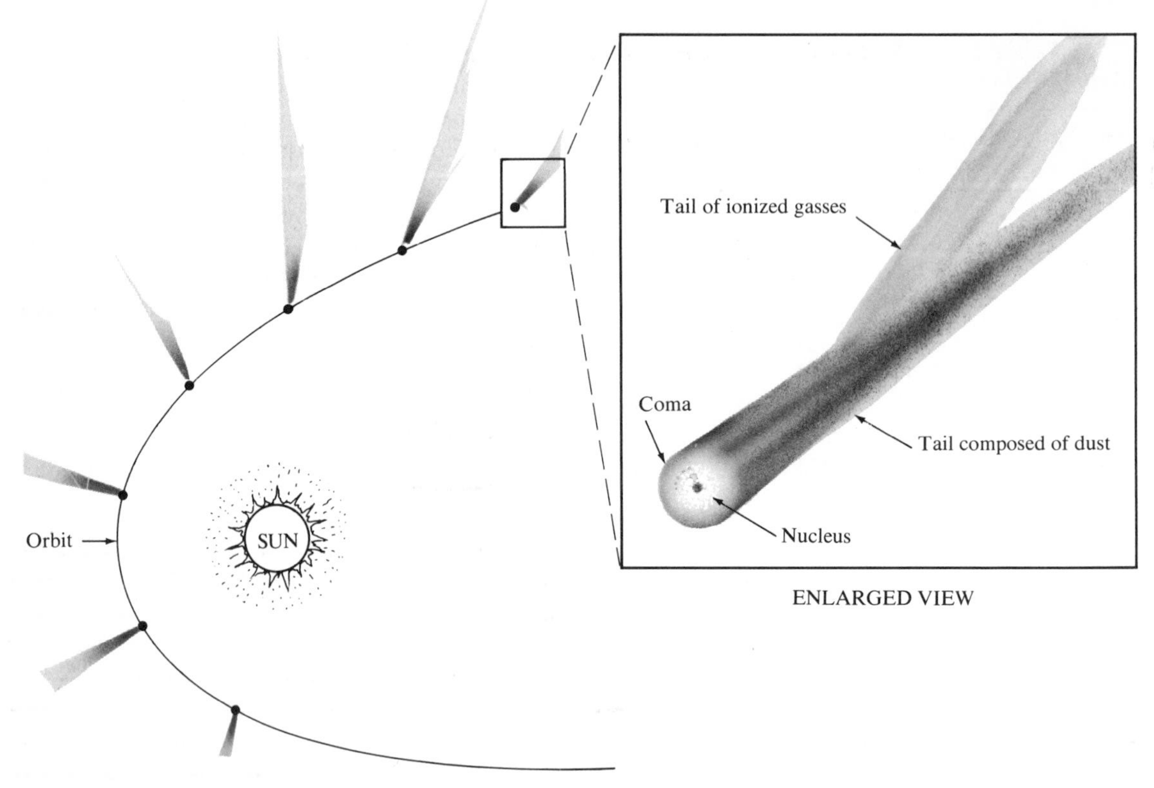

Figure 19.3 *Orientation of a comet's tail as it orbits the sun.*

tion. One, radiation pressure, pushes dust particles away from the coma, and the second, solar wind, is responsible for moving the ionized gasses, particularly carbon monoxide. Usually, a single tail composed of both types of materials is produced, but two somewhat separate tails like those of the comet Mrkos can form (Figure 19.4).

As a comet moves away from the sun, the gasses begin to condense, the tail disappears, and the comet once again returns to "cold storage" (Figure 19.5). The material that was blown from the coma to form the tail is lost to the comet forever. Consequently, it is believed that most comets cannot survive more than 100 close orbits of the sun. Once all the gasses are expended, the remaining material—a swarm of unconnected metallic and stony particles—continues the orbit without a coma or a tail.

Little is known about the origin of comets. The most popular theory considers them members of the solar system that formed at great distances from the sun. Accordingly, millions of comets are believed to orbit beyond Pluto with periods measured in hundreds of years. It is proposed that the gravitational effect of stars passing nearby send some of them into highly eccentric orbits which carry them toward the center of our solar system. Here, the gravitation of the larger planets, particularly Jupiter, alters their orbit and reduces their period of revolution. Many short-period comets of this type have been discovered. However, since they have a short life expectancy, we can be fairly certain that they are always being replaced by other long-period comets which are deflected toward the sun.

The most famous short-period comet is Halley's comet (Figure 19.5). Its period ranges between 74 and 79 years, and every one of its appearances since 240 B.C. has been recorded. When last seen, in 1910, Halley's comet had developed a tail nearly 1 million miles long and was visible during the daylight hours. It is ex-

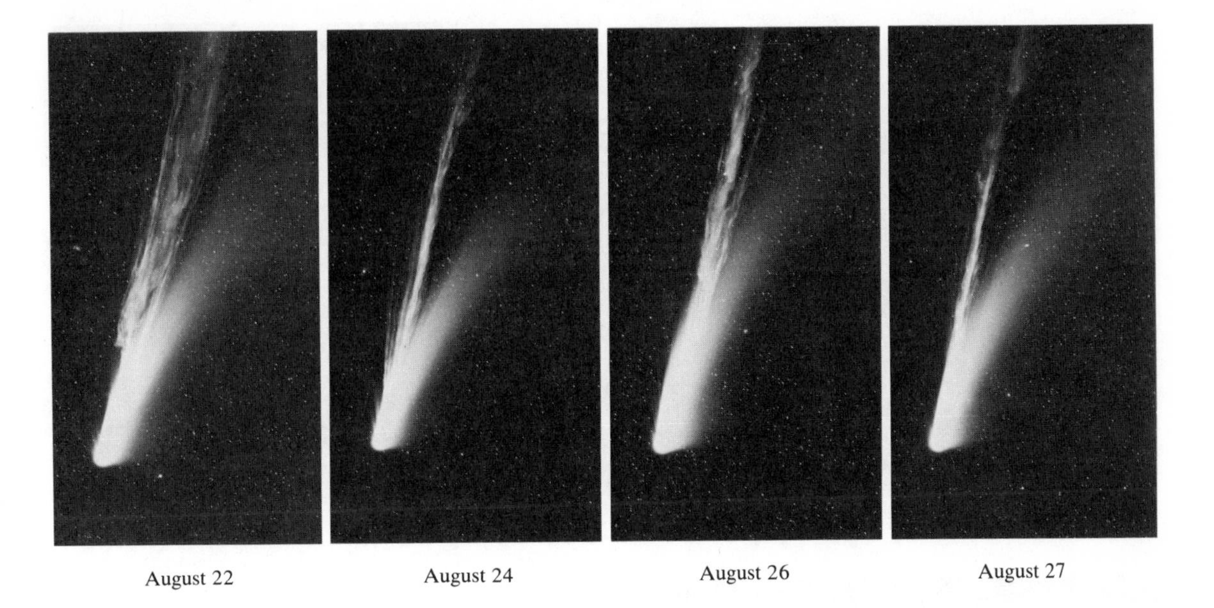

Figure 19.4 *Comet Mrkos photographed with the 48-inch Schmidt telescope on four nights in 1957. (Courtesy of Hale Observatories)*

pected again in February of 1986, when it should be visible for a few weeks. If it lives up to expectations, it should be a spectacular sight.

Meteoroids

Nearly everyone has seen a meteor, popularly called a "shooting star." This streak of light, which lasts for, at most, a few seconds, occurs when a small solid particle, a **meteoroid,** enters the earth's atmosphere from interplanetary space. The friction between the meteoroid and the air heats both and produces the light we see. Most meteors are about the size of sand grains and weight less than 1/100 of a gram. Consequently, they vaporize before reaching the earth's surface. Some, called **micrometeorites,** are so tiny that their rate of fall becomes too slow to allow them to burn up, so they drift down like dust. Each day, the total number of meteoroids that enter the earth's atmosphere must reach into the millions. From a single spot on earth, a half dozen or more are bright enough to be seen with the naked eye each hour.

Occasionally, the number of meteor sightings increases dramatically to 60 or more per hour. These spectacular displays, called **meteor showers,** result when the earth encounters a swarm of meteoroids traveling in the same direction and at nearly the same speed. The close association of these swarms to the orbits of some short-term comets strongly suggests that they represent material lost by these comets (Figure 19.6). Some swarms not associated with orbits of known comets are probably the remains of the nucleus of a defunct comet. The meteor showers that occur regularly each year around August 12 are believed to be the remains of the tail of Comet 1862 III, which has a period of 110 years. Meteoroids associated with comets are small and not known to reach the ground. Most meteoroids large enough to survive the fall are thought to originate in the belt of asteroids, where a chance collision sends them toward the earth. The earth's gravitational force does the rest. These remains, when found on the earth, are referred to as **meteorites.**

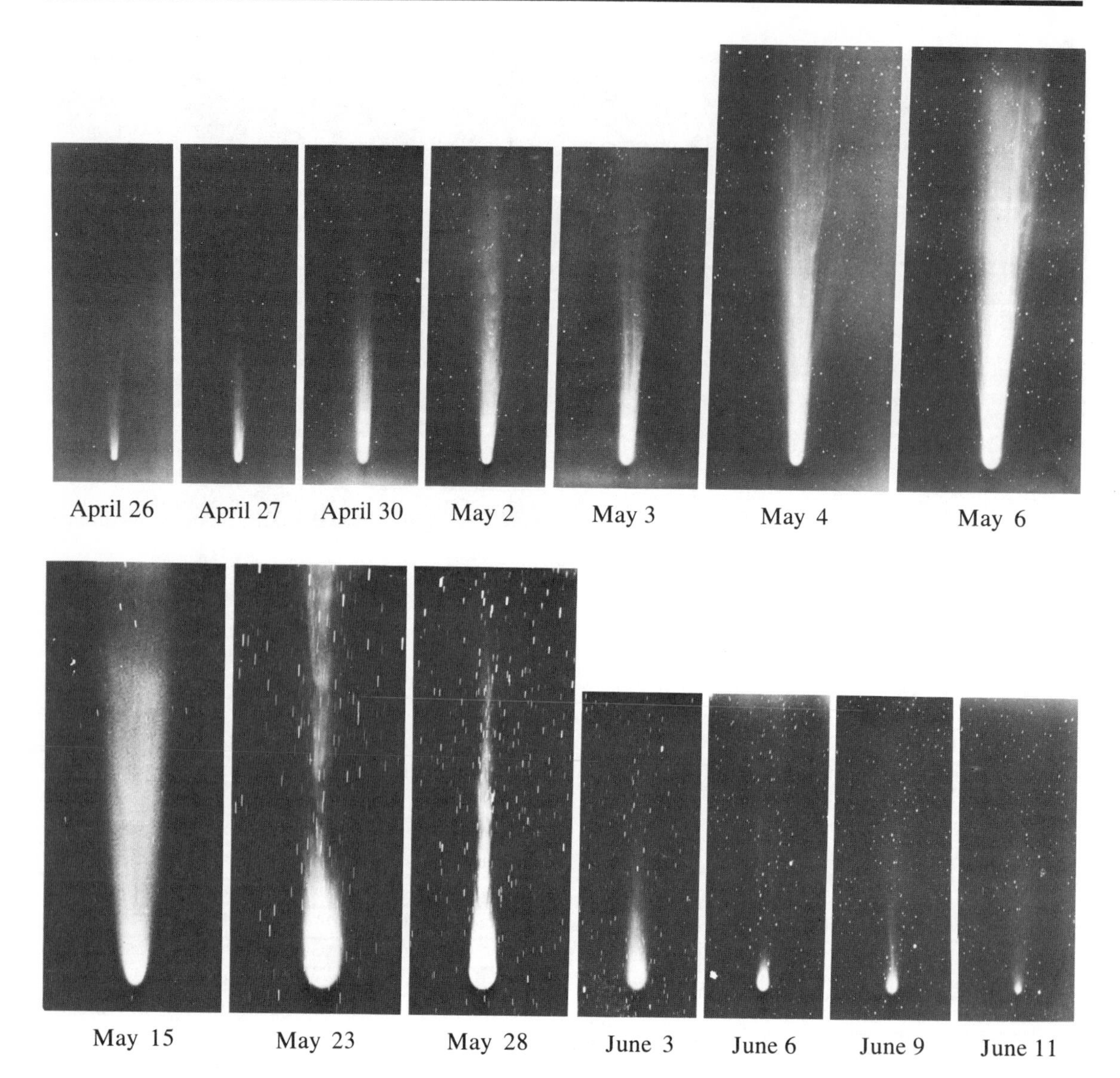

Figure 19.5 *Several views of the 1910 Halley's comet, illustrating its changing appearance as it moves near to and then away from the sun. (Courtesy of Hale Observatories)*

A few very large meteorites have blasted out craters on the earth's surface not too much different in appearance from those found on the lunar surface. The most famous is the Barringer meteorite crater, in Arizona (Figure 19.7). This crater is about 1.2 kilometers (0.80 mile) across, 170 meters (600 feet) deep, and has an upturned rim that rises 50 meters (150 feet) above the surrounding countryside. Over 30 tons of iron fragments have been found in the immediate area, but attempts to locate the main body have been unsuccessful. As evidenced from the amount of erosion, it appears that the impact occurred within the last 20,000 years.

One of the most spectacular meteoroid falls of modern times occurred in Siberia in 1908.

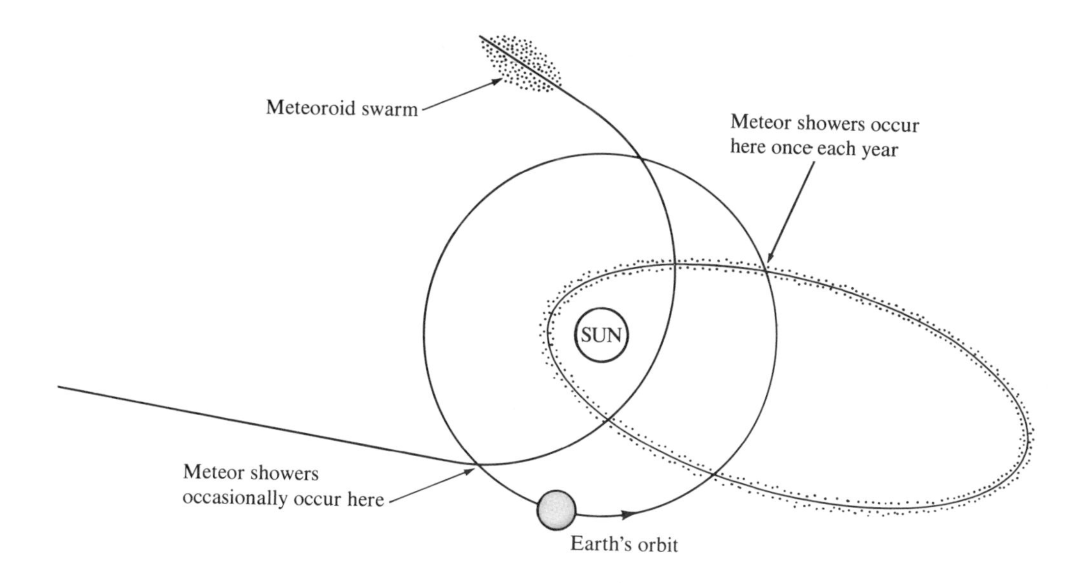

Figure 19.6 *Orbits of meteoroid swarms.*

A brilliant fireball exploded with a force strong enough to create a shock wave that rattled windows and was heard hundreds of kilometers away. Trees were scorched and flattened 30 kilometers from the impact zone, and 1500 raindeer were reportedly killed. Although numerous craters, some 50 meters (150 feet) wide, resulted, no metallic fragments were found. Recent speculation about the cause points to the impact of a comet nucleus.

Prior to lunar exploration, meteorites were the only samples of extraterrestrial material that could be directly examined (Figure 19.8). Depending upon their composition, they can gen-

Figure 19.7 *Barringer meteorite crater, about 32 kilometers (20 miles) west of Winslow, Arizona. (Courtesy of Meteor Crater Enterprises, Inc.)*

Figure 19.8 *Meteorite found near Barringer crater.*

erally be put into one of three categories: (1) **irons**—mostly iron with 5–20 percent nickel; (2) **stony**—silicate minerals with inclusions of other minerals; and (3) **stony-irons**—mixtures. Although stony meteorites are probably more common, most meteorite finds are irons. This is understandable, since irons tend to withstand the impact, weather more slowly, and are much easier for a lay person to distinguish from terrestrial rocks than are stony meteorites. Also, if the composition of meteorites is representative of the material that makes up the "earthlike" planets, as some astronomers believe, then the earth must contain a much larger percentage of iron than is indicated by the surface rock. This is one of the reasons why geologists suggest that the core of the earth may be mostly iron and nickel. In addition, the dating of meteorites has indicated that our solar system has an age in excess of 4700 million years. This "old age" has been confirmed by data obtained from lunar samples.

The Planets: An Inventory

Mercury: The Innermost Planet

Mercury, the smallest, innermost, and swiftest planet, is not much larger than the moon. Also like the moon, it absorbs most of the sunlight that strikes it, reflecting only 6 percent into space. This is characteristic of a terrestrial body without an appreciable atmosphere. By way of comparison, the earth reflects more than 30 percent of the light which it encounters, most of it from clouds.

Mercury's close proximity to the sun makes viewing from earthbound telescopes difficult at best. The first good glimpse of this planet came in the spring of 1974, when *Mariner 10* passed within 800 kilometers of its surface (Table 19.3).

Figure 19.9 *Photomosaic of Mercury taken by* Mariner 10 *at a distance of 200,000 kilometers (120,000 miles). (Courtesy of National Space Data Center)*

Its striking resemblance to the moon was immediately evident from the high-resolution photos that were radioed back (Figure 19.9). In fact, the similarity was so great that a project scientist remarked that the photos could be substituted for ones of the back side of the moon and most laymen would not suspect.

Mercury's period of revolution around the sun is 88 days, and until recently, its period of rotation was thought to be 88 days as well. It was therefore concluded that one side of Mercury perpetually faced the sun. Then in 1965 radar showed Mercury's rotational period to be only 59 days. Hence, Mercury rotates 1.5 times per revolution with every location on its surface experiencing the phenomenon of night and day. A typical period of daylight and darkness lasts 88 days. This causes the night temperatures to drop below −140° C (−220° F) and the day temperature to exceed 315° C (600° F), hot enough to melt tin and lead. Mercury, therefore, has the greatest temperature extremes of any planet. The odds of life existing on Mercury are nil.

Table 19.3 ***Summary of Significant Space Probes***

Apollo 8	1968	Astronauts circled the moon and returned to the earth
Apollo 11	1969	First astronaut landed on the moon
Apollo 17	1972	Last of six Apollo missions to carry people to the moon
Mariner 10	1974	Orbited the sun, allowing it to pass Mercury several times; fly-by mission to Venus
Pioneer 10,11	1973 1974	First close-up views of Jupitor
Venera 8, 9, 10	1972 1975 1975	Soviet landers on Venus (operative about one hour each)
Mariner 4, 6, 7	1965 1969 1969	Fly-by missions to Mars
Mariner 9	1971	Orbiter of Mars
Viking 1, 2	1976	Orbiters and landers of Mars

Venus: The Veiled Planet

Venus, second only to the moon in brilliance in the night sky, is appropriately named for the goddess of love and beauty. It orbits the sun in a nearly perfect circle once every 225 days. Because Venus is like the earth in size and mass, it has been referred to as the "earth's twin." However, the similarity ends there. Venus is shrouded with a thick cloud cover, which prohibits visual examination of its surface. Because of this, its period of rotation was not determined until the mid-1960s, when a technique using radar, to which its atmosphere is transparent, was developed. It was a surprise to learn that Venus rotates in a clockwise direction, opposite that of the other planets, and does so very slowly, having a period of 243 days. Radar also revealed that craters dot the Venusian surface, although a clear image of the planet still has not been obtained.

Before the advent of space vehicles, Venus and Mars were considered the most hospitable sites in the solar system (earth excluded) and the logical places to expect living organisms. Unfortunately, evidence from *Mariner* fly-by space probes and Soviet unmanned landings on Venus indicate differently. The surface of Venus reaches temperatures of 480° C (900° F), and the Venusian atmosphere is composed mostly of carbon dioxide (95 percent). Only minor amounts of water vapor and nitrogen have been detected. Its atmosphere contains an opaque cloud deck about 25 kilometers thick which begins about 70 kilometers from the surface. Although the unmanned Soviet *Venera 8* survived less than an hour on the Venusian surface, it determined that the atmospheric pressure on this planet is 90 times greater than that found on the earth's surface and is equivalent to the pressure at an ocean depth of 1000 meters (3000 feet). This hostile environment makes it unlikely that life as we know it exists on Venus and makes manned space flights to Venus improbable in the foreseeable future.

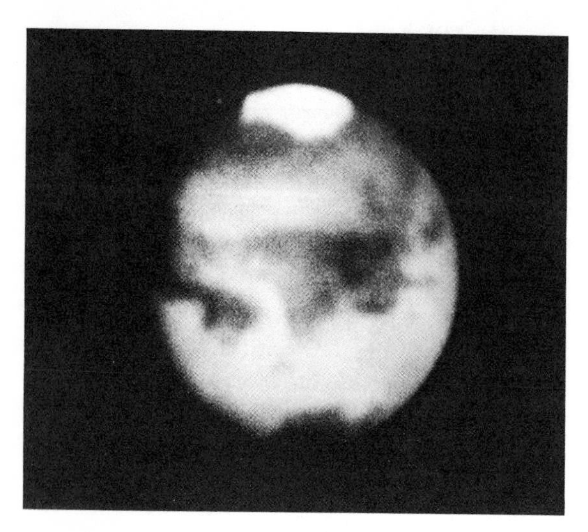

Figure 19.10 *Telescopic view of Mars showing its brilliant polar cap and dark markings. (Courtesy of Hale Observatories)*

Later *Venera* missions transmitted photographs before they succumbed to the tremendous heat and pressure. The landing site of *Venera 9* showed numerous angular rocks, a setting similar to what the Viking lander found on Mars. The *Venera 10* photographed rather flat, rounded rocks, possibly the result of longer exposure to erosion than those found by *Venera 9*. Soil measurements taken by the *Venera* landers indicated a composition similar to the basalts found on the earth, the moon, and Mars.

Because Venus is nearer the sun than the earth is, we would expect it to have a somewhat higher surface temperature. But, it is much higher than astronomers had predicted. This discrepancy may be explained by a phenomenon known as the **greenhouse effect** (see Chapter 12). This effect can be experienced on a warm, sunny day by anyone entering a greenhouse or a car with its windows rolled up. The dense Venusian atmosphere acts like the glass in a greenhouse, allowing visible solar energy to penetrate to the surface. It should be noted that as a planet absorbs energy, it must also lose energy, or its temperature would continually rise. It does so by reradiating heat energy to space. However, the carbon dioxide in the Venusian atmosphere, like the glass in the greenhouse, is rather opaque to the outgoing radiation. Thus, the temperature rises because the heat is temporarily trapped by the dense Venusian atmosphere.

Mars: The Red Planet

Mars has evoked greater interest than any other planet, for astronomers as well as for lay persons. When we think of intelligent life on other worlds, "little green Martians" may come to mind first. Interest in Mars stems mainly from this planet's accessibility to observation. All the other planets within telescopic range have their surfaces hidden by clouds, except for Mercury, whose nearness to the sun makes viewing difficult. Through the telescope, Mars appears as a reddish ball interrupted by some permanent dark regions that change in intensity during the Martian year (Figure 19.10). Early observers erroneously thought the dark areas were seas. The most prominent telescopic features of Mars are its brilliant white polar caps.

Mars rotates on its axis once in 24 hours and 37 minutes and is inclined at an angle of 24 degrees to the plane of its orbit. Since this closely approximates the earth's inclination, Mars has seasons much like the earth does. However, because of its greater distance from the sun, its seasons are cooler and twice as long as those on earth. Mars is nearest the sun (perihelion position) when its southern hemisphere is experiencing summer and farthest from the sun (aphelion position) when its southern hemisphere is experiencing winter, making seasons in the southern hemisphere more extreme than those of the northern hemisphere.

From earth-based telescopes, the most obvious seasonal change is the size of its white polar caps. With the onset of winter in a given hemisphere, the polar cap expands, occasionally extending halfway to the equator. As the polar cap diminishes in spring, a noticeable darkening of the gray areas occurs in that hemisphere. On occasion the dark regions have become blue green, but in autumn they redden again, blending somewhat into their surroundings. Prior to the *Mariner* space probes, this change in color had been interpreted as the result of vegetation that flourished in the summer when water was available and became dormant in the winter.

From the photo images that were radioed to earth by the *Mariner* space probes and the data collected by the two *Viking* landers, many of the controversies surrounding the nature of Mars were settled once and for all. However, although many of the old questions were answered, just as many new ones were raised. The pictures of Mars sent back from the first successful fly-by probe, *Mariner 4,* revealed a crater-pocked planet more akin to the moon than to the earth. As it turned out, these images were taken of the highly cratered southern hemisphere which exhibits a topography sharply different from that of the "younger" northern hemisphere. Follow-up missions revealed a dynamic planet which at some time in its history experienced volcanic activity, wind erosion, and crustal movements with associated ancient "marsquakes." Further, *Viking* orbital images provided unmistakable evidence for water erosion.

The Martian atmosphere is only 1 percent as dense as that of the earth, much less than that found at the top of Mt. Everest, and is composed primarily of carbon dioxide with only very small amounts of water vapor. Data confirmed speculations that the permanent polar caps are made of water ice covered by a relatively thin layer of carbon-dioxide ice. As winter nears, we see the growth of the ice cap, which is caused by equatorward deposition of additional carbon-dioxide ice. This finding is compatible with observed temperatures at the polar caps—about −125° C—cold enough for solid carbon dioxide to exist. Although the atmosphere of Mars is very thin, extensive dust storms do occur and may be responsible for the color changes observed from earth-based telescopes. Winds up to 270 kilometers (170 miles) per hour can persist for weeks. Sand dunes like those on the earth have been observed, and many Martian impact craters have flat bottoms because they have been partially filled with dust. Unlike those of the moon, the oldest craters on Mars are completely obscured by deposits of dust. Images made by the *Viking 1* lander reveal a Martian landscape remarkably similar to a rocky desert on the earth (Figure 19.11).

When *Mariner 9,* the first artificial satellite to orbit another planet, reached Mars in 1971, a large dust storm was raging, and only the ice caps were initially observable. It was spring in the southern hemisphere, and that polar cap was receding rapidly. About midsummer, when the rate of evaporation should have been at its greatest, there was little change in the size of the polar cap. A residual cap remained throughout the summer, providing strong evidence that the residual polar cap is made of water ice (Water ice has a much lower rate of evaporation than frozen carbon dioxide.) When the dust cleared, images of the northern hemisphere revealed numerous large volcanoes. The largest, called Olympus Mons, covers an area the size of the state of Ohio and is no less than 23 kilometers (75,000 feet) high, over twice as high as any mountain on the earth (Figure 19.12). These volcanoes most closely resemble the shield volcanoes found on earth, being similar to those which built Hawaii. Their extreme size is thought to be the result of the absence of plate movements on Mars. Therefore, rather than a chain of volcanoes forming as we find in Hawaii, one relatively large cone developed.

Impact craters are notably less abundant in the region where the volcanoes are the most numerous. This indicates that at least some of the volcanic topography formed more recently in that planet's history. Nevertheless, age determinations based on crater densities obtained from *Viking* photographs indicate that most of the Martian surface features are old by earth standards. The highly cratered Martian southern hemisphere is probably similar in age to the comparable Lunar highlands, being 3500–4000 million years old. The discovery of several highly cratered and weathered volcanoes on Mars further indicates that volcanic activity began early and had a rather long history. How-

Figure 19.11 *This spectacular picture of the Martian landscape by the* Viking 1 *lander shows a dune field with features remarkably similar to many seen in the deserts of the earth. The dune crests indicate that recent wind storms were capable of moving sand over the dunes in the direction*

ever, even the relatively fresh-appearing volcanic and lava plains of the northern hemisphere may all be older than 1000 million years. This fact, coupled with the absence of "marsquake" recordings by *Viking* seismographs, points toward a tectonically dead planet. Another surprising find made by *Mariner 9* was large canyons, which by comparison dwarf even the Grand Canyon. One of the largest of these canyons, Valles Marineris, is roughly 6 kilometers deep and 160 kilometers wide in places and winds for almost 5000 kilometers along the Martian equator (Figure 19.13). This vast chasm is thought to have formed by slippage of crustal material along great faults in the crustal layer. In this respect, it would be comparable to the rift valleys of Africa. This canyon also has tributaries which exhibit a dendritic pattern much like that of stream valleys found on the earth (see Chapter 3). In addition to these streamlike tributaries, *Viking* orbiter photographs have revealed landslides, gullies, and other evidence for water erosion on Mars. But how can water sculpture the land when no liquid water exists on the Martian surface? It is believed that large reserves of frozen water exist in the regolith on Mars, but present temperatures are far too cold for it to liquify. This has led to the speculation that heat from volcanic activity or meteoroid impact has caused local floods. Evaporation of the water released by these catastrophic events might generate brief, but torrential downpours. Other streamlike features may be the consequence of the slow melting of subsurface ice which results in the flow of material. These flows would be more akin to features formed by mass-wasting processes, so common on the earth.

Although no acceptable theory exists for the source of water needed to produce these streamlike features, most geologists are convinced that erosion by water is the only plausible explanation. With this in mind, along with the fact that the polar caps alternate in size every 50,000 years because of precession (see Chapter 17), it has been suggested that water moves in a 50,000-year cycle from one pole to the other. The hemisphere experiencing its summer season when Mars is farthest from the sun would contain most of the water ice. But as summers in that hemisphere continually warm, most of the water

from upper left to lower right. The large boulder at the left is about 10 meters from the space craft and measures 1 by 3 meters. (Courtesy of NASA)

would be released and possibly be available for rain and water erosion. Eventually, the other pole, which is experiencing progressively cooler summers, would accumulate most of the water as ice and the cycle would begin anew. This is an interesting piece of speculation, but at this time, it is just that—speculation.

Because of their small size, the two satellites of Mars, Phobos and Deimos, were not discovered until 1877. Phobos is closer to its parent than any other natural satellite (moon). This fact led to speculation that it was an artificial satellite. Only 5500 kilometers from the Martian surface, Phobos requires 7 hours and 39 minutes for one revolution. Deimos, which is smaller and 20,000 kilometers away, revolves in 30 hours and 18 minutes. *Mariner 9* revealed that both satellites are irregular in shape and have numerous impact craters, much like their parent (Figure 19.14). They may be captured asteroids. The maximum diameter of Phobos is 25 kilometers (15 miles), and the maximum diameter of Deimos is only about 15 kilometers (10 miles). One of the most interesting coincidences in astronomy is the close resemblance between

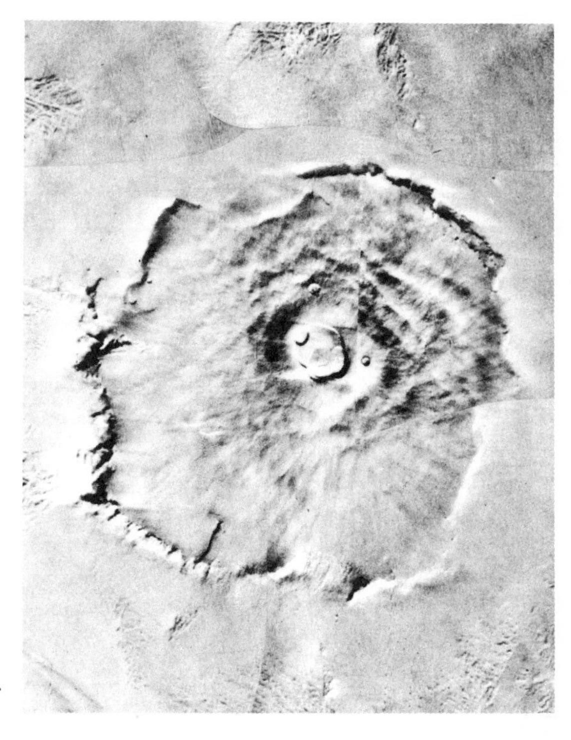

Figure 19.12 *Mons Olympus, a gigantic volcanic mountain on Mars. (Courtesy of NASA)*

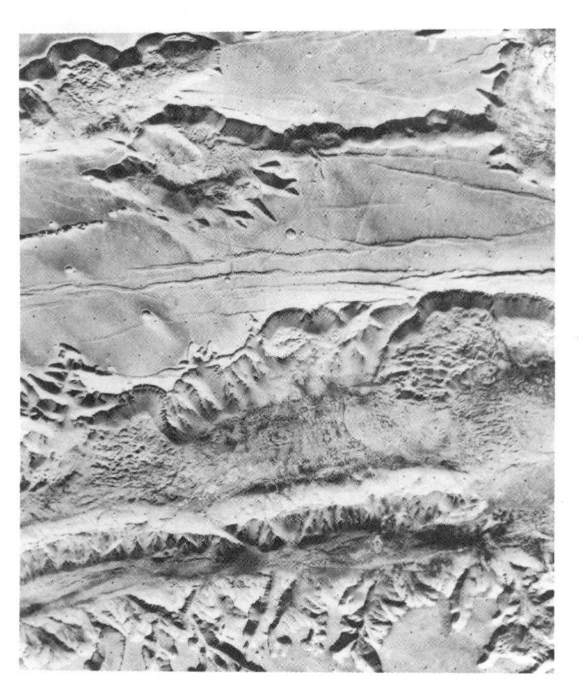

Figure 19.13 *Vast chasm with branching tributaries located 500 kilometers south of the Martian equator. (Courtesy of NASA)*

Phobos and Deimos and the two fictional satellites of Mars described by Jonathan Swift in *Gulliver's Travels,* written about 150 years before they were actually discovered.

Jupiter: The Giant Planet

Jupiter, the largest planet in the solar system, is twice as massive as the remaining planets combined. Despite its large size, however, it is only 1/1000 as massive as the sun. When viewed through a telescope or binoculars, its disk appears covered with alternate bands of various-colored clouds that run parallel to its equator (Figure 19.15). You can even observe the effects of Jupiter's rapid rotation: the equatorial region appears slightly bulged, while the polar dimension looks flattened. But the most striking feature is its great red spot (Figure 19.15). Several times larger than the earth, the red spot rotates with the planet, although sometimes faster and other times slower. It does, however, remain the same distance from the Jovian equator, although it changes intensity. On one occasion, at least, it became a brick red. The cause of the spot has been attributed to everything from volcanic activity to a large cyclonic-type storm. Images of Jupiter obtained by *Pioneer 11* as it moved to within 42,000 kilometers of its cloud tops in December of 1974 support the latter view. It appears that the great red spot is a counterclockwise rotating storm caught between two jet-streamlike bands of atmosphere flowing in opposite directions.

Jupiter's atmosphere is composed mostly of hydrogen and helium, with methane (CH_4) and ammonia (NH_3) as minor constituents. Its opaque cloud cover prohibits viewing the surface. These clouds are probably made of tiny frozen ammonia crystals afloat in the atmosphere. Atmo-

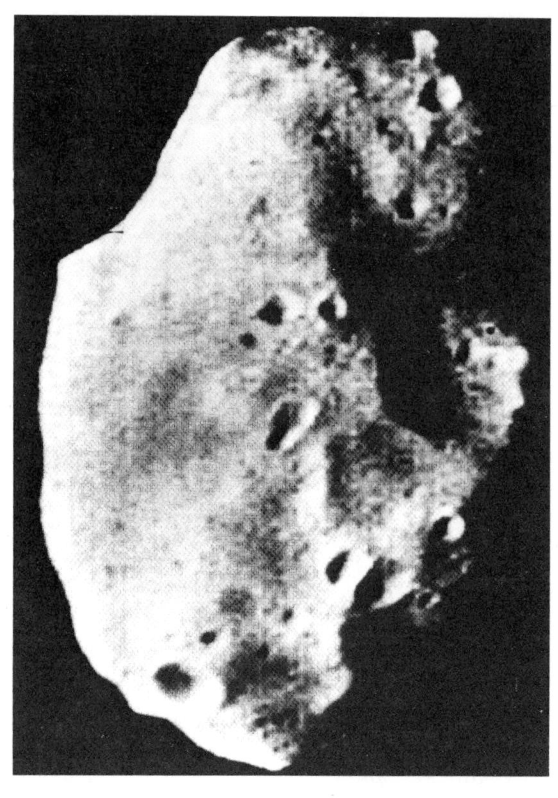

Figure 19.14 *Phobos, the larger of the two Martian satellites, photographed by* Mariner 9 *from 5500 kilometers. (Courtesy of NASA)*

Figure 19.15 *Jupiter, with its red spot and largest satellite, Ganymede, casting its shadow on Jupiter's surface. (Courtesy of Hale Observatories)*

spheric pressure at the top of the clouds is equal to sea level pressure on earth, but because of Jupiter's immense gravity, the pressure increases rapidly toward its surface. At 160 kilometers (100 miles) below the clouds, the pressure is great enough to liquify hydrogen. Consequently, the surface of Jupiter may be a gigantic ocean of liquid hydrogen. If the atmosphere of Jupiter is only 160 kilometers (100 miles) thick, as proposed, it makes up a rather small percentage of the total planet, which has a diameter of 140,000 kilometers. One model of the planet indicates that the ocean of hydrogen would be no more than 320 kilometers deep, at which depth the solid surface begins. However, other studies have shown that Jupiter may not have any truly solid part, but is rather a slushy mass of icy ammonia and liquid hydrogen. Jupiter is also believed to contain as much rocky and metallic material as found in the terrestrial planets, like the earth. Whether this "earthy" material makes up a central core or is scattered as small bits remains unknown, but the former idea has gained popularity. Because of Jupiter's dense atmosphere and its extreme pressure, it is difficult to imagine direct observation of its surface. Exploration of Jupiter in the near future will probably be conducted from one of its larger satellites.

Besides being the largest planet, Jupiter has the distinction of having the largest satellite system. The four largest of its thirteen known moons were all discovered by Galileo and travel in nearly circular orbits around the parent with periods of from two to seventeen days. The largest, Ganymede, is slightly bigger than Mercury, and the others approximate our moon in size. These Galilean moons can be observed with a small telescope and are interesting in their own right. Because their orbits are in the plane of the Jovian equator, they transit the parent, casting a shadow seen as a dark dot against its disk (Figure 19.15). On the other hand, the four outermost satellites are very small (20 kilometers in diameter) and have retrograde motion, and their orbits are inclined to the Jovian equator. It has been suggested that these satellites are asteroids that the giant Jupiter has captured.

Saturn: The Ringed Planet

Requiring 29½ years to make one revolution, Saturn is almost twice as far from the sun as Jupiter, yet its atmosphere, composition, and internal structure are thought to be very similar. The most spectacular feature of Saturn is its system of rings (see the chapter-opening photo). These rings were discovered by Galileo and appeared to him as two smaller bodies adjacent to the planet because he could not resolve them with his primitive telescope. Their ring nature was revealed 50 years later by the Dutch astronomer Christian Huygens. This system consists of three concentric rings. The inside ring is only 12,000 kilometers from the surface of Saturn and is very dim when viewed from earth. The middle, and brightest, ring is separated from the others by a gap on both sides. The **Cassini gap** is easily seen in a photo of Saturn (see chapter-opening photo). It is wide enough that the moon could fit without touching either ring.

From the earth, the ring system of Saturn can be viewed on edge once every 15 years, when it appears as an extremely fine line. Estimates of its thickness range from a few meters to a few kilometers. Thus it can be described as paper thin compared to its area. The rings are not solid, for occasionally stars can be seen through them. Studies of their rotational behavior also confirm this view, because if the rings were solid, their outermost edge would have the highest rotational speed. Measurement, however, indicates that the opposite is true. That portion of the ring closest to Saturn has the highest orbital velocity. The rings, no doubt, are made up of individual moonlets that orbit the planet in accordance with Kepler's laws. These moonlets are believed to be quite tiny and composed of ice (possibly frozen ammonia) and ice-covered dust. Recently, radar has been reflected from the rings, indicating that some of the particles must be a few meters in diameter, much larger than had originally been speculated.

The origin of the rings is undoubtedly related to their distance from the surface of Saturn. Being very close, the gravitational force of the parent will prevent two moonlets from joining if they happen to collide. In fact, if one of Saturn's larger satellites entered nearer the planet than the edge of the outer ring, it would be destroyed. The gravitational force of Saturn pulling on the near side of this satellite would be enough greater than the force pulling on its far side that it would pull the satellite apart. Although the rings could be the remains of a satellite destroyed in this fashion, these moonlets are most likely part of the original material from which Saturn was formed. Because of their close proximity to the planet, they were unable to grow into a "normal" satellite. Saturn has ten known satellites, and all of them lie beyond the outer edge of the last ring. The largest, Titan, is the only satellite in the solar system definitely known to have an atmosphere.

Uranus: The Green Planet

The honor of discovering the first planet that was unknown to the ancients belongs to William Herschel, the great German-born English astronomer. Herschel, a musician by training, originally made telescopes and studied astronomy as hobbies. On March 13, 1781, using a 7-inch reflecting telescope of his own making, he came upon a dim "star," which appeared as a small disk. Believing it was a comet, he recorded its position. After several months, a calculation of its orbit revealed that it was definitely a planet located beyond Saturn. For this discovery, King George III awarded Herschel a stipend which allowed him to pursue his interest in astronomy full-time. In response, Herschel named the planet "Georgium Sidus" (George's Star). In keeping with tradition, however, the name Uranus, a mythological Greek god and the grandfather of Jupiter, became universally adopted. Later, a review made of earlier records indicated that Uranus had been observed and recorded as a star on many occasions prior to Herschel's discovery.

Because of its great distance from the earth and rather featureless pale green disk, little is known about Uranus. Hydrogen and methane have been detected in its atmosphere, but ammonia, which is found in the atmospheres of Saturn and Jupiter, is absent. At the low temperatures of Uranus, which are never much above −170° C (−275° F), ammonia is undoubtedly frozen out. It has been proposed that Uranus is composed mostly of "ice" (frozen ammonia and methane), in which case it may closely resemble a comet in structure (but definitely not in size).

The unique feature of Uranus is that its axis of rotation lies only 8 degrees from the plane of its orbit (Figure 19.16). Its rotational motion, therefore, has the appearance of rolling, rather than spinning like a top as do the other planets. Because Uranus is inclined almost 90 degrees, the sun is nearly overhead at one of the poles once each revolution, and then half a revolution later, it is nearly overhead at the other.

A surprise discovery in 1977 revealed that Uranus is surrounded by rings, like those of Saturn. This find occurred as Uranus passed in front of a distant star and blocked its view, a process called **occulation.** Observers saw the star

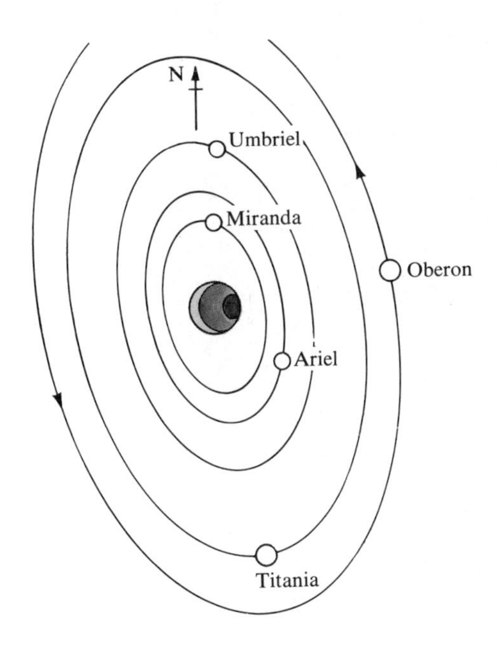

Figure 19.16 *Orientation of Uranus and its satellites. Note the "north" arrow. (After Wyatt and Kaler,* Principles of Astonomy, *Allyn, 1974, p. 183)*

"wink" briefly five times before the primary occulation and again five times afterward. Uranus, thus, has at least five distinct belts of moonlets orbiting its equatorial region.

Neptune: A Twin

While the discovery of Uranus was an accidental sighting by a very alert observer, the discovery of Neptune was truly a triumph for modern astronomy. A short while after an orbit had been calculated for Uranus, its motion began to show irregularities. Prior to 1822, it moved slightly ahead of its predicted position, and after 1822, it began lagging behind. Could it be that Newton's law of universal gravitation was not applicable for such great distances? Some people thought so.

A young Englishman, John Adams, approached the problem by considering the gravitational effect of a yet unknown planet. If this planet lay beyond Uranus, it would, in accordance with Kepler's laws, travel at a slower velocity. Consequently, Uranus would approach and pass this unknown planet during a revolution and be influenced by its gravity. When Uranus approached the unknown planet, its orbital speed would increase slightly, and as it receded from the unknown planet, its orbital speed would be reduced slightly. The amount of influence by the unknown planet would be dependent on its mass and its distance to Uranus. Adams made tedious calculations using different sized planets at various distances until he found a combination that worked. In October of 1845 he sent the location he had predicted for the unknown planet to the observatory at Greenwich, England. Undoubtedly, the Astronomer Royal had little faith in Adam's prediction, since a thorough search was never conducted. At this same time, a Frenchman, Urbain Leverrier, unaware of Adam's work, was tackling the same problem in a similar manner. Leverrier sent his data to the Berlin Observatory, and during the first night of observation, a likely candidate for the unknown planet was located. Since Adams and Leverrier independently predicted the position of Neptune, they are both credited with its discovery.

If any two planets in the solar system can be considered twins, Neptune and Uranus can. Besides being similar in size, they appear a pale green color, attributable to the methane in their atmospheres. Their structure and composition are believed to be similar, too, but because of its greater distance, Neptune experiences somewhat colder temperatures.

Pluto: Planet X

After the discovery of Neptune, there still remained discrepancies in the orbit of Uranus. This led some astronomers to suggest that another planet might lie beyond the orbit of Neptune. A vigorous search for this planet was begun in 1906 by Percival Lowell at the observatory in Flagstaff, Arizona, that now bears his

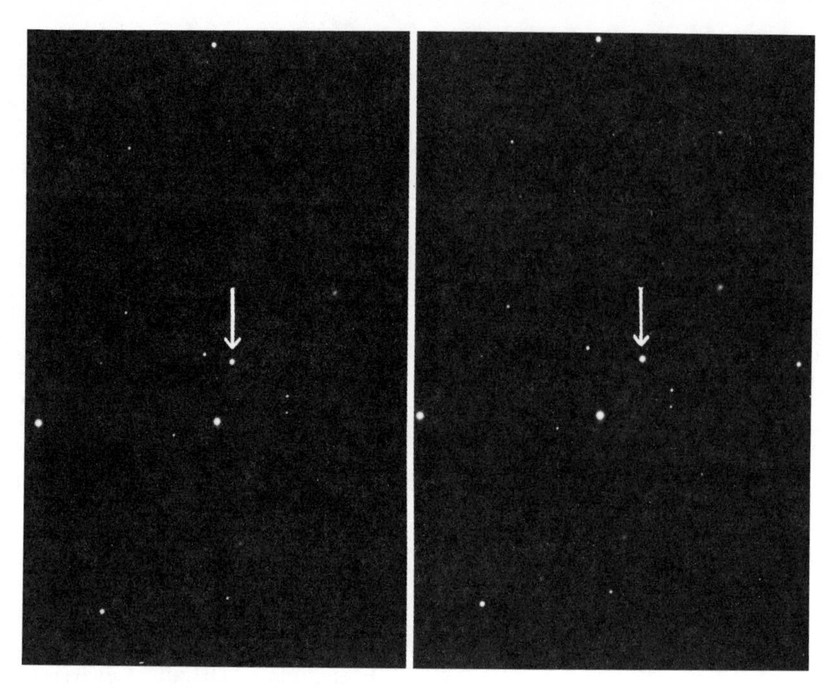

Figure 19.17 *Pluto appears starlike on a photograph but is revealed by its motion. (Courtesy of Hale Observatories)*

name. He unsuccessfully searched for the object he called "Planet X" until his death ten years later. Some years afterward, a new photographic telescope was installed at the Lowell Observatory and the search was resumed full scale. Two photographs of the same portion of the sky were taken at intervals of a week or so and then compared. In such short time intervals, stars appear stationary, but a planet shifts its position relative to its background. Twenty-five years had lapsed since Lowell's death, and 2 million stars were examined before Clyde Tombaugh discovered a body which had a shift that was about right for an object 1 billion miles beyond the orbit of Neptune (Figure 19.17). This newly discovered planet was named Pluto, for the Greek god of the underworld, because the first two letters were Percival Lowell's initials. Although the planet was located in the constellation Gemini as Lowell had calculated, its mass was much less than he had estimated. It seems unlikely that a planet of Pluto's size could have altered the orbit of Uranus measurably. Most probably, the discovery of Pluto can be credited to an extensive search of the correct portion of the sky, although for the wrong reasons.

Pluto lies on the fringe of the solar system almost 40 times farther from the sun than the earth. Because of this great distance and Pluto's slow orbital speed, it takes Pluto 248 years to orbit the sun. Since its discovery in 1930, it has completed less than ⅕ of a revolution. Pluto's orbit is noticeably elongated (highly eccentric), causing it to travel inside the orbit of Neptune. There is little likelihood that Pluto and Neptune will ever collide, because their orbits are inclined to each other and do not actually cross. Pluto's unusual orbit and small size suggest to some astronomers that it might originally have been a satellite of Neptune.

In June of 1978 a moon was discovered orbiting Pluto. Although this satellite is too small to be observed visually, it appears as an elongated bulge on photographic plates taken of Pluto. The satellite is about 20,000 kilometers from the parent planet, which is 20 times closer than our moon. This discovery greatly altered early estimations of the size of Pluto. Current data suggest that Pluto has a diameter of less than 3000 kilometers, less than that of our moon. Evidence from other studies indicates that its density is slightly greater than that of water. Consequently, Pluto might best be described as a large, dirty iceball made up of a mixture of frozen gasses and rock.

Other than its orbital and rotational motions, little else is known about Pluto. It is 10,000 times too dim to be visible with the unaided eye. We do know that Pluto fluctuates slightly in brightness with a period of 6.4 days. This variation is believed to be the result of its rotation, which causes regions of its surface that have different reflectibility to periodically face the earth. The average temperature of Pluto is estimated, considering its solar distance, at −210° C (−350° F), which is cold enough to solidify any gas that may be present. Pluto could not have an atmosphere.

After the discovery of Pluto, Tombaugh started a systematic search of the heavens for a tenth planet. His inability to find one makes it unlikely that any undiscovered large planets orbit the sun near the plane of the ecliptic. However, a planet with a very inclined orbit may exist. In fact, one has been predicted.

REVIEW

1. Compare the rotational periods of the terrestrial and Jovian planets.
2. How fast does a location on the equator of Jupiter rotate? Note that Jupiter's circumference equals 416,000 kilometers (260,000 miles).
3. By what criteria are the planets placed into either the Jovian or terrestrial group?
4. What evidence indicates that Saturn's rings are composed of individual moonlets rather than consisting of solid discs?
5. What do you think would happen if the earth passed through the tail of a comet?
6. What fact led to the search for asteroids? Why were they difficult to find?
7. Why has Mars been the planet most studied telescopically?
8. At one time it was suggested that the two "moons" of Mars were artificial. What characteristics do they have that would cause such speculation?
9. Describe the origin of comets according to the most widely accepted theory.
10. Compare a meteoroid, meteor, and meteorite.
11. What surface features does Mars have that are also common on earth?
12. Although Mars has valleys that appear to be the products of stream erosion, what fact makes it somewhat unlikely that they were eroded by water?
13. Examine the orbits of the satellites of Uranus in Figure 19.16. Do their orbits indicate that they were formed at the same time as Uranus or that they were captured at a later date?
14. Why are the four outer satellites of Jupiter thought to have been captured?
15. If distance were no object, which planet might be the most difficult for an *Apollo*-type expedition?
16. What are the three types of materials thought to make up the planets? How are they different? How does their distribution account for the density differences between the terrestrial and Jovian planetary groups?
17. Why are meteorite craters more common on the moon than on the earth, even though the moon is much smaller?

KEY TERMS

terrestrial planets
Jovian planets
escape velocity
asteroids (minor planets)
comets
coma
meteoroid
micrometeorite
meteor shower
meteorite
irons
stony
stony-irons
greenhouse effect
Cassini gap
occulation

Orion nebula. (Courtesy of Lick Observatory)

20

Properties of Stars
Stellar Classification
Hertzsprung-Russell Diagram
Unusual Stars
Stellar Evolution
Star Clusters
Interstellar Matter
The Milky Way
Galaxies
The Big Bang

Beyond Our Solar System

The star Proxima Centauri is about 4.3 light-years away, roughly 100 million times farther than the moon, yet next to our sun it is the closest star. The universe this fact suggests is incomprehensibly large. What is the nature of this vast cosmos beyond our solar system? Are the stars spread out at random, or are they organized into distinct clusters? Do stars move, or are they permanently fixed features, like lights strung out against the black cloak of outer space? Does the universe extend infinitely in all directions, or is it bounded? To consider these questions, this chapter will examine the universe by taking a census of the stars, which are the most numerous objects in the night sky.

Although the sun is the only star whose surface can be observed, a great deal is known about the universe beyond our solar system. In fact, more is known about the stars than about our outermost planet, Pluto. This knowledge hinges on the fact that stars, and even interstellar gasses, radiate energy in all directions into space. The only trick is to collect this radiation and unravel the secrets it holds. Astronomers have devised many ingenious methods to do just that. We will begin our discussion of the universe by examining some intrinsic properties of individual stars, including luminosity, temperature, size, color, and mass.

Properties of Stars

Distances to the Stars

As stated in Chapter 17, the annual back-and-forth shift of nearby stars with respect to very distant stars was considered proof of the earth's orbital motion. The measurement of this shift, called **stellar parallax,** is the only direct method of determining distances to stars. Figure 20.1 illustrates these shifts and the parallax angles determined from them. The nearest stars have the greatest parallax, while that of distant stars is imperceptible. Because the distance to even the nearest stars is thousands of times greater than the earth-sun distance, the triangles that astronomers work with are very "long and narrow," making the angles that are measured very small. The parallax angle of the nearest star, Proxima Centauri, is less than 1 second of arc, which equals 1/3600 of a degree. It should be apparent why Tycho Brahe was unable to observe stellar parallax without a telescope and why he therefore concluded that the earth did not revolve.

Astronomers generally express stellar distances in units called **parsecs.** One parsec is the distance at which a star would have a parallax of 1 second of arc. Simply stated, the distance in parsecs is equal to 1 divided by the parallax angle. For example, a star with a parallax of 0.5 second of arc is 1/0.5, or 2, parsecs away, and one with a parallax of 0.1 is 1/0.1, or 10, parsecs away. Another unit popularly used to express stellar distance is the **light-year,** which is the distance light travels in a year. For purposes of conversion, 1 parsec equals 3.26 light-years. The distances to some nearby stars are given in all three units in Table 20.1.

Stars more distant than 50 parsecs have such small shifts that accurate measurement is not

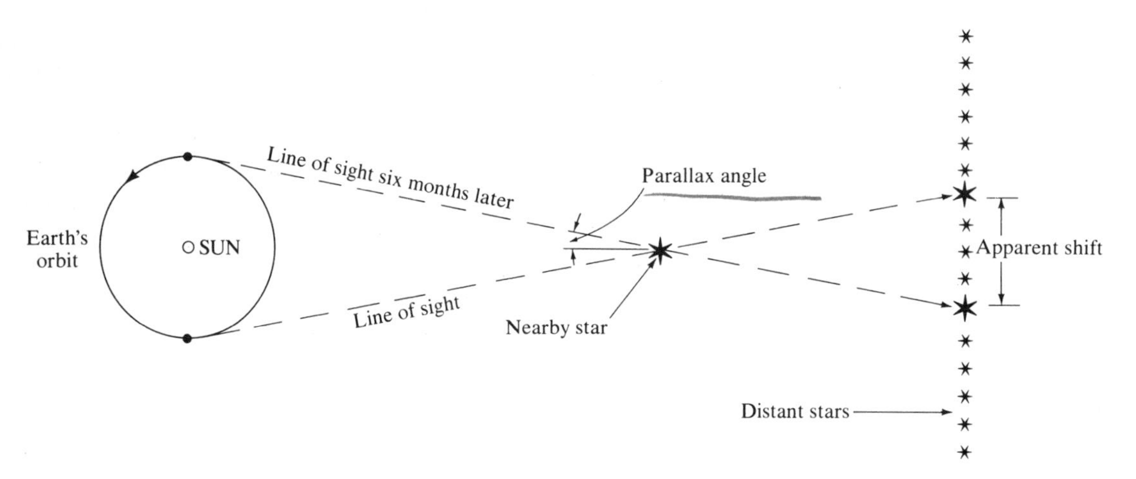

Figure 20.1 *Geometry of stellar parallax.*

Table 20.1 ***Distance to Some of the Nearest Stars***

Name	Parsecs	Parallax Angle	Light-years
Proxima Centauri	1.31	0.763	4.27
Alpha Centauri	1.33	0.752	4.34
Barnard's Star	1.83	0.545	5.97
Sirius	2.67	0.377	8.70
61 Cygni	3.42	0.292	11.15
Kapteyn's Star	3.98	0.251	12.97

possible. Fortunately, a few indirect methods exist so that distances to these stars can be estimated; however, it is important to check the accuracy of these methods on stars whose parallax is known.

In principle, the method used to measure stellar distances is rather elementary and was known to the ancient Greeks. In practice, the small angles involved and the fact that the sun, as well as the designated star, has actual motion greatly complicate the measurements. The first accurate stellar parallax was not determined until 1838, and today, parallaxes for only a few thousand of the nearest stars are known with certainty.

Stellar Magnitudes

Stars have been classified according to their brightness since the second century B.C., when Hipparchus placed about 1000 of them into six categories. The measure of a star's brightness is called its **magnitude.** Some stars may appear dimmer than others only because they are farther away. Therefore, a star's brightness, as it appears when viewed from the earth, has been termed its **apparent magnitude.** When numbers are employed to designate relative brightness, *the larger the magnitude number, the dimmer the star.* Stars that appear the brightest are of the first magnitude, while the faintest stars visible to the naked eye are of the sixth magnitude. With the advent of the telescope, many stars fainter than the sixth magnitude were discovered.

In the middle of the nineteenth century, a method was developed whereby magnitudes could be used to quantitatively compare the brilliance of stars. So, just as we can compare the brightness of a 50-watt light bulb to that of a 100-watt bulb, we can compare the brightness of stars having different magnitudes. It was determined that a first-magnitude star was about 100 times brighter than a sixth-magnitude star, so on the scale adopted, two stars which differ by 5 magnitudes have a ratio in brightness of 100 to 1. Hence, a seventh-magnitude star is 100 times brighter than a twelfth-magnitude star. It follows, then, that the brightness ratio of two stars differing by one magnitude is the fifth root of 100, or 2.512. Thus, a star of the first magnitude is 2.512 times brighter than a star of the second magnitude and $(2.512)^4$, or 40, times brighter than a star of the fifth magnitude (Table 20.2). Because some stars are brighter than the so-called first-magnitude stars, negative magnitudes were introduced. The brightest star in the night sky, Sirius, has an apparent magnitude of −1.4, about 10 times brighter than a first-magnitude star. On this scale, the sun has an apparent magnitude of −26.7, and when brightest, Venus has a magnitude of −4.3. At the other end of the spectrum, the 200-inch Hale telescope can view stars with an apparent magnitude of +23, which are approximately 100 million times dimmer than the average stars visible with the unaided eye.

Table 20.2 ***Ratios of Stellar Luminosity***

Difference in Magnitude	Brightness Ratio
0.5	1.6:1
1	2.5:1
2	6.3:1
3	16:1
4	40:1
5	100:1
10	10,000:1
20	100,000,000:1

Astronomers are also interested in the "true" brightness of stars, called their **absolute magnitudes.** Stars of the same luminosity or brightness usually do not have the same apparent magnitude, because their distances from us are not equal. In order to compare their true or intrinsic brightness, astronomers determine what magnitude the star would have if it were at a standard distance of 10 parsecs, or about 32.6 light-years. For example, the sun, which has an apparent magnitude of −26.7, would have a magnitude of 4.9 at the standard distance and would be one of the fainter stars visible in the night sky. Stars with absolute magnitudes greater than 4.9 (smaller numerical value) are intrinsically brighter than the sun, but because of their distance, they appear much dimmer. Table 20.3 lists the absolute and apparent magnitudes of some stars as well as their distances from the earth. Most stars have an absolute magnitude between −5 and +15, which puts the sun near the middle of this range.

For a star too distant for parallax measurements, knowing its absolute and apparent brightness provides astronomers with a tool for determining its distance. The apparent magnitude can be determined relatively easily with a photometer attached to a telescope. If we also know a star's true brightness, we can determine just how far away that star would have to be in order to have the apparent brightness we observe. You use this same principle when you drive at night and estimate the distance to an oncoming car simply from the brightness of its headlights. You know the true brightness of an automobile headlight, but how do astronomers determine the intrinsic brightness of a star? Fortunately, some stars have characteristics that provide the necessary data. One important group of these is called **cepheid variables.** These are pulsating stars that get brighter and fainter in a rhythmic fashion. The interval between two successive occurrences of maximum brightness of a pulsating variable is called its **light period.** In general, the longer the light period of a cepheid, the greater is its absolute magnitude (Figure 20.2). So, by determining the light period of a cepheid, its absolute magnitude is known. When this absolute magnitude is compared to the apparent magnitude, a good approximation of its distance can be made. Consequently, cepheid variables are truly informative stars.

Table 20.3 ***Distance, Absolute and Apparent Magnitudes of Some Stars***

Name	Distance (parsecs)	Apparent Magnitude	Absolute Magnitude
Sun		−26.7	4.9
Alpha Centauri A	1.3	0.0	4.4
Sirius A	2.7	−1.4	1.5
Epsilon Indi	3.4	4.7	7.0
Kruger 60B	3.9	11.2	13.2
Arcturus	11.0	−0.1	−0.3
Antares	120.0	1.0	−4.5
Betelgeuse	150.0	0.8	−5.5
Deneb	430.0	1.3	−6.9

Color Indexes

The next time you are out on a clear night, take a good look at the stars and note their color. Some that are quite colorful can be found in the constellation Orion (see Appendix E). One of the two brightest stars in Orion, Betelgeuse (α Orionis), is definitely red, while the other, Rigel (ß Orionis), appears blue.

Very hot stars with surface temperatures above 30,000 K emit most of their energy in the form of short-wavelength light and therefore appear blue. Red stars, on the other hand, are much cooler, generally less than 3000 K, and most of their energy is emitted as longer-wavelength red light. Stars with temperatures between 5000 and 6000 K appear yellow like the sun. Some stars are so cool that they radiate mostly in the infrared and are not visible. However, these can sometimes be detected by using infrared-sensitive photography. Since a star's color is primarily a factor of its temperature, it does more for the astronomer than just decorate the sky.

In order to obtain accurate stellar temperatures using color, astronomers have established a quantity called the **color index.** A method

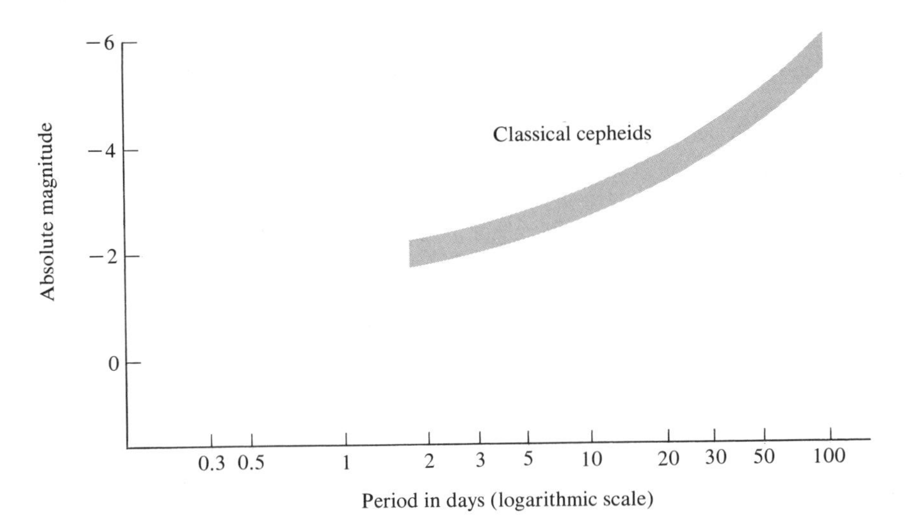

Figure 20.2 *Relationship between the light period and absolute magnitude of pulsating variable stars (cepheids).*

commonly employed to determine a star's color index uses two photometers, one sensitive to blue light, and the other sensitive to visual light. Visual light refers to the light to which our eyes are most sensitive. This light is often called yellow, since it is the light emitted by the sun, which is a yellow star. Using this system, a star that emits more of its light in blue than in yellow will have a smaller blue magnitude (seem brighter) and a larger visual magnitude (seem fainter). On the other hand, a red star emits more visible light than blue light. Hence, its visual magnitude will be smaller. The color index is then determined by subtracting the visual magnitude (V) from the blue magnitude (B); usually written as B − V. For example, Antares, in the constellation Scorpio, is a red star that has a magnitude of 2.7 in blue and 0.9 in visual, so its B − V index is 2.7 − 0.9, or +1.8. A blue star would have a negative color index, since its B value would be smaller (brighter). If a star has the same magnitude in both systems, it is a white star with a surface temperature of about 10,000 K and a color index of 0. The range of the color index is from −0.6 for the hottest blue stars to +2 for the coolest red stars. The sun is intermediate in color (yellow) and has an index of +0.6. In summary, *for most stars, their color index indicates their surface themperature.*

Stellar Classification

The vast majority of stars have continuous spectra on which a number of dark absorption lines are superimposed. Recall from our discussion of light that a given set of dark lines can be attributed to the presence of a particular element. Early studies of stellar spectra revealed that an element like calcium or helium produces strong absorption lines in the spectrum of some stars, but only weak lines in others, and no lines in still others. After the spectra of numerous stars were obtained, it was inevitable that someone would try to group stars according to their spectra. The first classification scheme was based on the relative strengths of selected absorption lines, and thus stars with similar-appearing spectra were placed in the same **spectral class.** This scheme assumed that stars with an abundance of a particular element, for instance calcium, would produce strong calcium lines. Elements for which no absorption lines appeared were considered absent or rare.

Later, it was learned that the spectral variations observed were primarily the result of temperature, not compositional, differences. For example, very cool red stars (class M) have strong absorption lines of molecular titanium oxide (TiO). However, in the spectrum of stars whose temperatures are above 5000 K, these lines do not appear. At these temperatures, the titanium oxide molecules are unstable and are split into atomic titanium and atomic oxygen. Hence, hotter stars probably contain as much titanium as the cooler stars, but the absorption lines of the titanium oxide molecule cannot be produced. Similarly, *the presence of most absorption lines was found to be dependent on the star's surface temperature, not its composition.*

The original spectral classes which had been labeled with letters of the alphabet were rearranged in order of decreasing temperatures (Table 20.4). This order is O, B, A, F, G, K, M. The O-type stars are hot, blue stars with temperatures up to 50,000 K, and the coolest red stars (temperatures less than 3500 K) belong to the M spectrum class. The sun is a yellow G-type star, and the somewhat cooler K-type stars appear orange. By now it should be obvious that *once a star's spectrum has been classified, its surface temperature has also been established.* As we

Table 20.4 ***Characteristics of Spectral Classes***

Spectral Class	Color	Approximate Temperature (K)	Principal Spectral Characteristics	Approximate Color Index (B − V)	Approximate Mass of Main-sequence Star (Sun = 1)	Stellar Example
O	Blue	>30,000	Strong lines of ionized helium; weak hydrogen lines	−0.6	40.0	10 Lacertae
B	Blue white	10,500–30,000	Stronger hydrogen lines than O type; lines of neutral helium dominate	−0.3	10.0	Rigel
A	White	7500–10,500	Hydrogen lines reach maximum strength; lines of ionized metals strong	0.0	3.0	Vega
F	Yellow white	6000–7500	Hydrogen lines weaken; lines of neutral metals strengthen; lines of ionized metals weaken	+0.3	1.5	Canopus
G	Yellow	5000–6000	Strong calcium lines; strong lines of neutral and ionized metals; weak hydrogen lines	+0.6	1.0	Sun
K	Orange	3500–5000	Lines of neutral metals dominate. Hydrogen lines weak but detectable	+1.4	0.8	Arcturus
M	Red	< 3500	Numerous lines of neutral metals; band of TiO molecules	+2.0	0.2	Antares

shall see, a star's temperature is very useful in determining its size.

Hertzsprung-Russell Diagram

Early in this century, two men, Einar Hertzsprung and Henry Russell, independently studied the relationship between a star's absolute magnitude (luminosity) and its spectral class (temperature). When nearby stars are plotted on a graph according to these two intrinsic properties as shown in Figure 20.3, they are not uniformly spread. Most of the stars, as you can readily see, fall along a band that runs from the upper-left corner to the lower-right corner, called the **main sequence.** These plots, which became known as **Hertzsprung-Russell diagrams,** or simply **H-R diagrams,** show that *the hottest main-sequence stars are intrinsically the brightest.* Although discovered independently, the luminosity of the main-sequence stars is also related to their mass. The hottest O-type stars are about 50 times more massive than the sun, while the coolest M-type stars are only 1/10 as massive. Therefore, *on the H-R diagram, the main-sequence stars appear in a decreasing order, from hotter, more massive blue stars to cooler, less massive red stars.* The position of the sun is shown by the *X*. The sun is a G-type, main-sequence star with an absolute magnitude of about +5. Because the magnitudes of a vast majority of main sequence stars lie between −5 and +15, and since the sun falls midway in this range, the sun is often considered an average star. However, more main-sequence stars are cooler and less massive than the sun.

Above and to the right of the main sequence lies a group of very luminous stars called **giants** or, on the basis of their color, **red giants.** The size of these giants can be approximated by comparing them with stars of known size that belong to the same spectral class and, therefore, have the same surface temperature. According to the Stefan-Boltzmann law (Chapter 18), stars having equal surface temperatures radiate the same amount of energy per unit area. Therefore, any difference in the luminosity (absolute magnitude)

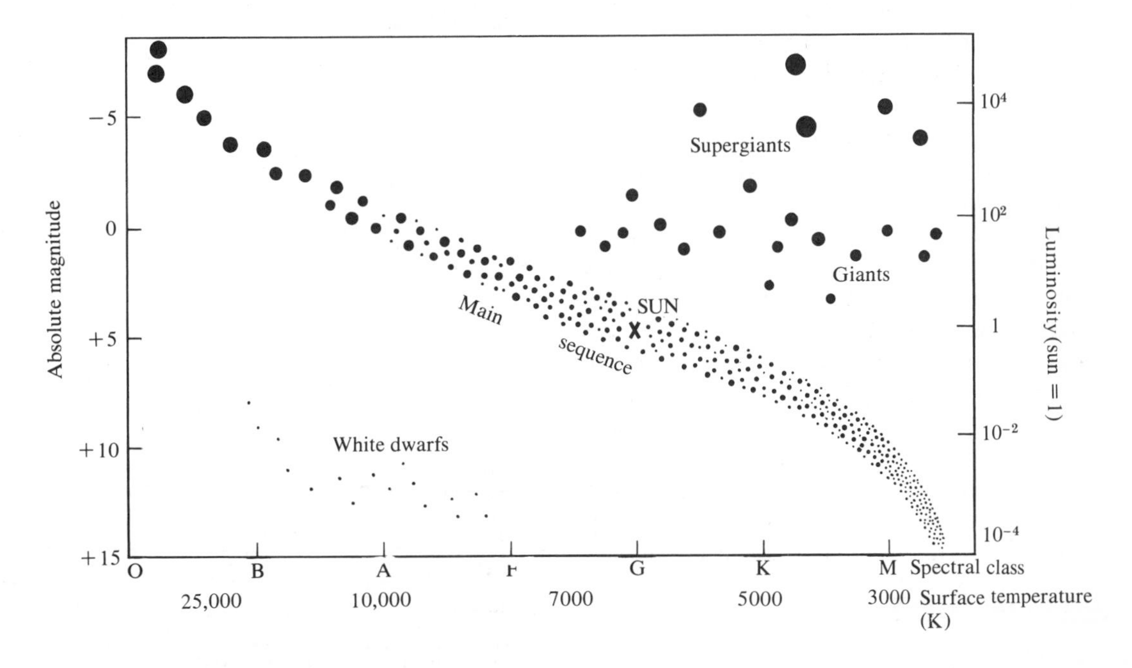

Figure 20.3 *Idealized Hertzsprung-Russell diagram, on which stars are plotted according to temperature and absolute magnitude.*

of stars belonging to the same spectral class is attributable to their relative sizes. As an example, let us compare the sun, which has an absolute magnitude of about +5, with another, more luminous G-type star that has an absolute magnitude of 0. Since their magnitudes differ by 5, their brightness ratio is 100 to 1. Because both stars radiate the same amount of energy per unit area, it follows that in order for the more luminous star to be 100 times brighter than the sun, it must have about 100 times more surface area. Note the position of a G-type star with a magnitude of 0 on the H-R diagram in Figure 20.3. It should be clear why stars whose plots fall in this portion of the graph are called giants.

Some stars are so large that they are called **supergiants.** Betelgeuse, a bright-red supergiant in the constellation Orion has a radius about 800 times that of the sun. If this voluminous star were at the center of the solar system, it would extend beyond the orbit of Mars, and the earth would find itself inside the star.

In the lower-left portion of the H-R diagram, the opposite situation occurs. These stars are much fainter than main-sequence stars of the same temperature, and by using the same reasoning as before, they must be much smaller. Some probably approximate the earth in size. This group has come to be called **white dwarfs,** although not all are white in color.

Soon after the first H-R diagrams were developed, astronomers realized their importance in interpreting stellar evolution. Just like living things, a star is born, ages, and dies. Because almost 90 percent of the stars lie on the main sequence, we can be relatively certain that stars spend most of their years as main-sequence stars. Only a few percent are giants, and perhaps 10 percent are white dwarfs. After considering some "unusual" stars and some star groups, we will return to this topic of stellar evolution.

Unusual Stars

Dwarfs

White dwarfs are extremely small stars with densities greater than that of any known terrestrial material. It is believed that white dwarfs were once normal sized stars whose internal heat energy was able to keep these gaseous masses from shrinking under their own gravitational force. Eventually, however, these stars depleted their nuclear fuel and collapsed to planetary size. Although some white dwarfs are no larger than the earth, they are nearly as massive as the sun. Thus, their densities may be a million times that of water. A spoonful of such matter would weigh several tons. Densities this great are only possible when electrons are displaced from their regular shells and pushed closer to the nucleus, allowing atoms to take up less space. Material in this state is called **degenerate matter.** Although atomic particles in degenerate matter are much closer together than usual, they are not yet "packed like sardines."

Once a star has contracted into a white dwarf, it is without a source of energy and can only become cooler and dimmer. Although none have been sighted, the terminal stage of a white dwarf must be a small, cold, nonluminous body called a **black dwarf.**

Black Holes

A study concerning white dwarfs produced what might at first appear to be a surprising conclusion. The *smallest* white dwarfs are the *most massive,* and the *largest* are the *least massive.* The reason for this is that a more massive star, because of its greater gravitational force, is able to squeeze itself into a smaller, more densely packed object than can a less massive star. Thus, the smallest white dwarfs were produced from the collapse of larger, more massive stars than were the larger white dwarfs.

Theoretical calculations indicate that collapsed stars slightly more massive than the sun would not become white dwarfs. Rather, they would form into the even smaller **neutron stars.** In a white dwarf, the electrons are pushed close to the nucleus, while in a neutron star, the electrons, under the much greater pressure, are forced to combine with the protons to produce *neutrons* (hence, the name). If the earth were to collapse to the density of a neutron star, it would have a diameter equivalent to the length of a football field.

Are neutron stars made of the most dense material possible? No, it seems that for any star more than three times the mass of the sun, the surface gravitation would be fantastically great. Upon collapse, such a star would form into a star still smaller and still denser than a neutron star. Even though these stars would be very hot, their immense gravitation would prevent light photons from leaving their surface, and consequently, they would literally disappear from sight. These hypothetical bodies have appropriately been named **black holes.** Anything which moved too near would be devoured and lost forever. Although no black holes have been positively identified, a few stars with invisible companions of great mass have been discovered. Further study of these companion stars will eventually reveal their true nature, but until then, the black hole remains an interesting possibility.

Variable Stars

Stars that fluctuate in brightness are known as variables. Some, called **pulsating variables,** fluctuate regularly in brightness by expanding and contracting in size. The importance of one member of this group (cepheid variables) in determining stellar distances was discussed earlier. Other pulsating variables have irregular periods and are of no value in this respect.

The most spectacular variables belong to a group known as **eruptive variables.** When one of these explosive events occurs, it appears as a sudden brightening of a star, called a **nova** (Figure 20.4). (The term *nova,* meaning "new," was used by the ancients since these stars were unknown to them before increasing in luminosity.) During the outburst, the outer layer of the star is ejected at high speeds. The "cloud" of ejected material is occasionally revealed photographically (Figure 20.5). A nova generally reaches maximum brightness in one day, remains bright for only a few weeks, then slowly returns in a year or so to its original brightness. Since the star returns to its prenova brightness, we can assume that only a small amount of its mass is lost during the flare-up. Some stars have experienced more than one such event. In fact, the process may occur over and over again, reducing the star's mass significantly. Because novae have only been observed in stars much hotter and larger than

Figure 20.4 *Photographs of Nova Herculis taken about two months apart showing the decrease in brightness. (Courtesy of Lick Observatory)*

Figure 20.5 *Expanding "cloud" around Nova Persei. (Courtesy of Hale Observatories)*

the sun, it is unlikely the sun will undergo such an outburst.

Possibly the most cataclysmic act of nature is the **supernova.** While a nova may increase its brightness several thousand times, a supernova becomes millions of times brighter than its prenova stage. If one of the nearer stars to the earth produced such an outburst, its brilliance would surpass that of the sun. Supernovae are rare; none have been observed in our galaxy since the advent of the telescope, but Tycho Brahe and Galileo each recorded one about 30 years apart. Probably an even larger one was recorded in 1054 by the Chinese. The remnant of this great outburst is the Crab nebula, shown in Figure 20.6. The amount of material ejected surely made up a considerable portion of the original star.

The current consensus of opinion is that supernovae occur when stars more massive than the sun exhaust their nuclear fuel. Then, just as in the formation of white dwarfs, these superstars collapse. Because of their great mass, the resulting implosion is of cataclysmic proportions, generating enormous amounts of energy, which eject the outer shell of the star. Theoretical work predicts that the remaining material is condensed into a very hot star possibly no larger than 16 kilometers in diameter. These incomprehensibly

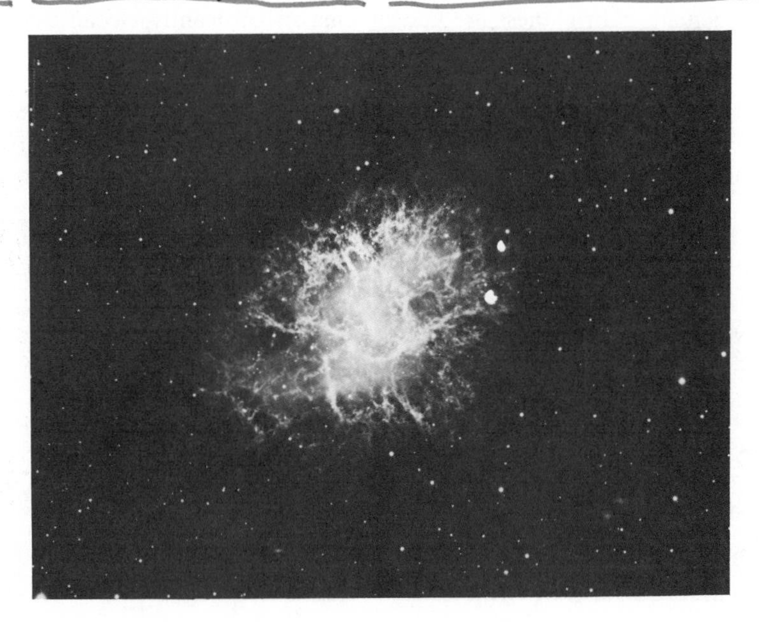

Figure 20.6 *Crab nebula in Taurus; the remains of the supernova of 1054* A.D. *(Courtesy of Hale Observatories)*

dense bodies have been named neutron stars. Although they have a high surface temperature, their small size would greatly limit their luminosity. Consequently, locating one would be extremely difficult.

At the beginning of the 1970s, a star that radiates short pulses of radio energy called a **pulsar** was discovered in the Crab nebula. Theory indicates that a rapidly rotating neutron star could produce the radiant energy observed for these pulsars. The remains of the supernova of 1054 may be the pulsar found in the Crab nebula. If it is, the first neutron star may have already been discovered.

Binary Stars

Persons with good eyesight can resolve the second star in the handle of the Big Dipper (Mizar) as two stars. Numerous star pairs were discovered by astronomers during the eighteenth century with their new tool, the telescope. One of the stars in the pair is usually fainter than the other, and for this reason, was considered farther away. In other words, the stars were not considered pairs, but were only thought to lie in the same line of sight. In the early nineteenth century, careful examination of numerous pairs by William Herschel revealed that these stars actually move about each other. The two stars are in fact united by their mutual gravitation. Newton's law of gravitation had been extended to the stars. These pairs of stars, in which the members are far enough apart to be resolved telescopically, are called **visual binaries.**

When binary stars are very close together, it is not possible to resolve them as two stars. However, some of these star systems have been detected from shifts in their spectral lines (Doppler effect) caused by their rapid rotation. For obvious reasons, these pairs are called **spectroscopic binaries.**

One of the most interesting groups of nonvisual binaries is detectable by changes in brightness. This change occurs when the plane of the orbits of two stars lies in our view such that the stars eclipse each other (Figure 20.7). When the fainter of the two stars eclipses the brighter, a major drop in intensity is noted, and when the brighter star eclipses the fainter one, a lesser drop is observed.

Binary stars are used to determine the property of a star most difficult to calculate—its mass. Recall that the mass of a body can be established if it is gravitationally attached to a partner, which is the case for any binary star system. Using Kepler's third law as redefined by Newton, the sum of the masses of two stars equals the cube of their mean distance (measured in AU) divided by their period squared. However, this gives information only about the combined masses of the two stars, not about the

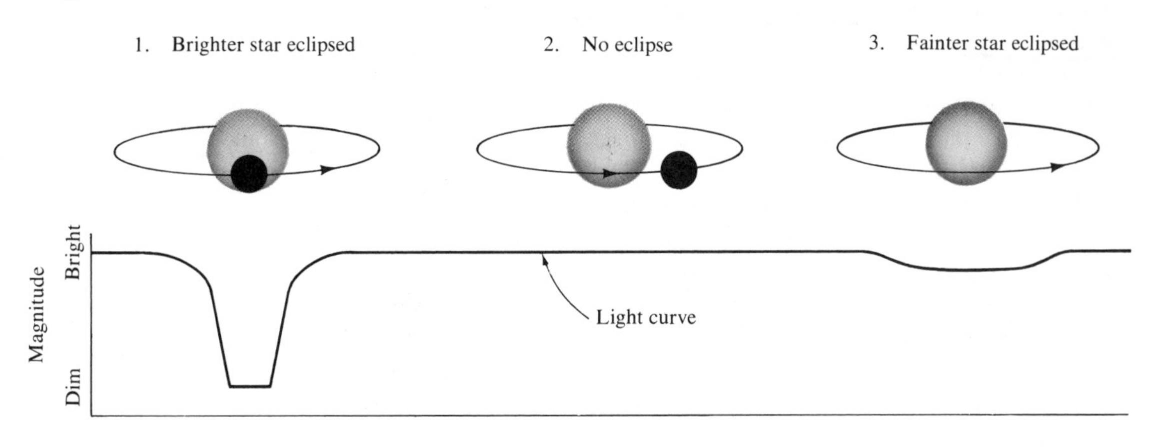

Figure 20.7 *Idealized light curve and geometry of an eclipsing binary star.*

individual masses. Fortunately, binary stars orbit each other around a common point called the **center of mass** in a manner dependent on their individual masses. For stars of equal mass, the center of mass lies exactly halfway between them. If one star is more massive than its partner, their common center will be located closer to the more massive one. Thus, if the sizes of their orbits can be observed, a determination of their individual masses can be made. You can experience this relationship on a seesaw by trying to balance a person having a much greater mass (Figure 20.8). For illustration, when one star has an orbit half the size of its companion, it is twice as massive. If their combined masses as determined by Kepler's law are equal to 3 times the mass of the sun, then the larger will be twice as massive as the sun, and the smaller will have a mass equal to that of the sun. Most stars have masses that range between 1/10 and 50 times the mass of the sun.

More than 50 percent of the nearby stars are known to be binaries or multiple stars. When the masses of main-sequence stars were compared to their absolute magnitude, the mass-luminosity relation described earlier was established. The intrinsically brighter main-sequence stars are the more massive ones.

Stellar Evolution

The idea of describing how a star is born, ages, and then dies may seem a bit presumptuous, for these objects have life spans that surely exceed tens of billions of years. However, by studying stars of different ages, astronomers have been able to piece together a rather acceptable model for stellar evolution. It should be pointed out that some parts of this story, the initial stage in particular, are rather sketchy.

The Hertzsprung-Russell diagrams have been very helpful in developing and testing models of stellar evolution. They are also useful for illustrating the changes that take place in an individual star during its life span. Figure 20.9 shows the evolution of a star about the size of the sun on an H-R diagram. We shall refer to this figure often in the following discussion. Keep in mind that the star does not physically move along this path, but rather that its position on the H-R diagram represents the color (temperature) and absolute magnitude (brightness) of the star at various stages in its evolution.

The birth of a star is thought to result when a mass of dust and gas becomes so dense that it begins to contract under its own gravitational attraction. As previously stated, contraction of a gaseous mass causes an increase in its temperature. Eventually, the temperature of the star will rise high enough that the contracting body will begin to radiate energy from its surface in the form of long-wavelength visible light. Because this large red object is not yet a star, the name **protostar** is generally applied to it. On the H-R diagram, a protostar would be located to the right and above the main sequence (Figure 20.9). It is found to the right because of its low temperature (red color) and above because it would be more luminous than a main-sequence star of the same color, a fact attributable to its large size. During the protostar phase, contraction continues, slowly at first, and then much

Figure 20.8 *Illustration of the center of mass using a seesaw.*

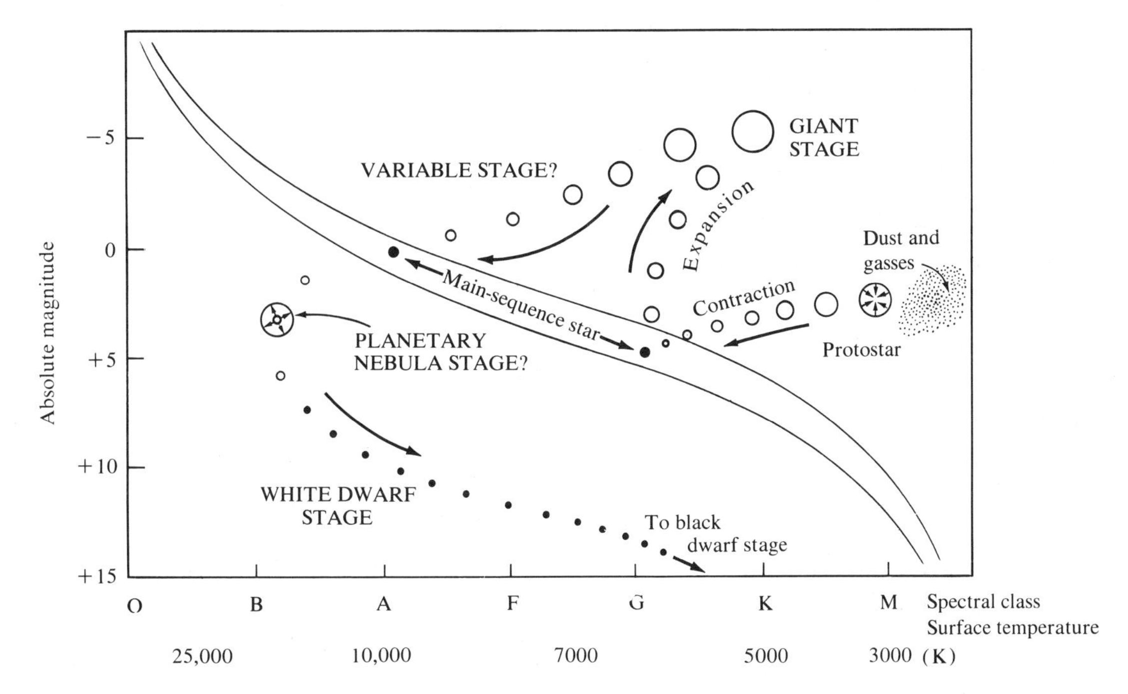

Figure 20.9 *Diagram of stellar evolution on a H-R diagram for a star about as massive as the sun.*

more rapidly, until its internal temperature has reached about 1 million K. At this time, the proton-proton reaction described earlier for the sun is initiated and begins converting hydrogen to helium. With this new heat source, the internal gas pressure steadily increases until it is sufficient to prevent further gravitational contraction. The star is now relatively stable and is located on the main sequence, where it will spend much of its life.

One difference between stellar evolution and evolution of the human population is that stars do not age at the same rate. As such, some stars remain on the main sequence much longer than others. The hot, massive blue stars radiate energy at such an enormously high rate that they substantially deplete their "fuel" in only a few million years. Once all of the hydrogen in the inner 10–20 percent of a star has been converted to helium, it leaves the main sequence to become a giant. For a yellow star, it takes about 10,000 million years before this change occurs. Since the solar system is approximately 5000 million years old, it is a comforting thought that the sun will remain stable for another 5000 million years. The very small (less massive) stars may remain stable for hundreds of thousands of millions of years.

A star leaves the main sequence because as its core becomes depleted of hydrogen, it again begins to contract. Contraction further heats the gaseous shell surrounding the core and raises its temperature high enough to trigger the proton-proton reaction there. This in turn heats and expands the star's outermost region enormously, and the star becomes a red giant or supergiant, depending on its original mass. The temperature in the interior of these superstars is thought to reach 100 million K, hot enough to initiate a complex nuclear reaction in which the helium in the core is converted to heavier elements. It has been predicted that this reaction is responsible for generating all the elements on the periodic table up to iron.

Astronomers are not clear on just what happens to a star after it leaves the red-giant stage.

It appears that a star about as massive as the sun would contract as its "fuel" supply dwindled and return to the main sequence. However, because of its increased surface temperature, it would move to a position on the main sequence slightly higher than its original position (Figure 20.9). On its return trip, it may go through an unstable stage, becoming a variable star, but this theory is uncertain. Whatever the case, it must eventually exhaust its "fuel" and contract to a white dwarf. Without a source of nuclear energy, a white dwarf radiates thermal energy, becoming cooler and dimmer, until it dies as a black dwarf.

A star having a mass greater than 1.2 times that of the sun probably goes through additional phases before reaching its terminal state. As previously mentioned, such a star is too large (massive) to become a white dwarf directly. It may lose mass during its red-giant phase, much as the sun does, in the form of solar wind, but at a much higher rate. Also, explosive events like novae may occur. Although a star loses only a small amount of mass during a nova, recurring novae have been observed and could account for greater losses. The existence of **planetary nebulae** suggests that some stars do, in fact, eject large quantities of material as they move toward the white-dwarf phase. Exemplified by the Ring nebula in the constellation Lyra, planetary nebulae are spherical shells of gas expanding from a central star (Figure 20.10).

A supernova may mark the end of a very massive star. The remnant of a supernova is the very dense neutron star. During the supernova implosion, the star's internal temperatures may reach 1000 million K, at which time very heavy elements like gold and uranium are thought to be produced. These heavy elements, plus the debris of novae and the planetary nebulae, are continually returning to interstellar space as dust and gas, which would then be available for the formation of other stars.

Astronomers believe that the original stars were made of pure hydrogen. They, in turn, produced the heavy elements, some of which were returned to space. Because the sun contains some heavy elements, and since it has not yet reached the stage in its evolution where it could have produced them, it must be a second-generation star. Thus, it, as well as the solar system, is believed to have formed, at least in part, from debris of preexisting stars. If so, the atoms in your body were produced thousands of millions of years ago inside a star, and the gold in your ring perhaps formed during a supernova event that occurred trillions of kilometers away.

Star Clusters

Many stars, like the members of ancient nomadic tribes, travel in groups called **clusters.** Two common types are the **open clusters** like Pleiades in the constellation Taurus (Figure 20.11) and the **globular clusters** like the cluster in the constellation Hercules (Figure 20.12). As their names suggest, the members of a globular cluster are arranged in a spherical shape, while the members of an open cluster are less well organized. Other differences between these groups include size, star density, and location within the galaxy.

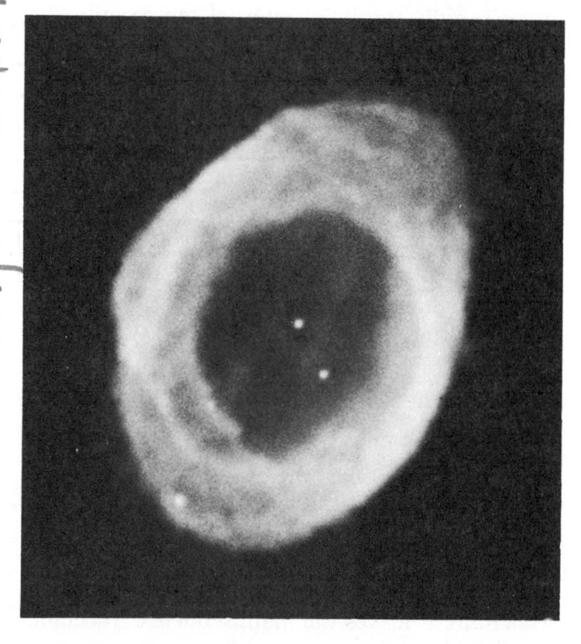

Figure 20.10 *Ring nebula, a planetary nebula in Lyra. (Courtesy of Hale Observatories)*

Figure 20.11 *Open-type star cluster in the constellation Libra. (Courtesy of Hale Observa-*

A large globular cluster may contain 100,000 member stars, which are packed near enough together to give the appearance of a solid when photographed. However, there is plenty of space, so the chance of stellar collision is almost nonexistent. Globular clusters are located in all directions from the center of our galaxy but appear in greater numbers nearest the center. The

Figure 20.12 *Globular star cluster in Hercules. (Courtesy of Hale Observatories)*

typical open clusters, on the other hand, contain tens to hundreds of members and are almost exclusively confined to the plane of our galaxy.

It is assumed that the members of a cluster have similar origins and compositions, and nearly the same birthday. Consequently, they are very beneficial in testing concepts of stellar evolution. At birth, each star cluster presumably contained members of every spectral type (color). Today, however, some clusters, like Pleiades, are notably missing the hot O- and B-type main-sequence stars, and others are nearly depleted of all members hotter than the G type. These differences can be explained if the clusters are of differing ages and the O-type stars do, in fact, use their fuel most rapidly, followed by B-type stars, and so on. Thus, studies of clusters support the proposal that the more massive (brighter) stars evolve to another stage, leaving the main sequences before less massive (dimmer) stars. In a similar manner, other aspects of stellar evolution have been tested using star clusters.

Interstellar Matter

The structure of the Milky Way galaxy was rather difficult to determine, since large amounts of interstellar dust and gas are found in its arms, obstructing the view. The name applied to concentrations of this intervening material is **nebula** (Latin for "cloud"), a term originally used for any fuzzy region in the heavens, including what we now know to be other galaxies. If this interstellar matter is in close proximity to a hot star, usually an O- or B-type star, it will glow and is then called a **bright nebula.** The two main types of bright nebulae are known as **emission nebulae** and **reflection nebulae.** Emission nebulae are gaseous in nature, containing a large percentage of hydrogen, which absorbs the *ultraviolet radiation* emitted by an embedded or nearby hot star. Because these gasses are under very low pressure, they reradiate this energy in the form of *visible light*. The process of converting ultraviolet light to visible light is known as **fluorescence,** an effect you have surely observed. A well-known emission nebula that can easily be observed with binoculars is located in the sword of the hunter in the constellation Orion (see the chapter-opening photo). Reflection nebulae, as the name implies, merely reflect the light of nearby stars. This fact was discovered when the spectra of these nebulae were found to closely match those of the benevolent stars. Reflection nebulae are thought to be composed of rather dense clouds of large particles called **interstellar dust.** This belief is based on the fact that atomic gasses with low densities could not reflect light sufficiently to produce the glow observed.

When a dense cloud of interstellar material is not close enough to a bright star to be illuminated, it is referred to as a **dark nebula.** Exemplified by the Horsehead nebula in Orion, dark nebulae appear as opaque objects silhouetted against a bright background (Figure 20.13). Dark nebulae can also easily be seen as starless regions—"holes in the heavens"—when viewing the Milky Way.

Although nebulae appear very dense, they actually approximate a good artificially made vacuum. However, because of their enormous size, their total mass may be many times that of the sun. Interstellar matter is of great interest to astronomers. It is from this material that the stars and the planets were formed. Although the original composition of this material is not known, it was probably mostly hydrogen. The heavier elements (dust) were added by supernovae and, less dramatically, by all stars. Nebulae hinder the work of astronomers, often impairing visual observations by dimming the light of stars or by completely blocking them from view.

The Milky Way

On a clear, moonless night away from the city lights, you can see a truly marvelous sight—the Milky Way galaxy (Figure 20.14). With his telescope, Galileo discovered that this band of light was produced by countless individual stars which the naked eye is unable to resolve. Today, we realize that the sun is actually part of this vast system of stars, which number about 100,000

Figure 20.13 *Horsehead nebula. (Courtesy of Hale Observatories)*

Figure 20.14 *Panorama of the Milky Way. Notice the dark bands caused by the presence of interstellar nebula. (Lund Observatory photograph)*

million. The "milky" appearance of our galaxy is a consequence of its rather flat disc shape. Thus, when it is viewed from the "inside," a higher concentration of stars appears in the direction of the galactic plane than in any other direction.

The Milky Way is a rather large spiral galaxy that spans some 100,000 light-years and is about 10,000 light-years thick at the nucleus (Figure 20.15). As viewed from the earth, the center of the Galaxy lies beyond the constellation Sagittarius. Although visual inspection has been difficult, radio telescopes reveal the existence of four, or possibly five, distinct spiral arms, with some showing splintering. The sun is believed positioned in one of these arms about two-thirds of the way from the center, at a distance of 30,000 light-years. The stars in the arms of the Galaxy rotate around the nucleus, with the most outward ones moving the slowest such that the ends of the arms appear to trail. The sun requires about 200 million years for each orbit. Surrounding the galactic disc is a nearly spheroidal **halo** made of very tenuous gas and numerous globular clusters. These star clusters do not participate in the rotating motion of the arms but rather have their own orbits that carry them through the disc. Although some clusters are very dense, they may pass among the stars of the arms with plenty of room to spare.

The Milky Way was probably formed from an enormous hydrogen cloud, which began to condense some 10,000 million years ago. As it contracted, its density increased, eventually becoming sufficient for stellar formation to begin. Initially, stars developed in scattered globular-type clusters. The globular clusters of our galaxy all lack hot O- or B-type stars, which testifies to their age. Astronomers call old stars like these, which formed early from nearly pure hydrogen, **population II stars.** At the time of the formation of population II stars, the Galaxy had the spheroidal shape and size which is presently outlined by the farthest extent of the globular clusters (Figure 20.15). As this gaseous mass continued to condense, its rotational speed increased, causing it to flatten. While this was taking place, the

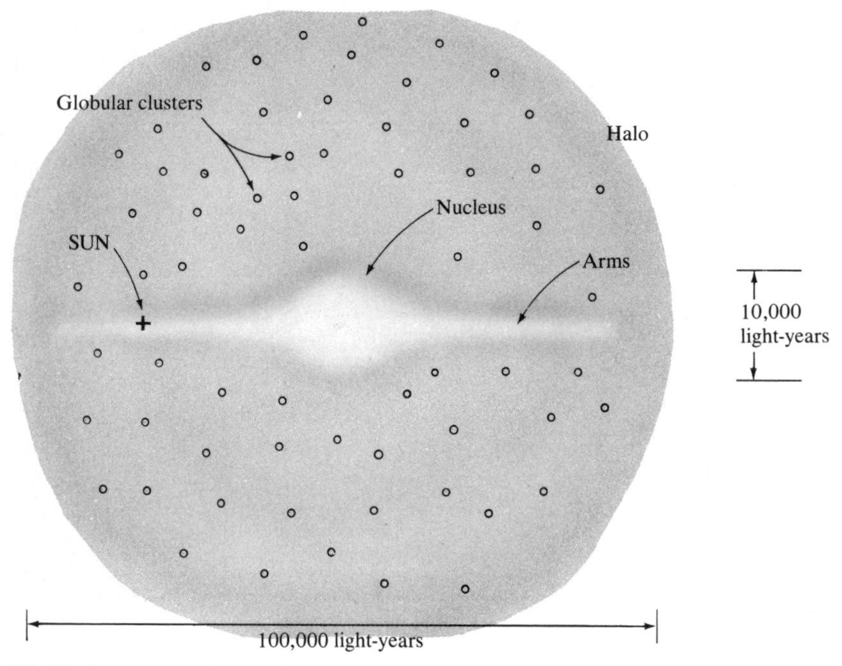

Figure 20.15 *Structure of the Milky Way.*

Figure 20.16 *Great Galaxy in the constellation Andromeda. (Courtesy of Hale Observatories)*

original stars were generating heavy elements, and more and more of this matter was being added to the condensing gasses. Therefore, the stars that formed later in the disc of the galaxy are richer in heavy elements. These young stars are said to be **population I stars.** *The sun is a population I star about 5000 million years old found in the galactic disc.* The arms of the galaxy still contain large quantities of dust and gasses, and stars are still forming there. These new stars will contain an even higher percentage of heavy elements.

Galaxies

In the mid-1700s Immanuel Kant proposed that the telescopically visible fuzzy patches of light that appeared scattered among the stars were actually distant galaxies like the Milky Way. Kant described them as "island universes." Each, he felt, contained billions of stars and, as such, was a universe in itself. The weight of opinion, however, favored the hypothesis that they were dust and gas clouds (nebulae) within our galaxy. This matter was not resolved until the 1920s when Edwin Hubble was able to identify some cepheid variables in the Great Galaxy in Andromeda (Figure 20.16). Hubble realized that since these intrinsically very bright stars had apparent magnitudes of + 18 (very dim), they must lie outside the Milky Way. Recall that cepheid variables can be used to determine stellar distances. When they were used to find the distance to Andromeda, astronomers determined it to be over a million light-years away. Hubble had extended the universe far beyond the limits of our imagination to consist of thousands of millions of galaxies, each of which contains millions to thousands of millions of stars. It has been said that a million galaxies are found in that portion of the sky bounded by the cup of the Big Dipper.

From the thousands of millions of galaxies, three basic types have been identified: elliptical, spiral, and irregular. Only a small percentage of the known galaxies lack symmetry and are classified as **irregular.** The best known are the Large and Small Magellanic Clouds, which are seen as the two bright areas in the lower half of Figure 20.14). They are easily visible with the unaided eye in the Southern Hemisphere and were named after the explorer Ferdinand Megellan, who observed them when he circumnavigated the earth. They are also our nearest galactic neighbors, only 150,000 light-years away.

The Milky Way and the famous Great Galaxy in Andromeda are examples of fairly large **spiral galaxies** (Figure 20.17). Andromeda can be seen with the naked eye as a fuzzy fifth-magnitude star. Typically, spiral galaxies are disc-shaped with a somewhat greater concentration of stars near their center, but there are numerous variations. Viewed broadside, arms are often seen extending from the central nucleus and sweeping

Figure 20.17 *Two views illustrating the idealized structure of spiral galaxies. (Courtesy of Hale Observatories)*

gracefully away. The outermost stars of these arms rotate most slowly, giving the galaxy the appearance of a fireworks pinwheel. However, one type of spiral galaxy has the stars arranged like a bar, which rotates as a rigid system. This requires that the outer stars move faster than the inner ones, a fact not easy for astronomers to reconcile with the laws of motion. Attached to each end of these bars are curved spiral arms. These have become known as **barred spiral galaxies** (Figure 20.18). Spiral galaxies are generally quite large, ranging from 20,000 to about 125,000 light-years in diameter. They make up about 70 percent of the known galaxies, but this figure is misleading, for they are probably not the most numerous group.

Figure 20.18 *Barred spiral galaxy. (Courtesy of Hale Observatories)*

The most abundant are probably the **elliptical galaxies.** These are generally smaller than spiral galaxies. Some are so much smaller, in fact, that the term *dwarf* has been applied. Since these dwarf galaxies are not visible at great distances, a survey of the sky reveals more of the conspicuous large spiral galaxies. Although most elliptical galaxies are small, the very largest known galaxies (200,000 light-years in diameter) are also elliptical. As their name implies, elliptical galaxies have an ellipsoidal shape that ranges to nearly spherical, and they lack spiral arms. The two dwarf companions of Andromeda shown in Figure 20.16 are elliptical galaxies. Some resemble the dense nucleus of a spiral galaxy, a fact which generated speculation that they may be the remains of a spiral galaxy whose arms wound around themselves. However, this has not been shown to be the case.

One of the major differences among the galactic types is the age of the stars which make them up. The irregular galaxies are composed mostly of young population I stars, while the elliptical galaxies contain old population II stars. The Milky Way and other spiral galaxies consist of both populations, with the youngest stars located in the arms. Attempts were made to develop an evolutionary scheme for galaxies based on their stellar populations and shape. It had been suggested that as irregular galaxies aged, they evolved into spiral galaxies, which became elliptical as their arms wound around their nucleus. Most evidence, however, indicates that this sequence does not occur and, in fact, that galaxies probably do not change from one form to another. However, since their component stars do evolve, galaxies themselves must also change with time.

The Big Bang

The universe—did it have a beginning? Will it have an end? Cosmologists are trying to answer this question, and that makes them a rare breed. Unlike other scientists in search for answers, cosmologists have little hope of finding a final solution.

First, and probably foremost, any viable theory on the origin of the universe must explain the so-called **red shift** obtained for most galaxies. Recall that light waves appear shortened or lengthened depending on whether their source is moving toward or away from the observer. The motion of galaxies, except for the very nearest, causes the lines in their spectra to shift toward the red (longer wavelengths), which means they are moving away from us. Since nearly all galaxies are moving away from us, are we in the center of the universe? If we are not in the center of the solar system and not in the center of the galaxy, it seems unlikely that we should be in the center of the universe. A more probable explanation exists for this fact. Imagine a rubber balloon with black dots painted on its surface equal distances apart. When the balloon is inflated, each dot spreads apart from every other dot. Similarly, if the universe is expanding, every galaxy would be moving away from all others, and observers in all galaxies would obtain a red shift in the spectra of every other galaxy.

This belief in the expanding universe led to the widely accepted **Big Bang** theory, according to which all of the material in the universe was at one time confined in a supermassive ball composed of hydrogen (Figure 20.19). Then, between 10,000 and 20,000 million years ago, a cataclysmic explosion occurred, hurling this material in all directions. The ejected masses of gas cooled and condensed, forming the stellar systems we now observe fleeing from their birthplace.

Can the galaxies expand forever? It has been suggested that after a certain point, perhaps 20,000 million years in the future, the galaxies will slow and begin to collapse under their own gravitational influence. But present estimates indicate that the amount of material in the universe is not enough to exert the needed

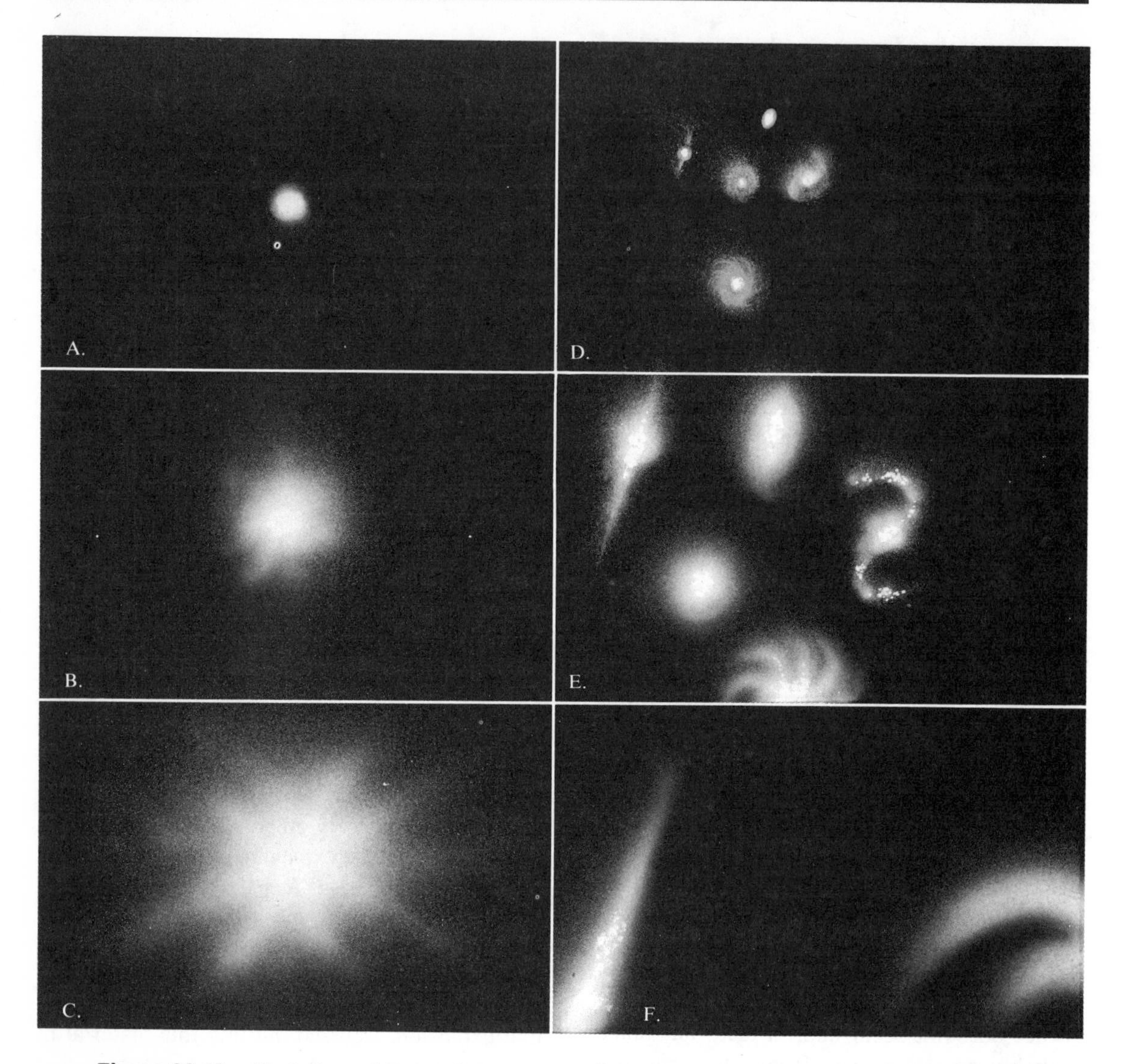

Figure 20.19 *Evolution of the universe according to the Big Bang theory.* **A.** *Supermassive ball.* **B.** *Explosion of the supermassive ball.* **C.** *Gas hurling into space from the exploding mass.* **D.** *Galaxies formed from gas moving outward into space.* **E.** *Galaxies spreading outward.* **F.** *Galaxies passing an observer who is looking into the blackness of space.*

gravitational pull to stop the outward flight of the galaxies. However, this subject does not end here. It has been proposed that heretofore undetected matter exists in large quantities in the universe, and if this is true, the galaxies could, in fact, collapse upon themselves. "Absence of evidence is not evidence of absence" (Anonymous).

REVIEW

1. What is the distance in parsecs to a star that has a parallax angle of 0.25 second of arc? What is its distance in light-years?
2. Explain the difference between a star's apparent and absolute magnitudes. Which one is an intrinsic property of a star?

3. What is the ratio of brightness between a twelfth- and fifteenth-magnitude star?
4. What information about a star can be determined from its color?
5. Why were the original spectral classes rearranged?
6. Where on an H-R diagram does a star spend most of its lifetime?
7. Why is the sun considered a second-generation star?
8. Make a generalization relating the mass and luminosity of main-sequence stars.
9. The disk of a star cannot be resolved telescopically. Explain the method that astronomers have used to estimate the diameter of stars.
10. What evidence supports the Big Bang theory?
11. Describe the general structure of the Milky Way.
12. Enumerate the steps thought to be involved in the evolution of stars larger than the sun.
13. Why are less massive stars thought to age slower than more massive ones, even though they have much less "fuel"?
14. Compare a black dwarf with a black hole.
15. Compare the three general types of galaxies.
16. How did Hubble determine that Andromeda is located beyond our galaxy?
17. Explain why astronomers consider elliptical galaxies more abundant than spiral galaxies, even though more spiral galaxies have been sighted.
18. What causes a star to become a giant?
19. How has the word *nebula* taken on different meanings?
20. Describe the different types of nebulae.

KEY TERMS

stellar parallax
parsecs
light-year
magnitude
apparent magnitude
absolute magnitude
cepheid variable
light period
color index
spectral class
main sequence
Hertzsprung-Russell (H-R) diagram
giant
red giant
supergiant
white dwarf
degenerate matter
black dwarf
neutron star
black hole
pulsating variable
eruptive variable
nova
supernova
pulsar
visual binaries
spectroscopic binaries
center of mass
protostar
planetary nebulae
cluster
open cluster
globular cluster
nebula
bright nebula
emission nebula
reflection nebula
fluorescence
interstellar dust
dark nebula
halo
population II stars
population I stars
irregular galaxy
spiral galaxy
barred spiral galaxy
elliptical galaxy
red shift
Big Bang

Metric and English Units Compared

UNITS

1 kilometer (km)	=	1000 meters (m)
1 meter (m)	=	100 centimeters (cm)
1 centimeter (cm)	=	0.39 inches (in.)
1 mile (mi)	=	5280 feet (ft)
1 foot (ft)	=	12 inches (in.)
1 inch (in.)	=	2.54 centimeters (cm)
1 square mile (mi^2)	=	640 acres (a)
1 kilogram (kg)	=	1000 grams (g)
1 pound (lb)	=	16 ounces (oz)
1 fathom	=	6 feet (ft)

CONVERSIONS

When you want to convert:	Multiply by:	To find:
Length		
inches	2.54	centimeters
centimeters	0.39	inches
feet	0.30	meters
meters	3.28	feet
yards	0.91	meters
meters	1.09	yards
miles	1.61	kilometers
kilometers	0.62	miles

When you want: to convert:	Multiply by:	To find:
Area		
square inches	6.45	square centimeters
square centimeters	0.15	square inches
square feet	0.09	square meters
square meters	10.76	square feet
square miles	2.59	square kilometers
square kilometers	0.39	square miles
Volume		
cubic inches	16.38	cubic centimeters
cubic centimeters	0.06	cubic inches
cubic feet	0.028	cubic meters
cubic meters	35.3	cubic feet
cubic miles	4.17	cubic kilometers
cubic kilometers	0.24	cubic miles
liters	1.06	quarts
liters	0.26	gallons
gallons	3.78	liters
Masses and Weights		
ounces	20.33	grams
grams	0.035	ounces
pounds	0.45	kilograms
kilograms	2.205	pounds

Temperature

When you want to convert degrees Fahrenheit (°F) to degrees Celsius (°C), subtract 32 degrees and divide by 1.8.

When you want to convert degrees Celsius (°C) to degrees Fahrenheit (°F), multiply by 1.8 and add 32 degrees.

When you want to convert degrees Celsius (°C) to kelvins (K), delete the degree symbol and add 273.

When you want to convert kelvins (K) to degrees Celsius (°C), add the degree symbol and subtract 273.

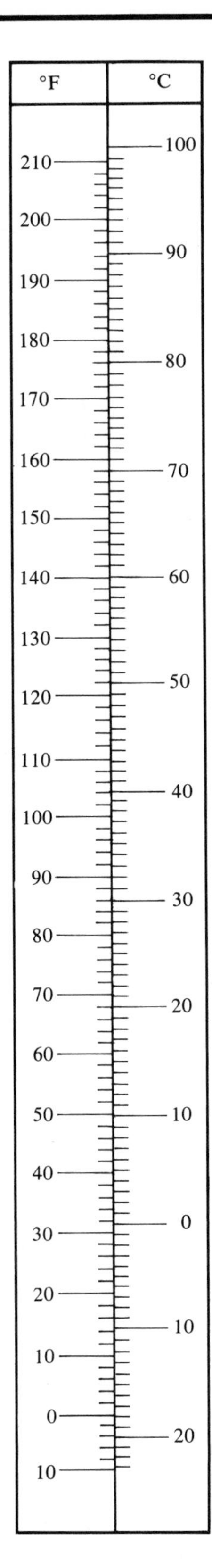
°F
°C
210
200
190
180
170
160
150
140
130
120
110
100
90
80
70
60
50
40
30
20
10
0
10
100
90
80
70
60
50
40
30
20
10
0
10
20

appendix B

Mineral Identification Key

Group I ***Metallic Luster***

Hardness	Streak	Other Diagnostic Properties	Chemical Composition
Harder than glass	Black streak	Black; magnetic; hardness = 6; specific gravity = 5.2; often granular	Magnetite (Fe_3O_4)
	Greenish black streak	Brass yellow; hardness = 6; specific gravity = 5.2; generally an aggregate of cubic crystals	Pyrite (FeS_2)—fool's gold
	Reddish streak	Gray or reddish brown; hardness = 5–6; specific gravity = 5; platy appearance	Hematite (Fe_2O_3)
Softer than glass	Greenish black streak	Brass yellow; hardness = 4; specific gravity = 4.2; massive	Chalcopyrite ($CuFeS_2$)
	Gray black streak	Silvery gray; hardness = 2.5; specific gravity = 7.6 (very heavy); good cubic cleavage	Galena (PbS)
	Yellow brown streak	Yellow brown to dark brown; hardness variable (1–6); specific gravity = 3.5–4; often found in rounded masses; earthy appearance	Limonite ($Fe_2O_3H_2O$)
	Gray black streak	Black to bronze; tarnishes to purples and greens; hardness = 3; specific gravity = 5; massive	Bornite (Cu_5FeS_4)
Softer than your fingernail	Dark gray streak	Silvery gray; hardness = 1 (very soft); specific gravity = 2.2; massive to platy; writes on paper (pencil lead); greasy feel	Graphite (C)

Group II ***Nonmetallic Luster (Dark Colored)***

Hardness	Cleavage	Other Diagnostic Properties	Chemical Composition
Harder than glass	Cleavage present	Black to greenish black; hardness = 5–6; specific gravity = 3.4; fair cleavage, two planes at nearly 90 degrees	Augite (Ca, Mg, Fe, Al silicate)
		Black to greenish black; hardness = 5–6; specific gravity = 3.2; fair cleavage, two planes at nearly 60 degrees and 120 degrees	Hornblende (Ca, Na, Mg, Fe, Al silicate)
	Cleavage not prominent	Red to reddish brown; hardness = 6.5–7.5; conchoidal fracture; glassy luster	Garnet (Fe, Mg, Ca, Al silicate)
		Gray to brown; hardness = 9; specific gravity = 4; hexagonal crystals common	Corundum (Al_2O_3)
		Dark brown to black; hardness = 7; conchoidal fracture; glassy luster	Smoky quartz (SiO_2)
		Olive green; hardness = 6.5–7; small grains	Olivine $(Mg, Fe)_2SiO_4$
Softer than glass	Cleavage present	Yellow brown to black; hardness = 4; good cleavage in six directions; light yellow streak that has the smell of sulfur	Sphalerite (ZnS)
	No cleavage	Generally tarnished to brown or green; hardness = 2.5; specific gravity = 9; massive	Native copper (Cu)
Softer than your fingernail	Cleavage present	Dark brown to black; hardness = 1; excellent cleavage in one direction; elastic in thin sheets; black mica	Biotite (K, Mg, Fe, Al silicate)
	Cleavage not prominent	Reddish brown; hardness = 1–5; specific gravity = 4–5; red streak; earthy appearance	Hemitite (Fe_2O_3)
		Yellow brown, hardness = 1–3; specific gravity = 3.5; earthy appearance; powders easily	Limonite ($Fe_2O_3 \cdot H_2O$)

Group III *Nonmetallic Luster (Light Colored)*

Hardness	Cleavage	Other Diagnostic Properties	Chemical Composition
Harder than glass	Cleavage present	Flesh colored or white to gray; hardness = 6; specific gravity = 2.6; two planes of cleavage at nearly right angles	Potassium feldspar ($KAlSi_3O_8$) Plagioclase feldspar ($NaAlSi_3O_8$ to $CaAl_2Si_2O_8$)
	Cleavage not prominent	Any color; hardness = 7; specific gravity = 2.65; conchoidal fracture; glassy appearance; varieties: milky, rose, smoky, amethyst (violet)	Quartz (SiO_2)
Softer than glass	Cleavage present	White, yellowish to colorless; hardness = 3; three planes of cleavage at 75 degrees (rhombohedral); effervesces in HCl; often transparent	Calcite ($CaCO_3$)
		White to colorless; hardness = 2.5; three planes of cleavage at 90 degrees (cubic); salty taste	Halite (NaCl)
		Yellow, purple, white; hardness = 4; white streak; translucent to transparent; four planes of cleavage	Fluorite (CaF_2)
Softer than your fingernail	Cleavage present	Colorless; hardness = 2; transparent and elastic in thin sheets; excellent cleavage in one direction; light mica	Muscovite (K, Al silicate)
		White to transparent; hardness = 2; when in sheets, is flexible but not elastic; varieties: selenite (transparent, three planes of cleavage; satin spar (fibrous, silky luster); alabaster (aggregate of small crystals)	Gypsum ($CaSO_4 . 2\ H_2O$)
	Cleavage not prominent	White, pink, green; hardness = 1–2; forms in thin plates; soapy feel; pearly luster	Talc (Mg silicate)
		Yellow; hardness = 1–2.5	Sulfur (S)
		White; hardness = 1 (very soft); smooth feel; earthy odor; when moistened, has typical clay texture	Kaolinite (Al silicate)
		Green; hardness = 2.5; fibrous; variety of serpentine	Asbestos (Mg, Al silicate)
		Pale to dark reddish brown; hardness = 1–3; dull luster; earthy; often contains spheroidal-shaped particles; not a true mineral	Bauxite (Hydrous Al oxide)

appendix C

The Earth's Grid System

A glance at any globe reveals a series of north-south and east-west lines that together make up the earth's grid system, a universally used scheme for locating points on the earth's surface. The north-south lines of the grid are called **meridians** and extend from pole to pole (Figure C.1). All are halves of great circles. A **great circle** is the largest possible circle that may be drawn on a globe; if a globe were sliced along one of these circles, it would be divided into two equal parts called **hemispheres.** By viewing a globe or Figure C.1, it can be seen that meridians are spaced farthest apart at the equator and converge toward the poles. The east-west lines (circles) of the grid are known as **parallels.** As their name implies, these circles are parallel to one another (Figure C.1). While all meridians are parts of great circles, all parallels are not. In fact, only one parallel, the equator, is a great circle.

Latitude and Longitude

Latitude may be defined as distance, measured in degrees, *north* and *south* of the equator. Parallels are used to show latitude. Since all points that lie along the same parallel are an identical distance from the equator, they all have the same latitude designation. The latitude of the equator is 0 degrees, while the north and south poles lie 90 degrees N and 90 degrees S, respectively.

Longitude is defined as distance, measured in degrees, *east* and *west* of the zero or prime meridian. Since all meridians are identical, the choice of a zero line is obviously arbitrary. However, the meridian that passes through the Royal Observatory

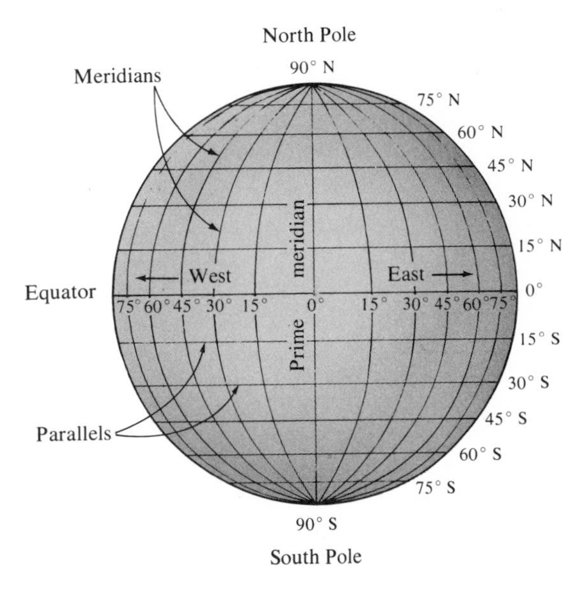

Figure C.1 *The earth's grid system. (After J.B. Hoyt,* Man and the Earth *3rd ed., © 1966. Adapted by permission of Prentice-Hall, Inc., Englewood Cliffs, N.J.)*

at Greenwich, England, is universally accepted as the reference meridian. Thus, the longitude for any place on the globe is measured east or west from this line. Longitude can vary from 0 degrees along the prime meridian to 180 degrees, halfway around the globe.

It is important to remember that when a location is specified, directions must be given, that is, north or south latitude and east or west longitude (Figure C.2). If this is not done, more than one point on the globe is being designated. The only exceptions, of course, are places that lie along the equator, the prime meridian, or the 180-degree meridian. It should also be noted that while it is not incorrect to use fractions, a degree of latitude or longitude is usually divided into minutes and seconds. A minute (′) is 1/60 of a degree, and a second (″) is 1/60 of a minute. When locating a place on a map, the degree of exactness will depend upon the scale of the map. When using a small-scale world map or globe, it may be difficult to estimate latitude and longitude to the nearest whole degree or two. On the other hand, when a large-scale map of an area is used, it is often possible to estimate latitude and longitude to the nearest minute or second.

Distance Measurement

The length of a degree of longitude depends upon where the measurement is taken. At the equator, which is a great circle, a degree of east-west distance is equal to approximately 111 kilometers (69 miles). This figure is found by dividing the earth's circumference—40,075 kilometers (24,900 miles)—by 360. However, with an increase in latitude, the parallels become smaller, and the length of a degree of longitude diminishes (see Table C.1). Thus, at about latitude 60 degrees N and S, a degree of longitude has a value equal to about half of what it was at the equator.

Since all meridians are halves of great circles, a degree of latitude is equal to about 111 kilometers (69 miles), just as a degree of longitude along the equator is. However, the earth is not a perfect sphere but is slightly flattened at the poles and

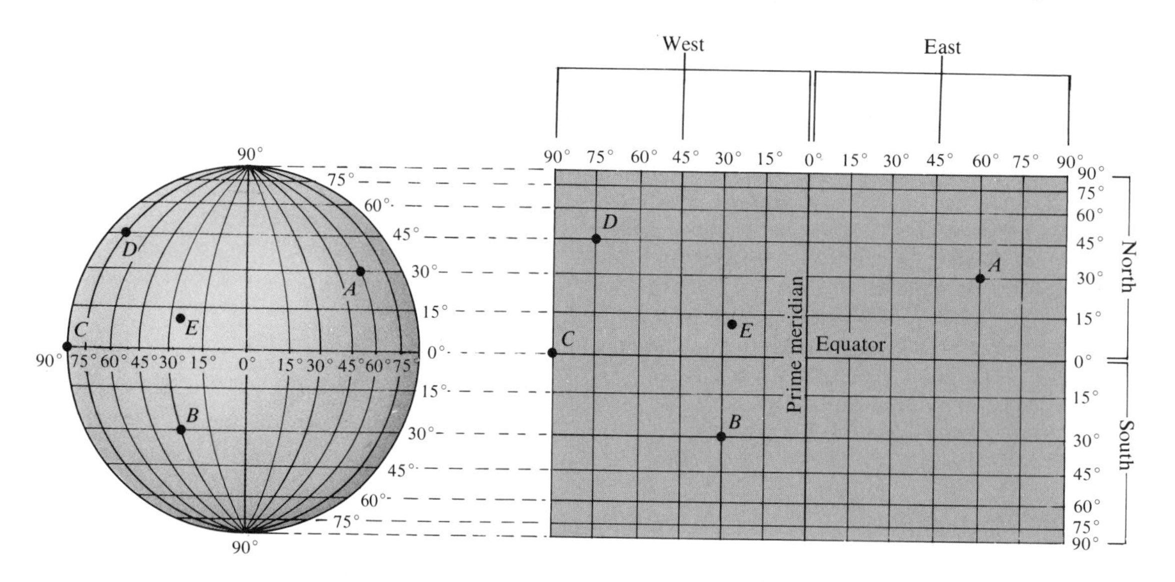

Figure C.2 *Locating places using the grid system. For both diagrams: Point* A *is latitude 30 degrees N, longitude 60 degrees E; Point* B *is latitude 30 degrees S, longitude 30 degrees W; Point* C *is latitude 0 degrees, longitude 90 degrees W; Point* D *is latitude 45 degrees N, longitude 75 degrees W; Point* E *is approximately latitude 10 degrees N, longitude 25 degrees W. (After J.B. Hoyt,* Man and the Earth *3rd ed., © 1966. Adapted by permission of Prentice-Hall, Inc., Englewood Cliffs, N.J.)*

Table C.1 ***Latitude and Longitude as Distance (Miles)***

° Lat.	Length of 1° Lat.	Length of 1° Long.	° Lat.	Length of 1° Lat.	Length of 1° Long.	° Lat.	Length of 1° Lat.	Length of 1° Long.
0	68.703	69.172	30	68.878	59.955	60	69.229	34.674
1	68.704	69.161	31	68.889	59.345	61	69.241	33.622
2	68.704	69.129	32	68.900	58.716	62	69.251	32.560
3	68.705	69.077	33	68.911	58.070	63	69.261	31.488
4	68.707	69.004	34	68.922	57.407	64	69.270	30.406
5	68.709	68.910	35	68.934	56.725	65	69.280	29.314
6	68.711	68.795	36	68.945	56.027	66	69.290	28.215
7	68.714	68.660	37	68.957	55.311	67	69.298	27.105
8	68.717	68.503	38	68.968	54.578	68	69.307	25.988
9	68.721	68.325	39	68.980	53.829	69	69.316	24.862
10	68.724	68.128	40	68.933	53.063	70	69.324	23.729
11	68.729	67.909	41	69.005	52.280	71	69.331	22.589
12	68.734	67.670	42	69.017	51.482	72	69.339	21.441
13	68.739	67.410	43	69.029	50.668	73	69.346	20.287
14	68.744	67.130	44	69.041	49.839	74	69.353	19.126
15	68.750	66.830	45	69.054	48.994	75	69.359	17.959
16	68.757	66.509	46	69.066	48.135	76	69.365	16.788
17	68.763	66.168	47	69.078	47.260	77	69.371	15.611
18	68.770	65.807	48	69.090	46.372	78	69.376	14.428
19	68.777	65.427	49	69.103	45.468	79	69.381	13.242
20	68.785	65.026	50	69.114	44.551	80	69.385	12.051
21	68.793	64.605	51	69.127	43.620	81	69.389	10.857
22	68.801	64.165	52	69.139	42.676	82	69.393	9.659
23	68.810	63.705	53	69.151	41.719	83	69.396	8.458
24	68.819	63.227	54	69.162	40.748	84	69.399	7.255
25	68.828	62.729	55	69.174	39.765	85	69.402	6.049
26	68.837	62.211	56	69.185	38.770	86	69.403	4.842
27	68.847	61.675	57	69.197	37.763	87	69.405	3.633
28	68.857	61.121	58	69.208	36.745	88	69.406	2.422
29	68.868	60.547	59	69.219	35.715	89	69.407	1.211
30	68.878	59.955	60	69.229	34.674	90	69.407	0.000

bulges slightly at the equator. Because of this, there are small differences in the length of a degree of latitude (see Table C.1).

Determining the shortest distance between two points on a globe can be done easily and fairly accurately using the "globe and string" method. It should be noted here that the arc of a great circle is the shortest distance between two points on a sphere. In order to determine the great circle distance (as well as observe the great circle route) between two places, stretch the string between the locations in question. Then, measure the length of the string along the equator (since it is a great circle with degrees marked on it) to determine the number of degrees between the two points. To calculate the distance in kilometers or miles, simply multiply the number of degrees by 111 or 69, respectively.

appendix D

Topographic Maps

A map is a representation on a flat surface of all or a part of the earth's surface drawn to a specific scale. Maps are often the most effective means for showing the locations of both natural and man-made features, their sizes, and their relationships to one another. Like photographs, maps readily display information that would be impractical to express in words.

While most maps show only the two horizontal dimensions, geologists, as well as other map users, often require that the third dimension, elevation, be shown on maps. Maps that show the shape of the land are called **topographic maps.** Although various techniques may be used to depict elevations, the most accurate method involves the use of contour lines.

Contour Lines

A **contour line** is a line on a map representing a corresponding imaginary line on the ground that has the same elevation above sea level along its entire length. While many map symbols are pictographs, resembling the objects they represent, a contour line is an abstraction that has no counterpart in nature. It is, however, an accurate and effective device for representing the third dimension on paper.

Some useful facts and rules concerning contour lines are listed as follows. This information should be studied in conjunction with Figure D.1.

(1) Contour lines bend upstream or upvalley. The contours form Vs that point upstream, and in the upstream direction the successive contours represent higher elevations. For example, if you were standing on a stream bank and wished to get to the point at the same elevation directly opposite you on the other bank, without stepping up or down, you would need to walk upstream along the contour at that elevation to where it crosses the stream bed, cross the stream, and then walk back downstream along the same contour.

(2) Contours near the upper parts of hills form closures. The top of a hill is higher than the highest closed contour.

(3) Hollows (depressions) without outlets are shown by closed, hatched contours. Hatched contours are contours with short lines on the inside pointing downslope.

(4) Contours are widely spaced on gentle slopes.

(5) Contours are closely spaced on steep slopes.

(6) Evenly spaced contours indicate a uniform slope.

(7) Contours usually do not cross or intersect each other, except in the rare case of an overhanging cliff.

(8) All contours eventually close, either on a map or beyond its margins.

(9) A single higher contour never occurs be-

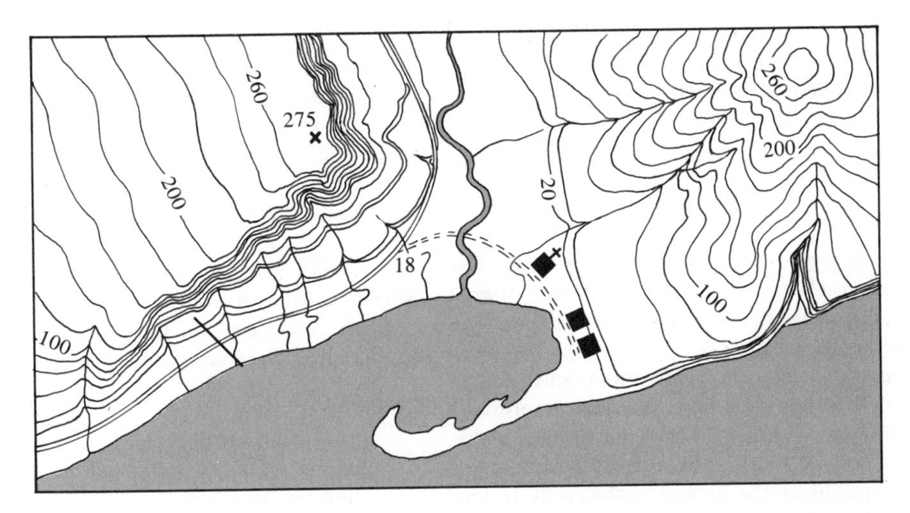

Figure D.1 *Perspective view of an area and a contour map of the same area. These illustrations show how various features are depicted on a topographic map. The upper illustration is a perspective view of a river valley and the adjoining hills. The river flows into a bay, which is partly enclosed by a hooked sandbar. On either side of the valley are terraces through which streams have cut gullies. The hill on the right has a smoothly eroded form and gradual slopes, whereas the one on the left rises abruptly in a sharp precipice, from which it slopes gently, and forms an inclined plateau traversed by a few shallow gullies. A road provides access to a church and two houses situated across the river from a highway that follows the seacoast and curves up the river valley. The lower illustration shows the same features represented by symbols on a topographic map. The contour interval (vertical distance between adjacent contours) is 20 feet. (After U.S. Geological Survey)*

tween two lower ones, and vice versa. In other words, a change in slope direction is always determined by the repetition of the same elevation either as two different contours of the same value or as the same contour crossed twice.

(10) Spot elevations between contours are given at many places, such as road intersections, hill summits, and lake surfaces. Spot elevations differ from control elevation stations, such as bench marks, in not being permanently established by permanent markers.

Relief

Relief refers to the difference in elevation between any two points. Maximum relief refers to the difference in elevation between the highest and lowest points in the area being considered. Relief determines the **contour interval,** which is the differ-

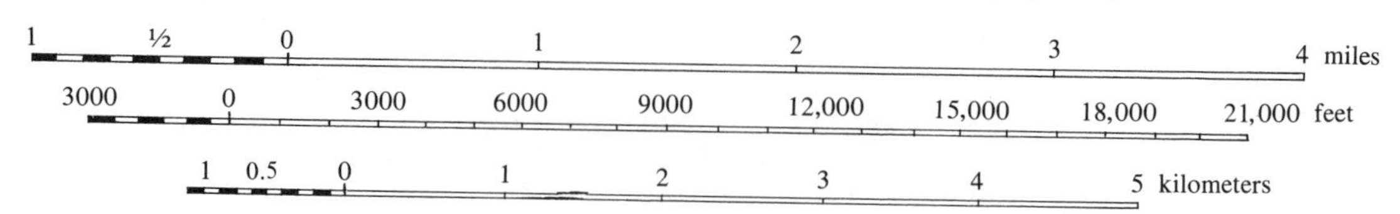

Figure D.2 *Graphic scale.*

ence in elevation between succeeding contour lines that is used on topographic maps. Where relief is low, a small contour interval, such as 10 or 20 feet, may be used. In flat areas, such as wide river valleys or broad, flat uplands, a contour interval of 5 feet is often used. In rugged mountainous terrain, where relief is many hundreds of feet, contour intervals as large as 50 or 100 feet are used.

Scale

Map **scale** expresses the relationship between distance or area on the map to the true distance or area on the earth's surface. This is generally expressed as a ratio or fraction, such as 1:24,000 or 1/24,000. The numerator, usually 1, represents map distance, and the denominator, a large number, represents ground distance. Thus, 1:24,000 means that a distance of 1 unit on the map represents a distance of 24,000 such units on the surface of the earth. It does not matter what the units are.

Often, the graphic or bar scale is more useful than the fractional scale, because it is easier to use for measuring distances between points. The graphic scale (Figure D.2) consists of a bar that is divided into equal segments, which represent equal distances on the map. One segment on the left side of the bar is usually divided into smaller units to permit more accurate estimates of fractional units.

Topographic maps, which are also referred to as quadrangles, are generally classified according to publication scale. Each series is intended to fulfill a specific type of map need. To select a map with the proper scale for a particular use, remember that large-scale maps show more detail and small-scale maps show less detail. The sizes and scales of topographic maps published by the U.S. Geological Survey are shown in Table D.1.

Each color used on the U.S. Geological Survey topographic maps has significance, as follows:

Color

Blue—water features
Black—works of man, such as homes, schools, churches, roads, and so forth
Brown—contour lines
Green—woodlands, orchards, and so forth
Red—Urban areas, important roads, public land subdivision lines

Table D.1 ***National Topographic Maps***

Series	Scale	1 Inch Represents	Standard Quadrangle Size (latitude-longitude)	Quadrangle Area (square miles)	Paper Size E-W Width × N-S Length (inches)
7½-minute	1:24,000	2000 feet	7½′ × 7½′	49–70	22 × 27 *
Puerto Rico 7½-minute	1:20,000	about 1667 feet	7½′ × 7½′	71	29½ × 32½
15-minute	1:62,500	nearly 1 mile	15′ × 15′	197–282	17 × 21 *
Alaska 1:63,360	1:63,360	1 mile	15′ × 20′–36′	207–281	18 × 21 †
U.S. 1:250,000	1:250,000	nearly 4 miles	1° × 2° ‡	4,580–8,669	34 × 22 §
U.S. 1:1,000,000	1:1,000,000	nearly 16 miles	4° × 6° ‡	73,734–102,759	27 × 27

SOURCE: U.S. Geological Survey.
* South of latitude 31 degrees, 7½-minute sheets are 23 × 27 inches; 15-minute sheets are 18 × 21 inches.
† South of latitude 62 degrees, sheets are 17 × 21 inches.
‡ Maps of Alaska and Hawaii vary from these standards.
§ North of latitude 42 degrees, sheets are 29 × 22 inches; Alaska sheets are 30 × 23 inches.

appendix E

Star Charts

These star charts can be used to locate and identify some of the brightest stars in the sky.* The months along the bottom of each map indicate the stars which will be over the observers meridian in the early evening during that month. The stars with a declination equal to the observer's latitude will be directly overhead (zenith).

*Reproduced from J.A. Ripley, Jr. and R.C. Whitten, *The Elements and Structure of the Physical Sciences,* 2nd ed.

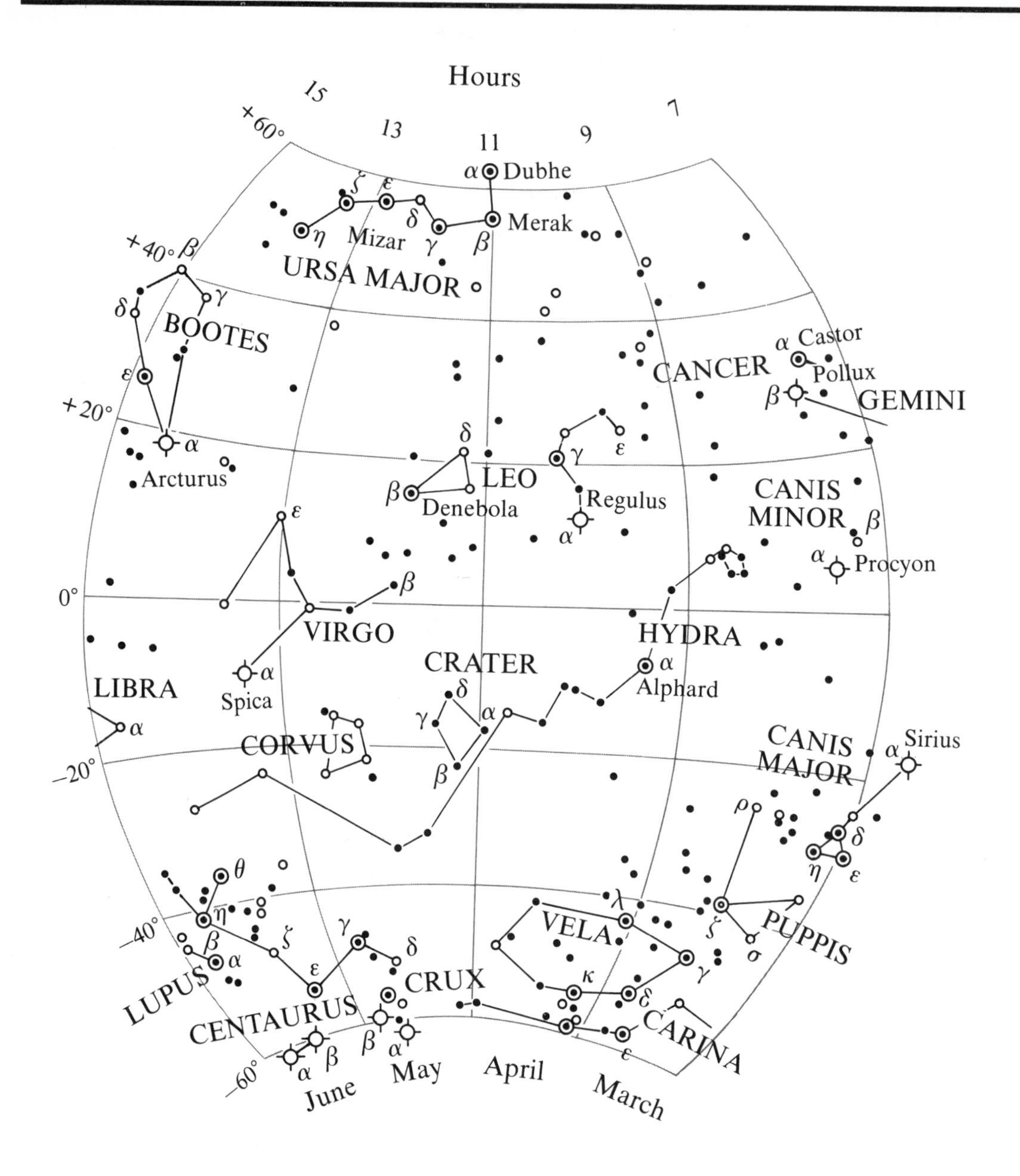

-✧- Brighter than magnitude 1.5

◉ Between magnitudes 1.5 and 2.5

○ Between magnitudes 2.5 and 3.5

• Fainter than magnitude 3.5

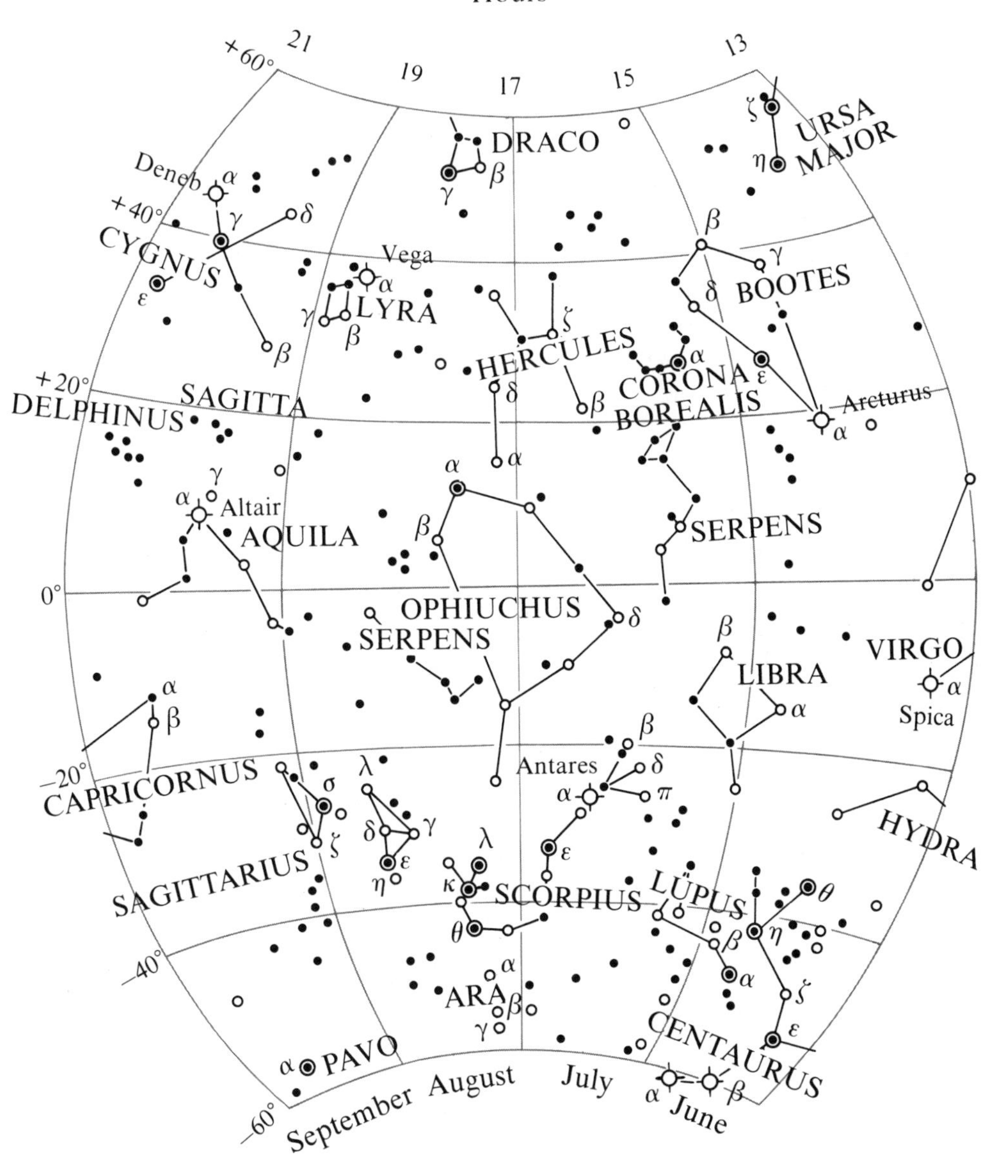

- Brighter than magnitude 1.5
- Between magnitudes 1.5 and 2.5
- Between magnitudes 2.5 and 3.5
- Fainter than magnitude 3.5

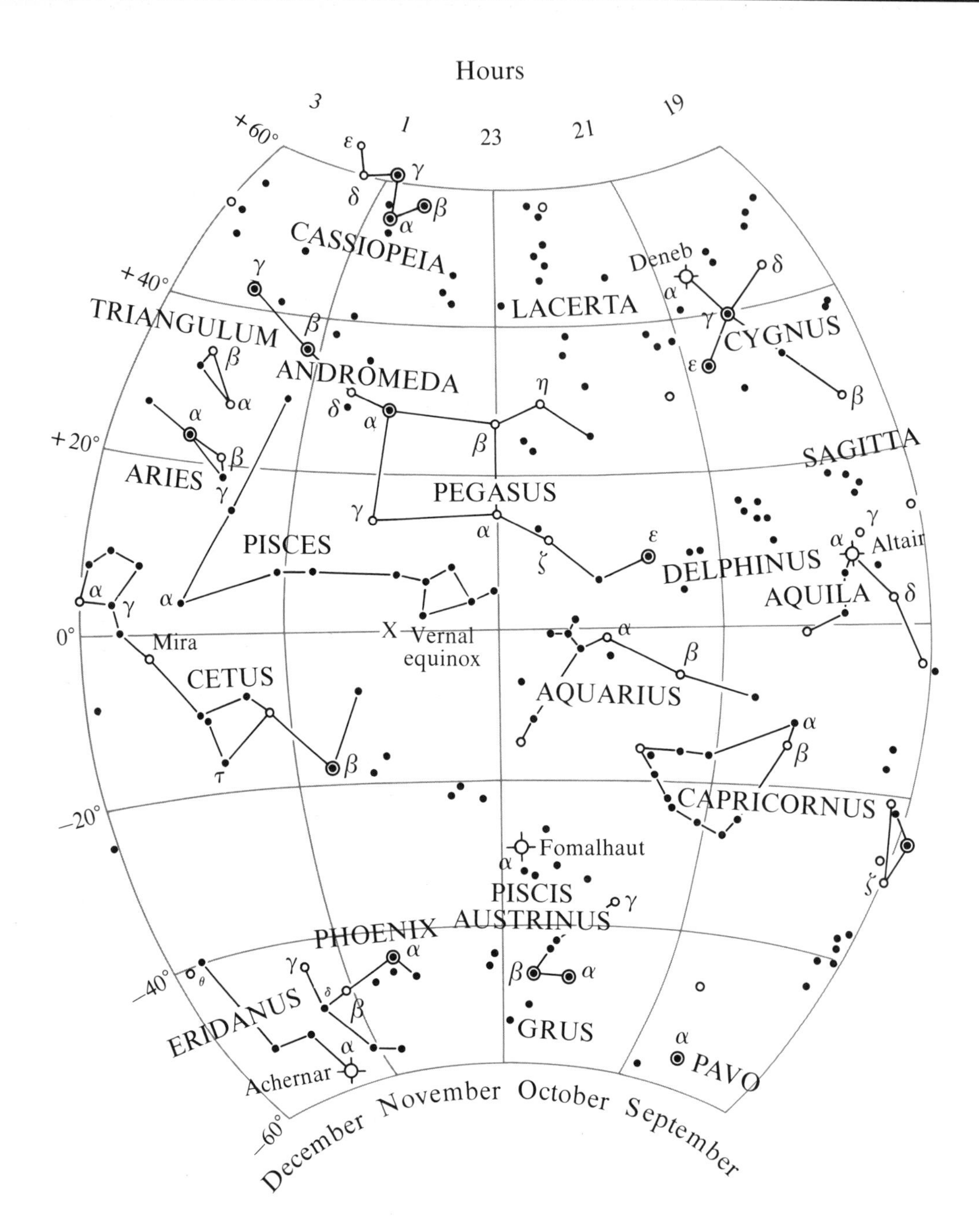
Hours
3
1
23
21
19
+60°
+40°
+20°
0°
–20°
–40°
–60°
CASSIOPEIA
LACERTA
Deneb
CYGNUS
TRIANGULUM
ANDROMEDA
ARIES
PEGASUS
SAGITTA
PISCES
DELPHINUS
Altair
AQUILA
Mira
X Vernal equinox
CETUS
AQUARIUS
CAPRICORNUS
Fomalhaut
PISCIS AUSTRINUS
PHOENIX
ERIDANUS
GRUS
PAVO
Achernar
December November October September
Brighter than magnitude 1.5
Between magnitudes 1.5 and 2.5
Between magnitudes 2.5 and 3.5
Fainter than magnitude 3.5

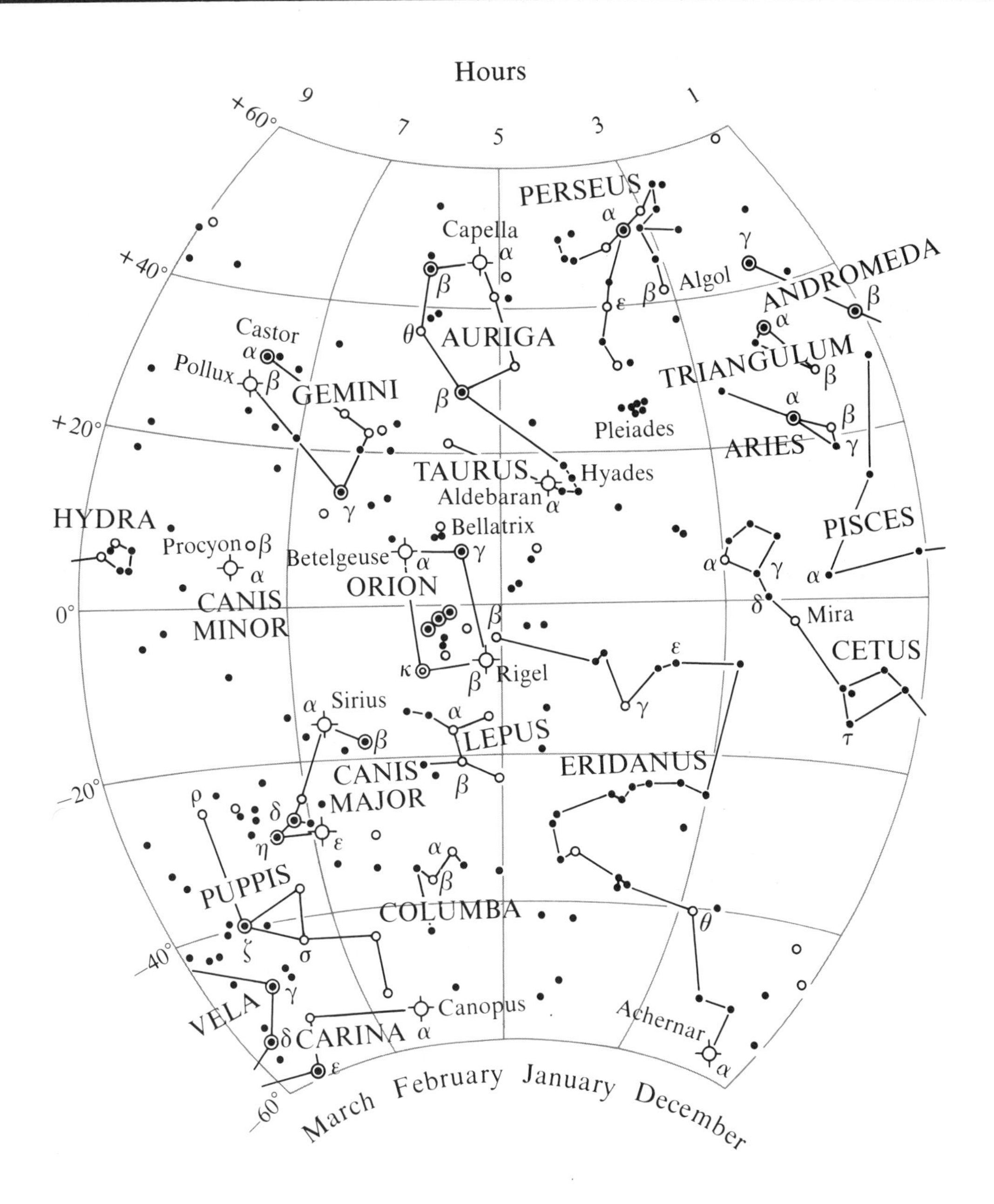

◆ Brighter than magnitude 1.5

◉ Between magnitudes 1.5 and 2.5

○ Between magnitudes 2.5 and 3.5

• Fainter than magnitude 3.5

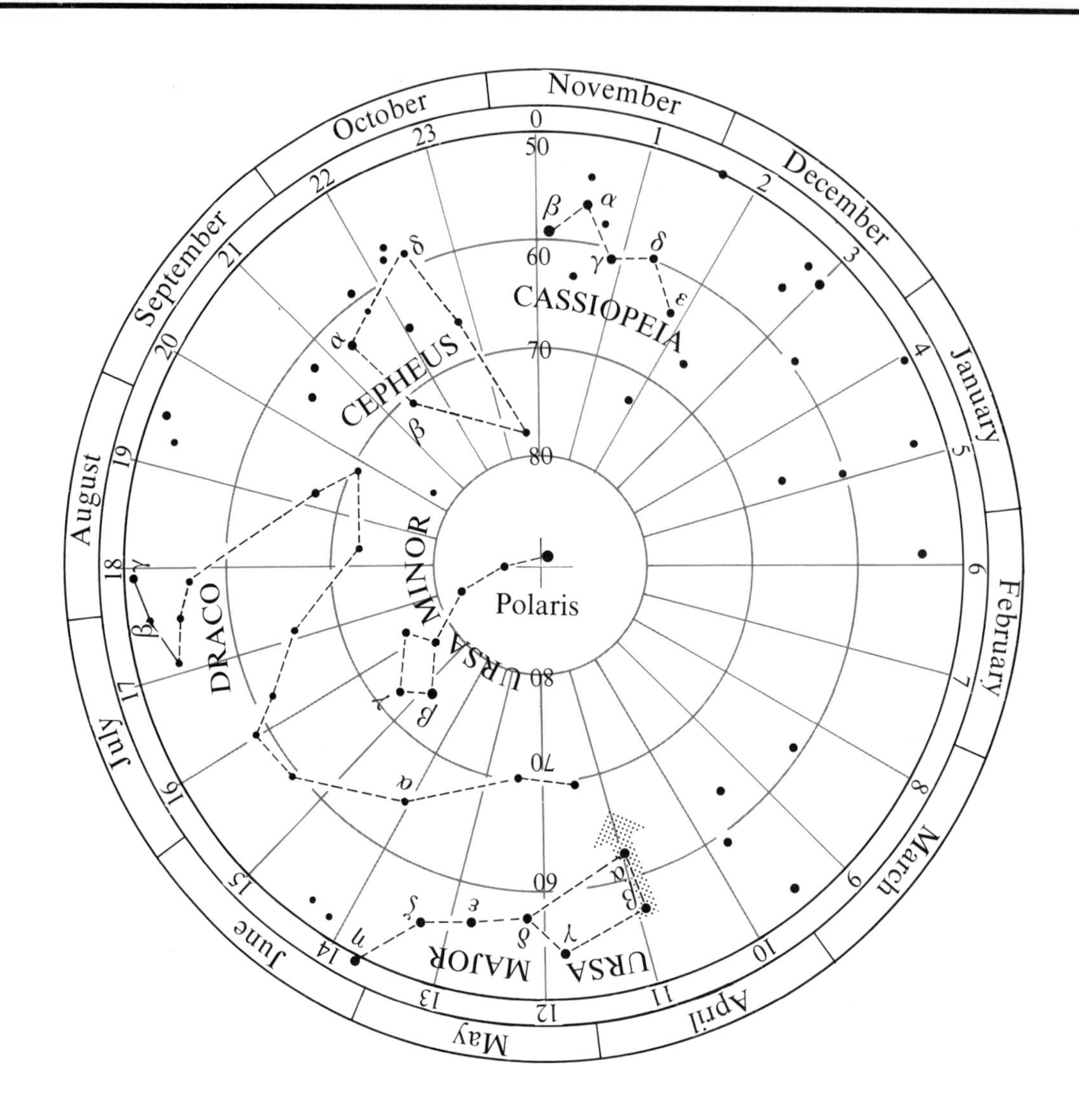
November
December
January
February
March
April
May
June
July
August
September
October
CASSIOPEIA
CEPHEUS
DRACO
URSA MINOR
URSA MAJOR
Polaris

appendix F

World Climates

The distribution of the major atmospheric elements is very complex. Because of the many variations from place to place and from time to time at the same place, it is unlikely that any two sites on the earth's surface experience exactly the same weather. Since the number of places on earth is almost infinite, the number of different climates must also be extremely large. In order to cope with such variety, it is essential to devise a means of classifying this vast array of data. By establishing groups consisting of items that have certain important characteristics in common, order and simplicity are introduced. Organizing large amounts of information not only aids comprehension and understanding but also facilitates analysis and explanation.

We should remember that the classification of climates (or of anything else) is not a natural phenomenon, but the product of human ingenuity. The value of any particular classification is determined largely by its intended use. Here we shall use a system devised by the Russian-born German climatologist Wladimir Koeppen (1846–1940). As a tool for presenting the general world pattern of climates, it has been the best known and most used system for more than fifty years. Koeppen believed that the distribution of natural vegetation was the best expression of the totality of climate. Therefore, the boundaries he chose were based largely on the limits of certain plant associations.

Five principal groups are recognized, with each group designated by a capital letter as follows:

A Humid tropical: winterless climates; all months have a mean temperature above 18° C.

B Dry: climates where evaporation exceeds precipitation; there is a constant water deficiency.

C Humid middle-latitude: mild winters; the average temperature of the coldest month is below 18° C but above −3° C.

D Humid middle-latitude: severe winters; the average temperature of the coldest month is below −3° C and the warmest monthly mean exceeds 10° C.

E Polar: summerless climates; the average temperature of the warmest month is below 10° C.

Notice that four of the major groups (A, C, D, E) are defined on the basis of temperature characteristics, and the fifth, the B group, has precipitation as its primary criterion. Each of the five groups is

Table F.1 ***Koeppen System of Climatic Classification***

Letter Symbol 1st	2nd	3rd	
A			Average temperature of the coldest month is 18° C or higher
	f		Every month has 6 centimeters of precipitation or more
	m		Short dry season; precipitation in driest month less than 6 centimeters but equal to or greater than $10 - R/25$ (R is annual rainfall in centimeters)
	w		Well-defined winter dry season; precipitation in driest month less than $10 - R/25$
	s		Well-defined summer dry season (rare)
B			Potential evaporation exceeds precipitation. The dry/humid boundary is defined by the following formulas as follows: (NOTE: R is average annual precipitation in centimeters and T is average annual temperature in °C) $R < 2T + 28$ when 70% or more of rain falls in warmer 6 months $R < 2T$ when 70% or more of rain falls in cooler 6 months $R < 2T + 14$ when neither half year has 70% or more of rain
	S		Steppe — The BS/BW boundary is 1/2 the dry/humid boundary
	W		Desert — The BS/BW boundary is 1/2 the dry/humid boundary
		h	Average annual temperature is 18° C or greater
		k	Average annual temperature is less than 18° C
C			Average temperature of the coldest month is under 18° C and above −3° C
	w		At least ten times as much precipitation in a summer month as in the driest winter month
	s		At least three times as much precipitation in a winter month as in the driest summer month; precipitation in driest summer month less than 4 centimeters
	f		Criteria for w and s cannot be met
		a	Warmest month is over 22° C; at least 4 months over 10° C
		b	No month above 22° C; at least 4 months over 10° C
		c	One to 3 months above 10° C
D			Average temperature of coldest month is −3° C or below; average temperature of warmest month is greater than 10° C
	s		Same as under C
	w		Same as under C
	f		Same as under C
		a	Same as under C
		b	Same as under C
		c	Same as under C
		d	Average temperature of the coldest month is −38° C or below
E			Average temperature of the warmest month is below 10° C
	T		Average temperature of the warmest month is greater than 0° C and less than 10° C
	F		Average temperature of the warmest month is 0° C or below

further subdivided by using the criteria and letter symbols presented in Table F.1.*

What follows is a brief summary of the major climatic types. While reading this summary, you should refer to Table F.2, which shows data for representative places of each climatic type.

Humid Tropics (A Climates)

Within the A group of climates, two main types are recognized—wet tropical climates (Af and Am) and tropical wet and dry (Aw).

Wet Tropical Climates (Af, Am)

Since places with an Af or Am designation are lowland areas that lie near the equator, temperatures are consistently high. Consequently, not only is the annual mean high, but the annual range is very small. Total annual precipitation is high, often exceeding 200 centimeters; and although precipitation is not evenly distributed throughout the year, places with this climate are generally wet in all months. If a dry season exists, it is very short. In response to the constantly high temperatures and year-round rainfall, the wet tropics are dominated by the most luxuriant vegetation found in any climatic realm—the tropical rain forest.

Tropical Wet and Dry (Aw)

In the latitude zone poleward of the wet tropics and equatorward of the tropical deserts lies a transitional climatic region called tropical wet and dry. Here the rain forest gives way to the savanna, a tropical grassland with scattered drought-tolerant trees. Since temperature characteristics among all A climates are quite similar, the primary factor which distinguishes the Aw climate from Af and Am is precipitation. Although the overall amount of precipitation in the tropical wet and dry realm is often considerably less than in the wet tropics, the most distinctive feature of this climate is not the annual rainfall total but the markedly seasonal character of the rainfall. As the equatorial low advances poleward in summer, the rainy season commences and features weather patterns typical of the wet tropics. Later, with the retreat of the equatorial low, the subtropical high advances into the region and brings with it intense drought conditions. In some Aw regions such as India, Southeast Asia, and portions of Australia, the alternating periods of rainfall and drought are associated with a pronounced monsoon circulation (see Chapter 14).

Dry (B) Climates

It is important to realize that the concept of dryness is a relative one and refers to any situation in which a water deficiency exists. Climatologists define a dry climate as one in which the yearly precipitation is not as great as the potential loss of water by evaporation. Thus, dryness is not only related to annual rainfall totals, but it is also a function of evaporation, which in turn is closely dependent upon temperature. To establish the boundary between dry and humid climates, the Koeppen classification uses formulas that involve three variables: average annual precipitation, average annual temperature, and seasonal distribution of precipitation. The use of average annual temperature reflects its importance as an index of evaporation. The amount of rainfall defining the humid-dry boundary increases as the annual mean temperature increases. The use of seasonal precipitation as a variable is also related to this idea. If rain is concentrated in the warmest months, loss to evaporation is greater than if the precipitation were concentrated in the cooler months.

Within the regions defined by a general water deficiency there are two climatic types: arid or desert (BW) and semiarid or steppe (BS). These two groups have many features in common; their differences are primarily a matter of degree. The semiarid is a marginal and more humid variant of the arid and represents a transition zone that surrounds the desert and separates it from the bordering humid climates.

The heart of the low-latitude dry climates (BWh and BSh) lies in the vicinity of the Tropic of Cancer and the Tropic of Capricorn. The existence and distribution of this rather extensive dry tropical realm is primarily a consequence of the subsidence and marked stability of the subtropical highs. Unlike their low-latitude counterparts, middle-latitude deserts and steppes are not controlled by the sub-

*When classifying climatic data by using Table F.1, you should first determine whether or not the data meet the criteria for the E climates. If the station is not a polar climate, proceed to the criteria for B climates. If your data do not fit into either the E or B groups, check the data against the criteria for A, C, and D climates, in that order.

Table F.2 ***Climatic Data for Representative Stations***

	J	F	M	A	M	J	J	A	S	O	N	D	Year
Iquitos, Peru (Af); lat. 3°39′S; 115m													
Temp (°C)	25.6	25.6	24.4	25.0	24.4	23.3	23.3	24.4	24.4	25.0	25.6	25.6	24.7
Precip. (mm)	259	249	310	165	254	188	168	117	221	183	213	292	2619
Rio de Janeiro, Brazil (Aw); lat. 22°50′S; 26m													
Temp. (°C)	25.9	26.1	25.2	23.9	22.3	21.3	20.8	21.1	21.5	22.3	23.1	24.4	23.2
Precip. (mm)	137	137	143	116	73	43	43	43	53	74	97	127	1086
Faya, Chad (BWh); lat. 18°00′N; 251m													
Temp. (°C)	20.4	22.7	27.0	30.6	33.8	34.2	33.6	32.7	32.6	30.5	25.5	21.3	28.7
Precip. (mm)	0	0	0	0	0	2	1	11	2	0	0	0	16
Salt Lake City, Utah (BSk); lat. 40°46′N; 1288m													
Temp. (°C)	−2.1	0.9	4.7	9.9	14.7	19.4	24.7	23.6	18.3	11.5	3.4	−0.2	10.7
Precip. (mm)	34	30	40	45	36	25	15	22	13	29	33	31	353
Washington, D.C. (Cfa); lat. 38°50′N; 20m													
Temp. (°C)	2.7	3.2	7.1	13.2	18.8	23.4	25.7	24.7	20.9	15.0	8.7	3.4	13.9
Precip. (mm)	77	63	82	80	105	82	105	124	97	78	72	71	1036
Brest, France (Cfb); lat. 48°24′N; 103m													
Temp. (°C)	6.1	5.8	7.8	9.2	11.6	14.4	15.6	16.0	14.7	12.0	9.0	7.0	10.8
Precip. (mm)	133	96	83	69	68	56	62	80	87	104	138	150	1126
Rome, Italy (Csa); lat. 41°52′N; 3m													
Temp. (°C)	8.0	9.0	10.9	13.7	17.5	21.6	24.4	24.2	21.5	17.2	12.7	9.5	15.9
Precip. (mm)	83	73	52	50	48	18	9	18	70	110	113	105	749
Peoria, Illinois (Dfa); lat. 40°45′N; 180m													
Temp. (°C)	−4.4	−2.2	4.4	10.6	16.7	21.7	23.9	22.7	18.3	11.7	3.8	−2.2	10.4
Precip. (mm)	46	51	69	84	99	97	97	81	97	61	61	51	894
Verkhoyansk, U.S.S.R. (Dfd); lat. 67°33′N; 137m													
Temp. (°C)	−46.8	−43.1	−30.2	−13.5	2.7	12.9	15.7	11.4	2.7	−14.3	−35.7	−44.5	−15.2
Precip. (mm)	7	5	5	4	5	25	33	30	13	11	10	7	155
Ivigtut, Greenland (ET); lat. 61°12′N; 129m													
Temp. (°C)	−7.2	−7.2	−4.4	−0.6	4.4	8.3	10.0	8.3	5.0	1.1	−3.3	−6.1	0.7
Precip. (mm)	84	66	86	62	89	81	79	94	150	145	117	79	1132
McMurdo Station, Antarctica (EF); lat. 77°53′S; 2m													
Temp. (°C)	−4.4	−8.9	−15.5	−22.8	−23.9	−24.4	−26.1	−26.1	−24.4	−18.8	−10.0	−3.9	−17.4
Precip. (mm)	13	18	10	10	10	8	5	8	10	5	5	8	110

siding air masses of the subtropical anticyclones. Instead, these dry lands exist principally because of their position in the deep interiors of large landmasses far removed from the oceans. In addition, the presence of high mountains across the paths of the prevailing winds further acts to separate these areas from water-bearing maritime air masses.

Humid Middle-latitude Climates with Mild Winters (C Climates)

Although the term *subtropical* is often used for the C climates, it can be misleading. While many areas with C climates do indeed possess some near-tropical characteristics, other regions do not. For example, we would be stretching the use of the term *subtropical* to describe the climates of coastal Alaska and Norway, which belong to the C group. Within the C group of climates, several subgroups are recognized.

Humid Subtropics (Cfa)

Located on the eastern sides of the continents, in the 25- to 40-degree latitude range, this climatic type dominates the southeastern United States, as well as other similarly situated areas around the world. In the summer the humid subtropics experience hot, sultry weather of the type one expects to find in the rainy tropics. Daytime temperatures are generally high, and since both specific and relative humidities are high, the night brings little relief. An afternoon or evening thunderstorm is also possible, for these areas experience such storms on an average of 40 to 100 days each year, the majority during the summer months. As summer turns to autumn, the humid subtropics lose their similarity to the rainy tropics. Although winters are mild, frosts are common in the higher-latitude Cfa areas and occasionally plague the tropical margins as well. The winter precipitation is also different in character from the summer. Some is in the form of snow, and most is generated along fronts of the frequent middle-latitude cyclones that sweep over these regions.

Marine West Coast Climate (Cfb, Cfc)

Situated on the western (windward) side of continents, from about 40 to 60 degrees north and south latitude, is a climatic region dominated by the onshore flow of oceanic air. The prevalence of maritime air masses means that mild winters and cool summers are the rule, as is an ample amount of rainfall throughout the year. Although there is no pronounced dry period, there is a drop in monthly precipitation totals during the summer. The reason for the reduced summer rainfall is the poleward migration of the oceanic subtropical highs. Although the areas of marine west coast climate are situated too far poleward to be dominated by these dry anticyclones, their influence is sufficient to cause a decrease in warm season rainfall.

Dry-summer Subtropics (Csa, Csb)

This climate is typically located along the west sides of continents, between latitudes of 30 and 45 degrees. It is unique because it is the only humid climate that has a strong winter rainfall maximum, a feature that reflects its intermediate position between the marine west coast on the poleward side and the tropical steppes on the equatorward side. In summer the region is dominated by the stable eastern side of the oceanic subtropical highs. In winter, as the wind and pressure systems follow the sun equatorward, it is within range of the cyclonic storms of the polar front. Thus, during the course of a year these areas alternate between being part of the dry tropics and being an extension of the humid middle latitudes. While middle-latitude changeability characterizes the winter, tropical constancy describes the summer.

Humid Middle-latitude Climates with Severe Winters (D Climates)

The D climates are land-controlled climates, the result of broad continents in the middle latitudes. Because continentality is a basic feature, D climates are absent in the Southern Hemisphere where the middle-latitude zone is dominated by the oceans.

Humid Continental (Dfa, Dfb, Dwa, Dwb)

This climate is located in the central and eastern portions of North America and Eurasia in the latitude range between approximately 40 and 50 degrees north latitude. Both winter and summer temperatures may be characterized as relatively severe. Consequently, annual temperature ranges are high throughout the climate. Precipitation is characteristically greatest in summer. A portion of the winter precipitation is in the form of snow, the proportion increasing with latitude. Although the precipitation is generally considerably less during

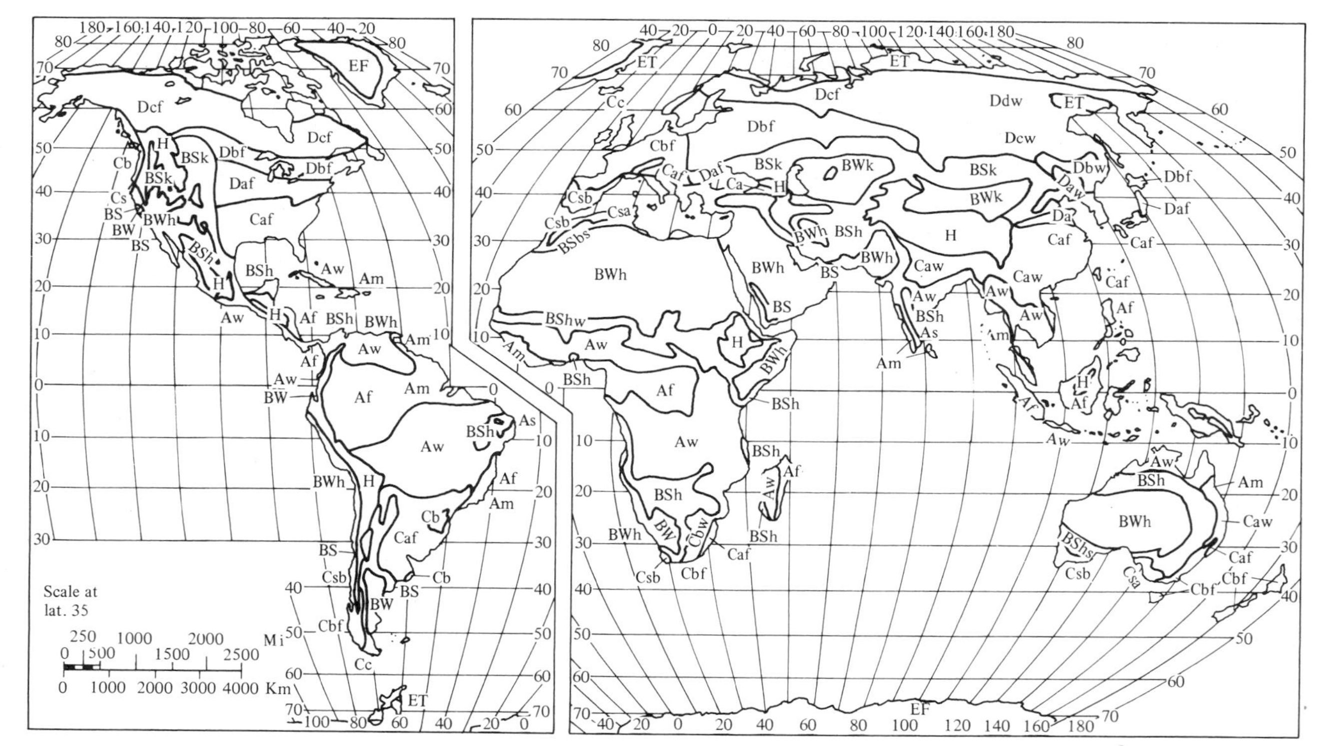

Legend. Af, Am: Tropical wet. Aw: Tropical wet, dry winters. BSh: Tropical and subtropical steppe. BSk: Middle latitude steppe. BWh: Tropical and subtropical desert. BWk: Middle latitude desert. Cs: Dry summer subtropical. Ca: Humid subtropical. Cb, Cc: Marine. Da: Humid continental, warm summer. Db: Humid continental, cold summer. Dc, Dd: Subarctic. ET: Tundra. EF: Ice cap. H: Undifferentiated highlands.

Figure F.1 *Climates of the world (Koeppen). (From Albert Miller and Jack C. Thompson,* Elements of Meteorology, *3rd ed., Columbus, Ohio: Charles E. Merrill, 1979)*

the cold season, it is usually much more conspicuous than the greater amounts which characterize the summer. An obvious reason is that snow remains on the ground, often for extended periods, and rain, of course, does not. Furthermore, while summer rains are often in the form of relatively short showers, winter snows generally occur over a prolonged period.

Subarctic Climate (Dfc, Dfd, Dwc, Dwd)

Situated north of the humid continental climate and south of the polar tundra is an extensive subarctic region. It is often referred to as the taiga climate because its extent closely corresponds to the northern coniferous forest of the same name. The outstanding feature in this realm is the dominance of winter. Not only is it long, but temperatures are also bitterly cold. By contrast, summers in the subarctic are remarkably warm, despite their short duration. However, when compared with regions farther south, this short season must be characterized as cool. The extremely cold winters and relatively warm summers combine to produce the highest annual temperature ranges on earth. Since these far northerly continental interiors are the source regions for cP air masses, there is very limited moisture available throughout the year. Precipitation totals are therefore small, with a maximum occurring during the warmer summer months.

Polar (E) Climates

Polar climates are those in which the mean temperature of the warmest month is below 10° C. Thus, just as the tropics are defined by their year-round warmth, the polar realm is known for its enduring cold. Since winters are periods of perpetual night, or nearly so, temperatures at most polar locations are understandably bitter. During the summer months temperatures remain cool despite the long days, because the sun is so low in the sky that its oblique rays are not effective in bringing about a genuine warming. Although polar climates are classified as humid, precipitation is generally meager. Evaporation, of course, is also limited. The scanty precipitation totals are easily understood in view of the temperature characteristics of the region. The amount of water vapor in the air is always small because low specific humidities must accompany low temperatures. Usually precipitation is most abundant during the warmer summer months when the moisture content of the air is highest.

Two types of polar climates are recognized. The tundra climate (ET) is a treeless climate found almost exclusively in the Northern Hemisphere. Because of the combination of high latitude and continentality, winters are severe, summers are cool, and annual temperature ranges are high. Further, yearly precipitation is small, with a modest summer maximum. The ice cap climate (EF) does not have a single monthly mean above 0° C. Consequently, since the average temperature for all months is below freezing, the growth of vegetation is prohibited and the landscape is one of permanent ice and snow. This climate of perpetual frost covers a surprisingly large area—more than 15.5 million square kilometers, or about 9 percent of the earth's land area. Aside from scattered occurrences in high mountain areas, it is confined to the ice caps of Greenland and Antarctica.

glossary

Aa A type of lava flow that has a jagged blocky surface.

Absolute dating Determination of the number of years since the occurrence of a given geologic event.

Absolute humidity The weight of water vapor in a given volume of air (usually expressed in grams per cubic meter).

Absolute instability Air that has a lapse rate greater than the dry adiabatic rate (about 5.5°F per 1000 feet).

Absolute magnitude The apparent brightness of a star if it were viewed from a distance of 10 parsecs (32.6 light-years). Used to compare the true brightness of stars.

Absolute stability Air with a lapse rate less than the wet adiabatic rate (about 2.5°F per 1000 feet).

Absorption spectrum A continuous spectrum with dark lines superimposed.

Abyssal plain Very level area of the deep ocean floor, usually lying at the foot of the continental rise.

Adiabatic temperature change Cooling or warming of air caused when air is allowed to expand or is compressed, not because heat is added or subtracted.

Advection Horizontal convective motion, such as wind.

Advection fog A fog formed when warm, moist air is blown over a cool surface.

Air mass A large body of air that is characterized by a sameness of temperature and humidity.

Albedo The reflectivity of a substance, usually expressed as a percentage of the incident radiation reflected.

Alluvial fan A fan-shaped deposit of sediment formed when a stream's slope is abruptly reduced.

Alluvium Unconsolidated sediment deposited by a stream.

Anemometer An instrument used to determine wind speed.

Angular unconformity An unconformity in which the strata below dip at an angle different from that of the beds above.

Anthracite A hard, metamorphic form of coal that burns clean and hot.

Anticline A fold in sedimentary strata that resembles an arch.

Anticyclone A high-pressure center characterized by a clockwise flow of air in the Northern Hemisphere.

Aphelion The place in the orbit of a planet where the planet is furthest from the sun.

Apparent magnitude The brightness of a star when viewed from the earth.

Aquifer Rock or soil through which groundwater moves easily.

Arête A narrow knifelike ridge separating two adjacent glaciated valleys.

Arkose A feldspar-rich sandstone.

Artesian well A well in which the water rises above the level where it was initially encountered.

Asteroids Thousands of small planetlike bodies,

ranging in size from a few hundred kilometers to less than a kilometer, whose orbits lie mainly between those of Mars and Jupiter.

Asthenosphere Layer of the earth found below the lithosphere that is thought to be partially molten.

Astronomical theory A theory of climatic change first developed by the Yugoslavian astronomer Milankovitch. It is based upon changes in the shape of the earth's orbit, variations in the obliquity of the earth's axis, and the wobbling of the earth's axis.

Atoll A continuous or broken ring of coral reef surrounding a central lagoon.

Atomic number The number of protons in the nucleus of an atom.

Aurora A bright display of ever-changing light caused by solar radiation interacting with the upper atmosphere in the region of the poles.

Barometer An instrument that measures atmospheric pressure.

Barred spiral A galaxy having straight arms extending from its nucleus.

Barrier island A low, elongate ridge of sand that parallels the coast.

Basalt A fine-grained igneous rock of mafic composition.

Base level The level below which a stream cannot erode.

Basin A circular downfolded structure.

Batholith A large mass of igneous rock that formed when magma was implaced at depth, crystallized, and was subsequently exposed by erosion.

Baymouth bar A sandbar that completely crosses a bay, sealing it off from the open ocean.

Beach drift The transport of sediment in a zigzag pattern along a beach caused by the uprush of water from obliquely breaking waves.

Bed load Sediment that is carried by a stream along the bottom of its channel.

Big Bang theory The theory that proposes that the universe originated as a single mass, which subsequently exploded.

Binary stars Two stars revolving around a common center of mass under their mutual gravitational attraction.

Biogenous sediment Sea-floor sediments consisting of material of marine-organic origin.

Bituminous The most common form of coal, often called soft, black coal.

Blowout (deflation hollow) A depression excavated by the wind in easily eroded deposits.

Bode's law A sequence of numbers that approximates the mean distances of the planets from the sun.

Braided stream A stream consisting of numerous intertwining channels.

Breccia A sedimentary rock composed of angular fragments that were lithified.

Bright-line spectrum The bright lines produced by an incandescent gas under low pressure.

Cactolith A quasi-horizontal chonolith composed of anastomosing ductoliths, whose distal ends curl like a harpolith, thin like a sphenolith, or bulge discordantly like an akmolith or ethmolith.

Caldera A large depression typically caused by collapse or ejection of the summit area of a volcano.

Capacity The total amount of sediment a stream is able to transport.

Catastrophism The concept that the earth was shaped by catastrophic events of a short-term nature.

Celestial sphere An imaginary hollow sphere upon which the ancients believed the stars were hung and carried around the earth.

Cepheid variable A star whose brightness varies periodically because it expands and contracts. A type of pulsating star.

Chemical weathering The processes by which the internal structure of a mineral is altered by the removal and/or addition of elements.

Chinook A wind blowing down the leeward side of a mountain and warming by compression.

Chromatic aberration The property of a lens whereby light of different colors is focused at different places.

Chromosphere The first layer of the solar atmosphere found directly above the photosphere.

Cinder cone A rather small volcano built primarily of pyroclastics ejected from a single vent.

Cirque An amphitheater-shaped basin at the head of a glaciated valley produced by frost wedging and plucking.

Clastic rock A sedimentary rock made of broken fragments of preexisting rock.

Climatic change A study that deals with variations in climate on many different time scales, from decades to millions of years, and the possible causes of such variations.

Climatic feedback mechanism One of the several different outcomes that may result if one of the many elements in the atmosphere's extremely complex interactive system is altered.

Cold front A front along which a cold air mass thrusts beneath a warmer air mass.

Color index The difference between the visual magnitude of a star and its magnitude when photographed using a selected color-sensitive film.

Columnar joints A pattern of cracks that forms during cooling of molten rock generating columns.

Comet A small body which generally revolves about the sun in an elongated orbit.

Competence A measure of the largest particle a stream can transport; a factor dependent on velocity.

Composite cone A volcano composed of both lava flows and pyroclastic material.

Condensation The change of state from a gas to a liquid.

Condensation nuclei Tiny bits of particulate matter that serve as surfaces on which water vapor condenses.

Conditional instability Moist air with a lapse rate between the dry and wet adiabatic rates (between about 2.5°F and 5.5°F per 1000 feet).

Conduction The transfer of heat through matter by molecular activity. Energy is transferred through collisions from one molecule to another.

Conglomerate A sedimentary rock composed of rounded gravel-sized particles.

Constellation An apparent group of stars originally named for mythical characters. The sky is presently divided into 88 constellations.

Contact metamorphism Changes in rock caused by the heat from a nearby magma body.

Continental drift theory A theory that originally proposed that the continents are rafted about. It has essentially been replaced by the plate tectonic theory.

Continental rise The gently sloping surface at the base of the continental slope.

Continental shelf The gently sloping submerged portion of the continental margin extending from the shoreline to the continental slope.

Continental slope The steep gradient that leads to the deep ocean floor and marks the seaward edge of the continental shelf.

Continuous spectrum An uninterrupted band of light emitted by an incandescent solid, liquid, or gas under pressure.

Convection The transfer of heat by the movement of a mass or substance. It can only take place in fluids.

Convergence zone An area where plates move together and either collide or one is subducted.

Coriolis force (effect) The deflective force of the earth's rotation on all free-moving objects, including the atmosphere and oceans. Deflection is to the right in the Northern Hemisphere and to the left in the Southern Hemisphere.

Corona The outer, tenuous layer of the solar atmosphere.

Correlation Establishing the equivalence of rocks of similar age in different areas.

Country breeze A circulation pattern characterized by a light wind blowing into a city from the surrounding countryside. It is best developed on clear and otherwise calm nights when the urban heat island is most pronounced.

Crater The depression at the summit of a volcano, or that which is produced by a meteorite impact.

Creep The slow downhill movement of soil and regolith.

Crevasse A deep crack in the brittle surface of a glacier.

Cross-cutting A principle of relative dating. A rock or fault is younger than any rock (or fault) through which it cuts.

Crystal An orderly arrangement of atoms.

Crystal form The external appearance of a mineral as determined by its internal arrangement of atoms.

Cyclone A low-pressure center characterized by a counterclockwise flow of air in the Northern Hemisphere.

Declination (stellar) The angular distance north or south of the celestial equator denoting the position of a celestial body.

Deflation The lifting and removal of loose material by wind.

Delta An accumulation of sediment formed where a stream enters a lake or ocean.

Dendritic pattern A stream system that resembles the pattern of a branching tree.

Density The weight per unit volume of a particular material.

Desalination The removal of salts and other chemicals from seawater.

Desert pavement A layer of coarse pebbles and gravel created when wind removed the finer material.

Dew point The temperature to which air has to be cooled in order to reach saturation.

Dike A tabular-shaped intrusive igneous feature that cut through the surrounding rock.

Discharge The quantity of water in a stream that passes a given point in a period of time.

Disconformity A type of unconformity in which the beds above and below are parallel.

Dissolved load That portion of a stream's load carried in solution.

Distributary A section of a stream that leaves the main flow.

Divergent zone A region where the rigid plates are moving apart, typified by the mid-oceanic ridges.

Divide An imaginary line that separates the drainage of two streams; often found along a ridge.

Dome A roughly circular upfolded structure similar to an anticline.

Doppler effect The apparent change in wavelength of radiation caused by the relative motions of the source and the observer.

Drumlin A streamlined asymmetrical hill composed of glacial till. The steep side of the hill faces the direction from which the ice advanced.

Dune A hill or ridge of wind-deposited sand.

Earthflow The downslope movement of water-saturated, clay-rich sediment. Most characteristic of humid regions.

Eccentricity The variation of an ellipse from a circle.

Eclipse The cutting off of the light of one celestial body by another passing in front of it.

Ecliptic The yearly path of the sun plotted against the background of stars.

Electromagnetic spectrum The distribution of electromagnetic radiation by wavelength.

End moraine A ridge of till marking a former position of the front of a glacier.

Entrenched meander A meander cut into bedrock when uplifting rejuvenated a meandering stream.

Epicenter The location on the earth's surface that lies directly above the focus of an earthquake.

Erosion The incorporation and transportation of material by a mobile agent, such as water, wind, or ice.

Esker Sinuous ridge composed largely of sand and gravel deposited by a stream flowing in a tunnel beneath a glacier near its terminus.

Evaporation The process of converting a liquid to a gas.

Evaporite A sedimentary rock formed of material deposited from solution by evaporation of the water.

Exfoliation A mechanical weathering process characterized by the splitting off of slablike sheets of rock.

Exotic stream A permanent stream that traverses a desert and has its source in well-watered areas outside the desert.

Extrusive Igneous activity that occurs outside the crust.

Fault A break in a rock mass along which movement has occurred.

Fault-block mountain A mountain formed by the displacement of rock along a fault.

Felsic The group of igneous rocks composed primarily of feldspar and quartz.

Fetch The distance that the wind has traveled across the open water.

Fiord A steep-sided inlet of the sea formed when a glacial trough was partially submerged.

Flare A sudden brightening of an area on the sun.

Floodplain The flat, low-lying portion of a stream valley subject to periodic inundation.

Fluorescence The absorption of ultraviolet light, which is reemitted as visible light.

Focus (earthquake) The zone within the earth where rock displacement produces an earthquake.

Focus (light) The point where a lens or mirror causes light rays to converge.

Foliated A texture of metamorphic rocks that gives the rock a layered appearance.

Fossil The remains or traces of organisms preserved from the geologic past.

Fractional crystallization The process that separates magma into components having varied compositions and melting points.

Front The boundary between two adjoining air masses having contrasting characteristics.

Frontal wedging Lifting of air resulting when cool

air acts as a barrier over which warmer, lighter air will rise.

Frost wedging The mechanical breakup of rock caused by the expansion of freezing water in cracks and crevices.

Geocentric The concept of an earth-centered universe.

Geostrophic wind A wind, usually above a height of 600 meters (2000 feet), that blows parallel to the isobars.

Geosyncline A large linear downwarp in the earth's crust in which thousands of meters of sediment have accumulated.

Geyser A fountain of hot water ejected periodically.

Glacial trough A mountain valley that has been widened, deepened, and straightened by a glacier.

Glacier A thick mass of ice originating on land from the compaction and recrystallization of snow that shows evidence of past or present flow.

Glaze A coating of ice on objects formed when supercooled rain freezes on contact.

Globular cluster A nearly spherically shaped group of densely packed stars.

Graben A valley formed by the downward displacement of a fault-bounded block.

Gradient The slope of a stream; generally measured in feet per mile.

Guyot A submerged flat-topped seamount.

Gyre The large circular surface current pattern found in each ocean.

Hail Nearly spherical ice pellets having concentric layers and formed by the successive freezing of layers of water.

Half-life The time required for one-half of the atoms of a radioactive substance to decay.

Halocline A layer of water in which there is a high rate of change in salinity in the vertical dimension.

Hanging valley A tributary valley that enters a glacial trough at a considerable height above its floor.

Heliocentric The view that the sun is at the center of the solar system.

Horn A pyramidlike peak formed by glacial action in three or more cirques surrounding a mountain summit.

Horst An elongate, uplifted block of crust bounded by faults.

H-R diagram A plot of stars according to their absolute magnitudes and spectral types.

Humus Organic matter in soil produced by the decomposition of plants and animals.

Hurricane A tropical cyclonic storm having winds in excess of 119 kilometers (74 miles) per hour.

Hydrogenous sediment Sea-floor sediments consisting of minerals that crystallize from seawater. The principal example is manganese nodules.

Igneous rock A rock formed by the crystallization of molten magma.

Inertia A property of matter that resists a change in its motion.

Infiltration The movement of surface water into rock or soil through cracks and pore space.

Intermediate composition The composition of igneous rocks lying between felsic and mafic.

Interstellar matter Dust and gasses found between stars.

Intrusive rock Igneous rock that formed below the earth's surface.

Ionosphere A complex zone of ionized gasses that coincides with the lower portion of the thermosphere.

Island arc A group of volcanic islands formed by the subduction and partial melting of oceanic lithosphere.

Isobar A line drawn on a map connecting points of equal atmospheric pressure, usually corrected to sea level.

Isostacy The concept that the earth's crust is "floating" in gravitational balance upon the material of the mantle.

Isotherms Lines connecting points of equal temperature.

Jet stream Swift (120–240-kilometer per hour 75–150-mile per hour), high-altitude winds.

Joint A fracture in rock along which there has been no movement.

Kame A steep-sided hill composed of sand and gravel originating when sediment collected in openings in stagnant glacial ice.

Karst A topography consisting of numerous depressions called sinkholes.

Kettle holes Depressions created when blocks of ice became lodged in glacial deposits and subsequently melted.

Laccolith A massive igneous body intruded between preexisting strata.

Land breeze A local wind blowing from land toward the water during the night in coastal areas.

Lapse rate (normal) The average drop in temperature (6.5° C per kilometer; 3.5° F per 1000 feet) with increased altitude in the troposphere.

Latent heat The energy absorbed or released during a change in state.

Lateral moraine A ridge of till along the sides of an alpine glacier composed primarily of debris that fell to the glacier from the valley walls.

Laterite A red, highly leached soil type found in the tropics that is rich in oxides of iron and aluminum.

Law of superposition In any undeformed sequence of sedimentary rocks, each bed is older than the one above it and younger than the one below.

Light-year The distance light travels in a year; about 6 trillion miles.

Lithification The process, generally cementation and/or compaction, of converting sediments to solid rock.

Lithogenous sediment Sea-floor sediments having their source as products of weathering on the continents.

Lithosphere The rigid outer layer of the earth, including the crust and upper mantle.

Loess Deposits of wind-blown silt, lacking visible layers, generally buff colored, and capable of maintaining a nearly vertical cliff.

Longshore current A near-shore current that flows parallel to the shore.

Low velocity zone *See* Asthenosphere.

Luminosity The brightness of a star. The amount of energy radiated by a star.

Mafic Igneous rocks with a low silica content and a high iron-magnesium content.

Magma A body of molten rock found at depth, including any dissolved gasses and crystals.

Magnitude (earthquake) The total amount of energy released during an earthquake.

Magnitude (stellar) A number given to a celestial object to express its relative brightness.

Mantle The 2900-kilometer (1800-mile) thick layer of the earth located below the crust.

Maria The Latin name for the smooth areas of the moon formerly thought to be seas.

Mass wasting The downslope movement of rock, regolith, and soil under the direct influence of gravity.

Meander A looplike bend in the course of a stream.

Mechanical weathering The physical disintegration of rock, resulting in smaller fragments.

Medial moraine A ridge of till formed when lateral moraines from two coalescing alpine glaciers join.

Mélange A highly deformed mixture of rock material formed in areas of plate convergence.

Melt The liquid portion of magma excluding the solid crystals.

Mesosphere The layer of the atmosphere immediately above the stratosphere and characterized by decreasing temperatures with height.

Metamorphism The changes in mineral composition and texture of a rock subjected to high temperature and pressure within the earth.

Meteoroid Small solid particles that have orbits in the solar system.

Mid-ocean ridge A continuous mountainous ridge on the floor of all the major ocean basins and varying in width from 500–5000 kilometers (300–3000 miles). The rifts at the crests of these ridges represent divergent plate boundaries.

Mineral A naturally occurring, inorganic crystalline material with a unique chemical structure.

Moho The boundary separating the crust from the mantle, discernible by an increase in seismic velocity.

Monsoon Seasonal reversal of wind direction associated with large continents, especially Asia. In winter, the wind blows from land to sea; in summer, from sea to land.

Mudflow The flowage of debris containing a large amount of water; most characteristic of canyons and gullies in dry, mountainous regions.

Natural levees The elevated landforms that parallel some streams and act to confine their waters, except during floodstage.

Nebula A cloud of interstellar gas and/or dust.

Normal fault A fault in which the rock above the fault plane has moved down relative to the rock below.

Nova A star that explosively increases in brightness.

Nuée ardente Incandescent volcanic debris buoyed up by hot gasses that moves downslope in an avalanche fashion.

Obliquity The angle between the planes of the earth's equator and orbit.

Obsidian A volcanic glass of felsic composition.

Occluded front A front formed when a cold front overtakes a warm front. It marks the beginning of the end of a middle-latitude cyclone.

Open cluster A loosely formed group of stars of similar origin.
Orbit The path of a body in revolution around a center of mass.
Orogenesis The processes that collectively result in the formation of mountains.
Orographic lifting Mountains acting as barriers to the flow of air force the air to ascend. The air cools adiabatically, and clouds and precipitation may result.
Outwash Stratified deposits dropped by glacial meltwater.
Oxbow lake A curved lake produced when a stream cuts off a meander.
Pahoehoe A lava flow with a smooth-to-ropy surface.
Pangaea The proposed supercontinent which 200 million years ago began to break apart and form the present land masses.
Parallax The apparent shift of an object when viewed from two different locations.
Parsec The distance at which an object would have a parallax angle of 1 second of arc (3.26 light-years).
Pedalfer Soil of humid regions characterized by the accumulation of iron oxides and aluminum-rich clays in the *B* horizon.
Pediment A sloping bedrock surface fringing a mountain base in an arid region, formed when erosion causes the mountain front to retreat.
Pedocal Soil associated with drier regions and characterized by an accumulation of calcium carbonate in the upper horizons.
Peridotite An igneous rock of ultramafic composition thought to be abundant in the upper mantle.
Perihelion The point in the orbit of a planet where it is closest to the sun.
Permeability A measure of a material's ability to transmit water.
Perturbation The gravitational disturbance of the orbit of one celestial body by another.
Phenocryst Conspicuously large crystals imbedded in a matrix of finer-grained crystals.
Photochemical reaction A chemical reaction in the atmosphere that is triggered by sunlight, often yielding a secondary pollutant.
Photon A discrete amount (quantum) of electromagnetic energy.
Photosphere The region of the sun that radiates energy to space. The visible surface of the sun.
Plage A bright region in the solar atmosphere located above a sunspot.
Planetary nebula A shell of incandescent gas expanding from a star.
Plate One of numerous rigid sections of the lithosphere that moves as a unit over the material of the asthenosphere.
Plate tectonics The theory which proposes that the earth's outer shell consists of individual plates which interact in various ways and thereby produce earthquakes, volcanoes, mountains, and the crust itself.
Playa A flat area on the floor of an undrained desert basin. Following heavy rain, the playa becomes a lake.
Porphyry An igneous texture consisting of large crystals embedded in a matrix of fine crystals.
Porosity The volume of open spaces in rock or soil.
Precession A slow motion of the earth's axis which traces out a cone over a period of 26,000 years.
Precipitation fog Fog formed when rain evaporates as it falls through a layer of cool air.
Pressure gradient The amount of pressure change occurring over a given distance.
Primary pollutants Those pollutants emitted directly from identifiable sources.
Principle of faunal succession Fossil organisms succeed one another in a definite and determinable order, and any time period can be recognized by its fossil content.
Prominence A concentration of material above the solar surface that appears as a bright archlike structure.
Pulsating star A star that fluxuates in brightness by periodically expanding and contracting.
P wave The fastest earthquake wave, which travels by compression and expansion of the medium.
Pyroclastic material The volcanic rock ejected during an eruption, including ash, bombs, and blocks.
Radial drainage A system of streams running in all directions away from a central elevated structure, such as a volcano.
Radiation The transfer of energy (heat) through space by electromagnetic waves.
Radiation fog Fog resulting from radiation heat loss by the earth.
Radioactivity The spontaneous decay of certain unstable atomic nuclei.
Radiocarbon (carbon-14) The radioactive isotope of carbon, which is produced continuously in the

atmosphere and is used in dating events as far back as 40,000 years.

Radiometric dating The procedure of calculating the absolute ages of rocks and minerals that contain radioactive isotopes.

Rainshadow A dry area on the lee side of a mountain range.

Recessional moraine An end moraine formed as the ice front stagnated during glacial retreat.

Reflecting telescope A telescope that concentrates light from distant objects by using a concave mirror.

Refracting telescope A telescope that employs a lens to bend and concentrate the light from distant objects.

Refraction The process by which the portion of a wave in shallow water slows, causing the wave to bend and tend to align itself with the underwater contours.

Regional metamorphism Metamorphism associated with the large-scale mountain-building process.

Regolith The layer of rock and mineral fragments that nearly everywhere covers the earth's land surface.

Rejuvenation A change, often caused by regional uplift, that causes the forces of erosion to intensify.

Relative dating Rocks are placed in their proper sequence or order. Only the chronologic order of events is determined.

Relative humidity The ratio of the air's water vapor content to its water vapor capacity.

Resolving power The ability of a telescope to separate objects that would otherwise appear as one.

Retrograde motion The apparent westward motion of the planets with respect to the stars.

Reverse fault A fault in which the material above the fault plane moves up in relation to the material below.

Revolution The motion of one body about another, as the earth about the sun.

Richter scale A scale of earthquake magnitude based on the motion of a seismograph.

Rift A region of the earth's crust along which divergence is taking place.

Right ascension An angular distance measured eastward along the celestial equator from the vernal equinox. Used with declination in a coordinate system to describe the position of celestial bodies.

Rime A thin coating of ice on objects produced when supercooled fog droplets freeze on contact.

Rock flour Ground-up rock produced by the grinding effect of a glacier.

Rockslide The rapid slide of a mass of rock downslope along planes of weakness.

Rotation The spinning of a body, such as the earth, about its axis.

Runoff Water that flows over the land rather than infiltrating into the ground.

Salinity The proportion of dissolved salts to pure water, usually expressed in parts per thousand ($^0/_{00}$).

Scoria Hardened lava which has retained the vesicles produced by the escaping gasses.

Sea arch An arch formed by wave erosion when caves on opposite sides of a headland unite.

Sea breeze A local wind blowing from the sea during the afternoon in coastal areas.

Sea-floor spreading The process of producing new sea floor between two diverging plates.

Seamount Isolated volcanic peak that rises at least 1000 meters (3000 feet) above the deep ocean floor.

Sea stack An isolated mass of rock standing just offshore, produced by wave erosion of a headland.

Secondary pollutants Pollutants that are produced in the atmosphere by chemical reactions that occur among primary pollutants.

Sedimentary rock Rock formed from the weathered products of preexisting rocks that have been transported, deposited, and lithified.

Seismic sea wave A rapid moving ocean wave generated by earthquake activity which is capable of inflicting heavy damage in coastal regions.

Seismograph An instrument that records earthquake waves.

Shield volcano A broad, gently sloping volcano built from fluid basaltic lavas.

Silicate Any one of numerous minerals that have the oxygen and silicon tetrahedron as their basic structure.

Sill A tabular igneous body that was intruded parallel to the layering of preexisting rock.

Sink hole A depression produced in a region where soluble rock has been removed by groundwater.

Sleet Frozen or semifrozen rain formed when raindrops freeze as they pass through a layer of cold air.

Slump The downward slipping of a mass of rock or unconsolidated material moving as a unit along a curved surface.

Snow A solid form of precipitation produced by sublimation of water vapor.

Soil A combination of mineral and organic matter, water, and air; that portion of the regolith that supports plant growth.

Soil horizon A layer of soil that has identifiable characteristics produced by chemical weathering and other soil-forming processes.

Soil profile A vertical section through a soil showing its succession of horizons and the underlying parent material.

Solar constant The rate at which solar radiation is received outside the earth's atmosphere on a surface perpendicular to the sun's rays when the earth is at an average distance from the sun.

Solar winds Subatomic particles ejected at high speed from the solar corona.

Solifluction Slow, downslope flow of water-saturated materials common to permafrost areas.

Source region The area where an air mass acquires its characteristic properties of temperature and moisture.

Specific humidity The weight of water vapor compared with the total weight of the air, including the water vapor.

Spit An elongate ridge of sand that projects from the land into the mouth of an adjacent bay.

Stalactite The iciclelike structure that hangs from the ceiling of a cavern.

Stalagmite The columnlike form that grows upward from the floor of a cavern.

Steam fog Fog having the appearance of steam; produced by evaporation from a warm water surface into the cool air above.

Strata Parallel layers of sedimentary rock.

Stratosphere The layer of the atmosphere immediately above the troposphere, characterized by increasing temperatures with height due to the concentration of ozone.

Stratovolcano *See* Composite cone.

Striations Scratches or grooves in a bedrock surface caused by the grinding action of a glacier and its load of sediment.

Strike-slip fault A fault along which the movement is horizontal.

Subduction The process of thrusting oceanic lithosphere into the mantle along a convergent zone.

Sublimation The conversion of a solid directly to a gas without passing through the liquid state, or vice versa.

Submarine canyon A seaward extension of a valley that was cut on the continental shelf during a time when sea level was lower, or a canyon carved into the outer continental shelf, slope, and rise by turbidity currents.

Sunspot A dark spot on the sun, which is cool by contrast to the surrounding photosphere.

Supernova An exploding star that increases in brightness many thousands of times.

Suspended load The fine sediment carried within the body of flowing water.

S wave An earthquake wave, slower than a P wave, that travels only in solids.

Swells Wind-generated waves that have moved into an area of weaker winds or calm.

Syncline A linear downfold in sedimentary strata; the opposite of anticline.

Talus An accumulation of rock debris at the base of a cliff.

Tarn A small lake in a cirque.

Tectonics The study of the large-scale processes that collectively deform the earth's crust.

Temperature inversion A layer in the atmosphere of limited depth where the temperature increases rather than decreases with height.

Terminal moraine The end moraine marking the farthest advance of a glacier.

Terrace A flat, benchlike structure produced by a stream, which was left elevated as the stream cut downward.

Texture The size, shape, and distribution of the particles that collectively constitute a rock.

Thermal gradient The increase in temperature with depth. It averages 1° C per 30 meters (1–2° F per 100 feet) in the crust.

Thermocline A layer of water in which there is a rapid change in temperature in the vertical dimension.

Thermohaline circulation Vertical movements of ocean water caused by density differences brought about by variations in temperature and salinity.

Thermosphere The region of the atmosphere immediately above the mesosphere and characterized by increasing temperatures due to absorption of very short-wave solar energy by oxygen.

Thrust fault A low-angle reverse fault.

Tide Periodic change in the elevation of the ocean surface.

Tombolo A ridge of sand that connects an island to the mainland or to another island.

Tornado A small, very intense cyclonic storm with exceedingly high winds, most often produced along cold fronts in conjunction with severe thunderstorms.

Transform fault A boundary where plates slide past each other, such as the San Andreas fault.

Transpiration The release of water vapor to the atmosphere by plants.

Travertine A form of limestone ($CaCO_3$) that is deposited by hot springs or as a cave deposit.

Trellis drainage A system of streams in which nearly parallel tributaries occupy valleys cut in folded strata.

Trench An elongate depression in the sea floor produced by bending of oceanic crust during subduction.

Troposphere The lowermost layer of the atmosphere. It is generally characterized by a decrease in temperature with height.

Tsunami The Japanese word for a seismic sea wave.

Turbidity current A downslope movement of dense, sediment-laden water created when sand and mud on the continental shelf and slope are dislodged and thrown into suspension.

Ultramafic Igneous rocks composed mainly of iron and magnesium-rich minerals.

Unconformity A surface that represents a break in the rock record, caused by erosion or nondeposition.

Uniformitarianism The concept that the processes that have shaped the earth in the geologic past are essentially the same as those operating today.

Upslope fog Fog created when air moves up a slope and cools adiabatically.

Upwelling The rising of cold water from deeper layers to replace warmer surface water that has been moved away.

Urban heat island Refers to the fact that temperatures within a city are generally higher than in surrounding rural areas.

Vadose water The unsaturated zone between the water table and the surface.

Variable sun theory One of the most persistent theories of climatic change which is based on the idea that the sun is a variable star and that its output of energy varies through time.

Ventifact A cobble or pebble polished and shaped by the sandblasting effect of wind.

Viscosity A measure of a fluid's resistance to flow.

Volcanic bomb A streamlined pyroclastic fragment ejected from a volcano while molten.

Volcanic dust theory A theory of climatic change that suggests that the injection of large quantities of dust and ash into the atmosphere during periods of explosive volcanism may have led to a global cooling.

Warm front A front along which a warm air mass overrides a retreating mass of cooler air.

Water table The upper level of the saturated zone of groundwater.

Wave-cut cliff A seaward-facing cliff along a steep shoreline formed by wave erosion at its base and mass wasting.

Wave-cut platform A bench or shelf in the bedrock at sea level, cut by wave erosion.

Wave height The vertical distance between the trough and crest of a wave.

Wave length The horizontal distance separating successive crests or troughs.

Wave period The time interval between the passage of successive crests at a stationary point.

Weathering The disintegration and decomposition of rock at or near the surface of the earth.

Wind vane An instrument used to determine wind direction.

Zodiac A band along the ecliptic containing the twelve constellations of the zodiac.

index